Perspectives on the History of Chemistry

Series Editor

Seth C. Rasmussen, Department of Chemistry and Biochemistry,
North Dakota State University, Fargo, ND, USA

Commonly described as the "central science", chemistry and the chemical arts have an extremely long history that is deeply intertwined with a wide variety of other historical subjects. Perspectives on the History of Chemistry is a book series that presents historical subjects covering all aspects of chemistry, alchemy, and chemical technology.

Potential topics might include:

- An updated account or review of an important historical topic of broad interest
- Biographies of prominent scientists, alchemists, or chemical practitioners
- Translations and/or analysis of foundational works in the development of chemical thought

The series aims to provide volumes that advance the historical knowledge of chemistry and its practice, while also remaining accessible to both scientists and formal historians of science. Volumes should thus be of broad interest to the greater chemical community, while still retaining a high level of historical scholarship. All titles should be presented with the aim of reaching a wide audience consisting of scientists, chemists, chemist-historians, and science historians.

All titles in the book series will be peer reviewed. Titles will be published as both printed books and as eBooks. Both solicited and unsolicited manuscripts are considered for publication in this series.

More information about this series at http://www.springer.com/series/16421

Carmen J. Giunta · Vera V. Mainz · Gregory S. Girolami
Editors

150 Years of the Periodic Table

A Commemorative Symposium

Springer

Editors
Carmen J. Giunta
Department of Chemistry
Le Moyne College
Syracuse, NY, USA

Vera V. Mainz
School of Chemical Sciences
University of Illinois at Urbana-Champaign
Urbana, IL, USA

Gregory S. Girolami
School of Chemical Sciences
University of Illinois at Urbana-Champaign
Urbana, IL, USA

ISSN 2662-4591 ISSN 2662-4605 (electronic)
Perspectives on the History of Chemistry
ISBN 978-3-030-67909-5 ISBN 978-3-030-67910-1 (eBook)
https://doi.org/10.1007/978-3-030-67910-1

This Springer imprint is published by the registered company Springer Nature Switzerland AG
The registered company address is: Gewerbestrasse 11, 6330 Cham, Switzerland

Contents

1 **Editors' Introduction** 1
Carmen J. Giunta, Vera V. Mainz, and Gregory S. Girolami

Part I Mendeleev and His Predecessors

2 **Dmitri Mendeleev and the Periodic System: Philosophy, Periodicity, and Predictions** 13
Ann E. Robinson

3 **The Trouble with Triads** 35
William B. Jensen

4 **Josiah Parsons Cooke, the Natural Philosophy of Sir John F. W. Herschel and the Rational Chemistry of the Elements** 43
Ronald Brashear and Gary Patterson

5 ***Vis Tellurique* of Alexandre-Émile Béguyer de Chancourtois** 61
Carmen J. Giunta

6 **Periodicity in Britain: The Periodic Tables of Odling and Newlands** 93
Carmen J. Giunta, Vera V. Mainz, and Julianna Poole-Sawyer

7 **Gustavus Hinrichs and His Charts of the Elements** 133
Gregory S. Girolami

8 **The Periodic Table of the Elements and Lothar Meyer** 195
Gisela Boeck

9 **Translation of §§ 91–94 of Lothar Meyer's *Modernen Theorien* (1864)** 215
Vera V. Mainz

Part II Discoveries of Elements: Successes and Challenges

10 **Discovery of Three Elements Predicted by Mendeleev's Table: Gallium, Scandium, and Germanium** 227
Mary Virginia Orna and Marco Fontani

11 The Rare Earths, a Challenge to Mendeleev, No Less Today 259
Simon A. Cotton

12 The History (and Pre-history) of the Discovery and Chemistry of the Noble Gases 303
Jay A. Labinger

13 Element Discovery and the Birth of the Atomic Age 329
Kit Chapman

14 Mary Elvira Weeks and *Discovery of the Elements* 343
Vera V. Mainz

Part III The Periodic Table from Other Perspectives

15 Astronomy Meets the Periodic Table, Or, How Much Is There of What, and Why? 387
Virginia Trimble

16 The Impact of Twentieth-Century Physics on the Periodic Table and Some Remaining Questions in the Twenty-First Century 409
Eric R. Scerri

17 An Essay on Periodic Tables 425
Pekka Pyykkö

18 The Periodic Table at 150: A Philatelic Celebration 439
Daniel Rabinovich

Index 451

About the Editors

Carmen J. Giunta is Professor Emeritus of Chemistry at Le Moyne College in Syracuse, New York. He received his B.S. degree in chemistry from the University of Scranton and an A.M. in physics and Ph.D. in chemical physics from Harvard University. He is editor of the *Bulletin for the History of Chemistry.*

Vera V. Mainz is retired from her position as the Director of the NMR Laboratory in the School of Chemical Sciences at the University of Illinois at Urbana-Champaign. She received her B.S. degrees in chemistry and mathematics from Kansas Newman University and her Ph.D. degree from the University of California at Berkeley working with Prof. Richard A. Andersen. She has been the Secretary/Treasurer of the Division of the History of Chemistry of the American Chemical Society since 1995 and she is Associate Editor of the *Bulletin for the History of Chemistry.*

Gregory S. Girolami is the William H. and Janet G. Lycan Professor of Chemistry at the University of Illinois at Urbana-Champaign. He received his B.S. degrees in chemistry and physics from the University of Texas at Austin and his Ph.D. degree from the University of California at Berkeley. Thereafter, he was a NATO post-doctoral fellow with Sir Geoffrey Wilkinson at Imperial College of Science and Technology, and joined the Illinois faculty in 1983. His research interests are primarily the synthesis, properties, and reactivity of new inorganic, organometallic, and solid-state species, especially mechanistic studies of reactions such as the activation of alkanes and the polymerization of alkenes, the synthesis of new "molecule-based" magnetic materials, the chemical vapor deposition of thin films from "designed" molecular precursors, and the chemistry of the f-elements.

Editors' Introduction

1

Carmen J. Giunta, Vera V. Mainz, and Gregory S. Girolami

Abstract

The declaration by UNESCO of 2019 as the International Year of the Periodic Table sparked celebrations and renewed study of this icon of science. Activities included exhibitions, symposia, and publications—including the present volume. A few of those events are reviewed, and the contents of the present volume are previewed.

1.1 International Year of the Periodic Table (IYPT2019)

Late in 2017, the United Nations (UN) General Assembly proclaimed 2019 to be the International Year of the Periodic Table of Chemical Elements. In so doing, it endorsed a resolution that UNESCO (the United Nations Educational, Scientific and Cultural Organization) had adopted earlier in the year. The proclamation was a recognition of the importance of chemistry to the UN's sustainable development agenda in contributing to "solutions to global challenges in energy, education, agriculture and health" [1]. The year 2019 was an appropriate one to celebrate the

C. J. Giunta (✉)
Department of Chemistry, Le Moyne College, Syracuse, NY, USA
e-mail: giunta@lemoyne.edu

V. V. Mainz · G. S. Girolami
School of Chemical Sciences, University of Illinois at Urbana-Champaign, Urbana, IL, USA
e-mail: mainz@illinois.edu

G. S. Girolami
e-mail: ggirolam@illinois.edu

C. J. Giunta et al. (eds.), *150 Years of the Periodic Table*, Perspectives on the History of Chemistry, https://doi.org/10.1007/978-3-030-67910-1_1

Fig. 1.1 Official logo of the International Year of the Periodic Table (https://iypt2019.org)

table because it was the 150th anniversary of the first table by the most influential discoverer of the periodic system, Dmitri Mendeleev.[1]

The idea for IYPT2019 did not originate with the UN, however. The germ of the proclamation appears to have come from a message from Martyn Poliakoff, Professor of Chemistry at Nottingham University in the United Kingdom, to Natalia Tarasova, Professor at the D. I. Mendeleev University of Chemical Technology in Russia in the middle of 2016 [1, p. 14]. Poliakoff is well known to chemists and chemistry students around the world as a presenter in Periodic Videos, a series of short videos about the elements [2]. Tarasova was president of the International Union of Pure and Applied Chemistry (IUPAC) at the time, and she had been on the management committee of the last UNESCO-sponsored year of chemistry, the International Year of Chemistry in 2011. The IUPAC Executive Committee approved exploring the idea, and before long Russian scientific and political support was lined up: the twentieth Mendeleev Congress on General and Applied Chemistry endorsed the idea in 2016, and soon afterward the Mendeleev Russian Chemical Society and the Russian Academy of Sciences followed suit.

IYPT2019 had a logo (Fig. 1.1) and a website [3], and it featured a large number of activities organized and run by local organizations (chemical societies, schools, museums, universities, etc.) around the world from Argentina (a nation named after an element) to Zambia. The opening ceremony was held at UNESCO headquarters in Paris in January 2019, and the closing ceremony at the Tokyo Prince Hotel in December. Both events featured dozens of speakers and attracted hundreds of attendees.

Among the notable conferences and symposia held in 2019 that focused on the periodic table and its history were "Setting their Table: Women and the Periodic Table of Elements" at the University of Murcia, Spain, in February; the Fourth International Conference on the Periodic Table, Mendeleev 150, in St. Petersburg, Russia, in July; and the 21st Mendeleev Congress on General and Applied

[1] As several chapters in this volume will make clear, Mendeleev is the best-known discoverer of the periodic system of the elements and of its chief embodiment, the periodic table, but he was not the first.

Chemistry, also in St. Petersburg, in September. The symposium out of which this volume grew was sponsored by the divisions of the History of Chemistry and Inorganic Chemistry of the American Chemical Society (ACS), held at the Fall 2019 National Meeting of ACS in San Diego, California, in August.

1.2 150 Years of the Periodic Table: Symposium at American Chemical Society San Diego Meeting

Vera Mainz, Gregory Girolami, and Carmen Giunta, the editors of this volume, began planning a symposium to commemorate the 150th anniversary of Mendeleev's table in the summer of 2017. The three of us have long been active in the ACS Division of the History of Chemistry, and we saw 2019 as an ideal time to mount a symposium exploring the origins of the periodic system of the elements, important episodes in its subsequent development, and even its future. Naturally enough, after the proclamation of IYPT2019, we added our symposium to the official list of IYPT2019 events.

The symposium program spanned three half-day sessions. The first session treated classifiers of elements who preceded Mendeleev or were contemporaries of his. Most of the second session dealt with developments in the periodic system in the nineteenth century after the publication of Mendeleev's first table. The final session treated topics mainly from the twentieth century and beyond. A list of symposium authors and the titles of their talks can be found in Table 1.1. We are grateful to all of the speakers who participated in the symposium: their knowledge about and interest in the periodic system made for lively presentations and discussions over the two days of the conference.

Most of the symposium speakers graciously agreed to contribute to the present volume; their chapters will be previewed in the next section. The remaining speakers also added greatly to the success of the symposium, despite being unable to contribute to the book. At our request, Michael Gordin spoke about Mendeleev's career apart from the work for which he is best remembered today. Gordin's biography of Mendeleev, *A Well-Ordered Thing*, was issued in a revised edition in 2019 [4]. Alan Rocke applied his expertise in nineteenth-century German chemistry to outline Lothar Meyer's pathway to periodicity; his research on the subject was published in *Ambix* [5]. Ana de Bettencourt-Dias is an inorganic chemist, specializing in the coordination chemistry and separation chemistry of the lanthanide elements. She spoke about the problems the "rare earth" elements presented to chemists and classifiers in the nineteenth century as well as on the present-day debate over the position of the lanthanides in the periodic table. Brigitte van Tiggelen compared views of the periodic system from Lise Meitner and Ida Noddack. Van Tiggelen spent much of 2019 engaged with the topic of women and the periodic table as co-organizer of the "Setting Their Table" conference mentioned above and as co-editor of a volume on women's contributions to the periodic system, *Women in their Element* [6]. Dawn Shaughnessy, leader of the Nuclear and

Table 1.1 Program for "150 Years of the Periodic Table" symposium held at the Fall 2019 ACS National Meeting in San Diego, CA

William Jensen	Trouble with triads
Carmen Giunta	Vis tellurique of Alexandre-Émile Béguyer de Chancourtois
Julianna Poole-Sawyer	Periodicity in Britain: Periodic tables of Odling and Newlands
Gregory Girolami	Gustavus Detlef Hinrichs and his charts of the elements
Michael Gordin	Mendeleev in St. Petersburg: Marginality of the periodic system
Alan Rocke	Lothar Meyer's path to periodicity
Mary Virginia Orna and Marco Fontani	Discovery of the elements predicted by Dmitri Mendeleev's table: Scandium, gallium, and germanium
Ana de Bettencourt-Dias	Rare earth elements
Jay Labinger	History (and pre-history) of the discovery and chemistry of the noble gases
Gary Patterson	Sir John F.W. Herschel and the concept of periodicity
Virginia Trimble	Hydrogen, helium, and metals: When astronomy met the periodic table
Daniel Rabinovich	Hydrogen to oganesson: Philatelic celebration of the periodic table
Eric Scerri	Impact of twentieth century physics on the periodic table and questions still outstanding in the twenty-first century
Brigitte Van Tiggelen	Uses of the periodic system after radioactivity and the discovery of the neutron: Contrasting views of Lise Meitner and Ida Noddack
Vera Mainz	Mary Elvira Weeks and *The Discovery of the Elements*
Kit Chapman	From neptunium to mendelevium: Element discovery and the birth of the atomic age[a]
Dawn Shaughnessy	Transactinide elements: How the 7th row of the periodic table was discovered
Pekka Pyykkö	Periodic table after period 7

[a]Scheduled but not given due to illness of the author

Chemical Sciences Division at Lawrence Livermore National Laboratory (LLNL) presented on the synthesis of transactinide elements—a topic in which she had firsthand experience as principal investigator in the LLNL Heavy Element Program.

1.3 *150 Years of the Periodic Table:* The Present Volume

1.3.1 Mendeleev and His Predecessors

The present volume begins with Mendeleev, the historical figure most closely associated with the periodic table and with the proclamation of IYPT2019, in a chapter by Ann Robinson titled "Dmitri Mendeleev and the Periodic System: Philosophy, Periodicity, and Predictions." Robinson briefly touches on efforts to classify the chemical elements before Mendeleev—efforts treated in greater detail in the chapters immediately following hers—before exploring how Mendeleev's understanding of elements shaped his periodic system. Mendeleev's distinction between simple substances and elements, and the definiteness and individuality of the latter, led him to eschew continuous representations of the properties of elements, and in particular to disfavor certain spiral representations of his periodic system.

After Robinson's chapter on Mendeleev, the first part of the book is organized roughly chronologically, featuring several chapters on attempts to classify elements before 1869. William Jensen's chapter is called "The Trouble with Triads." As Johann Wolfgang Döbereiner first observed early in the nineteenth century, triads are sequences of three similar elements for which the average of the atomic weights of the heaviest and lightest is approximately equal to that of the middle element. Jensen discusses the use of triads by later classifiers of chemical elements in the nineteenth century and examines the question whether Mendeleev based his well-known successful predictions of undiscovered elements on triads.

In the next chapter, Gary Patterson and Ronald Brashear treat a natural philosopher better known today for his contributions to astronomy than to chemistry, as well as an American who attempted to organize elements to teach chemistry. Their chapter is titled "Josiah Parsons Cooke, the Natural Philosophy of Sir John F. W. Herschel and the Rational Chemistry of the Elements." Herschel's chemical writings in his *Preliminary Discourse on the Study of Natural Philosophy* and his 1858 presidential address to the Chemical Section of the British Association for the Advancement of Science are examined in this chapter. In his address, Herschel pointed to an 1855 paper by Josiah Parsons Cooke Jr. on classifying the elements. Brashear and Patterson discuss that paper as well as Cooke's later classification in his 1868 textbook *First Principles of Chemical Philosophy*.

Carmen Giunta's chapter, "Vis tellurique of Alexandre-Émile Béguyer de Chancourtois" treats the helical arrangement of elements and radicals that the French geologist Béguyer de Chancourtois presented in 1862. Several historians of the periodic system identify the vis tellurique as the first periodic classification, and Giunta concurs. In addition to describing the arrangement itself, the chapter discusses its neglect by chemists until well after periodic classifications by Mendeleev and Meyer were well known.

Giunta joins Vera Mainz and Julianna Poole-Sawyer to present the work of two British classifiers of the elements in a chapter titled "Periodicity in Britain: The Periodic Tables of Odling and Newlands." William Odling and John Newlands, working independently in considerably different circumstances, published periodic arrangements of the elements in the middle 1860s. Neither was influential in the development of the periodic system, but Newlands received belated recognition for his work after pressing his claims.

Gregory Girolami delves into arguably the most obscure and least understood of the periodic systems that predate Mendeleev's in a chapter entitled "Gustavus Hinrichs and his Charts of the Elements." Hinrichs, a Danish-born American polymath included a spiral periodic arrangement (double-spiral, in fact) in an 1867 treatise on the structure of matter. That highly speculative *Programme der Atomechanik* explained similarity among analogous elements by similarity of the arrangements of their fundamental building blocks called pantatoms. Hinrichs also published a slightly different classification of the elements in tabular form in 1869.

Two chapters on the work of Lothar Meyer complete the book's first part. Gisela Boeck's chapter, "The Periodic Table of the Elements and Lothar Meyer," is a translation of a paper she originally had published in *Chemie in unserer Zeit*, a journal of the Gesellschaft Deutscher Chemiker [7]. During his lifetime, Meyer was recognized as a discoverer of the periodic system, sharing the 1882 Davy Medal of the Royal Society (London) with Mendeleev. Boeck discusses Meyer's incomplete periodic system of 1864 published in his short and influential book *Die modernen Theorien der Chemie* as well as his well-known periodic graph of atomic volumes published in 1870.

An introduction and translation by Vera Mainz of key passages from Meyer's *Modernen Theorien*, "Translation of §§ 91–94 of Lothar Meyer's *Modernen Theorien* (1864)," follows Boeck's chapter. The translated sections include several tables, including Meyer's table of 27 elements in a six-column periodic arrangement. They also contain Meyer's cogent discussion of the relationship that chemical theory and experiment ought to have vs. the relationship that they frequently do have.

1.3.2 Discoveries of Elements: Successes and Challenges

The next several chapters treat the discoveries of new elements, particularly after Mendeleev first formulated his periodic system in 1869. These chapters focus on how these new elements challenged Mendeleev's periodic system, forced it to change, and eventually served to convince the world of its merits. This section concludes with a study of the life and work of the foremost chronicler of these discoveries.

Mary Virginia Orna and Marco Fontani describe the discoveries of three elements predicted by Mendeleev and found within two decades of the predictions. Their chapter is called "Discovery of Three Elements Predicted by Mendeleev's Table: Gallium, Scandium, and Germanium." In addition to recounting how compounds of these elements and then the elements themselves were isolated, Orna and Fontani discuss the changing standards by which element discovery has been recognized since the 1870s and 1880s.

Simon Cotton's chapter, "The Rare Earths, A Challenge to Mendeleev, No Less Today," concentrates on the discovery of the rare earth elements: scandium, yttrium, and the lanthanides. He notes that the discoveries spanned about a century and a half from yttrium (or rather its oxide) in 1794 to promethium in 1947. Only five were known when Mendeleev made his first table. The questions of where to put them and even how many there were puzzled chemists until the advent of atomic number, and even now, just what elements are to be considered lanthanides is an unsettled question. Cotton's chapter concludes with a discussion of several aspects of recent lanthanide chemistry, including unusual coordination numbers and oxidation states.

Mendeleev may have known of only five rare earths when he made his first periodic table, but neither he nor anyone else knew of any noble gases at that time—and Mendeleev was reluctant to credit the discovery of the first few when they were found in the 1890s. Jay Labinger recounts how the elements of this group came to be known in "The History (and Pre-History) of the Discovery and Chemistry of the Noble Gases." One of the discoverers of argon, William Ramsay, speculated about the position of argon in the table even before it was isolated. Once characterized, it certainly did not belong in the periodic table where its atomic weight (40) would have placed it, between potassium (39.1) and calcium (40.1). Ramsay eventually solved the dilemma of its place, predicting and then finding most of the elements of the group. Labinger also recounts the flurry of noble gas compounds synthesized by several researchers within a few months in 1962 and 1963.

In "Element Discovery and the Birth of the Atomic Age," Kit Chapman describes the discovery of the first synthetic elements—acknowledging the philosophical question of whether synthesis really qualifies as discovery. Chapman begins by recounting early experiments by Enrico Fermi that were erroneously interpreted as resulting in the synthesis of elements more massive than uranium via the process of neutron capture. As it happens, the neutrons broke apart the uranium nuclei (nuclear fission) rather than sticking to them. Ironically, nuclear fission chain reactions ended up providing the high neutron fluxes that permitted neutron capture and the synthesis of elements 93 and 94 (neptunium and plutonium, respectively). The last elements produced by neutron capture were 99 and 100 (einsteinium and fermium, respectively), formed from the extremely high neutron fluxes in a thermonuclear explosion, the "Ivy Mike" test of the first hydrogen bomb. The discoveries of these elements were closely related to the development of nuclear power and nuclear weapons.

Vera Mainz concludes the focus on the discovery of elements with a chapter, "Mary Elvira Weeks and *Discovery of the Elements*," on a scholar who published accounts of practically all the elements discussed in this section and many more. Weeks was assistant professor of chemistry at the University of Kansas, the same university at which she had earned her Ph.D. in chemistry, when she began writing on how elements came to be recognized. She published 24 (!) papers on the subject in the *Journal of Chemical Education* in 1932–1933. The papers were collected into the classic book *Discovery of the Elements*, which went through seven editions between 1933 and 1968. Henry Marshall Leicester wrote a chapter on the elements of the atomic age for the 6th edition (1956) and was co-author with Weeks of the 7th and final edition (1968).

1.3.3 The Periodic Table from Other Perspectives

The final set of chapters examines aspects of the periodic system and its elements from perspectives of other disciplines—or at least using tools originating in other disciplines such as astronomy, quantum mechanics, and philately.

Virginia Trimble's chapter, "Astronomy Meets the Periodic Table. Or, How Much Is There of What, and Why?" chiefly addresses questions of nucleogenesis and the cosmic abundances of elements. Trimble recounts that early in the twentieth century, hydrogen was thought to account for only a small fraction of the mass of stars, and that estimate rose during the course of the century. The chapter begins, though, with observations of solar spectra leading to the proposal of three new elements in the 1860s—one of which (helium) can still be found on the periodic table.

Eric Scerri had an eventful IYPT2019—as did several other symposium speakers mentioned earlier. He was one of the organizers of the Fourth International Conference on the Periodic Table, Mendeleev 150, in St. Petersburg, and a revised edition of his book *The Periodic Table: its Story and its Significance* was published [8]. His chapter is titled "The Impact of Twentieth Century Physics on the Periodic Table and Some Remaining Questions in the Twenty-first Century." He begins with a brief review of physicists' work on the periodic table from the first half of the twentieth century, including J. J. Thomson's attempt to explain the table based on his electronic model of the atom, Henry Moseley's giving the table a better ordering principle than atomic weight, and quantum aspects of the atom proposed by Bohr, Pauli, and Schrödinger among others. Most of the chapter examines attempts over the years to apply concepts of symmetry and group theory to the periodic table, particularly the empirical Madelung rule for the filling order of atomic orbitals.

Pekka Pyykkö specializes in the theoretical study of the structure and chemistry of very heavy elements, including elements heavier than those that have yet been synthesized. His chapter, "An Essay on Periodic Tables," ranges from historical topics to theoretical limits to the periodic table (predicted for $Z = 172$), to physical effects (such as relativity and quantum electrodynamics) that cannot be neglected in the computational chemistry of heavy elements. Pyykkö's chapter was originally published in *Pure and Applied Chemistry*, the scientific journal of IUPAC [9].

Daniel Rabinovich's chapter, "The Periodic Table at 150: A Philatelic Celebration," returns us to where this introduction started, to IYPT2019. Algeria was the first nation to issue a stamp in honor of IYPT2019, featuring the IYPT logo. Several stamps depict Mendeleev through the IYPT logo, and one from Hungary also shows Mendeleev's handwritten draft of the periodic table. Sri Lanka issued a stamp that displays a full 118-element periodic table with color-coded groups. The stamps Rabinovich shows are visually engaging, celebratory, and often informative.

1.4 The End of the Beginning

The periodic table has been described as an icon of science, one that all scientists and students of science encounter at some point in their careers. The table's profile, its arrangement of orderly but unequal rows and columns of boxes, is a distinctive design that often appears in unexpected places in popular culture. How the periodic table came to be constructed is a fascinating story that rewards serious study and warrants celebration. We think you will find that the following book contains fascinating and occasionally surprising new insights into that story.

Finally, before plunging into the details previewed above, we hope you enjoy this alternative arrangement of the elements, courtesy of Tom Lehrer [10], updated to accommodate elements discovered since nobelium (1958)[2]:

There's antimony, arsenic, aluminum, selenium,
And hydrogen and oxygen and nitrogen and rhenium,
And nickel, neodymium, neptunium, germanium,
And iron, americium, ruthenium, uranium,

Europium, zirconium, lutetium, vanadium,
And lanthanum and osmium and astatine and radium,
And gold and protactinium and indium and gallium,
And iodine and thorium and thulium and thallium.

There's yttrium, ytterbium, actinium, rubidium,
And boron, gadolinium, niobium, iridium,
And strontium and silicon and silver and samarium,
And bismuth, bromine, lithium, beryllium, and barium.

There's holmium and helium and hafnium and erbium,
And phosphorus and francium and fluorine and terbium,
And manganese and mercury, molybdenum, magnesium,
Dysprosium and scandium and cerium and cesium.

And lead, praseodymium and platinum, plutonium,
Palladium, promethium, potassium, polonium,
And tantalum, technetium, titanium, tellurium,
And cadmium and calcium and chromium and curium.

There's sulfur, californium and fermium, berkelium,
And also mendelevium, einsteinium, nobelium,
And argon, krypton, neon, radon, xenon, zinc and rhodium,
And chlorine, carbon, cobalt, copper, tungsten, tin and sodium.

Rutherfordium, lawrencium, seaborgium, flerovium,
Darmstadtium, roentgenium, meitnerium, moscovium,
Copernicium, nihonium, oganesson, livermorium,
And tennessine, and hassium, and dubnium, and bohrium.

These are the only ones of which the news has reached Urbana
And there may be many others—for a chemist, that's Nirvana!

[2]A joint effort of one of the co-editors (GSG) and his University of Illinois colleague Alex Scheeline.

References

1. Sekeris F, Tarasova N, Reedijk J (eds) (2020) The International Year of the Periodic Table of Chemical Elements 2019, common language for science, global report 2019. https://iypt2019.org/wp-content/uploads/sites/4/2020/10/IYPT2019-Final-Report_150dp1-Oct23.pdf. Accessed 11 Nov 2020
2. Periodic videos. http://www.periodicvideos.com. Accessed 11 Nov 2020
3. International Year of the Periodic Table (IYPT), a common language for science. https://iypt2019.org. Accessed 11 Nov 2020
4. Gordin MD (2004) A well-ordered thing: Dmitrii Mendeleev and the shadow of the periodic table. Basic Books, New York. Revised ed. (2019) Princeton University Press, Princeton, NJ
5. Rocke A (2019) Lothar Meyer's pathway to periodicity. Ambix 66:265–302
6. Lykknes A, van Tiggelen B (2019) Women in their element: selected women's contributions to the periodic system. World Scientific
7. Meyer JL (2019) Das Periodensystem der Elemente und Lothar Meyer. Chem. unserer Zeit. 53:372–381
8. Scerri ER (2007) The periodic table: its story and its significance, 2nd ed. Oxford, New York (2019)
9. Pyykkö P (2019) An essay on periodic tables. Pure Appl Chem 91:1959–1967
10. Lehrer T. The elements. https://tomlehrersongs.com/wp-content/uploads/2018/12/the-elements.pdf. Accessed 13 Nov 2020

Part I
Mendeleev and His Predecessors

Dmitri Mendeleev and the Periodic System: Philosophy, Periodicity, and Predictions

2

Ann E. Robinson

Abstract

One of the projects undertaken by chemists during the first half of the nineteenth century was organizing the elements in a meaningful way. By the time of the Karlsruhe Conference in 1860, it was clear that atomic weight was likely the key to creating an organizational scheme that could encompass most, if not all, of the elements rather than small groupings. When Dmitri Mendeleev developed his periodic law in 1869, others had already created systems in which the elements were organized by atomic weight, in which it was postulated that the atomic weights of some elements should be adjusted, and in which gaps were left for elements that had not yet been discovered. These are also hallmarks of Mendeleev's system, yet it is his system which gained wide attention and acceptance. This paper looks at Mendeleev's system in relation to those of his contemporaries. Three areas that Mendeleev emphasized as important in his writings will be explored in greater depth: his philosophical understanding of elements, which assisted in the development of the periodic law; his detailed predictions for elements not yet discovered, which showcased the utility of the periodic system; and his stress on the finiteness of periods, which influenced what he saw as the best forms of the periodic table.

A. E. Robinson (✉)
Harvard University, Widener Library, Cambridge, MA 02138, USA
e-mail: ann_robinson@harvard.edu

C. J. Giunta et al. (eds.), *150 Years of the Periodic Table*, Perspectives on the History of Chemistry, https://doi.org/10.1007/978-3-030-67910-1_2

2.1 Introduction

In a lecture on the history of chemistry, German chemist Alfred Ladenburg informed his audience that natural laws "do not originate suddenly in the head of a single individual" [1]. Rather, over time and through the work of many, the facts are determined and the fundamental ideas mature. It is then that someone, or several someones, announce the law. The history of the periodic law follows this pattern. Chemists spent much of the nineteenth century creating different ways to organize and classify the elements. None of these schemes was entirely satisfactory, however, as many failed to encompass all of the elements and most lacked a single fundamental property upon which a system could be based. When Dmitri Mendeleev developed his periodic law in 1869, he built upon what he referred to as "the stock of generalisations and established facts which had accumulated by the end of the decade 1860–1870" [2].

This paper begins with an overview of the organization and classification of the elements before Mendeleev and then considers three areas Mendeleev emphasized in his writings which shaped his understanding of the periodic system. His philosophical understanding of elements assisted in the development of the periodic law. His detailed predictions for elements not yet discovered and reasons for changing the atomic weights of some elements showcased the utility of the periodic system. His stress on the finiteness of periods influenced what he saw as the best forms of the periodic table.

2.2 Classification Before Mendeleev

Over the course of the first half of the nineteenth century, chemists searched for meaningful ways to organize and classify the chemical elements. In his *A New System of Chemical Philosophy*, John Dalton had tied atomic weight to the identity of an element [3]. While not all of Dalton's chemical atomic theory was accepted, atomic weight quickly became the defining characteristic of an element and the determination of atomic weights became an important field of research. It was noticed early on that there was an arithmetic relationship between small groupings of elements. These so-called triads were composed of three elements with similar chemical properties in which the atomic weight of the center element was the mean of the weights of the other two elements.[1] For example, in the grouping of calcium, strontium, and barium, the atomic weight of strontium was the mean of that of calcium and barium. These triads, however, were just that, groups of three. There was no obvious way to connect all of the elements in a numerical relationship. There were few attempts to do so, in part because there was a lack of accurate atomic weights. Chemists knew that atomic weights were highly uncertain. Multiple bases for calculating weight and multiple understandings of the term atomic

[1]For more on triads, see Chapter 3 of this volume.

weight made a single system based on that criterion difficult. A better method for the determination of atomic weight became well-known in 1860, but until then other means of organization needed to be used.

2.2.1 Before 1860

One of the most commonly used methods of classifying the elements was to divide them into two groups, metals and metalloids (or non-metals). This was a seemingly straightforward system that nonetheless caused confusion. As Leopold Gmelin wrote in his noted handbook, "No exact line of demarcation can be drawn between metals and metalloids" [4]. For example, some chemists placed iodine with the metals as it had a visible metallic luster while others considered it to be a metalloid based on other characteristics. It was a system based on only one characteristic but one which was not always the most important characteristic of an element and one which was open to interpretation. Other methods for classifying and organizing the elements were sought.

2.2.1.1 Gmelin

Leopold Gmelin had been studying the numerical relationships between the elements since at least 1827. His arrangement displays triads, the word he used in place of Döbereiner's triplets [5]. In 1843, Gmelin published a system of the elements in the fourth edition of his *Handbook of Chemistry*. Gmelin divided the known elements into metals and metalloids. But, as noted, he was clear that it was not an easy division to make. Based on triads, he then arranged the elements in groups according to their chemical and physical properties. This was shown in a table with a vee shape (Fig. 2.1). It was an "imperfect attempt" to arrange the elements and he noted it would be better shown in three dimensions [4]. Across the top of the vee were oxygen, hydrogen, and nitrogen, as none had a known analogous element. The electro-negative elements were placed on the left and the electro-positive elements on the right [6].

2.2.1.2 Gladstone

In 1853, in an ordering that was unusual at the time, British chemist John Hall Gladstone placed all of the elements in order according to their atomic weights.[2] However, he did not see any obvious relationships. According to Francis Preston Venable, the American chemist who wrote the first history of the periodic law, this was because "the numbers used by Gladstone are too faulty to show any noteworthy regularity" [7]. Gladstone then grouped the elements as Gmelin had done but replaced the symbols of the elements with their atomic weights. He discerned some numerical relationships in doing this, but they were largely triads. Why this was the case, he did not know, but he was sure that these numerical relationships were not

[2]Venable states Gladstone was the first to do so, but this is incorrect. Marc Antoine Gaudin did so in 1833, and before that Dalton published several partial lists organized in this way.

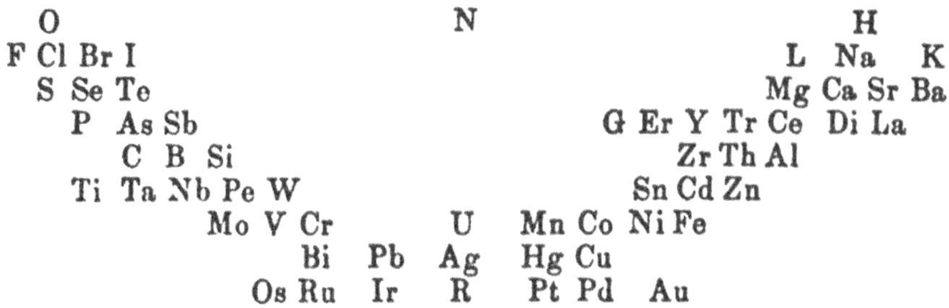

Fig. 2.1 Gmelin's vee shape classification. Reproduced from Gmelin (1844) [4]

mere chance. Further research was required, but he was hopeful an answer would soon be found: "We can scarcely imagine that the intimate constitution of these related elementary bodies will long remain an unfruitful field of investigation" [8].

2.2.1.3 Cooke[3]

Josiah Parsons Cooke, Jr., who taught chemistry at Harvard College, found the use of only one characteristic as the basis of a classification system to be ridiculous. He wrote in 1855 [9]:

> For a zoologist to separate the ostrich from the class of birds because it cannot fly, would not be more absurd, than it is for a chemist to separate two essentially allied elements, because one has a metallic lustre and the other has not.

Rather a "correct" classification system should be based upon a "fundamental property common to all the elements, the law of whose variation is known." However, it was not clear what that common fundamental property was. Such a system also needed to encompass all of the elements, not just groups of triads. Cooke created a classification in which the elements were placed into six series, each containing elements that formed similar compounds and produced similar reactions, had the same crystalline forms, and whose properties varied in a regular manner [10].

2.2.2 The 1860s

A seminal event in the history of chemistry took place in 1860: an international congress of chemists was held in Karlsruhe to discuss important aspects of chemistry.[4] Among the matters discussed were the definitions of the terms atom, molecule, and equivalent; a uniform chemical notation and nomenclature; and the

[3]For more on Cooke, see Chapter 4 of this volume.
[4]The Karlsruhe Congress was international in that chemists came from 11 European countries, with a lone representative from Mexico. No chemists from the United States or Asia were present.

question of equivalent weights and formulae [11, 12]. As Venable had noted about the atomic weights used by Gladstone, there were many ways to determine the atomic weights of the elements but the conversion of chemical analyses to atomic weight remained uncertain. Perhaps the most important thing to happen at the Karlsruhe Congress occurred at the end, when Italian chemist Angelo Pavesi distributed copies of an article written by his colleague Stanislao Cannizzaro. In it, Cannizzaro described a method for determining atomic weight using the hypothesis of Amedeo Avogadro and André-Marie Ampère [13]. Julius Lothar Meyer later recalled that after reading it, "It was as though the scales fell from my eyes, doubt vanished, and was replaced by a feeling of peaceful certainty" [14].

2.2.2.1 Chancourtois[5]

The French geologist Alexandre-Émile Béguyer de Chancourtois presented a series of papers before the Académie des Sciences in 1862. He described the natural classification of the elements he had developed. Although he had prepared a full diagram of his vis tellurique, it was not published in the *Comptes Rendus* but rather in a self-published pamphlet [15]. Chancourtois's system was designed to be shown on a three-dimensional helix. The elements were placed on the helix in order of their atomic weight. His key insight was that, "The properties of the bodies are the properties of the number" [16]. In other words, the properties of the elements were tied to their atomic weights. Despite the importance of this insight, Chancourtois's work received little attention until after Mendeleev's periodic law had gained acceptance [17].

2.2.2.2 Hinrichs[6]

Danish-American Gustavus Detlef Hinrichs first hinted at his classification in an 1866 paper, stating that a series of papers was forthcoming that would show "the properties of the elements as functions of their atomic weights" [18]. These papers were not published and Hinrichs self-published his classification system in 1867 [19]. In his arrangement, the elements were placed in a spiral in order of atomic weight. The lightest elements were closest to the center of the spiral while the heaviest were furthest away. This spiral arrangement was not easy to read or to print and Hinrichs later produced a tabular form which he used in his textbooks and other publications [20].

2.2.2.3 Odling[7]

English chemist William Odling made several attempts to classify the elements. In his 1857 attempt, he noted, much like Cooke, that although it was an acknowledged fact that "the groupings of the elements are as real and certain as the natural families of plants and animals," in the usual systems "bodies manifesting the strongest analogies are widely separated from one another." Starting with triads, he arranged

[5]For more on Chancourtois, see Chapter 5 of this volume.
[6]For more on Hinrichs, see Chapter 7 of this volume.
[7]For more on Odling and Newlands, see Chapter 6 of this volume.

the elements into 13 groups that shared important properties, emphasizing the use of "fundamental" characteristics rather than "superficial" ones [21]. In his 1864 work, Odling organized the elements into a table. He left blank spaces as he felt that the discovery of new elements "is not by any means improbable." He also noted that the numerical relations between the elements must "depend upon some hitherto unrecognized general law" [22]. Odling's table appeared as an appendix in the 1865 edition of his textbook, *A Course of Practical Chemistry Arranged for Use of Medical Students* [23] and its French translation [24].

2.2.2.4 Newlands

John Alexander Reina Newlands, an English chemist, published several short works in the weekly magazine *Chemical News* during the 1860s as he developed what he referred to as the Law of Octaves [25]. Newlands arranged the elements by atomic weight and discerned that "the numbers of analogous elements generally differ by 7 or by some multiple of seven," meaning that the same characteristics reappeared every eight elements. He arranged the elements in the order of increasing atomic weight in a table of eight columns and seven rows [26]. Given that a number of new elements had recently been discovered, Gladstone wondered if "the finding of one more would throw out the whole system" but Newlands believed that the finding of new elements or the revision of atomic weights would not "upset, for any length of time, the existence of a simple relation among the elements, when arranged in the order of their atomic weights" [27, 28]. For his Law of Octaves, Newlands was awarded the Royal Society's Davy Medal in 1887 [29].

2.2.2.5 Meyer[8]

Although Mendeleev is often credited as the sole discoverer of the periodic law, German chemist Julius Lothar Meyer has a good claim for the title: he was, with Mendeleev, awarded the Royal Society's Davy Medal in 1882 [30]. After attending the Karlsruhe Congress and reading Cannizzaro's article, Lothar Meyer began work on a chemistry textbook that would become a classic text, *Die modernen Theorien der Chemie* [31, 32]. As Meyer revised his textbook, he also continued to develop his classification of the elements. The first edition of 1864 contained a table with only 28 of the known elements [33]. By 1868, the table contained 52 elements, the vast majority, arranged by atomic weight and organized in 15 families [34] —but this table was not published until 1895.

In response to Mendeleev's first announcement of the periodic law, Meyer published another table that he described as being "essentially identical with that given by Mendeleev." This table included 56 elements in nine columns. There were gaps that he felt would be filled either by already known elements, once their atomic weights had been more accurately determined, or by yet unknown elements. He also included a second figure, one that would become very well-known in the following decades. The atomic volume curve illustrated the variation of atomic volume of solid elements when plotted against atomic weight. It clearly showed, he

[8]For more on Meyer, see Chapters 8 and 9 of this volume.

wrote, "that the atomic volume of the elements, like their chemical properties, is a periodic function of their atomic weight." Although he believed the curve showed there were errors in the accepted atomic weights of several elements, he stated that it would be "premature" to make changes [35].

2.3 Mendeleev's Periodic Law

Much has been written about Mendeleev's discovery of the periodic law so only a very brief overview will be given here [36–39]. Just as Meyer was working on a classification system for the elements while writing and revising his textbook *Modernen Theorien*, Mendeleev was working on a textbook of inorganic chemistry, *Principles of Chemistry*, having already written an organic chemistry text. He was seeking "some system of simple substances in order to be guided in their classification—not by arbitrary or subjective reasons, but by some exact, definite principle" [40]. Mendeleev began arranging the elements by their atomic weights and realized, much as Newlands had, that properties of the elements recurred on a regular basis. The periodic law, as he stated it in his important paper of 1871, was thus: "The properties of the elements (and of the simple and compound substances which they form) show a periodic dependence on their atomic weights" [41].

2.3.1 Mendeleev's Writings on the Periodic Law

Mendeleev wrote many papers relating to the periodic system [42]. The very first paper published in 1869 served to announce and explain his discovery [43]. The seminal 1871 paper, originally published in German, based on earlier Russian papers, and later translated into French and English,[9] provided an in-depth look at the periodic law, how it was applied to create a system of the elements, and how it could be used to discover new elements, correct atomic weights, and otherwise complete our knowledge of the elements [44]. In 1889, Mendeleev was asked by the Chemical Society of London to deliver the Faraday Lecture and he used the opportunity to reflect on the discovery and use of the periodic law and to comment on more recent trends in chemistry [2]. Each edition of his textbook, *The Principles of Chemistry*, also included thoughts about the periodic system.

These writings emphasize three areas which were important to Mendeleev's conception of the periodic system: his philosophical understanding of elements, which assisted in the development of the periodic law; his detailed predictions for elements not yet discovered and changes in atomic weights of known elements,

[9]The French translation was made eight years after the original German publication and Mendeleev included a brief letter in which he provided some new thoughts on the periodic law. The English translation, serialized in the weekly *Chemical News*, was hastily done; see Jensen [41] for a more accurate version.

which showcased the utility of the periodic system; and his stress on the finiteness of periods, which influenced what he saw as the best forms of the periodic table.

2.4 Philosophical Conception of the Chemical Element

In 1810, Dalton defined the term element in this way [45]:

> By elementary principles, or simple bodies, we mean such as have not been decomposed, but are found to enter into combination with other bodies. We do not know that any one of the bodies denominated elementary, is absolutely indecomposable; but it ought to be called simple, till it can be analyzed.

For most of the nineteenth century, the terms "simple body" and "simple substance" were generally used interchangeably with the term "element." Mendeleev, however, saw a distinction between elements and simple substances [46]. In a lecture given at St. Petersburg University in 1867, he gave a definition for simple substances which is almost identical with Dalton's definition of elements. A simple body, Mendeleev said, is a substance "which taken individually, cannot be altered chemically by any means produced up until now or be formed through the transformation of any other kinds of bodies."

Mendeleev described elements as something else altogether—an abstract concept. An element was "the material that is contained in a simple body and that can, without any change in weight, be converted into all the bodies that can be obtained from this simple body" [47]. The frequently given example is that carbon is an element while graphite and diamond are simple bodies. We cannot see the element carbon in the diamond, but it is still present in the simple body that is the diamond.

The clearest statement of this distinction can be found in his first major paper of 1869 [40]:

> everybody understands that in all changes in the properties of simple substances, something remains unchanged and that, in the transformation of the elements into compounds, this material something determines the characteristics common to the compounds formed by a given element. In this regard only a numerical value is known, and this is the atomic weight appropriate to the element.

Continuing with the example of carbon, Mendeleev states that atomic weight does not belong to coal or diamond but to carbon. He thus tied atomic weight to this abstract concept of element.

By 1860, the theories of Charles Frédéric Gerhardt and Stanislao Cannizzaro as well as advances in experimental analysis resulted in increasingly more reliable atomic weights. For this reason, Mendeleev felt confident in basing his classification system for the elements upon atomic weight. It was the fundamental property common to all the elements that Cooke had been looking for and it was the basis for the unrecognized general law that Odling was sure explained the numerical relations between the elements.

Historian Helge Kragh stated that Mendeleev's periodic law "was about both elements and simple substances, but in different ways" [48]. The periodic law was primarily about atomic weight—Mendeleev's elements—as it was the basis for the law. The properties of the elements—Mendeleev's simple substances—showed a periodic dependence on the atomic weight, which was an important aspect of the law.

Mendeleev clearly had a philosophical point of view when it came to the elements. However, unlike previous systems, "the periodic law furnishes facts and emphasizes that philosophical question which highlights the mysterious nature of the elements" [49]. The philosophical question was the nature of the elements. For Mendeleev, the nature of the elements depended on their atomic weight. The periodic law was based "on the solid and wholesome ground of experimental research" whereas other supposedly philosophical ideas about the nature of the elements were "relic[s] of the torments of classical thought," remnants of an ancient time when our ancestors concocted hypotheses to explain the universe [2].

The periodic law had, of course, not yet been proved when he first wrote about it in 1869, but Mendeleev was certain that "new interest will be awakened in the determination of atomic weights, in the discovery of new simple substances, and in the detection of new analogies among the elements" [40]. With this statement, Mendeleev suggested that the periodic system could assist in future research in ways that he explicated in greater detail in his 1871 paper.

2.5 Predictions and Adjustments

Mendeleev discussed the application of the periodic law to the following areas in his 1871 paper: the system of the elements, the determination of atomic weights of insufficiently studied elements, the determination of the properties of presently unknown elements, the correction of the magnitude of atomic weights, and the completion of our knowledge of the forms of chemical combinations [41]. In regard to the periodic system, Mendeleev noted that it held not only "purely pedagogic importance as a means of learning more easily various facts" but also scientific importance as it "paves the way for new methods of investigating the elements." Most famously, Mendeleev left gaps in his periodic tables that he was confident would be filled by as-yet-undiscovered elements. He also corrected the weights of several elements and suggested some others were incorrect. The work of other chemists would prove that Mendeleev was correct: the periodic system was a useful tool for research.

2.5.1 Leaving Gaps and Predicting Characteristics

Mendeleev was far from the first to leave gaps in his table for new elements. Newlands, in his Law of Octaves, had not, although he admitted that new elements would be able to slide into his system. Odling did leave blank spaces in his tables as he felt new elements would be discovered. Meyer also left gaps that, he later explained, would be filled by already known elements after their atomic weights had been more accurately determined or by elements yet unknown. None of them was as bold as Mendeleev who not only left gaps but predicted the atomic weights and chemical and physical properties of the elements that would fill those gaps. As these gaps were filled with newly discovered elements that, more or less, fit Mendeleev's predictions, chemists increasingly began to view the periodic system as a useful tool for both research and pedagogy.[10]

2.5.1.1 Gallium

The French chemist Paul-Émile Lecoq de Boisbaudran announced the discovery of a new element in 1875, which he decided to call gallium in honor of France. After reading of this discovery, Mendeleev composed a note for the *Comptes Rendus* in which he reminded his audience that he had proposed the periodic law in 1869. The note also included a table which had blank spaces, one of which was for an element Mendeleev had called eka-aluminum as it should be analogous to aluminum. The characteristics he had predicted for eka-aluminum were more or less in agreement with those of the newly discovered gallium. He concluded by stating, "If subsequent researches confirm the identity of the properties of gallium with those which I have pointed out as belonging to eka aluminum, the discovery of this element will furnish an interesting example of the utility of the periodic law" [50].

Boisbaudran denied that he was aware of Mendeleev's periodic law or of his predicted element eka-aluminum. He was skeptical that gallium was eka-aluminum and of the utility of Mendeleev's system. He said, "I will even add that this ignorance may perhaps have been advantageous to me, for I should have experienced serious delays" [51]. M. M. Pattison Muir, the British chemist who translated Mendeleev's note on gallium into English, expressed some skepticism himself, believing that further research was required before accepting that eka-aluminum was gallium. However, he declared, "Mendelejeff's hypothesis is at least of much value as a guide to future research" [52].

2.5.1.2 Scandium

Four years after the discovery of gallium, the Swedish chemist Lars Fredrik Nilson discovered a new element among minerals found only in Scandinavia. He called this new element scandium. Nilson was also apparently unaware of Mendeleev's predictions; however, Swedish chemist Per Cleve was aware and he explicitly made the connection in his publication reporting on his research which confirmed the discovery of scandium. He wrote: "The great interest of scandium is that its existence has been predicted. Mendeleef, in his memoir on the law of periodicity, had foreseen the existence of a metal which he named ekabor[on], and whose characteristics agree very fairly with those of scandium" [53].

[10]For more on discoveries of Mendeleev's predicted elements, see Chapter 10 of this volume.

Mendeleev saw that interest in his periodic system had been strengthened and he had a French translation of his 1871 paper sent to the journal *Le Moniteur Scientifique*. It was accompanied by a letter in which he expressed his gratification that his law was under scrutiny and proving itself through experimentation. These recent discoveries were but "the first fruits of the periodic law" and he hoped they would lead to "a new philosophical order, by securing it with pillars strengthened by new experiments, so as to give greater stability to the edifice already begun" [49].

2.5.1.3 Germanium

The discovery of eka-silicon came 15 years after its prediction. In 1886, Clemens Winkler announced the discovery of a new element he called germanium. Initially, he believed it would fit into the periodic system between the elements antimony and bismuth [54]. Further experiments revealed an atomic weight of 72.32 and many of the properties of the new element correlated with those Mendeleev had predicted for eka-silicon [55]. Winkler was not immediately convinced that germanium was eka-silicon but after discussions with Victor von Richter, Lothar Meyer, and Mendeleev, Winkler changed his mind. Germanium fit in the periodic system between gallium, the first of Mendeleev's predicted elements to be discovered, and arsenic. This discovery served to further strengthen the acceptance of Mendeleev's periodic law.

2.5.2 Changing Atomic Weights

Mendeleev is renowned for changing the atomic weights of some elements in order to make them better fit into his periodic system. The most well-known of these changes are the pair reversals. Pairs of elements, such as tellurium and iodine, were flipped, giving precedence to their chemical properties rather than to their accepted atomic weights. Odling, Newlands, and Meyer had also flipped tellurium and iodine, but Mendeleev flipped more than one set of elements. Mendeleev is also well-known for doubling the atomic weight of uranium, from the accepted weight of 120 to 240. Perhaps one of Mendeleev's lesser known adjustments was to the atomic weight of beryllium. In a memoir on beryllium, the American chemist Charles Lathrop Parsons noted that between 1873 and 1885, "a long, earnest and interesting discussion ... regarding the valency of beryllium and its place in the periodic system" took place [56]. By the end of this period, many chemists who had remained skeptical about the utility of the periodic system had changed their positions.

The assignment of the atomic weight of beryllium was a test of the accuracy of the periodic system. Chemists believed the atomic weight of beryllium was close to either 9 or 13.5, but there was no consensus. Mendeleev believed that beryllium, or glucinum as it was also called, was divalent and had an atomic weight of approximately 9. Its characteristics made it analogous to magnesium. Odling, Newlands, and Hinrichs had also assigned the atomic weight of 9 to beryllium in their systems. However, others believed that beryllium was trivalent with an atomic weight closer to 13.5, making it analogous to aluminum. In an 1880 paper on their experiments, Swedish chemists Lars Fredrik Nilson and Sven Otto Pettersson announced that their results showed beryllium to have an atomic weight of 13.65. "In consequence of what has been indicated here," they wrote, "the periodic law in its present condition cannot be said to be quite an adequate expression of our knowledge of the elements." However, they expected that "the periodic law may be so modified and developed that it can embrace and explain every fact stated by experiment" [57].

The English chemist Thomas Samuel Humpidge, professor of chemistry at the University College of Wales, surveyed the field of beryllium research in an 1880 paper. He preferred to accept fact over theory, writing, "I am not arguing for the rejection of the periodic law, but only wish to show that if facts are discovered which are incompatible with it, it must of necessity receive some modification" [58]. Humpidge obtained a grant from the Royal Society of £50, which went toward materials and apparatus, and began his own experiments.

In his 1883 report, Humpidge described his experimental results which resulted in a specific heat measurement that was "nothing near" what it should be if the atomic weight of beryllium was 9. "The result is unfortunate for the periodic law, and is the first serious rebuff which this useful generalisation of facts has received," he concluded [59]. Humpidge continued his experiments on the vapor-density of several compounds of beryllium. The results now showed that beryllium was divalent with an atomic weight of 9.1. In his second paper, he declared [55]:

> The long disputed question of the atomic weight of glucinum is thus definitely and finally decided in the favour of that number which satisfies the requirements of the periodic law, and another element is added to the long list of those whose atomic weights have been corrected by this important generalisation.

Over the course of his research, Humpidge changed his opinion on the utility of the periodic system. He wrote, "In all future determinations of the atomic weight of an element, the position which the element should occupy in the periodic arrangement must receive due importance."

Regarding the controversy over the atomic weight of beryllium, Mendeleev stated that the confirmation of the bivalency of beryllium was "as important in the history of the periodic law as the discovery of scandium." And, he observed, "It is most remarkable that the victory of the periodic law was won by the researches of the very observers who previously had discovered a number of facts in support of the tri-valency of beryllium" [60].

2.6 The Importance of Periodicity

One important aspect of Mendeleev's periodic system is the way it is represented graphically. In other words, the periodic table. To date, probably over 1000 different forms of the periodic table have been drawn [61–63]. Mendeleev himself drew more than 60 different tables during his lifetime [64]. Whereas he was bold in his predictions of new elements and in changing the atomic weights of already known elements, Mendeleev was, as historian Bernadette Bensaude-Vincent noted, "more hesitant about the best visualization of the periodic system that he discovered" [65]. He may have been hesitant about what the best form of the periodic table might be but he definitely had opinions about what the best forms were not. Mendeleev frequently noted that the important aspect of the periodic law was just that, periodicity. In *The Principles of Chemistry*, he wrote, "The elements, if arranged according to their atomic weights, exhibit an evident *periodicity* of properties." He further emphasized this point in a footnote about the representation of the periodic law, stating that the law [66]:

> *above all*, depends on there being but few types of chemical compounds, which are arithmetically simple, *repeat themselves*, and offer no uninterrupted transitions, and therefore each period can only contain a definite number of members.

In other words, periods were finite, not continuous, and this fact influenced Mendeleev's views on the graphic representation of the periodic law.

2.6.1 Spiral Forms

By far, the most popular forms of the periodic table are spirals and tables. Mendeleev only drew a handful of spiral forms and only one was these was published, though it looks more like a table than a spiral (Fig. 2.2) because it omits the elements that would furnish transitions that would connect the bottom of some columns to the top of the next ones [40]. It was, perhaps, meant to be a three-dimensional or screw-shaped spiral rather than a flat, two-dimensional spiral [67]. In 1870, after reading the works of Mendeleev and Meyer, the Swiss chemist Heinrich Baumhauer suggested that the periodic law could be represented in the form of a spiral. By arranging the elements in order of atomic weight, with hydrogen in the center of the spiral, a clear view of the elements could be obtained, he said [68]. Mendeleev was dismissive of Baumhauer's spiral, claiming it was the spiral table from his paper and that Baumhauer's arrangement had "little application" and was "artificial" [69].

John Russell Smith suggested that Mendeleev's dismissiveness toward Baumhauer's spiral was due to his "failure in early 1871 to draw up a satisfactory spiral arrangement of the elements." Mendeleev also mentioned spiral arrangements several times, so he was not entirely against them. However, he only found certain types of spirals to be useful ones. In lectures in 1889–90, he said that the periodic law "may be represented in the form of a spiral, where each turn will express a definite period. It may also be represented in the form of a screw-shaped line, where each turn of the screw will represent a period" [70]. And that was, of course, the important aspect—the periods were definite.

Li	Na	K	Cu	Rb	Ag	Cs	—	Tl
7	23	39	63,4	85,4	108	133		204
Be	Mg	Ca	Zn	Sr	Cd	Ba	—	Pb
B	Al	—	—	—	Ur	—	—	Bi?
C	Si	Ti	—	Zr	Sn	—	—	—
N	P	V	As	Nb	Sb	—	Ta	—
O	S	—	Se	—	Te	—	W	—
F	Cl	—	Br	—	J	—	—	—
19	35,5	58	80	190	127	160	190	220.

Fig. 2.2 The spiral table from Mendeleev's 1869 paper. Reproduced from Mendeleev (1869) [40]

Mendeleev did not look upon other curved forms as favorably. Of Lothar Meyer's popular atomic volume curve, Mendeleev commented [71, 72][11]:

> This method, although graphic, has the theoretical disadvantage that it does not in any way indicate the existence of a limited number of elements in each period. … The actual periodic law does not correspond with a continuous change of properties, with a continuous variation of atomic weight.

Forms that did not represent the periods—curves, spirals, two-dimensional, or three-dimensional—were not, in Mendeleev's view, true graphic representations of the periodic law.

2.6.2 Tabular Forms

Tabular forms, that is forms with columns and rows, had an advantage in that the periods were easily shown. Aside from a handful of attempts at spirals, all of the periodic tables that Mendeleev drew were tabular. He drew tables in which the periods were represented in vertical columns and the groups in horizontal rows, and vice versa. The famous first attempt of 1869 is an example of a table with vertical periods while the table from his 1871 paper has horizontal periods.

Mendeleev also drew short-form or long-form tables. In short-form tables, the periods double back (as in the 1871 table) whereas in long-form tables, the periods extend across the table (as in most of the periodic tables you will find in a textbook or on a wall chart today). Until the 1920s, the short-form was the most popular type of periodic table. That began to change with the more complete understanding of the structure of the atom and the change to arranging the periodic table by atomic

[11]Mendeleev had noticed periodicity in atomic volume; see Girolami and Mainz [72].

Gruppen	Reihen: 1	2	4	6	8	10	12
I.		Li = 7	K (39)	Rb (85)	Cs (133)	—	—
II.		Be = 9.2	Ca (40)	Sr (87)	Ba (137)	—	—
III.		B = 11	? Sc ^1)	Yt (89 ?)	? Di (139 ?)	Er (175 ?)	—
IV.		C = 12 } Typische Elemente	Ti (48)	Zr (90)	Ce (141)	? La (180 ?)	Th (231)
V.		N = 14	V (51)	Nb (94)	(? ^2)	Ta (182)	—
VI.		O = 16	Cr (52.5)	Mo (96)	?	W (184)	Ur (240)
VII.		F = 19	Mn (55)	—	—	—	—
			Fe (56)	Ru (103)	—	Os (194 ?)	—
VIII.			Co (58.6)	Rh (104)	—	Jr (195 ?)	—
			Ni (58.6)	Pd (106)	—	Pt (197)	—
I.	H = 1	Na = 23	Cu (63.5)	Ag (108)	—	Au (197)	—
II.		Mg (24)	Zn (65)	Cd (112)	^3)	Hg (200)	—
III.		Al (27.3)	Ga (69)	Jn (113)	—	Tl (204)	—
IV.		Si (28)	? ? ?	Sn (118)	—	Pb (204)	—
V.		P (31)	As (75)	Sb (120) ^4)	—	Bi (208)	—
VI.		S (32)	Se (79)	Te (125 ?)	—	—	—
VII.		Cl (35.5)	Br (80)	Jod (127)	—	—	—
	Reihen	3	5	7	9	11	

St. Petersburg, 4./16. April 1880.

Fig. 2.3 The long-form table from Mendeleev's 1880 paper on the history of the periodic law. Reproduced from Mendelejeff (1880) [76]

number rather than atomic weight. By the 1950s, the long-form table had overtaken the short-form in textbooks [73], though short-form tables remained in some classrooms at least into the 1970s.

Mendeleev's opinion regarding which form—short or long—was the best way to represent the periodic system changed. According to Smith, Mendeleev initially preferred the long-form arrangement but after 1869 began to favor the short-form. However, after a decade, Mendeleev again showed a preference for long-form tables [74]. In a short history of the periodic law written in 1880, Mendeleev included a long-form table. This sort of table, he wrote, was the tabular arrangement that he considered "to be the best and most complete expression of the harmony of the elements or of the periodic law (and the most convenient with respect to typography)" (Fig. 2.3) [75]. However, Mendeleev did not abandon the short-form and continued to draw updated versions, such as the one in the 7th edition of the *Principles of Chemistry* that incorporated the newly discovered noble gases (Fig. 2.4) [76].

Short-form and long-form tables had different advantages. Mendeleev believed that the advantage of the long-form table lay in its ability to better show the periodicity of physical properties, such as atomic volume, and to better show analogies between elements. Short-forms, on the other hand, better illustrated the valency of the elements and brought together sub-groups on the basis of the similarity of their compounds. But, as Bensaude-Vincent put it, "Mendeleev never considered one single representation because none of them was totally satisfying" [66]. The only form Mendeleev showed a definite preference for was the tabular form as it showcased the essence of the periodic law—periodicity.

PERIODIC SYSTEM OF THE ELEMENTS IN GROUPS AND SERIES.

Series	GROUPS OF ELEMENTS 0	I	II	III	IV	V	VI	VII	VIII
1	—	Hydrogen H 1·008	—	—	—	—	—	—	
2	Helium He 4·0	Lithium Li 7·03	Beryllium Be 9·1	Boron B 11·0	Carbon C 12·0	Nitrogen N 14·04	Oxygen O 16·00	Fluorine F 19·0	
3	Neon Ne 19·9	Sodium Na 23·05	Magnesium Mg 24·3	Aluminium Al 27·0	Silicon Si 28·4	Phosphorus P 31·0	Sulphur S 32·06	Chlorine Cl 35·45	
4	Argon Ar 38	Potassium K 39·1	Calcium Ca 40·1	Scandium Sc 44·1	Titanium Ti 48·1	Vanadium V 51·4	Chromium Cr 52·1	Manganese Mn 55·0	Iron Cobalt Nickel Fe Co Ni (Cu) 55·9 59 59
5		Copper Cu 63·6	Zinc Zn 65·4	Gallium Ga 70·0	Germanium Ge 72·3	Arsenic As 75	Selenium Se 79	Bromine Br 79·95	
6	Krypton Kr 81·8	Rubidium Rb 85·4	Strontium Sr 87·6	Yttrium Y 89·0	Zirconium Zr 90·6	Niobium Nb 94·0	Molybdenum Mo 96·0	—	Ruthenium Rhodium Palladium Ru Rh Pd (Ag) 101·7 103·0 106·5
7		Silver Ag 107·9	Cadmium Cd 112·4	Indium In 114·0	Tin Sn 119·0	Antimony Sb 120·0	Tellurium Te 127	Iodine I 127	
8	Xenon Xe 128	Cæsium Cs 132·9	Barium Ba 137·4	Lanthanum La 139	Cerium Ce 140	—	—	—	— — —
9		—	—	—	—	—	—	—	
10	—	—	—	Ytterbium Yb 173	—	Tantalum Ta 183	Tungsten W 184	—	Osmium Iridium Platinum Os Ir Pt (Au) 191 193 194·9
11		Gold Au 197·2	Mercury Hg 200·0	Thallium Tl 204·1	Lead Pb 206·9	Bismuth Bi 208	—	—	
12	—	—	Radium Rd 224	—	Thorium Th 232	—	Uranium U 239		
HIGHER SALINE OXIDES	R	R_2O	RO	R_2O_3	RO_2	R_2O_5	RO_3	R_2O_7	RO_4
HIGHER GASEOUS HYDROGEN COMPOUNDS					RH_4	RH_3	RH_2	RH	

Fig. 2.4 The short-form table from the 7th edition of Mendeleev's *Principles of Chemistry*. Reproduced from Mendeléeff (1905) [76]

2.7 Conclusion

The periodic law, as formulated by Mendeleev, was: "The properties of the elements (and of the simple and compound substances which they form) show a periodic dependence on their atomic weights." As seen, periodicity was an important aspect and it influenced Mendeleev's opinions on the best ways to represent the law graphically. Tabular forms were best to illustrate the periods, whereas most spiral forms did not show definite periods. Another important aspect of Mendeleev's periodic system was its use in changing atomic weights of already known elements and in predicting the characteristics of yet-to-be discovered ones. The changes and predictions showcased the utility of the periodic law. And Mendeleev's philosophical conception of chemical elements assisted in the development of the periodic law by allowing him to consider atomic weight as the essential part of the element upon which the law rested.

Just as Ladenburg had told his audience about natural laws, the periodic law did not suddenly spring out of the mind of Mendeleev. He was assisted by "the stock of generalisations and established facts" that had accumulated over decades as chemists sought different ways of determining relationships between the elements, determining atomic weights, and organizing and classifying the elements. These established facts combined with Mendeleev's own understanding of the elements to develop an organizational scheme that encompassed all of the elements, that was flexible enough to survive changes in chemical and physical knowledge, and was useful in both research and pedagogy.

References

1. Ladenburg A (1900) Lectures on the history of the development of chemistry since the time of Lavoisier. Edinburgh, Alembic Club, p 3
2. Mendeléeff D (1889) The periodic law of the chemical elements. J Chem Soc Trans 55:634–656. https://doi.org/10.1039/CT8895500634
3. Dalton J (1808–10) A new system of chemical philosophy. R. Bickerstaff, Manchester
4. Gmelin L (1844) Handbuch der Chemie, band 2. Karl Winter, Heidelberg. English edition: Gmelin L (1849) Handbook of chemistry, vol 2. (trans: Watts H). Cavendish Society, London, p 1
5. van Spronsen JW (1969) The periodic system of chemical elements: A history of the first hundred years. Elsevier, Amsterdam, pp 69–71
6. For more on Gmelin's system, see Scerri E (2020) The periodic table: Its story and its significance, 2nd edn. Oxford University Press, New York, pp 48–51
7. Venable FP (1896) The development of the periodic law. Chemical Publishing Co., Easton, p 39
8. Gladstone JH (1853) On the relations between the atomic weights of analogous elements. Phil Mag ser 4(5):313–320. https://doi.org/10.1080/14786445308562716
9. Cooke JP (1855) The numerical relation between the atomic weights, with some thoughts on the classification of the chemical elements. Mem Am Acad Arts Sci 5:235–256. https://doi.org/10.2307/25058181
10. For more on Cooke's system, see Venable FP (1896) The development of the periodic law. Chemical Publishing Co., Easton, pp 41–43

11. deMilt C (1951) The congress at Karlsruhe. J Chem Ed 28:421–425. https://doi.org/10.1021/ed028p421
12. Wurtz A (1984) Account of the sessions of the international congress of chemists in Karlsruhe, on 3, 4, and 5 September 1860. In: Nye MJ (ed) The question of the atom: From the Karlsruhe congress to the first Solvay conference, 1860–1911. Tomash, Los Angeles, pp 5–28
13. Cannizzaro S (1858) Lettera del Prof. Stanislao Cannizzaro al Prof. S. De Luca; sunto di un corso di filosofia chimica, fatto nella r. Università di Genova. Nuovo Cimento 7:321–366. https://doi.org/10.1007/bf02827711 English edition: Cannizzaro S (1911) Sketch of a course of chemical philosophy. Alembic Club, Edinburgh
14. Cannizzaro S (1891) Abriss eines Lehrganges der theoretischen Chemie vorgetragen an der k. Universität Genua (trans: Miolati A). Ostwalds Klassiker der exakten Wissenschaften 30. Wilhelm Engelmann, Leipzig, p 59
15. For the full diagram, see Béguyer de Chancourtois A-E (1863) Vis tellurique: classement naturel des corps simples ou radicaux, obtenu au moyen d'un système de classification helicoïdal et numérique. Mallet-Bachelier, Paris. A three-dimensional version of the Vis tellurique can be seen at https://patrimoine.mines-paristech.fr/document/Vis_tellurique. Accessed 1 Nov 2020
16. Béguyer de Chancourtois A-E (1863) Suite du mémoire de la vis tellurique, du 7 Avril 1860, adressé à propos du thallium. C R Acad Sci 56:479–482
17. Hartog PJ (1889) A first foreshadowing of the periodic law. Nature 41:186–188. https://doi.org/10.1038/041186a0
18. Hinrichs G (1866) On the spectra and composition of the elements. Am J Sci Arts ser 2 (42):350–368
19. Hinrichs G (1867) Programme der Atomechanik oder die Chemie eine Mechanik der Panatome. Privately printed, Iowa City, IA
20. Hinrichs G (1869) Natural classification of the elements. Pharmacist 2:10–12
21. Odling W (1857) On the natural groupings of the elements. Phil Mag ser 4(13):423–439, 480–497. https://doi.org/10.1080/14786445708642323, https://doi.org/10.1080/14786445708642334
22. Odling W (1864) On the proportional numbers of the elements. Q J Sci 1:642–648
23. Odling W (1865) A course of practical chemistry arranged for the use of medical students, 2nd edn. Longmans, Green, London, p 226
24. Odling W (1869) Cours de chimie practique (analytique, toxocologique, animale) a l'usage des étudiants en médicine, 3rd edn (trans: Naquet A). F. Savy, Paris, p 261
25. Newlands collected his articles into a single volume: Newlands JAR (1884) On the discovery of the periodic law, and on relations among the atomic weights. Spon, London
26. Newlands JAR (1865) On the law of octaves. Chem News 12:83
27. Anon (1866) Chemical Society. Chem News 13:113–114
28. Newlands JAR (1866) On the "law of octaves". Chem News 13:130
29. For more on Newlands, see Gordin, M (2018) Paper tools and periodic tables: Newlands and Mendeleev draw grids. Ambix 65:30–51. https://doi.org/10.1080/00026980.2017.1418251
30. For more on the priority dispute between Lothar Meyer and Mendeleev, see Gordin, M (2012) The textbook case of a priority dispute: D. I Mendeleev, Lothar Meyer, and the periodic system. In: Biagioli M, Riskin J (ed) Nature engaged: Science in practice from the Renaissance to the present. Palgrave Macmillan, New York, pp 59–82
31. Bedson PP (1901) Lothar Meyer memorial lecture. Memorial lectures delivered before the Chemical Society, 1893-1900. Gurney and Jackson, London, pp 1402–1439
32. Rocke A (2019) Lothar Meyer's pathway to periodicity. Ambix 66:265–302. https://doi.org/10.1080/00026980.2019.1677976
33. Meyer L (1864) Die modernen Theorien der Chemie und ihre Bedeutung für die chemische Statik. Breslau, Maruschke & Berendt, p 137
34. Seubert K (1895) Zur Geschichte des periodischen Systems. Z Anorg Chem 9:334–338. https://doi.org/10.1002/zaac.18950090130

35. Meyer L (1870) Die Natur der chemischen Elemente als Function ihrer Atomgewichte. Ann Chem suppl 7:354–364. English translation: Meyer JL (1952) The nature of the chemical elements as a function of their atomic weights. In: Leicester HM, Klickstein HS (ed) A source book in chemistry, 1400–1900. Harvard University Press, Cambridge, pp 434–438
36. Gordin MD (2019) A well-ordered thing: Dmitrii Mendeleev and the shadow of the periodic table, rev edn. Princeton University Press, Princeton, chapter 2
37. Kaji M (2018) The origin of Mendeleev's discovery of the periodic system. In: Scerri E, Restrepo G (eds) Mendeleev to oganesson: A multidisciplinary perspective on the periodic table. Oxford University Press, New York, pp 219–244
38. Brooks NM (2000) Dmitrii Mendeleev's *Principles of chemistry* and the periodic law of the elements. In: Lundgren A, Bensaude-Vincent B (eds) Communicating chemistry: Textbooks and their audiences, 1789-1939. Science History Publications, Canton, pp 295–309
39. Bensaude-Vincent B (1986) Mendeleev's periodic system of chemical elements. Br J Hist Sci 19:3–17
40. Mendeleev D (2005) On the correlation between the properties of the elements and their atomic weights. In: Mendeleev DI (2005) Mendeleev on the periodic law: Selected writings, 1869–1905 (trans: Jensen WB). Dover, Mineola, pp 18–37. Russian original: Mendeleev D (1869) Sootnoshenie svoistv s atomnym vesom elementov. Zh Russ Khim O-va 1:60–77
41. Mendeleev D (2005) On the periodic regularity of the chemical elements. In: Mendeleev DI (2005) Mendeleev on the periodic law: Selected writings, 1869–1905 (trans: Jensen WB). Dover, Mineola, pp 38–109. German original: Mendelejeff D (1871) Die periodische Gesetzmäßigkeit der chemischen Elemente. Ann Chem suppl 8:133–229
42. For English translations of many of Mendeleev's papers, see Mendeleev DI (2005) Mendeleev on the periodic law: Selected writings, 1869–1905 (trans: Jensen WB). Dover, Mineola
43. Mendeleev D (1869) Sootnoshenie svoistv s atomnym vesom elementov. Zh Russ Khim O-va 1:60–77. German abstract: Mendelejeff D (1869) Ueber die Beziehungen der Eigenschaften zu den Atomgewichten der Elemente. Z Chem 5:405–406
44. Mendelejeff D (1871) Die periodische Gesetzmäßigkeit der chemischen Elemente. Ann Chem suppl 8:133–229. French translation: Mendeleeff D (1879) La loi périodique des éléments chimiques. Monit Sci ser 3(9):691–737
45. Dalton J (1808–10) A new system of chemical philosophy. R. Bickerstaff, Manchester, pp 221–222
46. For a deeper look at Mendeleev's philosophical views about elements, see Bensaude-Vincent B (2019) Reconceptualizing chemical elements through the construction of the periodic system. Centaurus 61:299–310. https://doi.org/10.1111/1600-0498.12228
47. Mendeleev D (1949) Lektsii po obshchei khimii 1867/68 g. In: Sochineniia, vol 15. Izd. Akademii Nauk SSR, Leningrad. Quoted in Kaji M (2002) D. I. Mendeleev's concept of chemical elements and *The Principles of Chemistry*. Bull Hist Chem 27:4–16
48. Kragh H (2019) The periodic system and the idea of a chemical element: From Mendeleev to superheavy elements. Centaurus 61:329–344. https://doi.org/10.1111/1600-0498.12231
49. Mendeleev D (2005) The periodic law of the chemical elements. In: Mendeleev DI (2005) Mendeleev on the periodic law: Selected writings, 1869–1905 (trans: Jensen WB). Dover, Mineola, pp 138–141. French original: Mendeleef D (1879) La loi périodique des éléments chimiques. Monit Sci ser 3(9):691–737
50. Mendelejeff D (1876) Remarks on the discovery of gallium. Phil Mag ser 5, 1:542–546. French original: Mendeleeff D (1875) Remarques à propos de la découverte du gallium. C R Acad Sci 81:969–972
51. Lecoq de Boisbaudran (1877) On the new metal—gallium. Chem News 35:148–151, 157–160, 167–170. French original: Lecoq de Boisbaudran (1875) Sur un noveau métal, le gallium. Ann Chim Phys 10:100–141
52. Pattison Muir MM (1877) Observations on gallium. Phil Mag ser 5(3):281–283

53. Clève P (1879) On scandium. Chem News 40:159–160. French original: Clève P (1879) Sur le scandium. C R Acad Sci 89:419–422
54. Winkler C (1886) Germanium, Ge, ein neues, nichtmetallisches Element. Ber Dtsch Chem Ges 19:210–211
55. Humpidge TS (1885) On the atomic weight of glucinum (beryllium). Second paper. Proc R Soc London 39:1–19
56. Parsons CL (1909) The chemistry and literature of beryllium. Chemical Publishing Co., Easton, p 2
57. Nilson LF, Pettersson O (1880) On the essential properties and chemical character of beryllium (glucinum). Chem News 42:297–299
58. Humpidge TS (1880) On the atomic weight of beryllium. Chem News 42:261–263
59. Humpidge TS (1883) On the atomic weight of glucinum (beryllium). Philos Trans R Soc London 174:601–613
60. Mendeléeff D (1891) The principles of chemistry, vol 2. (trans: Kamensky G). Longmans, Green, London, pp 449, 448. Russian original: Mendeleev DI (1889) Osnovy khimii, 5th edn. St. Petersburg
61. van Spronsen JW (1969) The periodic system of chemical elements: A history of the first hundred years. Elsevier, Amsterdam
62. Quam GN, Quam MB (1934) Types of graphic classifications of the elements. J Chem Educ 11:27–32, 217–223, 288–297. https://doi.org/10.1021/ed011p27; https://doi.org/10.1021/ed011p217; https://doi.org/10.1021/ed011p288
63. Mazurs EG (1957) Graphic representations of the periodic system during one hundred years, rev edn. University of Alabama Press, University
64. For a collection of tables drawn by Mendeleev, see Smith JR (1976) Persistence and periodicity: A study of Mendeleev's contribution to the foundations of chemistry. Dissertation, University of London, chapter 4. https://ethos.bl.uk/OrderDetails.do?uin=uk.bl.ethos.473179. Accessed 6 Nov 2020
65. Bensaude-Vincent B (2001) Graphic representations of the periodic system of chemical elements. In: Klein U (ed) Tools and modes of representation in the laboratory sciences. Kluwer, Dordrecht, pp 133–161
66. Mendeléeff D (1891) The principles of chemistry, vol 2. (trans: Kamensky G). Longmans, Green, London, pp 16, 19. Russian original: Mendeleev D (1889) Osnovy khimii, 5th edn. St. Petersburg
67. Smith JR (1976) Persistence and periodicity: A study of Mendeleev's contribution to the foundations of chemistry. Dissertation, University of London, p 292. https://ethos.bl.uk/OrderDetails.do?uin=uk.bl.ethos.473179. Accessed 6 Nov 2020
68. Baumhauer H (1870) Die Beziehungen zwischen dem Atomgewichte und der Natur der chemischen Elemente. Friedrich Vieweg und Sohn, Braunschweig, pp 21–22
69. Mendeleev D (2005) On the question concerning the system of elements. In: Mendeleev DI (2005) Mendeleev on the periodic law: Selected writings, 1869–1905 (trans: Jensen WB). Dover, Mineola, pp 115–119. German original: Mendelejeff D (1871) Zur Frage über das System der Elemente. Ber Dtsch Chem Ges 4:348–352
70. Smith JR (1976) Persistence and periodicity: A study of Mendeleev's contribution to the foundations of chemistry. Dissertation, University of London, pp 292, 293. https://ethos.bl.uk/OrderDetails.do?uin=uk.bl.ethos.473179. Accessed 6 Nov 2020
71. Mendeléeff D (1891) The principles of chemistry, vol 2. (trans: Kamensky G). Longmans, Green, London, p 19. Russian original: Mendeleev DI (1889) Osnovy khimii, 5th edn. St. Petersburg
72. Girolami GS, Mainz VV (2019) Mendeleev, Meyer, and atomic volumes: an introduction to an English translation of Mendeleev's 1869 article "On the atomic volume of simple bodies". Bull Hist Chem 44:100–108
73. Robinson AE (2019) Chemical pedagogy and the periodic system. Centaurus 61:360–378. https://doi.org/10.1111/1600-0498.12229

74. Smith JR (1976) Persistence and periodicity: A study of Mendeleev's contribution to the foundations of chemistry. Dissertation, University of London, pp 288–289. https://ethos.bl.uk/OrderDetails.do?uin=uk.bl.ethos.473179. Accessed 6 Nov 2020
75. Mendeleev D (2005) On the history of the periodic law. In: Mendeleev DI (2005) Mendeleev on the periodic law: Selected writings, 1869–1905 (trans: Jensen WB). Dover, Mineola, pp 142–151. German original: Mendelejeff D (1880) Zur Geschichte des periodischen Gesetzes. Ber Dtsch Chem Ges 13:1796–1804
76. Mendeléeff D (1905) The principles of chemistry, 3rd English edn, vol 1. (trans: Kamensky G). Longman, Greens, London, p xviii. Russian original: Mendeleev D (1902) Osnovy khimii, 7th edn. St. Petersburg

The Trouble with Triads

3

William B. Jensen

Abstract

Both Döbereiner's original atomic weight triads and the recent atomic number triads are critically evaluated and found to be neither necessary nor sufficient to determine the degree of relatedness among the chemical elements. Mendeleev's revision of the triad concept to include both vertical and horizontal triads and to interpolate various properties rather than to determine chemical relatedness is also reviewed.

"They do seem to be all over."
The Trouble with Tribbles
Star Trek 1967

3.1 Origins

As many readers of this symposium volume are aware, the concept of triads—or the concept that, for a consecutive sequence of three chemically analogous elements,[1] the average of the atomic weights of the heaviest and lightest members of the

[1]By chemically analogous we mean elements that we today would place in the same group of the periodic table by virtue of having identical valence electron counts and analogous electron configurations. This excludes the transition element triplets created by Mendeleev in order to force all elements into just eight groups.

W. B. Jensen (✉)
Department of Chemistry, University of Cincinnati, Cincinnati, OH 45221, USA
e-mail: jensenwb@ucmail.uc.edu

C. J. Giunta et al. (eds.), *150 Years of the Periodic Table*, Perspectives on the History of Chemistry, https://doi.org/10.1007/978-3-030-67910-1_3

sequence is approximately equal to that of the intermediate element—was first introduced by the German chemist, Johann Wolfgang Döbereiner (Fig. 3.1), in a paper published in 1829 [1], though he claimed to have mentioned the idea several years earlier in his chemical lectures.[2] To the modern chemist, this paper is a mess in that it contains inconsistent data, inconsistent symbolism, and a failure to explicitly state its underlying assumptions.

Döbereiner first illustrated his approach by applying it to four triads involving elements that were already widely recognized by chemists and textbooks of the time as being chemically analogous. These are shown in a slightly reformatted form in Table 3.1 and served as his learning set or test cases. However, as can be seen, for the nonmetallic elements he used their atomic weights as measured on an H = 1.00 scale, whereas for the metals he used the equivalent weights of their oxides as measured on an O = 100 scale. In addition, the formulas for the oxides of the alkaline metals are now known to be incorrect and he is under the illusion that in 1829 it was mathematically meaningful to report his atomic and equivalent weights to the nearest thousandth.

Next Döbereiner reversed his tactics and used the observed formation of triads as an argument for the classification of hitherto unclassified metals into chemically analogous groups, again using the equivalent weights of their oxides as measured on an O = 100 scale (Table 3.2). As may be seen, in light of our current knowledge of the periodic table—which was, of course, unavailable to Döbereiner—all of these proposed groupings are incorrect.

These results lead us to formulate two pertinent questions:

1. Do elements known to be chemically analogous via other criteria automatically form triads?
2. Can the formation of triads be used to predict which elements are chemically analogous?

Based on Döbereiner's results, we can tentatively answer the first of these questions in the affirmative, though, as we will see later, this answer must be modified. Likewise, we can answer the second question in the negative. Indeed, I would like to dub this second, incorrect, proposition as "Döbereiner's fallacy."

As is well known to historians of the periodic table, both triads and Döbereiner's fallacy were widely used prior to the 1860s by chemists interested in the classification of the chemical elements, often with less than satisfactory results. Lest you doubt this, let me cite an 1857 paper by the German chemist, Ernst Lenssen [2], that is in many ways the culmination of the triad fad. In this paper, Lenssen attempted to classify the 58 elements known at the time into 20 triads (Table 3.3). Those triads considered correct by today's standards are in italics and all but one of these are identical to those given by Döbereiner 28 years earlier. The remaining triads do not involve elements that we today would consider chemically analogous. This is a 75% error rate!

[2]See Gisela Boeck's article, Chapter 8 in this volume, which cites an 1816 letter from Döbereiner to Goethe in which he describes triads.

Fig. 3.1 Johann Wolfgang Döbereiner (1780–1849) (Courtesy of the Oesper Collections in the History of Chemistry, University of Cincinnati)

Table 3.1 Döbereiner's Learning Set (Reformatted)

Cl-Br-I:	$(AW_{Cl} + AW_{I})/2 = (35.470 + 126.470)/2 = 80.470 = AW_{Br}$
S-Se-Te:	$(AW_{S} + AW_{Te})/2 = (32.239 + 129.243)/2 = 80.741 = AW_{Se}$
CaO-SrO-BaO:	$(EW_{CaO} + EW_{BaO})/2 = (356.019 + 956.880)/2 = 656.449 = EW_{SrO}$
LiO-NaO-KO:	$(EW_{LiO} + EW_{KO})/2 = (195.310 + 589.916)/2 = 392.613 = EW_{NaO}$

Table 3.2 Döbereiner's Predictions (Reformatted)

Fe_2O_3-Cr_2O_3-Mn_2O_3:	$(EW_{Fe2O3} + EW_{Mn2O3})/2 = (979.426 + 1011.574)/2 = 995.000 = EW_{Cr2O3}$
FeO-MnO-CoO:	$(EW_{FeO} + EW_{CoO})/2 = (439.213 + 468.911)/2 = 452.102 = EW_{MnO}$
NiO-CuO-ZnO:	$(EW_{NiO} + EW_{ZnO})/2 = (469.675 + 503.226)/2 = 486.450 = EW_{CuO}$
PtO-IrO-OsO:	$(EW_{PtO} + EW_{OsO})/2 = (1233.260 + 1244.210)/2 = 1238.735 = EW_{IrO}$

Table 3.3 Lenssen's 20 Triads

I	*Li–Na-K*	XI	C-N-O
II	*Ca-Sr-Ba*	XII	B-F-Si
III	Mg-Zn-Cd	XIII	*Cl-Br-I*
IV	Mn–Fe-Co	XIV	*S-Se-Te*
V	La-Ce-Nb	XV	*P-As-Sb*
VI	Y-Tb–Er	XVI	Ti-Sn-Ta
VII	Al–No-Th	XVII	V-Mo-W
VIII	Be-Zr-U	XVIII	Ru-Rh-Pd
IX	Cr-Ni-Cu	XIX	Os–Ir-Pt
X	Ag-?-Pb	XX	Bi-?-Au

3.2 The Modern Revival of Triads

One would have thought that these abysmal results would have been sufficient to consign the triad concept to the dustbin of history, but, rather remarkably, in recent years it has once more been revived by two well-known authors on the periodic table [3–6]. However, these newer triads involve atomic numbers rather than atomic weights. As may be seen by applying this revised concept to the four triads in Döbereiner's learning set, these new atomic number triads work perfectly:

Cl-Br-I: $(Z_{Cl} + Z_{I})/2 = (17 + 53)/2 = 35 = Z_{Br}$
S-Se-Te: $(Z_{S} + Z_{Te})/2 = (16 + 52)/2 = 34 = Z_{Se}$
Ca-Sr-Ba: $(Z_{Ca} + Z_{Ba})/2 = (20 + 56)/2 = 38 = Z_{Sr}$
Li–Na-K: $(Z_{Li} + Z_{K})/2 = (3 + 19)/2 = 11 = Z_{Na}$

This is hardly surprising given the extremely high correlation coefficient between atomic numbers and atomic weights (Fig. 3.2)—a correlation that allowed Mendeleev, in the absence of the former, to discover the periodic law using the latter. This newer version also has the advantage of using only whole numbers, rather than numbers with messy fractions, and of giving exact rather than approximate agreement between the predicted and actual values.

However, these newer triads also allow us to discover that chemically analogous elements do not always form triads. Thus, for consecutive group 3 elements, the average of the atomic numbers of Ga and Tl gives a value of 56 for In rather than its actual value of 49:

Ga-In-Tl: $(Z_{Ga} + Z_{Tl})/2 = (31 + 81)/2 = 56 \neq 49 = Z_{In}$

Likewise, for consecutive noble gases, the average of the atomic numbers of Kr and Rn gives a value of 61 for Xe rather than the actual value of 54:

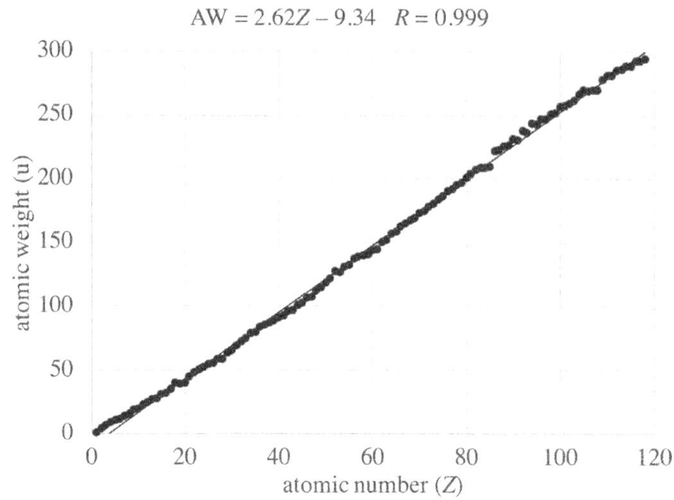

Fig. 3.2 The correlation between atomic weights and atomic numbers

Kr-Xe–Rn: $(Z_{Kr} + Z_{Rn})/2 = (36 + 86)/2 = 61 \neq 54 = Z_{Xe}$

What is going on here may be understood using a simple mathematical proof in which we predict the atomic number of an element A by averaging the atomic number of an element y units greater than that of A and the atomic number of an element x units less than that of A:

$$Z_A = [(Z_A + y) + (Z_A - x)]/2 = [2Z_A + y - x]/2$$

Obviously the right side of this equation is equal to Z_A if and only if $x = y$. In other words, to form an atomic number triad the two elements being averaged must be equally spaced on the atomic number scale above and below the element in the center of the triad.

This explains our earlier exceptions, since the full periodic table—due to the insertion of the *d*- and *f*-blocks—is not a rectangle but rather a triangle (Fig. 3.3) and therefore not all elements within a given group are necessarily equally or even approximately spaced with respect to their atomic numbers and atomic weights.

Indeed we can go further, since our proof shows that there are no restrictions on how far the elements must be spaced that are being averaged as long as the spacing for the largest and smallest about the center element are equal. This means that there are literally hundreds of potential atomic number triads within the periodic table. Thus we can form a triad between Al, As, and I—three elements that are equally spaced by 20 units but are neither in the same group nor the same row of the periodic table:

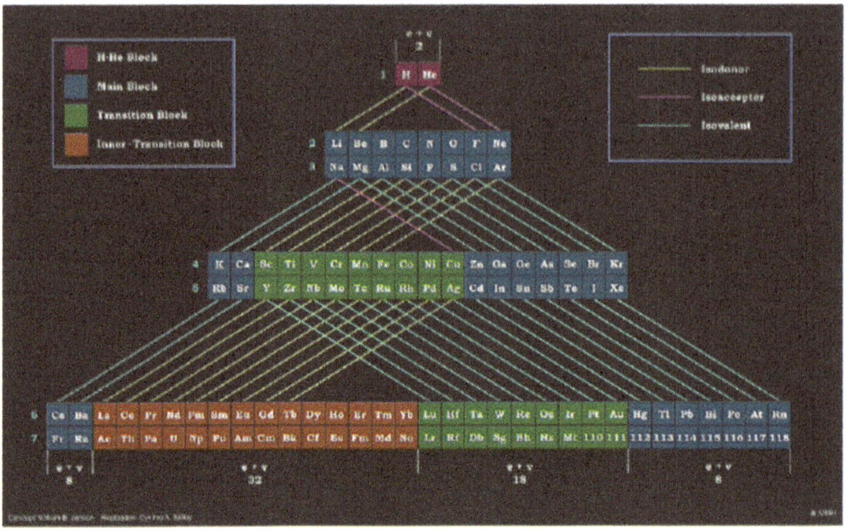

Fig. 3.3 A step-pyramid periodic table illustrating unequal row lengths (Courtesy of the Oesper Collections in the History of Chemistry, University of Cincinnati)

$$\text{Al-As-I: } (Z_{Al} + Z_I)/2 = (13 + 53)/2 = 33 = Z_{As}$$

or a triad involving a transition element, a group 3 main-block element, and a noble gas that are equally spaced by 5 units:

$$\text{Fe-Ga-Kr: } (Z_{Fe} + Z_{Kr})/2 = (26 + 36)/2 = 31 = Z_{Ga}$$

But, advocates of the new triads will object that this as an abuse, since triads are intended to be formed only between chemically analogous elements. However, if you are restricting their use to chemically analogous elements in order to predict which elements are chemically analogous, then this is a circular argument, since you are assuming the very thing you are trying to prove.

So, in summary—while remembering that what is true of atomic number triads is also approximately true of atomic weight triads—the answers to the two questions posed in the previous section are:

1. Chemically analogous elements do not automatically form triads.
2. Triad formation is neither necessary nor sufficient to establish chemical relatedness.

3.3 Mendeleev and Triads

To return once more to the history of chemistry, we now ask the question "Did Mendeleev use triads to make his famous predictions of the properties of eka-boron or scandium, eka-aluminum or gallium, and eka-silicon or germanium?" To the best of my knowledge, Mendeleev never used the term triad in his various writings on the periodic law, nor did he ever explicitly explain how he made his famous predictions. However, in his comprehensive review of 1871 [7] [3] he indirectly implied that he made them by averaging the values for both an element's vertical and horizontal nearest neighbors, a procedure that he illustrated using the nearest neighbors (S, Te, As and Br) of Se:

$$AW_{Se} = (AW_{S} + AW_{Te} + AW_{As} + AW_{Br})/4 = (32 + 125 + 75 + 80)/4 = 78$$

This gives an average value of 78, and is the value reported by Mendeleev, who at the time rounded most of his atomic weights to the nearest whole number or, at best, to the nearest tenth. Mendeleev viewed this averaging procedure as one of interpolation, which is what triad formation really is, mathematically speaking, and not as a procedure for predicting chemical relatedness. In other words, he introduced a new use for the triad concept, both by expanding it to include horizontal as well as vertical triads and by rejecting Döbereiner's fallacy.

When this procedure is applied to the case of Sc there is no lighter vertical nearest neighbor, since Sc is the first member of its group and the atomic weight of Y, its heavier vertical nearest neighbor, was questionable at the time. This leaves us with its two horizontal nearest neighbors, Ca and Ti:

$$AW_{Sc} = (AW_{Ca} + AW_{Ti})/2 = (40 + 48)/2 = 44$$

which gives an average value of 44 for eka-boron, as reported by Mendeleev in 1871.

In the case of Ga there were only three nearest neighbors known at the time, Al, In and Zn, since the space for Ge was blank:

$$AW_{Ga} = (AW_{Al} + AW_{In} + AW_{Zn})/3 = (27.3 + 113 + 65)/3 = 68.4$$

This gives an average value of 68.4 for eka-aluminum as compared with the value of 68 reported by Mendeleev when rounded to the nearest whole number.

In the case of Ge we once again have only three nearest neighbors, Si, Sn, and As, since the space for Ga was blank at the time:

[3] Mendeleev dated his review as August of 1871, though it appeared in an 1871-72 supplement volume of Liebig's journal. Historians are divided over which date to cite. The atomic weights predicted in 1871 differ from those predicted in 1869.

$$AW_{Ge} = (AW_{Si} + AW_{Sn} + AW_{As})/3 = (28 + 118 + 75)/3 = 73.7$$

This gives an average value of 74 for eka-silicon when rounded to the nearest whole number and does not agree with the value of 72 reported by Mendeleev. However, if we also include the value predicted above for Ga:

$$AW_{Ge} = (AW_{Si} + AW_{Sn} + AW_{As} + AW_{Ga})/4 = (28 + 118 + 75 + 68)/4 = 72.3$$

we get the value reported by Mendeleev when it is rounded to the nearest whole number.

3.4 Other Applications

There are several other important topics related to triads, including their use in predicting properties other than atomic weights, and their use in extrapolating rather than interpolating properties. Unfortunately, limitations of both time and space preclude coverage of these topics in this chapter, though I hope to have the opportunity to deal with them in a future paper.

References

1. Döbereiner JW (1829) Versuch zu einer Gruppirung der elementaren Stoffe nach ihrer Analogie. Ann Phys Chem 15:301–307. English translation In: Leicester HM, Klickstein HS (eds) (1952) A source book in chemistry, 1400–1900, Harvard, Cambridge, MA, pp 268–272
2. Lenssen E (1857) Ueber die Gruppirung der Elemente nach ihrem chemisch-physikalischen Character. Ann Chem Pharm 103:121–131
3. Bent HA (2000) New ideas in chemistry from fresh energy for the periodic law. authorHouse, Bloomington, IN
4. Scerri ER (2008) The role of triads in the evolution of the periodic table: past and present. J Chem Educ 85:585–589
5. Scerri ER (2010) Explaining the periodic table and the role of chemical triads. Found Chem 12:69–83
6. Scerri ER (2011) The periodic table: a very short introduction. Oxford University Press, Oxford, UK
7. Mendeleev, DI (1872) Die periodische Gesetzmässigkeit der chemischen Elemente. Ann Chem Pharm 8(suppl):133–229. English translation in: Jensen WB (ed) (2005) Mendeleev on the periodic law: Selected writings, 1869–1905. Dover, Mineola, NY, pp 38–109

Josiah Parsons Cooke, the Natural Philosophy of Sir John F. W. Herschel and the Rational Chemistry of the Elements

4

Ronald Brashear and Gary Patterson

Abstract

The natural philosophy of chemistry grew slowly until the articulation of the axiom of chemical atoms by John Dalton in the early nineteenth century. Sir John F. W. Herschel was one of the leading natural philosophers of the nineteenth century and published his own monograph, in which he listed the ten key axioms of chemistry. Later in that decade, he gave the President's address for the Chemistry Section at the 1858 British Association for the Advancement of Science meeting. He identified significant progress in understanding the relationships between the known chemical elements by Josiah Parsons Cooke of Harvard. Examination of the long paper by Cooke published in the *Memoirs of the American Academy of Arts and Sciences* in 1855 gives his natural groupings of the known chemical elements. Further progress contained in Cooke's 1868 textbook, *First Principles of Chemical Philosophy*, documents just how important his work was in the development of the periodic table, soon to be published by Mendeleev. I. Bernhard Cohen lauded Cooke as the first really significant academic chemist in America.

4.1 Introduction

The science we know as Chemistry is composed of many parts. An empirical base of technological knowledge existed in antiquity, and by the sixteenth century, chemistry was being taught in medical schools by physicians such as Oswald Croll

R. Brashear · G. Patterson (✉)
Science History Institute, Philadelphia, PA, USA
e-mail: gp9a@andrew.cmu.edu

G. Patterson
Carnegie Melon University, Pittsburgh, PA, USA

C. J. Giunta et al. (eds.), *150 Years of the Periodic Table*, Perspectives on the History of Chemistry, https://doi.org/10.1007/978-3-030-67910-1_4

(1580–1609) [1] and Andreas Libavius (1550–1616) [2]. A philosophical foundation for chemistry was attempted by alchemists such as Paracelsus (1493–1541) [3], but this paradigm was not a way forward. The seventeenth century saw great progress in the overall field of Natural Philosophy. The empirical data of astronomy was reduced to a coherent system by Copernicus, Kepler and Newton. While such an elegant mathematization of the data is not necessary for a science to achieve a coherent natural philosophy, some axioms must be created that are empirically verified and will serve as the foundation for further experimentation and articulation. For chemistry, this level of scientific sophistication needed to wait until the nineteenth century. 2019 was the UNESCO International Year of the Periodic Table, and celebrated the formulation of a coherent arrangement of the known chemical elements in 1869.

The central underlying concept that leads to the Periodic table is the existence of a class of substances called **elements**. Robert Boyle (1627–1691) is credited with early cogent thoughts on this subject [4–6]. Chemistry could be based on many different coherent sets of fundamental substances, and in the seventeenth century many common chemicals were viewed in this light, even though we now consider them to be compounds.

The eighteenth century was dominated by the coherent version of chemistry created by Herman Boerhaave (1668–1738) [7]. (Boerhaave was a Professor of Medicine, Botany, and Chemistry at the University of Leyden. His Latin lectures on chemistry were published and used for more than 100 years.) Chemistry started with the substances found in Nature, and continued through the use of chemical forms of transformation, especially the furnace. Classes of reactions were developed, and a form of systematic chemistry was created. But, in the absence of a truly microscopic perspective, the task was too hard. Attempts to articulate the Newtonian paradigm of particles interacting through complicated potentials were carried out by geniuses such as Roger Boscovich (1711–1787) [8], but such an opaque theory was not the way forward. (Boscovich was a Jesuit priest and mathematician from Ragusa (now Dubrovnik, Croatia) who revolutionized natural philosophy throughout Europe [9].)

The number of substances that were viewed as elements increased throughout the eighteenth century. A good compilation of elements can be found in the book by Antoine Lavoisier (1734-1794) published in 1789 [10]. A constructed Table based on this book is presented in *The Chemical Tree* by William Brock [11] (see Fig. 4.1). While the number of substances was increasing, the elementary clarity was primitive.

From a pedagogical perspective, the nineteenth century was dominated by the synoptic view of chemistry instantiated in *The Elements of Experimental Chemistry* by William Henry (1774–1836) [12]. Following the concepts created by John Dalton (1766–1844) [13], Humphry Davy (1778-1829) [14] and William Wollaston (1766–1828) [15], Henry compiled a list of atomic elements and compounds. There were 48 known substances in this list. Henry also noted that there were a few groups of elements that had similar chemistries. Oxygen, nitrogen, hydrogen, and chlorine were observed at room temperature as gases. (Fluorine had not yet been

TABLE 3.2 *Lavoisier's 'elements' or 'simple substances'.*

Light	Sulphur	Antimony	Mercury	Lime
Caloric	Phosphorus	Arsenic	Molybdena	Magnesia
Oxygen	Charcoal	Bismuth	Nickel	Barytes
Azote	Muriatic radical	Cobalt	Platina	Argilla
Hydrogen	Fluoric radical	Copper	Silver	(alumina)
	Boracic radical	Gold	Tin	Silex
		Iron	Tungsten	(silica)
		Lead	Zinc	
		Manganese		

Fig. 4.1 Table of simple substances compiled from Lavoisier (1789) [10] (Reproduced from *The Chemical Tree* [11] by permission)

isolated as a gas.) Sodium and potassium were obviously related as the bases of *alkaline salts* with chlorine as the corresponding element. Barium, strontium, calcium, magnesium, silicon, aluminum, zirconium, glucinum (beryllium), and yttrium were observed as the bases of *alkaline earths.* The earths were all oxides. Carbon, boron, sulfur, nitrogen, and phosphorus were all the bases of oxygen acids. Chlorine, fluorine, and iodine were the bases of hydrogen acids. And, of course, there were many *metals.* Even long after the formulation of the periodic table, elementary descriptive chemistry was predominantly a repetition of Henry. The story of the development of the rational chemistry of the elements is told by Alan Rocke in *Chemical Atomism in the Nineteenth Century: From Dalton to Cannizzaro* [16].

One of the most important precursors of the atomic theory was developed by Benjamin Richter (1762–1807) of Prussia. He considered the combinations of many compounds, especially salts, acids, and bases, and concluded that fixed weights of each substance combined with different fixed weights of other members of these classes. He formulated the theory of stoichiometry in essentially the form in which it exists today. But, he did not yet know about atoms. He expressed his specific weights in terms of equivalents for each substance.

The importance of the work of John Dalton is that he articulated an axiom of chemistry that forms the basis for all further work: All matter is composed of ***chemical atoms***, and there is a unique such particle associated with each element. While Dalton's understanding of these fictive objects was imperfect, and some of the basic chemistry of compounds was incorrect, this paradigm inspired the right kind of experimental program, since it turned out to be essentially correct. Further articulations occurred as new experiments were carried out and many surprises were yet to appear, but chemistry now had a way forward.

The story of the chemistry being pursued in Great Britain during the first third of the nineteenth century is well told by Thomas Thomson (1773–1852, FRS), the Regius Professor of Chemistry in the University of Glasgow [17]. In addition, he was a superb laboratory chemist who made many contributions to the field of experimental chemistry. He was memorialized in an Alembic Club reprint of the key papers by himself, Dalton and Wollaston concerning the Foundations of the Atomic Theory [18]. In his monumental *History of Chemistry* (1835), he introduces the subject [17]:

> Like all other great improvements in science, the atomic theory developed itself by degrees, and several of the older chemists ascertained facts which might, had they been aware of their importance, have led them to conclusions similar to those of the moderns.

Thomson traveled around Great Britain in his attempt to improve his understanding of chemistry. In 1804 he spent time with Dalton in Manchester. During their extensive discussions, Dalton revealed his thoughts about the microscopic basis of the properties of chemicals. Dalton opined that "The ultimate particles of all simple bodies are ***atoms*** incapable of further division. These atoms (at least as viewed along with their atmospheres of heat) are all spheres, and are each of them possessed of particular weights, which shall be denoted by numbers" [17]. The paradigm that all elements are defined by actual physical objects with a definite mass remains the basis of microscopic chemistry today. Many articulations have occurred, and many surprises have been observed, but the fundamental idea led to a series of fruitful experiments that continue today.

Thomson was a prolific author and actually introduced Dalton's system to the public before its originator, with full credit being given [19]. One of the best stories in all the history of chemistry involves Thomson, Davy, Wollaston, and Davies Gilbert (1767–1839) (the President of the Royal Society). Thomson scripts it this way:

> In the autumn of 1807 I had a long conversation with him [Davy] at the Royal Institution, but could not convince him that there was any truth in the hypothesis [Dalton's chemical atoms]. A few days after, I dined with him at the Royal Society Club, at the Crown and Anchor, in the Strand. Dr. Wollaston was present at the dinner. After dinner every member of the club left the tavern, except Dr. Wollaston, Mr. Davy, and myself, who staid behind and had tea. We sat about an hour and a half together, and our whole conversation was about the atomic theory. Dr. Wollaston was a convert as well as myself; and we tried to convince Davy of the inaccuracy of his opinions; but, so far from being convinced, he went away, if possible, more prejudiced against it than ever. Soon, after, Davy met Mr. Davis [*sic*] Gilbert, the late distinguished president of the Royal Society; and he [Davy] amused him with a caricature description of the atomic theory, which he exhibited in so ridiculous a light, that Mr. Gilbert was astonished how any man of sense or science could be taken in with such a tissue of absurdities. Mr. Gilbert called on Dr. Wollaston (probably to discover what could have induced a man of Dr. Wollaston's sagacity and caution to adopt such opinions), and was not sparing in laying the absurdities of the theory, such as they had been represented to him by Davy, in the broadest point of view. Dr. Wollaston begged Mr. Gilbert to sit down, and listen to a few facts which he would state to him. He then went over all the principal facts at that time known respecting the salts; mentioned the alkaline carbonates and bicarbonates, the oxalate, binoxalate, and quadroxalate of potash, carbonic oxide and carbonic acid, olefiant gas, and carburetted hydrogen; and doubtless many other similar compounds, in which the proportion of one of the constituents increases in a regular ratio. Mr. Gilbert went away a convert to the truth of the atomic theory; and he had the merit of convincing Davy that his former opinions on the subject were wrong.

This is science at its best! Four of the leading scientists of the nineteenth century carrying out vigorous arguments based on real facts and sound arguments. The state of the atomic theory in 1807 was still very tenuous, and many aspects of experimental chemistry were in a bad way, but as better data was obtained and clearer arguments were formulated, the atomic theory continued to look better and better. Progressive paradigms can incorporate new data and point the way to better arguments.

When Dalton published his own *New System of Chemical Philosophy* in 1808, he included a Table of atoms and their proposed atomic weights [20] (see Fig. 4.2 for a reproduction of this Table). One of the first things to notice is the uniformly integral values for the atomic weights. Most of the values are different from modern atomic weights because little knowledge of the true atomic formulae for the compounds on which the atomic weights were based had been gleaned. The whole theory could have been discarded at this point with no tears shed. But, a few disciples continued to collect better data, and new ideas about gases and solids helped to clarify the actual compositions of many substances. Progress in measuring atomic weights occurred when Michael Faraday and Davy started electrodepositing metals, such as sodium and potassium. Faraday's Law allowed measured deposition currents to be related in a one-to-one basis with the number of atoms deposited. (The Faraday constant is equal to Avogadro's number times the charge on an electron: 96500 C/mol.) Progress was also obtained by considering the crystalline salts of these metals. The concept of stoichiometry, presented first in a coherent form by Richter, allowed the anionic partner atomic weight to be determined from the density of the salt and knowledge of its composition. Similar crystal forms helped to identify other salts with identical atomic crystal structure. While the theory of macroscopic crystallography was derived on the basis of purely mathematical concepts, the atomic theory of Dalton allowed a microscopic realization of these symmetries in terms of actual arrangements of atoms. The consilience of both mathematical and physical arrangements helped chemistry to "come of age."

Fig. 4.2 Dalton's initial Table of atomic weights reproduced from Thomson's *History of Chemistry* [17]. (scanned from personal copy)

	Weight of atom.		Weight of atom
Hydrogen . .	1	Strontian. .	46
Azote . . .	5	Barytes . .	68
Carbon . .	5	Iron . . .	38
Oxygen . .	7	Zinc . . .	56
Phosphorus .	9	Copper . .	56
Sulphur . .	13	Lead . . .	95
Magnesia . .	20	Silver . . .	100
Lime . . .	23	Platinum . .	100
Soda . . .	28	Gold . . .	140
Potash . . .	42	Mercury . .	167

4.2 The Natural Philosophy of Sir John F. W. Herschel

By the 1830s, lists of elements were lengthening, with the help of chemists such as Baron Jöns Jacob Berzelius (1779–1848) of Sweden [21]. Some atomic weights were being determined with values within experimental error of current values. One of Thomson's best students was Thomas Graham (1805–1869) at University College London. In his *Elements of Chemistry* (1850), he summarized all the data obtained by that date on the elements and their atomic weights [22]. Even so, the total number had only grown to 59. Atomic weights were now reported to four significant digits, but were still often wrong by multiples or divisors of 2 or 3. While better gravimetric data were obtained, there was still uncertainty about the actual empirical formula for many salts and molecules. Some of the best atomic weights include: chlorine (35.50), fluorine (18.70), iodine (126.36), potassium (39.00), silver (108.00), and sodium (22.97). This was the state of chemistry at the time that Sir John F. W. Herschel (1792–1871) published the new edition of his *Preliminary Discourse on the Study of Natural Philosophy* (1851) [23] (see Fig. 4.3).

Chapter IV of Part III in the Discourse is entitled "Of the Examination of the Material Constituents of the World." Herschel was a master geologist and one of the Fellows of the Geological Society of London. The study of minerals was very important in the history of nineteenth-century chemistry. The sheer number of

Fig. 4.3 Title page of Herschel's *Preliminary Discourse on the Study of Natural Philosophy* (1851) (scanned from personal copy)

crystalline minerals provides excellent grist for the empirical mill of the natural philosopher. Crystals had definite shapes characterized by precise three-dimensional geometries. When the macroscopic theory of the Abbé Haüy (1743–1822) was joined with the atomic theory of Dalton, these shapes could inspire insight into specific arrangements of atoms in the crystals. (Haüy was a French priest, mathematician, and mineralogist who formulated the theory of crystallography.)

Herschel also reflected on the properties of liquid mixtures of substances. He observed that two homogeneous solutions can produce a heterogeneous mixture when combined, even though no visible structure exists in the initial solutions. This implies that the size of fundamental chemical structures is very small (atomic even). The fact that the mixture produces a new substance that was not present in the original solutions is evidence of chemical forces between the elementary chemical particles in solution. (This subject was extensively studied by Graham.)

Herschel admired nineteenth-century chemistry because it focused largely on experimentally observable phenomena that could be explained in terms of other known generalizations from experience. But, he yearned for the day when chemists could predict phenomena not yet observed using precise concepts: the *axioms* of chemistry. He enumerated the known axioms in 1851 as:

1. The discovery of the proximate, if not the ultimate, elements of all bodies, and the enlargement of the list of known elements to its present extent of between 50 and 60 substances.
2. The development of the doctrine of latent heat by Black, with its train of important consequences, including the scientific theory of the steam-engine.
3. The establishment of Wenzel's law of definite proportions on his own experiments, and those of Richter, a discovery subsequently merged in the more general wording and better development of Dalton's atomic theory.
4. The precise determination of the atomic weights of the different chemical elements, mainly due to the astonishing industry of Berzelius, and his unrivalled command of chemical resources, as well as to the researches of the other chemists of the Swedish and German school.
5. The assimilation of gases and vapours, by which we are led to regard the former, universally, as particular cases of the latter, a generalization resulting from the experiments of Faraday on the condensation of the gases, and of those of Gay-Lussac and Dalton, on the laws of their expansion by heat compared with that of vapours.
6. The establishment of the laws of the combination of gases and vapours by definite volumes, by Gay-Lussac.
7. The discovery of the chemical effects of electricity, and the decomposing agency of the Voltaic pile, by Nicholson and Carlisle; the investigation of the laws of such decomposition, by Berzelius and Hisinger: the decomposition of the alkalies by Davy, and the consequent introduction into chemistry of new and powerful agents in their metallic bases.

8. The application of chemical analysis to all the objects of organized and unorganized nature, and the discovery of the ultimate constituents of all, and the proximate ones of organic matter, and the recognisance of the important distinctions which appear to divide these great classes of bodies from each other.
9. The application of chemistry to innumerable processes in the arts, and among other useful purposes to the discovery of the essential medical principles in vegetables, and to important medicaments in the mineral kingdom.
10. The establishment of the intimate connection between chemical composition and crystalline form, by Haüy and Vauquelin, with the successive rectifications the statement of that connection has undergone in the hands of Mitscherlich, Rose, and others, with the progress of chemical and crystallographic knowledge.

While many natural philosophers were still reticent to embrace the concept of distinct chemical atoms in 1851, and remained so until 1911, Herschel warmly endorsed this notion. He remained open to the discovery of yet deeper facts of atomic structure, but reminded his readers that the paradigm of the chemical atom made sense out of all known compounds and chemical reactions. He considered the chemical atomic doctrine to rank with the laws of Newtonian mechanics in terms of their importance to natural philosophy.

Herschel understood how difficult it was to measure the atomic weights accurately, and in 1851 many of the reported values were in serious error, but he knew how important it would be to obtain a truly reliable set. Early values of atomic weights were all given as integers, as expected by William Prout (1785–1850) and Thomson, but the better values from Graham reported above prove that the actual values are not pure integers. Exactly why the atomic weight of chlorine is 35.5 was not yet known, but it was not an integral value.

Herschel envisioned the day when all the chemical atoms could be comprehended in terms of a precise mathematical expression. The Law of Moseley that relates the frequency of the principal X-rays emitted by atoms to their atomic number is the realization of this dream. But, in his own time, he continued to follow developments in chemistry with great interest. He was chosen as President of the Chemistry Section of the British Association for the Advancement of Science (BAAS) in 1858. In his address as President, Herschel pointed forward to the time "when, from a knowledge of the family to which a chemical element belongs, and its order in that family, we may be able to predict with confidence the system of groups into which it is capable of entering, and the part it will play in the combination" [24].

In fact he believed that Josiah Parsons Cooke (1827–1894) of Harvard University had made major progress in this area and had published it in 1855 in the *Memoirs of the American Academy of Arts and Sciences* [25]. Cooke identified families of elements that shared common properties. The first group consisted of oxygen, fluorine, chlorine, bromine, and iodine. Such a group was identified by Thomson in his *History*. The second group included nitrogen, phosphorus, arsenic, antimony, and bismuth. The third group encompassed hydrogen, lithium, sodium, and potassium. The fourth group listed calcium, strontium, barium, and lead. Herschel implored his expert audience to cogently discuss what Cooke had found. In addition, he speculated that the newly emerging science of optical spectroscopy would be very helpful in unraveling the mysteries of the atom. Yes, Sir John Herschel, FRS, was instrumental in founding the field of chemical spectroscopy.

4.3 Josiah Parsons Cooke and the Relationships Between the Elements

With the pointer from Sir John Herschel, it is time to examine in detail the paper by Cooke [25]. Josiah Parsons Cooke was the Erving Professor of Chemistry at Harvard University. This professorship was established in 1791, but Aaron Dexter had been teaching a regular course in Chemistry since 1783 [26]. Cooke graduated from Harvard in 1848 and was appointed to the Erving Chair in 1850. Rather than start teaching a subject in which he was unprepared, he traveled to Europe and attended lectures by Jean-Baptiste Dumas (1800–1884) and Henri Victor Regnault (1810–1878). Upon his return in 1851, he was made M.A. and introduced both lecture and laboratory instruction in Chemistry and Materia Medica. He also helped raise the money to build Boylston Hall, which would later be devoted entirely to chemistry. He was elected to the National Academy of Sciences in 1872 [27]. Cooke went on to carry out many measurements of atomic weight, and one of his greatest students, Theodore Richards (1868–1928), went on to win the 1914 Nobel Prize in Chemistry for his work on atomic weights.

Cooke starts off by acknowledging his debt to M. Dumas of Paris. He cites the paper presented by Dumas at the 1851 meeting of the British Association for the Advancement of Science that discusses the "numerical relations between the atomic weights of the chemical elements." Dumas reminded the BAAS worthies of the triads of Johann Döbereiner (1780–1849). (Döbereiner noticed that for several groups of elements ({Cl, Br, I}, {Li, Na, K}, {Ca, Sr, Ba} and {S, Se, Te}), the atomic weight of the middle-most member was the arithmetic average of the first and last element [28].) But, Cooke concluded that a focus on triads was not the best way forward. He proposed to construct groups of elements that had similar chemistries, and then examine the relations between their atomic weights. His proposed numerical relations are more a part of the history of chemistry than a contribution to natural philosophy; they did not produce new insights and were not a way forward. The Karlsruhe conference in 1860 helped the scientific community to see the need to adopt an experimental technique that could yield reliable particle weights; the ideal gas law and Avogadro's Law allowed gas densities to be converted to particle weights [16]. The leading natural philosopher at the Karlsruhe Conference was Stanislao Cannizzaro (1826–1910).

Cooke's natural motivation for this work was his desire to present a coherent version of chemistry to his students at Harvard. Most extant chemistry texts presented a qualitative survey of the elements and their compounds in an arbitrary order. For example, a chapter on oxygen, nitrogen, and hydrogen depended on the fact that these were all gases, rather than on any similarity in their chemistries. It did make sense in 1817 [12] to discuss the many compounds of oxygen, nitrogen, and hydrogen. But, since many of the proposed molecules were incorrectly understood, only confusion resulted. Cooke created groups of elements, based first on their similar chemistries. The first group was the now-familiar halogens, but with two additional members: oxygen, which in 1855 [25] was considered very similar to chlorine, and the molecule cyanogen (NCCN), which is still considered a pseudo-halogen. The proposed numerical relations are fanciful, and depend on bad atomic and molecular weights, but the grouping was brilliant. Since water was still considered to be the molecule HO, this choice of elements was rational, albeit incorrect. (The prototypical acids are HO, HF, HCN, HCl, HBr, and HI). This grouping led to correct atomic weights for the halogens and disproved Prout's Law.

In addition to the main group, Cooke presented an "affiliated" group that included chromium, manganese, osmium, and gold. Good atomic weights were listed for all but osmium, which was half its modern value. The compounds were oxides of the elements. The composition of the gold oxides was not yet known. (They are now known to be aurous oxide (Au_2O) and auric oxide (Au_2O_3).) Cooke does list the correct formula for osmium tetroxide (OsO_4).

The next grouping consisted of oxygen, sulfur, selenium, and tellurium. The proposed numerical relationships work for this group, since all the "accepted" values were wrong by a factor of two (8, 16, 40, 64). But, sulfur does not work as a triad with oxygen and selenium. Just as with the halogens, this grouping has remained as a fact. It was noted that two oxides of sulfur, selenium, and tellurium existed, of the form XO_2 and XO_3. This inspired Cooke to create a group of affiliated elements that also displayed oxides of this form: molybdenum, vanadium, tungsten, tantalum, and manganese.

Cooke definitely liked oxygen, and included it in a third grouping: oxygen, nitrogen, phosphorus, arsenic, antimony, and bismuth. Ignoring the obviously errant inclusion of oxygen (Cooke noted that oxygen did not share the chemistry of this group but was chosen for numerical reasons), the remaining five elements continue to be considered a true group. Of even more interest, the atomic weights chosen by Cooke were all close to modern values, except for the ancient stibium (antimony). One of the reasons for this is that ammonia and phosphine were correctly identified as NH_3 and PH_3. The basis for this was the observation that three volumes of hydrogen combined with one volume of nitrogen to produce two volumes of ammonia, noted by Dumas and Joseph Louis Gay-Lussac (1778–1850). Why was this data accepted when the analogous data for oxygen and hydrogen (two volumes of hydrogen combine with one volume of oxygen to make two volumes of water) was rejected?! The incorrect atomic weight for oxygen ($O = 8$) caused the empirical formulas for some oxides to be incorrect: NO_5 instead of N_2O_5. And there are actually six known oxides of nitrogen. The oxide chemistry of antimony is especially rich and many famous colors are based on these compounds.

Cooke chose to create a group with carbon, boron, and silicon. His discussion is highly flawed with respect to the underlying compounds. He lists bad atomic weights for carbon (6) and silicon (21), but a correct value for boron (11). This leads to incorrect oxides of boron and silicon: BO_3 rather than B_2O_3, and SiO_3 rather than SiO_2. It is hard to be coherent when so much error is being organized. Cooke emphasizes the similarities of boron with silicon. Both form crystals and glasses, and hydrogen acids. Creating such a group may have helped to teach his class, but when better atomic weights and correct compositions were obtained, it was time to reconsider (see next section).

The apparent wealth of oxide minerals provided a basis to organize many metals into a coherent arrangement. Cooke further subdivided this collection into three subgroups. He used a matrix of potential oxides from M_2O through MO_4. The first subgroup consisted of titanium, palladium, tin, platinum, iridium, osmium, and gold. It was claimed that all members had univalent oxides of the form MO. This eliminates gold, which we have already noted was Au_2O and Au_2O_3. All the atomic weights were incorrect (half their modern values due to O = 8), except for gold which had a modern value (197). Titanium has a very rich oxide chemistry from Ti_2O to TiO_2, with many non-stoichiometric compositions. The attempt to use the oxide matrix as an organizing principle was fully appropriate in 1855, but inconsistencies still haunted the fundamental composition data.

The next subgroup consisted of aluminum, chromium, manganese, iron, cobalt, nickel, and uranium. The atomic weights are all wrong, especially uranium. The only oxide listed for aluminum is the correct Al_2O_3, with none of the other plethora of compounds. It would be a long time before boron and aluminum were recognized as brothers. The other members of the group do share many compounds in common, from MO to M_2O_7. They form the familiar sequence of transition metals: Cr, Mn, Fe, Co, and Ni!

The final small subgroup consisted of copper and mercury. The atomic weights are half of the modern values. The compounds of mercury are the familiar Hg_2O and HgO. Cooke lists nonexistent oxides of copper: Cu_2O_3 and CuO_2. Not much insight in this grouping.

The last large grouping is also divided into three subgroups. The first subgroup consisted of magnesium, zinc, and cadmium. All the atomic weights are half of their modern values. All of the elements have oxides of the form MO. Cooke lists oxides for zinc and cadmium of the form M_2O, but these are not known today. Not a great grouping, but zinc and cadmium are highly related.

The second subgroup consisted of calcium, strontium, barium, and lead. The atomic weights are half of the modern values. The characteristic compound is of the form MO, just like the last group, but additional compounds are listed in the form MO_2. Strontium and barium do not exhibit such compounds. The oxide chemistry of lead includes compounds from PbO (litharge and massicot) to PbO_2. Red lead (minium) is a mixture of lead(II) and lead(IV). Cooke lists a Pb(I) compound that is not currently acknowledged. When Cooke's organization is cleaned up, and fictive compounds are eliminated, the combined subgroup would be Mg, Ca, Sr, and Ba: the familiar alkaline earths.

The final subgroup, based on the compounds MO, was composed of hydrogen, lithium, sodium, potassium, and silver. Also of interest is the fact that the atomic weights listed for this group are also in excellent agreement with modern values. (1, 6.5, 23, 39.5, 108.5) Inclusion of silver made complete sense in 1855, based on AgCl. It was univalent and a metal. But Cooke's justification for this group is based on nonexistent compounds: OH, LiO, NaO, KO, and AgO. The error occurs for the same reason the compounds of mercury were misrepresented: good metal atomic weights and incorrect oxygen atomic weight. Correcting the compounds to the real H_2O, Li_2O, Na_2O, K_2O, and Ag_2O yields a sound grouping for 1855. The alkali metals remain as a grouping that makes sense, even though hydrogen is not usually considered a metal at atmospheric pressure. (At sufficient temperature and pressure it is predicted to become a metal.)

Cooke engaged in a monumental program to make sense of the known elements. He succeeded enough to teach a course in elementary chemistry. He tried to carry out the natural philosophical analysis of the known compounds of the elements, mostly oxides. But, his knowledge of the atomic weights was imperfect and his knowledge of the compositions of his prototypical compounds was also flawed. He never stopped trying to improve both his database and his analysis, and by 1868 he had published a textbook that took advantage of better atomic weights and better descriptive chemistry.

4.4 Josiah Parsons Cooke and *First Principles of Chemical Philosophy*

Josiah Parsons Cooke taught chemistry at Harvard throughout the 1860s. He benefitted from the Karlsruhe Conference in 1860, as did all the chemists in the world [16]. Many of his atomic weights were now within experimental error of modern values, but not all. As in 1855, he tried to organize the now 63 known elements into groups with similar chemistries. His efforts produced Table II at the back of his textbook: *First Principles of Chemical Philosophy* (1868) [29] (see Fig. 4.4).

Cooke divided the elements into two large groups: (1) Perissad elements with odd quantivalence, from 1 to 5, and (2) Artiad elements with even quantivalence from 2 to 6 [30].[1] (Quantivalence is a nineteenth-century concept that expresses the number of bonded neighbors of an atom.) He was aware that many elements displayed several valences. And not all his quantivalence assignments were correct, by modern standards. He provided an extensive discussion of the known properties of each element and many of its known compounds. Without an accurate knowledge of the chemistry of each element, further arrangement into subgroups would be futile.

[1]William Odling previously used these terms [30].

TABLE II.

ELEMENTARY ATOMS.

Perissad Elements.	Atomic Weights.	Symbols of Molecules.	Quantivalence.
Hydrogen	1.0	*H-H*	I
Fluorine	19.0	*F-F*	“
Chlorine	35.5	*Cl-Cl*	“
Bromine	80.0	*Br-Br*	“
Iodine	127.0	*I-I*	“
Lithium	7.0	*Li-Li*	“
Sodium	23.0	*Na-Na*	“
Potassium	39.1	*K-K*	“
Rubidium	85.4	*Rb-Rb*	“
Cæsium	133.0	*Cs-Cs*	“
Silver	108.0	*Ag-Ag?*	“
Thallium	204.0	*Tl-Tl?*	I or III
Gold	197.0	*Au≡Au?*	III
Boron	11.0	*B≡B?*	“
Nitrogen	14.0	*N≡N*	III or V
Phosphorus	31.0	P_2≣P_2	“
Arsenic	75.0	As_2≣As_2	“
Antimony	122.0	Sb_2≣Sb_2?	“
Bismuth	210.0	Bi_2≣Bi_2?	“
Vanadium	51.37	*V≡V?*	“
Uranium	120.0	*U≡U?*	“
Columbium	94.0	*Cb≣Cb?*	V
Tantalum	182.0	*Ta≣Ta?*	“
Artiad Elements.			
Oxygen	16.0	*O=O*	II
Sulphur	32.0	*S=S*	II or VI
Selenium	79.4	*Se=Se*	“
Tellurium	128.0	*Te=Te*	“
Molybdenum	96.0	*Mo?*	VI
Tungsten	184.0	*W?*	“

Artiad Elements.	Atomic Weights.	Symbols of Molecules.	Quantivalence.
Copper	63.4	*Cu?*	II
Mercury	200.0	*Hg*	“
Calcium	40.0	*Ca?*	“
Strontium	87.6	*Sr?*	“
Barium	137.0	*Ba?*	“
Lead	207.0	*Pb?*	“
Magnesium	24.0	*Mg?*	“
Zinc	65.2	*Zn?*	“
Indium	72.0	*In?*	“
Cadmium	112.0	*Cd*	“
Glucinum	9.3	*G?*	“
Yttrium	61.7	*Y?*	“
Erbium	112.6	*E?*	“
Cerium	92.0	*Ce?*	“
Lanthanum	93.6	*La?*	“
Didymium	95.0	*D?*	“
Nickel	58.8	*Ni?*	“
Cobalt	58.8	*Co?*	“
Manganese	55.0	*Mn?*	II or IV
Iron	56.0	*Fe?*	“
Chromium	52.2	*Cr?*	“
Aluminum	27.4	*Al?*	“
Ruthenium	104.4	*Ru?*	“
Osmium	199.2	*Os?*	“
Rhodium	104.4	*Rh*	“
Iridium	196.0	*Ir?*	“
Palladium	106.6	*Pd?*	“
Platinum	197.4	*Pt?*	“
Titanium	50.0	*Ti?*	“
Tin	118.0	*Sn?*	“
Zirconium	89.6	*Zr?*	IV
Thorium	231.4	*Th?*	“
Silicon	28.0	*Si?*	“
Carbon	12.0	*C?*	“

Fig. 4.4 Table II from *First Principles of Chemical Philosophy* (1868) by Josiah Parsons Cooke [29] (scanned from personal copy)

He chose to discuss hydrogen (H = 1) in a group by itself, as one of the Perissad elements. He seemed unaware of hydride compounds. (Sodium (NaH) and potassium hydride (KH) were isolated by Henri Moissan (1852–1907) in 1902.) He did note that hydrogen seems able to react with most of the other elements. The current fashion to include hydrogen with the alkali metals, because it heads column I of the periodic table, ignores the rich chemistry of hydrogen compared with sodium.

The next group is the familiar halogens: F, Cl, Br, I. Excellent atomic weights are listed and many compounds are cited. He was aware of many oxides of chlorine; this indicates that a quantivalence of 1 is not the only possibility for the heavier halogens. He knew about fluorspar (CaF_2) and other fluorine-containing minerals, but it was not until 1886 that Moissan isolated pure fluorine gas.

The third group is the collection of alkali metals: Li, Na, K, Rb, and Cs. Again, excellent atomic weights are given. They are indeed univalent and many compounds are discussed. Except for the radioactive halogen, astatine (At), and the radioactive alkali metal, francium (Fr)—among the naturally occurring elements—these two foundational groups of elements are established. They allow good atomic weights to be determined for any element that reacts with them.

The coinage metals are all discussed (Cu, Ag, Au), but they are not linked as a group. Silver is included as a univalent metal with many compounds such as silver chloride (AgCl) and silver nitrate ($AgNO_3$). Gold is listed as a trivalent metal in its own group as well. Gold trichloride ($AuCl_3$) is the key compound. Copper is listed as an Artiad metal with a quantivalence of 2. Copper(I) compounds are ignored. The insights that link these three elements were to be found elsewhere. Inorganic chemistry continued its rapid development throughout the nineteenth century. (And the wonders to be discovered in the twentieth century were beyond imagination in 1868.)

Boron (B) is the crux of a problem. Much was known about this element and its compounds. Cooke listed it as trivalent, and an obvious lightest element of some group. But, alas, it is placed in a group of one. It was prepared in a solid state by Davy in 1808. It occurs in the earth's crust primarily as some form of borate $(BO_3)^{3-}$. The pure oxide is B_2O_3. Cooke knew about the remarkable compound boron nitride (BN(s)). He also knew that pure boron could be alloyed with pure aluminum. He even knew that a polymer of boron, oxygen and hydrogen existed. (Commercial "Borax" can still be purchased in a grocery store.) The chloride (BCl_3) and bromide (BBr_3) were known. With such a rich chemistry, it should have been a simple matter to match boron up with other elements with the same chemistry. But, alas, aluminum (Al) was listed as an Artiad element with quantivalences of 2 and 4! Cooke was confused by the enormous number of complex minerals containing aluminum. One of the most common is alum ($KAl(SO_4)_2.12H_2O$) If he had focused on pure alumina (Al_2O_3) or $AlCl_3$, he would have seen the relationship more easily. But, pure aluminum was only commonly available after 1855, and Cooke may not have known of the new developments. It only became a commercial product after 1886 with the discovery of the Hall process. And the next element in the modern group, gallium (Ga), had not yet been discovered.

Nitrogen (N) was recognized as the first member of a group with quantivalence III or V. The other members included phosphorus (P), arsenic (As), antimony (Sb) and bismuth (Bi). Good atomic weights were known. There are many compounds of nitrogen (and all the other members of this group) with oxygen. Davy would be proud. (He studied many oxides of nitrogen, including breathing more nitrous oxide than anyone.) There are many forms of hydrogen acids formed with nitrogen and oxygen. And there are hydrogen compounds such as ammonia (NH_3), phosphine (PH_3), etc. Ammonium ion (NH_4^+) forms salts with many anions, such as *sal ammoniac* (NH_4Cl), which was well-known in antiquity. Cooke was even aware of the highly explosive compounds formed from nitrogen and halogens, such as NI_3. With such a rich chemistry, and so many similar elements known to the chemist in 1860s, this group was natural.

While uranium (U) was known, Cooke listed a bad atomic weight and did not associate it with anything else. He did know about the uranyl yellow dyes, for which uranium is famous. (The yellow wall tiles in the Mellon Institute at Carnegie Mellon University create quite a racket when a Geiger counter is held nearby.) But, it was the extremely rich chemistry of uranium that prevented him from assigning it to either major group, since it can adopt very many quantivalences. While he does mention one chloride of uranium, there are at least five. One of the most important modern compounds of uranium is the hexafluoride (UF_6). This gaseous compound is used to fractionate the isotopes of uranium to obtain the enriched form needed for fission reactors. This embarrassment of riches made classification based on chemistry difficult.

The Artiad elements were characterized by even quantivalence. The *prima materia* for this group is oxygen (O). Cooke was aware of a bewildering array of compounds containing oxygen (and some that are no longer accepted as real). While oxygen is now placed at the head of a group containing sulfur (S), selenium (Se), and tellurium (Te), it merited a special place in Cooke's Table because of the richness of its chemistry. But, the higher members of the group also introduce new richness. Any scheme of organization faces the dilemma of the lumper and the splitter. The more you know, the more the "differences" between members of the group are magnified. Where do you draw the line?!

The second major subgroup in this list comprises the alkaline earths: calcium (Ca), strontium (Sr), and barium (Ba). Cooke also included lead (Pb). It is not that there is no overlap between lead and the other elements in this group, but the existence of PbO_2 should have ruled it out. All these elements do have oxides of the form MO. Magnesium (Mg) is placed in a small group with zinc; and glucinum (Be) in yet another small group. At the end of the discussion of lead, Cooke suggests that it should not really be placed with calcium. In his discussion of magnesium, which has all the compounds associated with calcium, he decides to emphasize the differences, rather than the similarities. This group helps to illustrate the inability of organizing all the elements on the basis of known chemistry alone. Another measurable property is necessary to make further progress.

The Table of Artiad elements is dominated by many metals. The discussion above [25] related the known chemistries for many of these substances, but the quantivalences are not correctly given for most of them in Table II. Many groups of two are “split” by Cooke’s analysis, but elements such as chromium are linked with obviously different elements such as aluminum. Chromium can have a very large quantivalence, while aluminum is actually a Perissad element. Cooke’s attempt to make the quantitative organization of the elements has foundered on the rocks of bad data and incorrect analysis. However, the progress displayed in Table II is impressive.

At the bottom of the Table we find silicon and carbon in groups of one. The quantivalence is listed as IV, which by 1868 was the notion of Kekule. There are enough differences between the pure elements and their compounds to make the similarities less obvious.

Fortunately, Dmitri Mendeleev (1834–1907) was also concerned with making chemistry transparent to his students [31]. He now had the same table of atomic weights as Cooke; and he had the same extensive descriptive inorganic chemistry. The obvious choice of using the measured atomic weights had been there since Dalton, but the values were incorrect until Karlsruhe. Ordering the elements by atomic weight was a brilliant step and helped Mendeleev create his famous initial version of the Periodic table. There were still many ambiguities and problems, but the leap from Table II in Cooke to the Table developed by Mendeleev was a major advance.

Cooke went on to devote his career to making precise measurements of atomic weight. He succeeded and was elected to the National Academy of Sciences in 1872. Among the American chemists of the nineteenth century, I. Bernard Cohen ranked Cooke as “the first university chemist to do truly distinguished work in the field of chemistry” [32]. He set a standard for precise work that propelled both the development of inorganic chemistry and the teaching of coherent principles of chemical philosophy.

The sheer volume of chemical facts can be both discouraging and confusing. But, the goal of Chemical Natural Philosophy is to create sound principles that can guide daily thoughts and inspire new explorations. Both Sir John F. W. Herschel, FRS, and Josiah Parsons Cooke, NAS, were exemplars of natural philosophy. Their role in the story of the development of a rational chemistry of the elements is worth knowing by all chemists and other scientists.

References

1. Hannaway O (1975) The chemists and the word: the didactic origins of chemistry. Johns Hopkins University Press, London
2. Moran BT (2007) Andreas Libavius and the transformation of alchemy: separating chemical cultures with polemical fire. Science History Publications, Sagamore Beach, MA
3. Webster C (2008) Paracelsus, medicine, magic and mission at the end of time. Yale University Press, New Haven, CT
4. Hall MB (1965) Robert Boyle on natural philosophy: an essay with selections from his writings. Greenwood Press, Westport, CT
5. Principe LM (1998) The aspiring adept: Robert Boyle and his alchemical quest. Princeton University Press, Princeton, NJ
6. Newman WR (2006) Atoms and alchemy: Chymistry and the experimental origins of the scientific revolution. University of Chicago Press, Chicago

7. Powers JC (2012) Inventing chemistry: Herman Boerhaave and the reform of the chemical arts. University of Chicago Press, Chicago
8. Whyte LL (1961) Roger Joseph Boscovich, S.J., F.R.S., 1711–1787: studies of his life and work on the 250th anniversary of his birth. Fordham University Press, New York
9. Boscovich RJ (1763) Theoria philosophiae naturalis. Venetiis
10. Lavoisier A (1789) Traité élémentaire de chimie, présenté dans un ordre nouveau et d'après les découvertes modernes. Chez Cuchet, Paris
11. Brock WH (2000) The chemical tree: a history of chemistry. W.W. Norton & Co., New York
12. Patterson G (2018) William Henry and *The Elements of Experimental Chemistry*. In: Patterson G (ed) Preceptors in chemistry. ACS Symposium Series 1273, American Chemical Society, Washington, DC, pp 67–82
13. Thackray A (1972) John Dalton: critical assessments of his life and science. Harvard University Press, Cambridge, MA
14. Golinski J (2016) The experimental self: Humphry Davy and the making of a man of science. University of Chicago Press, Chicago
15. Usselman MC (2015) Pure intelligence: the life of William Hyde Wollaston. University of Chicago Press, Chicago
16. Rocke AJ (1984) Chemical atomism in the nineteenth century: from Dalton to Cannizzaro. Ohio State University Press, Columbus
17. Thomson T (1835) The history of chemistry, 2nd edn. Henry Colburn and Richard Bentley, London
18. Alembic Club (1911, originally 1893) Foundations of the atomic theory comprising papers and extracts by John Dalton, William Hyde Wollaston, M.D. and Thomas Thomson, M.D. (1802–1808). University of Chicago Press, Chicago
19. Thomson T (1807) A system of chemistry, 3rd ed., vol 3. Bell & Bradfute and E. Balfour, Edinburgh
20. Dalton J (1808) A new system of chemical philosophy. Bickerstaff, Manchester
21. Berzelius JJ (1829) Traité de chimie. F. Didot Frères, Paris
22. Graham T (1850) Elements of chemistry, 2nd edn. Hippolyte Baillière, London
23. Herschel JFW (1851) Preliminary discourse on the study of natural philosophy, new ed. Longman, Brown, Green & Longmans, London
24. Herschel JFW (1858) Address. Report of the 28th Meeting of the British Association for the Advancement of Sciences, Transactions of the Sections, 41–45
25. Cooke JP (1855) The numerical relation between the atomic weights, with some thoughts on the classification of the chemical elements. Memoirs of the American Academy of Arts and Sciences 5:235–256
26. Elliott CA, Rossiter MW (1992) Science at Harvard University: historical perspectives. Lehigh University Press, Bethlehem, PA
27. Jackson CL (1902) Memoir of Josiah Parsons Cooke. Biographical Memoirs of the National Academy of Sciences 4:175–183
28. Döbereiner JW (1829) Versuch zu einer Gruppirung der elementaren Stoffe nach ihrer Analogie. Annalen der Physik und Chemie 15:301–307
29. Cooke JP (1868) First principles of chemical philosophy. Sever, Francis and Co., Boston
30. Odling W (1864) On the hexatomicity of ferricum and aluminium. Philos Mag 27:115–119
31. Mendeleev D (1869) Osnovy khimii. St. Petersburg, Russia
32. Cohen IB (1959) Some reflections on the state of science in America during the nineteenth century. Proc Nat Acad Sci USA 45:666–677

Vis Tellurique of Alexandre-Émile Béguyer de Chancourtois

5

Carmen J. Giunta

Abstract

The *vis tellurique*, the arrangement of elements and radicals that the French geologist Alexandre-Émile Béguyer de Chancourtois presented in 1862, is described. Reactions to the arrangement by contemporaries, by later chemists of the nineteenth century, and by selected historians are examined. It is argued that Béguyer de Chancourtois discovered the law of chemical periodicity, albeit in a form too flawed and indefinite to appeal to contemporary chemists.

5.1 Introduction

On April 7, 1862, Alexandre-Émile Béguyer de Chancourtois[1] (1820–1886) made the first of six presentations over the span of a year to the French Académie des Sciences on a natural classification of simple bodies and radicals he called the *vis tellurique*. The *vis tellurique* is considered by many chemists and historians to be the first formulation and embodiment of the periodic law, that is, of the recognition

[1] I will refer to him as Béguyer de Chancourtois, following the guidance of the Bibliothèque Nationale de France, or simply Béguyer. The most common nineteenth-century usage was Chancourtois or de Chancourtois. The Notice de personne in the catalogue général of the Bibliothèque says that Béguyer is the start of the surname.

C. J. Giunta (✉)
Department of Chemistry, Le Moyne College, Syracuse, NY, USA
e-mail: giunta@lemoyne.edu

C. J. Giunta et al. (eds.), *150 Years of the Periodic Table*, Perspectives on the History of Chemistry, https://doi.org/10.1007/978-3-030-67910-1_5

that chemical and physical properties of elements repeat when one orders elements by atomic weight.[2] The memoirs, or rather extracts from them, were published in the Académie's *Comptes rendus* [1–6]. Famously **not** included in the journal publications was the chart that Béguyer described and that embodied the classification. He privately published a booklet in 1862 containing the chart as a six-panel fold-out, along with reprints of the memoirs, some of them containing a small amount of additional material left out of the publications in *Comptes rendus* [7]. The booklet and chart were revised in 1863 [8].

After a brief bit of biographical information [9, 10] on Béguyer de Chancourtois to complete this introductory section, this chapter will summarize his presentations and describe the *vis tellurique*. Reactions to the work during Béguyer's lifetime will be examined. An account of the posthumous rediscovery of his work and claims of priority made on his behalf, as well as the treatment of the *vis tellurique* by later chemists and historians, will follow. The chapter will conclude with my assessment of the extent to which Béguyer de Chancourtois deserves to be considered a discoverer of the periodic law.

Béguyer de Chancourtois (Fig. 5.1) was a reasonably well-established academic and professional geologist at the Paris École des Mines in 1862. At age 42, he was assistant to Léonce Élie de Beaumont, France's preeminent geologist, who held the chair of geology at the École des Mines and elsewhere in Paris. Béguyer was also assistant director (also under Élie) of the French geological survey, and his volume on the stratigraphy of the Haute-Marne [11], co-authored with Élie, would come out later in 1862.

Born in Paris in 1820, Béguyer de Chancourtois was a Parisian throughout his life, and he was affiliated with the École des Mines for most of it. He entered the École Polytechnique in 1838 and the École des Mines two years later. As a student, he traveled to Eastern Europe and Western Asia, and an excerpt from one of his letters [12] (to Élie de Beaumont) became his first publication in the *Comptes rendus*. He returned to the École des Mines as an instructor in 1848 and became *professeur suppléant*[3] to Élie in 1852.

Béguyer de Chancourtois worked with a senior colleague at the École des Mines, Frédéric Le Play, in organizing the Exposition Universelle held in Paris in 1855, a project which led both to connections in high places and to similar activities in the future. Prince Napoleon, also known as Napoléon-Jérôme Bonaparte, cousin of Emperor Napoleon III (Louis Napoleon), had noticed Béguyer's work on the exposition. The Prince invited him to a voyage in the polar North Atlantic on the royal yacht *Reine Hortense* in the next year. Later in the 1850s, Béguyer was chief of staff to the Prince during his brief stint as Minister of Algerian and Colonial Affairs. The Prince also tapped him for the commission organizing the 1867 Paris Exposition Universelle. Béguyer located the Pavilion of Measures and Currency,

[2]Now, of course, the proper order is by atomic number.

[3]At this time, it was not unusual for prominent French scientists such as Élie de Beaumont to hold chairs simultaneously at multiple institutions or for such professors to hire more or less permanent substitutes to undertake their teaching duties.

Fig. 5.1 Portrait of Alexandre-Émile Béguyer de Chancourtois (1820–1886) taken by Alexandre Quinet around 1882 (Courtesy of the Bibliothèque nationale de France)

whose dome was inscribed with the words "Omnia, o Deus, fecisti ex numero, mensura et pondere,"[4] in the center of the exposition. Béguyer later organized the French geological exhibits at international expositions in Venice in 1881 and Madrid in 1883.

Élie de Beaumont proved to have a lasting effect on Béguyer de Chancourtois, as both a patron and an intellectual influence. Béguyer was described as Élie's "scientific heir and successor" [9]. He was his successor in the chair of geology at the École des Mines from 1875, the year after Élie's death. He was also the primary advocate of Élie's hypothesis of the "pentagonal network," a geometric idea of the distribution of relief characteristics of the earth. We will see Élie invoked in Béguyer's thinking behind the *vis tellurique*.

Béguyer de Chancourtois died in Paris in 1886, not long after being named president of the French geologic mapping service. In the year of his death he arranged to have two seismological observatories set up in France. At the time he was Inspector General of mines and still held his chair at the École des Mines.

5.2 The *Vis Tellurique*

The *vis tellurique* chart (Fig. 5.2) is about 20 cm wide by nearly 146 cm long. The graph is a grid marked out in what for our purposes are atomic weight units (see below for discussion of the units.), 16 units wide (the atomic weight of oxygen),

[4]"You have made everything, O God, from number, measure, and weight." See Wisdom 11:21.

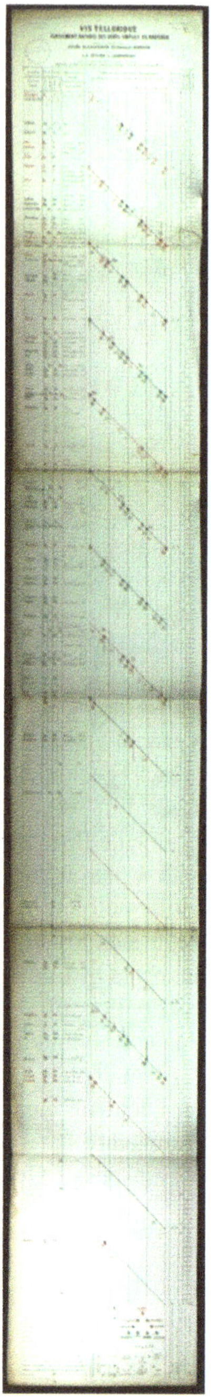

Fig. 5.2 *Vis tellurique* chart [7]. At this scale, one can really see only the 45° line segments with points clustered on them and the fact that the graph is much taller than it is wide (Courtesy of the Master and Fellows of St Catharine's College, Cambridge)

and 240 units tall. Béguyer de Chancourtois plotted points representing atomic weights on a descending 45° line proceeding from the upper left of the graph to the lower right and starting again on the left every time it reached 16 or a multiple of 16. The graph was intended to be wrapped around a circular cylinder, so that the parallel 45° line segments would connect to form a continuous descending helix.

In his first presentation of the *vis tellurique* to the Académie des Sciences, on April 7, 1862, Béguyer de Chancourtois described how he constructed the chart, saying of the resulting arrangement [13][5]

> The relations between the properties of different bodies are manifested by simple geometrical relations between the positions of their characteristic points.[6]

The first such relationship he described was that related bodies fall on or near the same vertical lines. His first example was oxygen, sulfur, selenium, and bismuth. On the opposite side of the cylinder magnesium, calcium, iron, strontium, uranium, and barium line up. Just to the left of the first set fall hydrogen and zinc; just to its right fall bromine, iodine, copper, and lead. In another vertical line are lithium, sodium, potassium, manganese, and more (See Fig. 5.3, the top portion of the chart). Today one can see definite similarities in chemical and physical properties among the elements within some of these groups, while other groups appear more disparate. We shall see that Béguyer was rather expansive in what he considered to be related elements, partly because his criteria were not just chemical and partly, I believe, because his faith in order led him to perceive it in some places where it was absent.

In this first paper, Béguyer did not limit himself to vertical relationships [13]:

> Each helix drawn through two characteristic points and passing through several other points or only near them, brings out relations of a certain kind between their properties; likenesses and differences being manifested by a certain numerical order in their succession, for example, immediate sequence or alternation at various periods.

This statement explicitly mentions periodicity in properties, but it asserts relationships practically at will or imagination: join a pair of elements and follow the helix that connects them. Some non-vertical groupings will be examined below.

Near the end of this first memoir, Béguyer indulged in some numerological speculations. Perhaps the set of characteristic numbers is the set of natural numbers. In fact he plotted all points on whole numbers, in accord with Prout's hypothesis. Furthermore, he included some "compound radicles" [13] such as cyanogen and ammonium (the latter visible in Fig. 5.3) on some natural numbers that did not correspond to elements. But he went further: "In this natural series, the bodies which are really simple, or at least irreducible by the ordinary means at our disposal, would be represented by the *prime numbers*" [13] (italics in original). The rightmost three columns of the *vis tellurique* chart (Fig. 5.3, to the right of the

[5]The English translation here is by Philip J Hartog whose 1889 paper [13] includes a translation of Béguyer's first paper on the *vis tellurique* [1].

[6]For characteristic points, we would say atomic weights. See below for Béguyer's description of characteristic points. Both Béguyer and Hartog used italics for this statement.

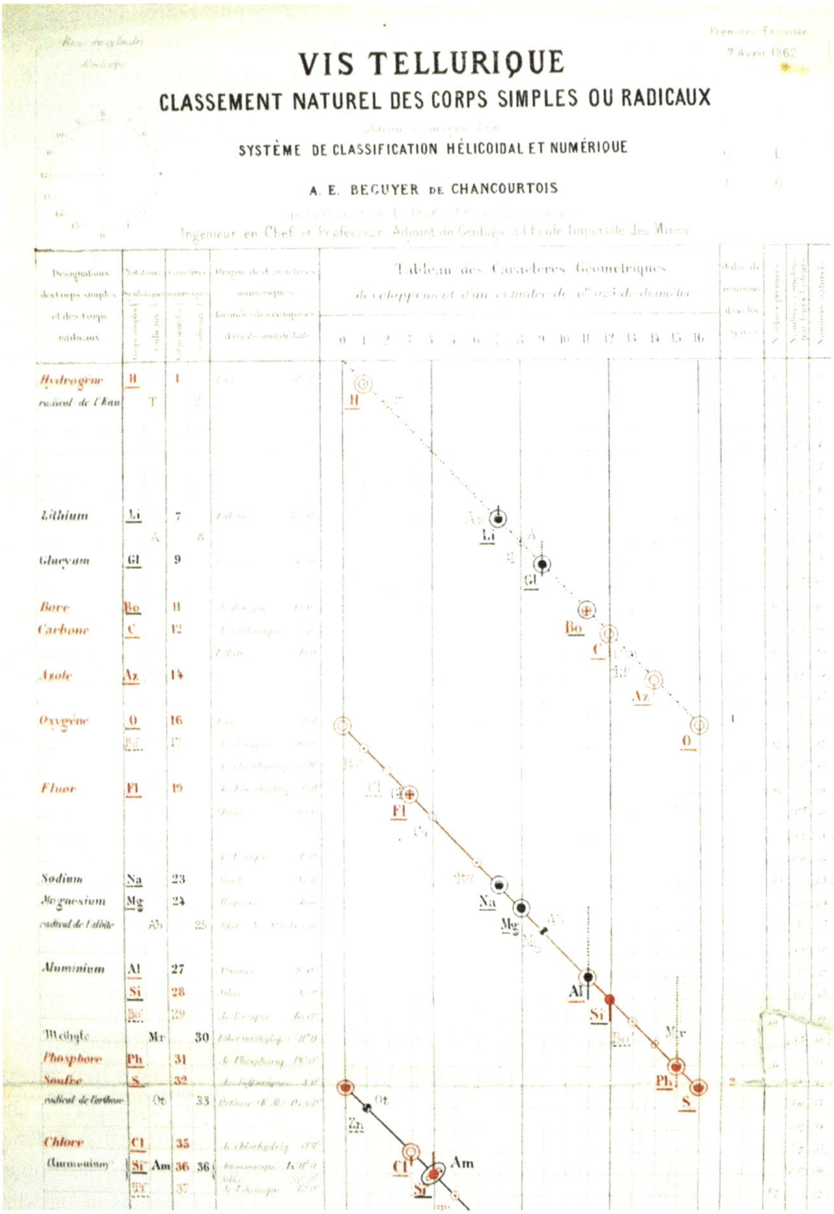

Fig. 5.3 The first two turns of the *vis tellurique* [7]. Note Li above Na, C above Si, O above S, and Fl (fluorine) above Cl. Note also Bo (boron) in light print at three points on the second turn in addition to where one would expect it on the first turn and Am (ammonium) near the bottom of the figure (Courtesy of the Master and Fellows of St Catharine's College, Cambridge)

actual graph) list, respectively, prime numbers, the factorization of compound numbers, and natural numbers.

Writers about the *vis tellurique* in the twentieth and twenty-first centuries usually state that it was ignored for decades. That is not quite true: in fact, it was discussed at the very next session of the Académie, April 14, 1862, by Charles Joseph Sainte-Claire Deville [14]. If the name Sainte-Claire Deville looks familiar in history of chemistry, it is not because of this one, but because of his brother Étienne-Henri, who devised a way of obtaining aluminum in metallic form. Both brothers were members of the Académie des Sciences in the mineralogical section. (Béguyer de Chancourtois was not a member of the Académie, then or thereafter.)

Charles wished to assert a claim of priority for the classification of simple bodies according to their roles in lithology. He had first published such a classification in 1855, having organized the material as part of the lithology course he taught at the Collège de France, where he was an assistant of the omnipresent Élie de Beaumont. An updated version of the table was included in the *Comptes rendus* of this session. Sainte-Claire Deville's table displays some familiar groupings of chemical elements, ones that have similar reactivity and stoichiometry manifest in the composition of minerals. No one who has seen a periodic table would see anything like it in Sainte-Claire Deville's table—which should come as no surprise, as he was not trying to construct such a system.

It was Béguyer de Chancourtois' turn to respond at the next session of the Académie, on April 21 [2]. He also invoked teaching for Élie (at the École des Mines) as the inspiration for his arrangement, which was also based in lithology and geognosy (classification of rocks and rock formations). The helical plot struck him "as a means of joining in a fundamental series all the elements scattered on my chart, then to manifest the correspondences of properties of all kinds."[7] He argued that his chart is much more flexible than a double-entry table, for it can draw on a practically infinite number of possible relationships, as helices can be drawn at practically any inclination joining any two points. Although Béguyer portrayed this extreme flexibility as an advantage of the *vis tellurique*, it is really a deficiency, for he offers no guidance on how to distinguish helices that connect related elements from those that connect arbitrary elements.

Figure 5.4, which is a detail of the 1863 edition of the *vis tellurique* [8, 15], illustrates these additional helices. The diagonal lines other than the main 45° descending line on which the points are plotted are other helices that connect sets of plotted points to which Béguyer attached some significance. If the graph is wrapped around a cylinder, all of the diagonal lines would wrap around to make helices of different pitch. Some of the helices are labeled with numbers such as −5/11. That helix descends five units as it moves right to left (signified by the minus sign) 11 units. The numerical labels also appear at the left and right margins of the graph, where the helices would wrap around. The 1863 edition is better for illustrating

[7]Quotations in English from French-language publications, are by the present author unless otherwise stated.

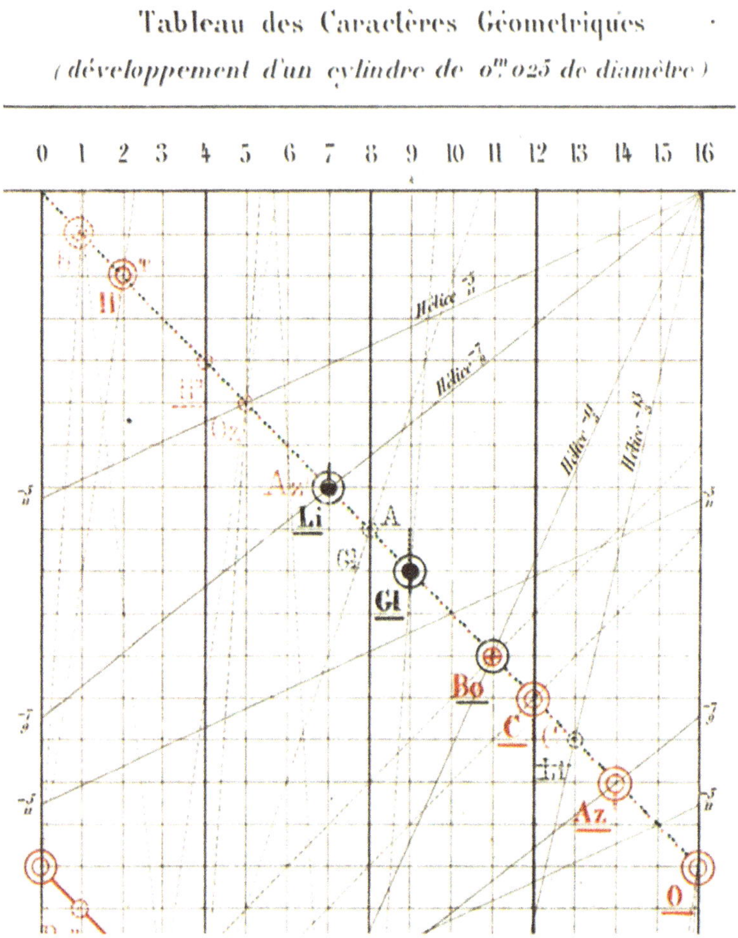

Fig. 5.4 Detail of the first turn of the 1863 edition of the *vis tellurique* chart [8, 15] illustrating helices as diagonal lines in addition to the main helix on which the points are plotted (Courtesy of the Bibliothèque patrimoniale numérique, MINES ParisTech)

these secondary helices because it shows these numerical labels, which are not present in the 1862 edition [7], and because the lines depicting these helices are darker.

Sainte-Claire Deville made another presentation on his classification on April 28 [16]. In tone, it was more conciliatory than the previous one, noting that he and Béguyer must arrive at the same place if they were correct, even if they traveled different paths to get there. He praised the helical design of the *vis tellurique* for displaying the continuous change in chemical properties, something that his own table did not do. (By the way, Sainte-Claire Deville mused about making a 24-faced right prism to illustrate his classification. If the *vis tellurique* was a challenge for printers, imagine producing such a prism!) This memoir was practically exclusively geological, about how different elements entered into different classes of minerals.

Béguyer de Chancourtois was on the program at the next session of the Académie on May 5 [3]. This paper looks like a natural follow-up to the April 7 paper had Sainte-Claire Deville not spoken up. Once again Béguyer pointed out that helices of various inclinations would illustrate relationships among the elements, and here he gave another example. One such helix joins sulfur and iron, also passing through tellurium and gold, "explaining the association of common iron and gold in gold-bearing pyrite" [3]. This is an example of a geognostic association, an observation quite different from properties such as valence or electrical polarity or isomorphism, emphasized by chemists who classified elements.

Speaking of tellurium, it is in this memoir that Béguyer explained the name he gave his system. *Vis tellurique* has usually been translated into English as telluric screw, although telluric helix [13] better conveys the abstract geometry of the arrangement. Both *tellurique* and its English cognate telluric can refer to the element tellurium or to the earth (*tellus* in Latin), for which tellurium is named. Béguyer called the helix telluric partly for the element, which lies about halfway down the chart (albeit near the bottom of the part that is densely covered with entries), and partly to emphasize the earthy or geognostic origins of the system.

In this paper Béguyer gave a bit more information about the set of characteristic points or numerical characters he used. We can regard these numbers as atomic weights, although it is difficult to discern exactly how Béguyer thought of them. He insisted that his classification was independent of preconceived theoretical ideas such as atoms [2]. Leaving aside the characters he based on specific heat, the numbers he used were, for a specified set of elements, the "proportional numbers" found in chemical treatises or, for the remaining elements, twice those proportional numbers [3]. Now the "proportional numbers" available in most French chemical treatises available at the time were equivalents. Doubling equivalents of elements with even valences basically produced the atomic weight system advocated by French chemists such as Charles Gerhardt, Auguste Laurent, and Adolphe Wurtz. At about this time, Stanislao Cannizzaro famously elucidated theoretical bases for deriving atomic weights [17]. Béguyer did not, nor did he mention Cannizzaro, Gerhardt, Laurent, or Wurtz. Recalling that Béguyer was not a chemist, we cannot assume that he was familiar with the debates on atoms, molecules, and equivalents that engaged chemists around 1860. Regardless of the conceptual significance Béguyer attached to them, his numerical characters are the atomic weights one would obtain by chemical analysis from compounds whose formula he specified based on a system where hydrogen is one and oxygen 16.

One important respect in which Béguyer's characteristic numbers differed from contemporary chemists' concepts of atomic weights is that Béguyer believed that many elements have more than one numerical character [3]:

> Most of the known simple bodies then offer several distinct numerical characters corresponding to different physical states and by which one must reconcile, as a case of isomerism, the apparently contradictory results of chemical and physical experiments. We can say that instead of one body there are several associates, one of which is the type and the others are immediate derivatives.

The reference to a type in this last sentence illustrates a key difference between Béguyer's view of elements and that of contemporary chemists. In mineralogy, a mineral species is often defined by a physical exemplar, the "type specimen." Other specimens of the same mineral can coexist with the type despite differences from it. Chemists, by contrast, regarded atoms of the same element as identical.

The detail of the *vis tellurique* shown in Fig. 5.3 shows silicon at 28 (on the second turn of the helix) and again at 36 (on the third turn); silicon appears again on the chart at 43. These numbers are based on differing assumptions about the formula of silica: SiO_2, Si_2O_5, and SiO_3, respectively. The column in the lithographed chart just to the left of the actual graph lists the compound and formula Béguyer used to justify the atomic weight plotted on that line.

It is worth emphasizing that Béguyer did not regard alternative values of atomic weight as a matter of uncertainty over a unique property whose value was one of a discrete set of possibilities depending on which of several assumptions was correct. Chemists of this time would have assigned one value to the atomic weight of silicon if its oxide was SiO_3 and a different value if it was SiO_2: they may not have known which stoichiometry (and corresponding atomic weight) was correct, but they did not believe both could be correct. But Béguyer did not use the language of uncertainty in the passage quoted above: multiple values represented different physical states rather than alternative estimates of a unique value. Furthermore, the legend of the *vis tellurique* chart (Fig. 5.5) employs the language of type and derivative used in the passage above. Of the three characters for silicon, 43 (based on silica as SiO_3) is designated the type; all three characters are shown as "definite" according to the symbols in the legend.[8]

As noted, most of the numerical characters on Béguyer's chart were based on "proportional numbers" from chemical treatises based on particular stoichiometry of specified compounds as described above. Some, however, were based on the law of Dulong and Petit and specific heats reported by Victor Regnault. Béguyer used the principle that the product of atomic weight and specific heat was the same for all elements [1]; indeed, this is the sole occurrence of the phrase atomic weight in the set of *vis tellurique* memoirs in the *Comptes rendus*. Cannizzaro had described how to use such heat capacity information to select among possible atomic weights based on uncertain stoichiometry [17]. Béguyer, however, placed numerical characters derived from heat capacities alongside those from proportional numbers. For

[8]Each silicon point is also shown with an experimental uncertainty of one unit, depicted as a bold line segment above or below the plotted point. Figure 5.3 also illustrates multiple characters for boron (Bo) with a primary character of 11 and three secondary characters (17, 22, and 29). These values are based on the formula of "boric acid" (the compound we would call boric oxide or boric anhydride) as Bo_2O_3, Bo_2O_5, BoO_3, and BoO_4, respectively. In this case, only the type character of 11 is designated as definite; the others, plotted in dotted symbols, are characterized as "probable."

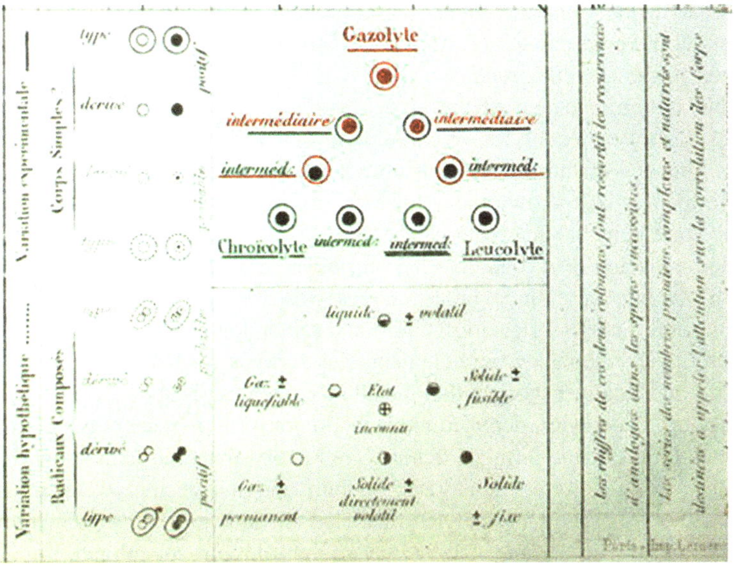

Fig. 5.5 The legend of the *vis tellurique* from Ref [7]. The triangle of circled dots in the upper part of the legend represents three classes of elements, gazolytes (top, red), leucolytes (lower right, black), and chroicolytes (lower left, green) (Courtesy of the Master and Fellows of St Catharine's College, Cambridge)

example, the chart shows antimony at 129 based on thermal measurements alongside the type character at 121; those two values are reported as definite, and characters plotted at 123 and 242 are depicted as probable.

In a passage left out of the *Comptes rendus* but included in the later pamphlet version of this paper [8], Béguyer went on to discuss considerations that led him to some of the characters he selected for some specific elements. (And when he reported more than one character for one element, he described them as "a first" and "a second," etc.—not as alternatives). In the same section, he mentioned just a few compound (that is, polyatomic) radicals he included on the chart, namely cyanogen, which he formulated as C_2N_2, ammonium (N_2H_8), and a mineralogical radical of orthoclase. Orthoclase is a potassium aluminosilicate, and Béguyer gave this radical a character of 33, the average of the primary characters of potassium and aluminum.

Several months passed before another presentation to the Académie on the *vis tellurique*. In the meantime, the first edition of his booklet and chart was published [7]. The booklet is 12 pages long, reprinting the first three memoirs from the *Comptes rendus* [1–3].

Béguyer de Chancourtois presented the fourth paper in the series on October 13, 1862 [4]. At that time, he presented the Académie with a copy of the lithographed chart. He hoped that it would promote understanding of the pieces published earlier in the year in the *Comptes rendus* and would facilitate critiques from interested scientists. It did not seem to accomplish this aim—but then again, the chart was not published in the *Comptes rendus*.

Within his arrangement of the *vis tellurique*, Béguyer continued to class elements as gazolytes, leucolytes, or chroicolytes, or some combination of these main classes. This classification had been introduced by André-Marie Ampère in 1816 [18].[9] Ampère is best known today for his work in electricity; his name is the base unit for current in the International System of Units (SI). In chemistry, he also published a version of Avogadro's hypothesis in 1814. At any rate, the division of elements among these three broad classes was still employed in the 1860s in textbooks and treatises of chemistry [19] and mineralogy [20]. Gazolytes, depicted in red on the *vis tellurique*, were elements that form permanent gases (nitrogen, for example), and Béguyer thought that metalloids (such as phosphorus) also belonged to that group. Leucolytes, depicted in black, do not form permanent gases with each other, but they do not form particularly refractory materials either; in colorless acids, they form colorless solutions. Sodium and magnesium were among the elements in this class. Chroicolytes, depicted in dark green, *do* form refractory materials, and if their oxides dissolve in acid, the solutions are colored. Iron, cobalt, and copper were examples. The rather elaborate key Béguyer included at the bottom of the chart (Fig. 5.5), permits entries to be classed as one of these types or as some combination of them.

The fifth of Béguyer de Chancourtois' memoirs in the *Comptes rendus* was presented on February 9, 1863 [5]. This application of the *vis tellurique* to the theory of steel is the one paper of his six that he would not reproduce in the 1863 edition of his booklet [8]. Béguyer observed that many elements that are either hard themselves or that are used to harden alloys fall on a certain helix that connects bodies whose character is a multiple of 11. These include boron (11 and 22), carbon (44, based on the law of Dulong and Petit and the heat capacity of diamond), manganese, zinc, arsenic, antimony, tungsten, and iridium. Note that some of these points are secondary characters and others depend on error bars assigned to atomic weights. He concluded that "The number 11 therefore seems characteristic of a certain hardness" [5]. There are also a number of elements whose atomic weights are multiples of 7 that are hard or can act as hardeners: they include nitrogen, silicon (28 and 42), titanium, iron, arsenic, iodine, vanadium, and tungsten. Absent from this paper, and apparently superseded, is an aside from the May 1862 paper [4] asserting that practically all of the bodies used to make steel are grouped on or near the vertical line down from atomic weight 12.

Béguyer de Chancourtois presented the last of his six papers [6] on the *vis tellurique* at the Académie's session of March 16, 1863. This paper was largely about how the recently discovered element thallium fit into his system. It contains some of the most explicit statements of Béguyer's numerological ideas. It also

[9]Ampère's long article was serialized in four consecutive monthly issues of the *Annales de chimie et de physique*. The terms gazolytes, leucolytes, and chroicolytes are explained in the last six pages of the final installment.

happens to be the paper whose version in the 1863 booklet [8] shows the most additions compared to the *Comptes rendus*.

A recent report of the heat capacity of thallium led Béguyer to adopt 103 as the element's primary numerical character, which would place it on the same vertical line as lithium, sodium, potassium, manganese, and rubidium. In fact, the accepted atomic weight of thallium[10] is nearly double 103, and it is not classified as an alkali metal, although it commonly takes on an oxidation number of +1. Still, Béguyer's willingness to group thallium with the metals above it strikes me as the most chemically astute example given in the whole set of six papers. That is, this vertical line connects many elements that exhibit similar chemical behavior while omitting few elements that also exhibit that behavior and including few that do not belong. The other observation he made about thallium in this paper is that it lies on a helix consisting only of "rare, singular or virtual bodies, such as arsenic, the radical of orthoclase, and fluorine" [6]. He proposed to add to this helix a point at atomic weight 5, which he assigned to ozone (one-sixth of the sum of oxygen and nitrogen). Why oxygen and nitrogen, why one-sixth, and what ozone has in common with thallium and the other singular bodies mentioned (other than fluorine) was not explained.

In this paper Béguyer repeated his belief that prime numbers are important in the structure of matter. Almost all of the numerical characters on the chart, he noted, are either prime numbers, one unit away from prime numbers, or half of "one unit away from prime numbers."[11] If this observation sounds like a suggestion that prime numbers are surprisingly connected to the structure of matter, the following observation (not made by Béguyer) puts the statement into context: nearly 90% of the natural numbers up to 200 (roughly the range of numerical characters in the *vis tellurique*) fall into one of these categories. So the observation that nearly all of the atomic weights plotted at the nearest natural number falls into one of these categories is neither surprising nor physically significant.

Béguyer's guiding philosophy was, perhaps, summarized by the statement that "The properties of bodies are the properties of numbers,"[12] a statement emphasized in the booklet version of this paper by being printed in small capital letters [6].

The 1863 edition of the booklet (Fig. 5.6) and chart [8] contains a postscript dated April 6, 1863—one day short of a year after his first presentation on the *vis tellurique*. The postscript, on the 21st and final page of the booklet, describes the preceding pages as the abridged papers on his work that had appeared in the *Comptes rendus*, supplemented by a few additional words (shown in smaller print) needed to convey the sequence of ideas. (Béguyer did not mention that the booklet contains only five of the six papers [1–4, 6].) Most of the postscript describes

[10]A value in the low 200s was also used by other pioneers of the periodic system in the 1860s, including John Newlands, William Odling, Dmitri Mendeleev, and Lothar Meyer.

[11]That is, if we represent a prime number by p, numbers that can be represented as $(p \pm 1)/2$. Since all primes greater than 2 are odd numbers, the corresponding numbers $(p \pm 1)/2$ are all integers.

[12]*Les propriétés des corps sont les propriétés des nombres.*

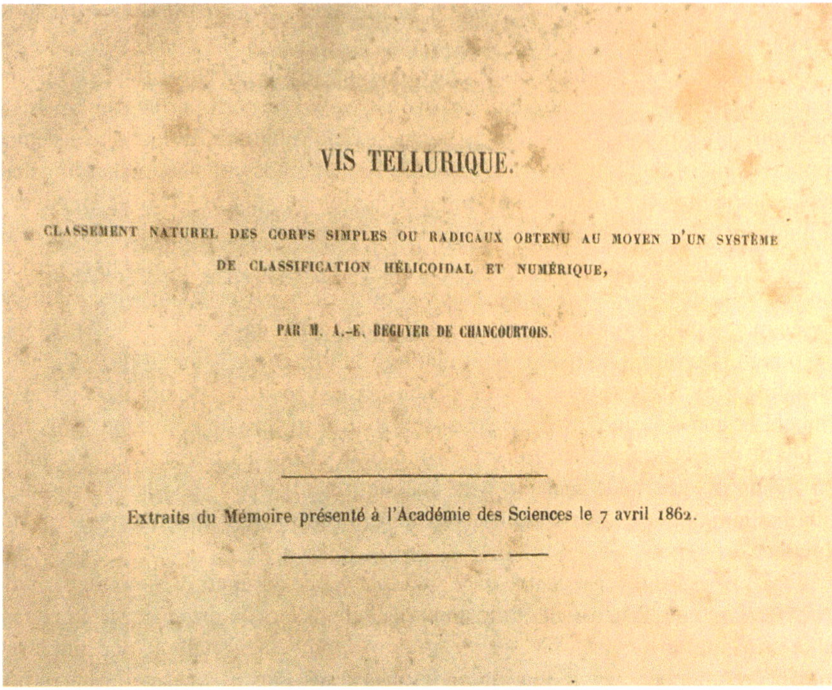
VIS TELLURIQUE.

CLASSEMENT NATUREL DES CORPS SIMPLES OU RADICAUX OBTENU AU MOYEN D'UN SYSTÈME DE CLASSIFICATION HÉLICOIDAL ET NUMÉRIQUE,

PAR M. A.-E. BEGUYER DE CHANCOURTOIS.

Extraits du Mémoire présenté à l'Académie des Sciences le 7 avril 1862.

Fig. 5.6 Front cover of the 1863 edition of the *vis tellurique* booklet [8]. This copy was once owned by Stanislao Cannizzaro (Courtesy of the private collection that holds it)

differences between the second and third printings of the chart.[13,14] The most important change in the chart was the subject of the final paper, namely the addition of the recently discovered thallium to the system. The other changes were minor, such as adding some secondary characters and plotting some points in dots (indicating probable values) rather than solidly (indicating definite ones).

[13]Reference to the second printing is a bit mysterious. The copies of the chart currently in libraries are either the first printing or the third printing, according to responses to my inquiries. The upper right of both of these printings bears the words "Première Esquisse, 7 Avril 1862." In the chart that accompanied the 1863 booklet, the phrase "3e Tirage 16 Mars 1863" follows those words. "Première Esquisse" was evidently intended to indicate when the chart was first drafted or presented, and not to label a given version.

[14]At least one version of the booklet included a second fold-out chart, "une seconde planche muette, du développement du cylindre disposée pour l'étude et l'extension du système" according to nineteenth-century advertisements and twenty-first-century library catalog descriptions. The planche muette appears to be a blank version of the chart, displaying the grid and some secondary helices but lacking plotted points. I have not been able to examine a physical example of the planche muette or a high-resolution scan of it.

5.3 Reaction to the *Vis Tellurique*

As mentioned above, the *vis tellurique* was ignored by chemists in the 1860s, or at least it made little or no impression in print. It was not as though Béguyer's presentations were published in an obscure journal: they appeared in one of the most important scientific publications in Europe, the Académie's *Comptes rendus*. To be sure, Béguyer de Chancourtois was a geologist, not a chemist; however, the Académie published all six of his presentations under the heading of general chemistry.

To be sure, the papers make little sense without the chart. The Académie did not publish the chart, but Béguyer did. And he sent the chart and booklet to other scholars (*savants*), as he noted in the postscript to the 1863 booklet [8]: he said that his list of changes from the previous version was addressed to those scholars whose attention he was soliciting and to whom he had sent the earlier version. It would be interesting to know how widely he diffused the booklets, especially by discipline: were the recipients mostly chemists? mostly geologists?

The *vis tellurique* does not appear to have been reported on much outside France in 1862 or 1863. A short paragraph abstract of it appeared under the heading of general chemistry in the *Jahresbericht über die Fortschritte der Chemie und verwandter Theile anderer Wissenschaften* for 1862 [21]. Another short paragraph on it appeared in the *American Journal of Science and the Arts* in 1863 [22], in which the French chemist Jerome Nicklès intimated that it appeared to be the key to a fundamental law.[15]

At the end of Béguyer's fourth memoir in the *Comptes rendus* [4], it is noted that Jean-Baptiste Dumas, then dean of French chemistry, was named as one of the *commissaires* the Académie assigned to this work. It was normal practice for the Académie to name a few members as *commissaires* for memoirs presented by non-members. The *commissaires* originally named for Béguyer's work, listed under the title of the first publication [1], were Henri Hureau de Sénarmont, Gabriel Delafosse, and Gabriel-Auguste Daubrée, all members of the mineralogical section of the Académie. De Sénarmont died in June 1862, and Dumas was listed as his replacement at the end of Béguyer's October 1862 memoir.

Dumas had published on relations among the equivalents of the elements just a few years earlier, including an 82-page memoir in 1858 [23]. Dumas' memoir is frequently mentioned among the work of chemists in the 1850s who were seeking patterns and relationships among the atomic weights of the elements—as forerunners of the periodic law who did not, however, observe chemical periodicity [24, 25]. We know, therefore, that Dumas was at least exposed to the *vis tellurique* and

[15]Jerome Nicklès was professor of chemistry at Nancy, France, and a regular correspondent to the journal more in the capacity of a reporter than a researcher. His contributions were mainly collection of short reports and blurbs on current scientific developments in France. It is in this capacity that he mentioned the *vis tellurique*. Nicklès is best known for his research on fluorine, which ended his life tragically in 1869 in an attempt to isolate that element.

that he was interested in classification of the elements. I know of no reaction by Dumas to Béguyer's work, nor of any report on his work by the *commissaires*.[16]

It is worthwhile to mention one other famous nineteenth-century chemist who apparently left no reaction to the *vis tellurique*. Cannizzaro is best known today for rationalizing atomic weight measurements into a self-consistent system [17], a development that is widely considered to be a prerequisite to the discovery of chemical periodicity [26, 27]. Cannizzaro owned a copy of Béguyer de Chancourtois' booklet and chart (1863 version, Fig. 5.6). That copy bears no inscription or notes, however. It is not known when or how Cannizzaro acquired the copy or what he thought of it.

If few of Béguyer's contemporaries saw his lithographed chart, few likewise saw it subsequently. It is a rarity among book dealers and collectors: only four copies of the item have been offered for sale over the last 50 years [28]. WorldCat lists only a handful of copies in its member institutions around the world, most of them in France, including at the Bibliothèque nationale de France and MINES ParisTech (successor of the École des Mines). In addition, there are copies at the Science Museum of London and St Catharine's College, Cambridge. At least two were placed on exhibit [29, 30] in 2019 during the International Year of the Periodic Table. High-resolution scans of the chart or of a significant portion of it are now available online [15, 31].

5.4 Later Attention and Priority Claims

We pick up the story of the *vis tellurique* some 15 years later, in the late 1870s. The intervening time was very eventful for the development of the periodic law, as can be seen in several other chapters of this book. Periodic systems of the elements were proposed by several chemists. This time also includes the discovery of gallium, whose resemblance to Mendeleev's predicted eka-aluminum was noted in short order. In sum, the periodic system was beginning to become widely known in chemistry during this time [32]. But the *vis tellurique* had to wait still longer to capture the attention of many chemists.

The next published mention of the *vis tellurique* by a chemist appears to be an implicit one in passing by the French chemist Wurtz. In his 1879 book *La Théorie atomique*, he described Mendeleev's system of classification of elements. Periodicity was observed, he wrote, after putting the elements in order by atomic weight. Atomic weight order, though simple, was a key idea. In a footnote, Wurtz added that Mendeleev's idea was not unlike one put forward earlier by "M. de Chancourtois" [33]. The mention does not name the *vis tellurique* or provide any bibliographic pointers. In context, it seems to me that Wurtz was claiming priority for

[16]I inquired whether the archives of the Académie des Sciences contained any items related to the *vis tellurique*—manuscripts or charts from the author or a report of the *commissaires*, for example. The answer was that there was nothing. Appointment of a commission did not imply that a report was produced, let alone published.

Béguyer not for periodicity, but for arranging elements in order by atomic weight—although he was not actually the first to do so.

Six years later (1885), Wurtz's *Introduction à l'étude de chimie* contained two mentions of Béguyer in connection with the periodic law [34]. The first, in the preface by Charles Friedel and Georges Salet, lists "de Chancourtois" along with Dumas, John Newlands, and Mendeleev as investigators of ingenious relationships between the atomic weights of elements and their properties. In the body of the text, Wurtz discussed Dumas' work on relationships among the atomic weights of the elements, then listed Béguyer along with John Hall Gladstone, Josiah Parsons Cooke, Newlands, Max von Pettenkofer, William Odling, and Peter Kremers as other investigators of relationships among atomic weights. Also in 1885, in his book *Les origines de l'alchimie*, Marcellin Berthelot credited first Béguyer, then Newlands, Lothar Meyer, and Mendeleev for parallel periodic series [35]. Although Berthelot credited Béguyer with what later chemists would consider a more important advance than Wurtz did, both of these French chemists mentioned his work in passing, providing neither the name *vis tellurique* nor explanation nor bibliographic information about it.

Béguyer de Chancourtois died in Paris in November 1886, and the obituaries that appeared early in 1887 in French geological periodicals inevitably mentioned the *vis tellurique*, sometimes in detail. Edmond Fuchs, a colleague of Béguyer at the École des Mines, highlighted the *vis tellurique* in his long obituary published in the *Annales des Mines* [9] and in the *Bulletin of the French Geological Society* [36]: it must rank at the top of a series of first-rate works of Béguyer. Fuchs described the *vis tellurique* in some detail, and noted that chemists had later developed similar ideas independently.

> Thus Lothar Meyer, Mendeleef, & c., have established a general classification of simple bodies according to their atomic weights; but they did it in a much less felicitous way and in the rudimentary form of a double-entry table.

Fuchs also quoted at length from Béguyer's inscription of a presentation copy of the *vis tellurique* to Prince Napoleon. The inscription is idealistic, religious, and mystical, expressing a faith in science for the benefit of humanity.

Mineralogist Ernest Mallard, also of the École des Mines, also highlighted the *vis tellurique* in the much shorter obituary he published in the *Bulletin of the French Society of Mineralogy* [37], in effect calling it Béguyer's best work. "We know ... what has been the success of M. Mendelejeff's ideas," he wrote, and one can see that the same ideas inspired Béguyer several years earlier. Mallard attributed to Béguyer what has become the classic statement of the periodic law, that physical and chemical properties are a periodic function of atomic weight.

Mendeleev himself mentioned Béguyer in his 1889 Faraday lecture [38]:

> The idea of seeking for a relation between the atomic weights of all the elements was foreign to the ideas then current, so that neither the *vis tellurique* of De Chancourtois, nor the *law of octaves* of Newlands, could secure anybody's attention.

In effect, Mendeleev said that the two were ahead of their time. He added:

> And yet both De Chancourtois and Newlands, like Dumas and Strecker, more than Lenssen and Pettenkofer, had made an approach to the periodic law and had discovered its germs.

Later in 1889 came a milestone in making Béguyer's work known in the chemical community, namely the first graphical representation of his system ever published in a major journal (Fig. 5.7) and the first anywhere since his own booklet with fold-out chart. This came in Philip Hartog's article in *Nature*, "A First Foreshadowing of the Periodic Law" [13].

Hartog's paper includes a translation of Béguyer's first paper, Hartog's own assessment of the significance of the *vis tellurique*, and a highly simplified version of the first two turns of the graph, leaving out radicals, secondary characters, and

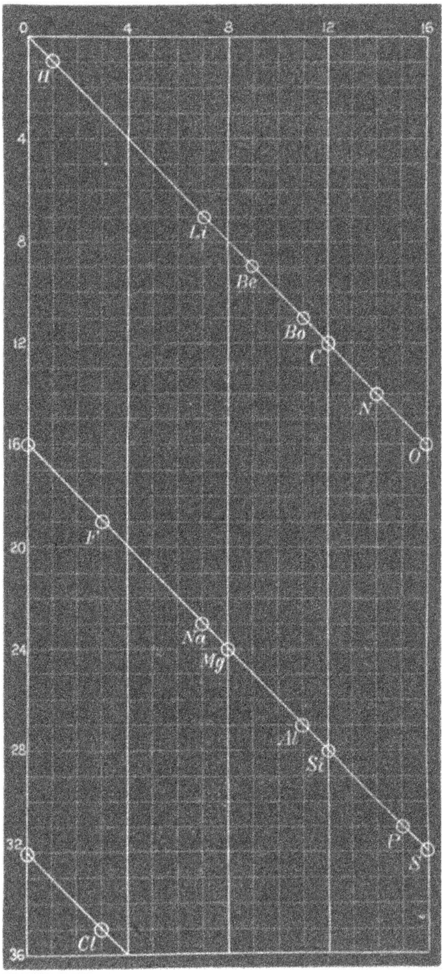

Fig. 5.7 Figure illustrating Philip Hartog's report on the *vis tellurique* [13]. Digitized by Google Books

additional helices intended to show relationships among elements. Although the graph published in Hartog's article is much abbreviated and simplified compared to Béguyer's own, no caption or explanatory text tells the reader that it is simplified, let alone **how** it is simplified. Hartog did not claim for Béguyer "the discovery of Newlands and Mendeleev." He asserted (incorrectly) that Béguyer was the first to publish a list of all the known elements in order of atomic weight, an explicit claim that echoes Wurtz's 1879 passing mention. But Hartog did not see octaves or periods in Béguyer's work.

In 1891, two French scientists made a priority claim on behalf of Béguyer de Chancourtois, namely François Lecoq de Boisbaudran,[17] who had discovered gallium, and Albert de Lapparent, a French geologist who attended the École des Mines while Béguyer was Élie's assistant there. They published in the *Comptes rendus* [39], and a translation of their article appeared [40], along with an editorial [41], in the *Chemical News*. The person whose priority these authors wished to contest was not Mendeleev or Meyer, but Newlands, for Newlands had, they noted, established priority with respect to Mendeleev. Near the beginning of the article, they noted that Newlands had claimed to be the first to list the known elements in order of atomic weight [42, 43]; they correctly pointed out that Béguyer had done so before him, but mistakenly asserted that he was the first.

They described the *vis tellurique* and included a simplified and abridged version of the chart (Fig. 5.8), which they called a "reduction." It is considerably simplified, but it does include secondary characters for some elements (an issue discussed in their text) and a secondary helix. The authors judged that the *vis tellurique* contains both atomic weight order and evidence of periodicity, even though it has flaws as well.

Near the end of the article, they asked how it was possible that Béguyer's work, published in the most widely circulated scientific periodical in the world, could have escaped the notice of Newlands "whose good faith cannot be doubted" [40]. They answered this question in a way echoed by many writers on the *vis tellurique* after them: because the diagram was not included in the publication, and that the privately published diagram did not receive wide circulation.

William Crookes published a translation of Lecoq de Boisbaudran and Lapparent's paper in the *Chemical News* (without the figure, by the way), but he preceded it with a decidedly skeptical editorial. Béguyer's memoirs were in the *Comptes rendus* for all to see over all these years, and no one had interpreted them in this way before: when Lecoq discussed the predictions of Mendeleev while reporting the discovery of gallium, he made no mention of the *vis tellurique*; when Mendeleev and Meyer had something of a priority dispute, no one brought it up; and when the matter was reexamined in light of Newlands's priority claims, no one mentioned it. Crookes's judgment of Béguyer's papers was [41]:

[17]According to the Notice de personne in the catalogue général of the Bibliothèque national de France, the international form of the name of the discoverer of gallium is Lecoq de Boisbaudran, François. His name is also given as Paul-Émile Lecoq de Boisbaudran in many sources.

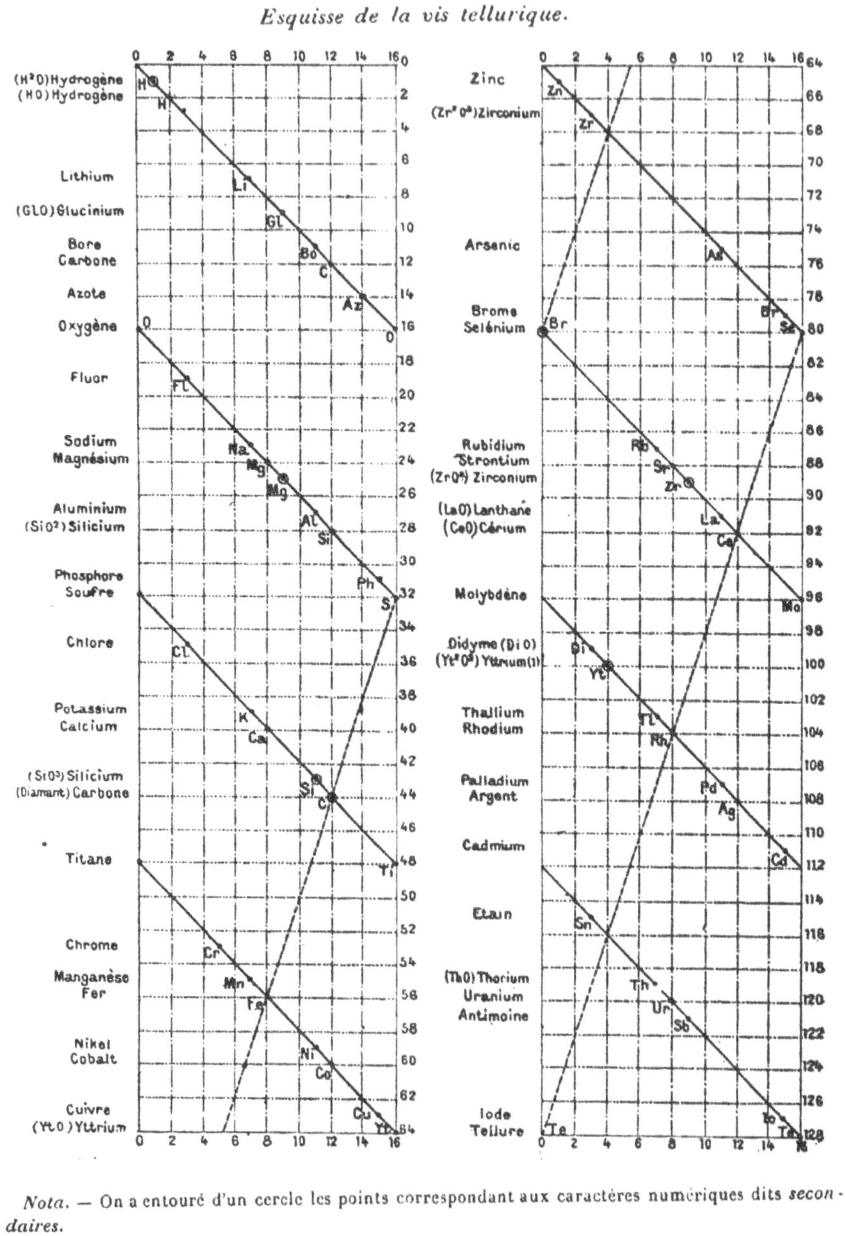

Fig. 5.8 Reduction of the *vis tellurique* illustrating the report by Lecoq de Boisbaudran and Lapparent [39]. Digitized by Google Books

They certainly contain a proposal to classify the elements with reference to their atomic weights. But we may be permitted to doubt whether they can be fairly considered as the germ of the Periodic Law.

5.5 Historical Treatments

By 1891, Béguyer de Chancourtois' work was rediscovered, and writers on the origins of the periodic table would nearly always mention him, either as a precursor or as a discoverer of chemical periodicity. And they would nearly always include an illustration, albeit always abbreviated and nearly always simplified.

In 1895, Karl Seubert collected and edited a group of writings on the periodic table for the 68th number in the series of Ostwalds *Klassiker*.[18] Along with primary texts by Meyer and Mendeleev were notes by Seubert [44]. Those notes offered the interesting suggestion that chemists were too focused on organic chemistry in the early 1860s to pay attention to Béguyer's work. The suggestion is simply an assertion, though, not supported by argument. The emergence of the structural theory of organic chemistry at about this time may have appeared, from the perspective of the 1890s, to be the major development in chemistry during the late 1850s and early 1860s. The careers of Dumas and Cannizzaro, however, as well as the Karlsruhe Congress of 1860, would seem to be counterexamples to the assertion. Both Dumas and Cannizzaro were organic chemists and both wrote important papers on atomic weights in the late 1850s [17, 23]. Both attended the Karlsruhe Congress, where atoms, molecules, and equivalents were the central topics of discussion, as did many prominent chemists from throughout Europe.

Seubert included an illustration of the *vis tellurique* (Fig. 5.9) which had the advantage of being able to be set in type rather than needing to be engraved or otherwise reproduced as artwork. Technically it is a table rather than a figure. In fact, Seubert's tabular version is an only slightly simplified depiction of Lecoq and Lapparent's graphic. But Seubert stated (incorrectly) that Lecoq and Lapparent gave a faithful copy of Béguyer's figure.

Francis Venable's 1896 book on the development of the periodic law [45] includes the statement that Béguyer de Chancourtois "may in some measure be regarded as the originator of the periodic law." Venable illustrated the *vis tellurique* in an abbreviated but quite faithful way (Fig. 5.10), if not in perfect French. It is abbreviated in that it shows only the first three turns of the main helix, but it shows secondary characters (mentioned also in the text), radicals, and even the columns of prime numbers, factorized numbers, and natural numbers. Aside from abbreviation,

[18]Ostwalds *Klassiker der exakten Wissenschaften* was a long-running series of reprints in German of important texts in the natural sciences and mathematics. The series is named after its originator, the physical chemist Wilhelm Ostwald, and it was published for many years by Wilhelm Engelmann in Leipzig. The series was launched in 1889 with a reprint of *Über die Erhaltung der Kraft* by Hermann Helmholtz. Nearly 200 volumes were published over the next 30 years, and the series continued at a slower pace after that.

121

Esquisse de la vis tellurique.

			0		2		4		6		8		10		12		14		16
(H_2O)	Wasserstoff	1	.	H															
(OH)	Wasserstoff	2	.	.	H														
	Lithium	7	.	.	.	.	.	.	.	Li									
(GlO)	Glucinium	9	.	.	.	.	.	.	.	.	.	Gl							
	Bor	11	.	.	.	.	.	.	.	.	.	.	.	Bo					
	Kohlenstoff	12	.	.	.	.	.	.	.	.	.	.	.	.	C				
	Stickstoff	14	.	.	.	.	.	.	.	.	.	.	.	.	.	.	N		
	Sauerstoff	16	O	.	.	.	.	.	.	.	.	.	.	.	.	.	.	.	O
	Fluor	19	.	.	.	Fl													
	Natrium	23	.	.	.	.	.	.	.	Na									
(25)	Magnesium	24	.	.	.	.	.	.	.	.	Mg	Mg							
	Aluminium	27	.	.	.	.	.	.	.	.	.	.	.	Al					
(SiO_2)	Silicium	28	.	.	.	.	.	.	.	.	.	.	.	.	Si				
	Phosphor	31	.	.	.	.	.	.	.	.	.	.	.	.	.	.	.	P	
	Schwefel	32	S	.	.	.	.	.	.	.	.	.	.	.	.	.	.	.	S
	Chlor	35	.	.	.	Cl													
	Kalium	39	.	.	.	.	.	.	.	K									
	Calcium	40	.	.	.	.	.	.	.	.	Ca								
(SiO_3)	Silicium	43	.	.	.	.	.	.	.	.	.	.	.	Si					
(Diamant)	Kohlenstoff	44	.	.	.	.	.	.	.	.	.	.	.	.	C				
	Titan	48	Ti	.	.	.	.	.	.	.	.	.	.	.	.	.	.	.	Ti
	Chrom	53	.	.	.	.	.	Cr											
	Mangan	55	.	.	.	.	.	.	.	Mn									
	Eisen	56	.	.	.	.	.	.	.	.	Fe								
	Nickel	59	.	.	.	.	.	.	.	.	.	.	.	Ni					
	Kobalt	60	.	.	.	.	.	.	.	.	.	.	.	.	Co				
	Kupfer	63	.	.	.	.	.	.	.	.	.	.	.	.	.	.	.	Cu	
(YtO)	Yttrium	64	Yt	.	.	.	.	.	.	.	.	.	.	.	.	.	.	.	Yt
	Zink	65	.	Zn															
(Zr_2O_3)	Zirkonium	67	.	.	.	Zr													
	Arsen	75	.	.	.	.	.	.	.	.	.	.	.	As					
(79)	Brom	80	Br	.	.	.	.	.	.	.	.	.	.	.	.	.	.	Br	
	Selen	80	Se	.	.	.	.	.	.	.	.	.	.	.	.	.	.	.	Se
	Rubidium	87	.	.	.	.	.	.	.	Rb									
	Strontium	88	.	.	.	.	.	.	.	.	Sr								
(ZrO_2)	Zirkonium	89	.	.	.	.	.	.	.	.	.	Zr							
(LaO)	Lanthan	91	.	.	.	.	.	.	.	.	.	.	.	La					
(CeO)	Cerium	92	.	.	.	.	.	.	.	.	.	.	.	.	Ce				
	Molybdän	96	Mo	.	.	.	.	.	.	.	.	.	.	.	.	.	.	.	Mo
(DiO)	Didym	99	.	.	.	Di													
(Yt_2O_3)	Yttrium	100[1]	.	.	.	.	Yt												
	Thallium	103	.	.	.	.	.	.	.	Tl									
	Rhodium	104	.	.	.	.	.	.	.	.	Rh								
	Palladium	107	.	.	.	.	.	.	.	.	.	.	.	Pd					
	Silber	108	.	.	.	.	.	.	.	.	.	.	.	.	Ag				
	Kadmium	111	.	.	.	.	.	.	.	.	.	.	.	.	.	.	.	Cd	
	Zinn	115	.	.	.	Sn													
(ThO)	Thorium	119	.	.	.	.	.	.	.	Th									
	Uran	120	.	.	.	.	.	.	.	.	Ur								
	Antimon	121	.	.	.	.	.	.	.	.	.	Sb							
	Jod	127	.	.	.	.	.	.	.	.	.	.	.	.	.	.	.	J	
	Tellur	128	Te	.	.	.	.	.	.	.	.	.	.	.	.	.	.	.	Te

[1] Irrthümlich für 96.

Fig. 5.9 Karl Seubert's representation of the *vis tellurique* [44]. Public domain image from Universitat de Barcelona, Biblioteca Patrimonial Digital

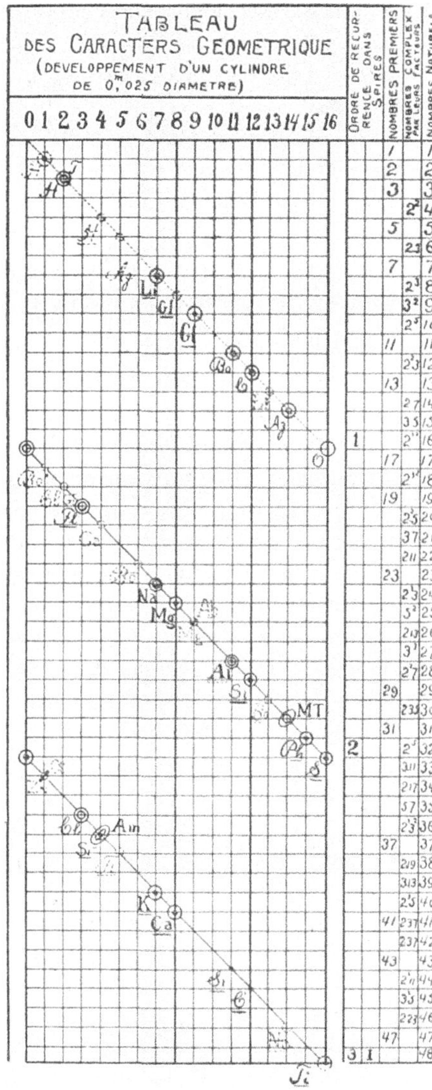

Fig. 5.10 Depiction of the *vis tellurique* in Venable's book on the periodic law [45]. Digitized by the Internet Archive

this illustration of the *vis tellurique* is a simplification only in that it does not show helices other than the main one.

Venable also pronounced on the priority of publishing a table of elements in atomic weight order [46]. He credited John Hall Gladstone for this innovation in 1853 [47], nearly a decade before Béguyer. In the twenty-first century it seems remarkable that a table of elements ordered by atomic weight was such an innovation, particularly in light of the primacy that determining atomic and molecular weights had in the chemical research programs of the first half of the nineteenth century. My own unsystematic perusal of tables of weights in papers and textbooks of the time showed alphabetical order to be usual. Often the tables listed both elements and compounds. John Dalton's famous table in *A New System of Chemical Philosophy* [48] lists 20 elements in atomic weight order, and then several compounds.[19]

Let us consider one more example from near the turn of the twentieth century. George Rudorf, a 19-year-old student of William Ramsay's at University College London wrote of Béguyer [52]

> Being a mathematician and a geologist, he did not consider his spiral from the point of view of chemical facts, and thus the "vis tellurique," as it was called, received no attention, and fell entirely into oblivion until unearthed in 1889.

When I first read Rudorf's words, they struck me as brash and not entirely accurate; for example, there is a difference between a mathematician and a number mystic, and Béguyer appears to have been the latter rather than the former. On reflection, though, I think he is right about chemical considerations not having been the prime consideration in Béguyer's thinking about the relationships of elements. Rudorf's treatment of the *vis tellurique* is very brief, but it does mention that Béguyer used helices other than the main one to illustrate relationships. Rudorf's illustration of the *vis tellurique* is more simplified even than Hartog's [13].

This review of Béguyer's place in historical treatments of the periodic law moves ahead now to the work of the historian who first introduced me to the *vis tellurique*, Jan van Spronsen. His classic history of the first 100 years of the periodic system of the elements [53], published about 50 years ago, named Béguyer de Chancourtois one of the independent discoverers of the periodic law. That book raised more questions with me about the *vis tellurique* than it answered, for it included both a highly simplified graphic and an image of the first three turns of Béguyer's lithographed chart.[20] Although the fragment of the original chart was neither large enough nor extensive enough to study in detail, it was clear that the chart was much more complex than the simplified version. The simplified figure illustrated well most of what van Spronsen wrote about the *vis tellurique*, and he did note that the redrawn figure was simplified. Put another way, van Spronsen wrote little or nothing about the portions of the original chart that were omitted from the simplification. I wondered why, for instance, I could see points representing boron in several places on the original. I wondered about the radicals mentioned in the title of the chart. The discrepancy between the two charts was at the back of my mind for perhaps 20 years before I saw a high-quality scan of the original (and then later an actual paper copy).

[19]Dalton published a partial list of the relative weights of the ultimate bodies of gases and volatile bodies even earlier [49, published in 1805, read in 1803]. A notable but not comprehensive list of elements only arranged by atomic weight appears in an 1833 paper by Marc-Antoine Gaudin [50]. In another paper read in 1831 and published in 1833, Gaudin included a more complete table of elements in atomic weight order, but the table was not published. See Theron Cole's paper on Gaudin [51].

[20]The simplified graphic in Ref. [53] is identical to that displayed as Fig. 5.11 in the present work except for the typeface and the size of plotted points. The image of the original in van Spronsen's book is comparable to Figs. 5.3 and 5.4 of the present work.

As far as I have been able to trace, van Spronsen's simplified *vis tellurique* (Fig. 5.11) first appeared in a 1951 article on priority for the periodic table in the Royal Dutch Chemical Society's member magazine [54]. In the text, van Spronsen noted that reprinting the original would have been impractical so he presented the graphic somewhat simplified, but preserving the original as much as possible. This article named Béguyer as the first of the independent discoverers of the periodic law. Van Spronsen regretted the fact that no graphic was published with Béguyer's memoirs. He noted that Meyer, Mendeleev, and especially Newlands claimed credit for their formulations of the periodic law once it became widely known, but that Béguyer did not. He speculated that Béguyer no longer cared about his system, possibly because he was not a chemist. At any rate, van Spronsen dated the discovery of the periodic system to 1862.

5.6 Did Béguyer de Chancourtois Discover the Periodic Law?

Having examined the *vis tellurique* memoirs and chart and much of what has been written about them since, I turn to the question of whether or not Béguyer de Chancourtois ought to be counted among the discoverers of the periodic law. I choose to engage this question understanding that "Who discovered *x*?" is both philosophically and historically vexed. See Michael Gordin's chapter on the pedagogical origins of Mendeleev's and Meyer's classification systems for a discussion of the problems with "who discovered" questions in general and their application to the periodic system in particular [55]. Among the problems is identifying the essence of the discovery in question and drawing a sharp boundary around it, conceptually, linguistically, and chronologically isolating it from what was not the discovery and from subsequent development of the discovery. Identifying the essence of a discovery is always anachronistic, always done after the fact (whenever that was) and having in mind a version of the discovery that emerged from whatever historical processes produced it.

Nevertheless, "who discovered it" is a tantalizing question for scientists of the present day interested in the development of the theories, practices, and concepts they now use. Trying to identify the essence of a discovery can promote reflection on what is important and what superfluous to a concept as currently employed. Looking at the past of the discipline from the perspective of the present can satisfy the curiosity of present-day scientists. So I offer my answer to the question "Did Béguyer de Chancourtois discover the periodic law?" for the consideration of the readers of this chapter, understanding full well that "discovering the periodic law" is not what Béguyer set out to do, not what he thought he was doing while working on the *vis tellurique*, and not what any other scientist at the time understood him to have done.

Fig. 5.11 Depiction of the *vis tellurique* in van Spronsen's 1951 article on the periodic system [54]. (Courtesy of *C2W* and J. W. van Spronsen)

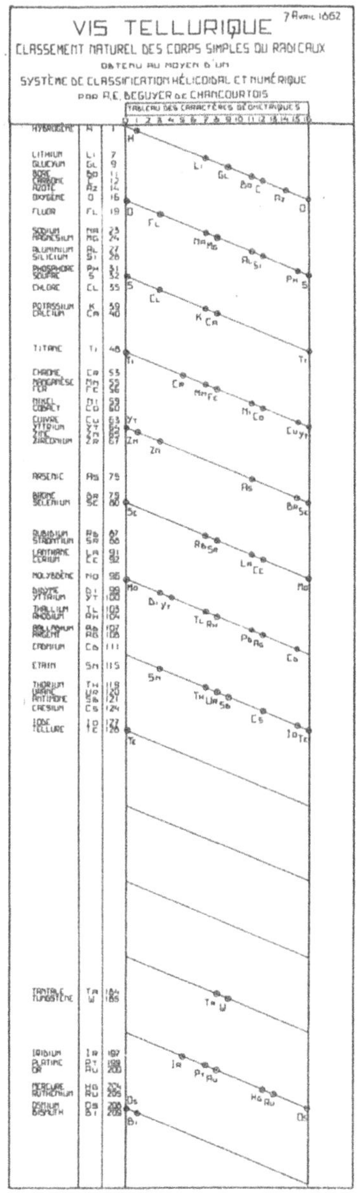

Fig. 1. Systeem van de Chancourtois uit 1862.

I agree with those who count Béguyer among the discoverers of the periodic law. His classification was comprehensive, including all known elements, and it related properties of elements to atomic weights. Béguyer both explicitly and implicitly asserted periodicity (that is recurrence) for related elements. (A comprehensive classification based on atomic weight order that asserts recurrence of properties is what I take to be the essence of the periodic law when that law became widely accepted by chemists in the later nineteenth century.) He presented very little, though, that would make chemists of the time think that there was anything useful for them in the arrangement. Looking back from 150 years, it is obvious that there was much dross in the *vis tellurique*, but chemical periodicity seems to be there as well.

The first geometrical relationship among related bodies that Béguyer described in the *vis tellurique* was that they fall on or near the same vertical lines. In effect, because of the way he constructed the *vis tellurique*, this amounts to saying that related bodies have atomic weights that differ by 16 or multiples of 16 (or have differences near to multiples of 16). This certainly seems to be the case among the lighter elements, as a glance of Fig. 5.7 (Hartog's simplification) shows. Among the perfect vertical alignments visible here are Li with Na, Bo with Al, C with Si, O with S, and F with Cl; and the alignments of Be with Mg and N with P are off by just one unit. The earliest embodiments of the periodic law worked very well for the lightest elements, and even those of the later discoverers of the law, Mendeleev and Meyer, struggled with the heaviest elements.

It is very easy to understand how a chemist of the late nineteenth century who already knew about chemical periodicity would see it in Fig. 5.7. In addition to this apparent display of periodicity, there is an explicit statement of alternation of properties at various periods in Béguyer's first paper on the *vis tellurique* [1, 13]. Béguyer both displayed periodicity in the *vis tellurique* (albeit more complexly and less perfectly than does Fig. 5.7) and described it.

But I would not call the *vis tellurique* a periodic **system** because it is too flexible to yield definite predictions or classifications: it is highly underdetermined. Comparing the simplified Fig. 5.7 to an image of the same part of the original chart (Fig. 5.3) reveals three sources of its indeterminacy: secondary characters, radicals, and diagonal lines intended to show relationships among points other than on vertical lines. As previously mentioned, there are three secondary characters for boron on this section of the chart (three on the second turn of the helix). Also visible on this portion of the chart are secondary characters for lithium, calcium, and silicon based on alternative formulas of key compounds. This section of the chart also shows several radicals: ammonium (as noted above), methyl, and mineralogical "radicals" of orthoclase and albite. To be fair, Béguyer did not include radicals in most of his examples of related bodies, but he did include them on the chart and in at least one example of relationships [6]. The diagonal lines denoting relationships among bodies that do not line up vertically are barely visible in Fig. 5.3, but they are more prominent and more plentiful in Fig. 5.4, taken from the 1863 version of the lithographed chart [8].

One other factor contributes to the underdetermination of the *vis tellurique*: the basis for asserting relationships among bodies is not always clear and when clear, not always based on chemistry. Some of the diagonal relationships described in the later memoirs in the series are geognostic [3] or metallurgical [5] relationships. With so many points to connect and so many ways of connecting them, the *vis tellurique* offers too little guidance to constitute a useful system.

The *vis tellurique* is perhaps the best example from the career of Béguyer de Chancourtois that suggests he was something of a latter-day Pythagorean, but it is not the only one. We have already noted his belief that "The properties of bodies are the properties of numbers," coming near the conclusion of his final paper on the *vis tellurique* [6], as well as his suspicion that the prime numbers were of fundamental significance to understanding the elements [1, 6, 13]. His colleague and obituarist, Fuchs, stated several times that organization and system were hallmarks of Béguyer: his inscription on and positioning of the Pavilion of Measures and Currency at the 1867 exposition, his advocacy of Élie de Beaumont's pentagonal network of geological features, and his advocacy (along with other geodesists) of a decimal system of angles[21] were all examples [9]. Clearly Béguyer saw and valued patterns, a laudable trait shared with scientists before and after him. Equally clearly, some of the patterns that he saw were more illusory than real, including some he pointed out in the *vis tellurique*.

Having given my assessment of the *vis tellurique*, I make a few observations on the opinions of the chemists who brought the *vis tellurique* to the attention of other chemists in the late nineteenth century. Hartog's assessment [13] puzzles me. As already noted, his illustration of Béguyer's chart (Fig. 5.7) is an oversimplification. It seems to me that if one had wished to illustrate the *vis tellurique* in the best possible light to make the case that it embodies the periodic law—glossing over its complexities and deficiencies and maximizing the appearance of periodicity—one could hardly have done better than Fig. 5.7. (I do not suggest that Hartog had any such deceptive intent, and after all, Hartog's figure illustrates the main idea of the *vis tellurique*, which had previously not been illustrated in any widely available publication).

And yet Hartog did not credit Béguyer with discovery of the periodic law. The title of his article refers to a "first foreshadowing" of that law, and his paper concludes with the statement [13]

> But the discovery of the "octaves" or "periods" cannot be ascribed to our author, although it seems almost impossible that chemists should not have perceived their existence on looking at his table.

Hartog seems to be saying that although it did not announce the periodic law, the *vis tellurique* would have led chemists to it if they had only paid attention.

Lecoq de Boisbaudran and Lapparent did a better job than Hartog both in illustrating and in describing the *vis tellurique*, and in my opinion they arrived at a sounder appraisal. They described and depicted both the significance and flaws of the *vis tellurique* [39, 40]:

> We are far from pretending that the theory of the screw is free from faults, and that the author has not grafted upon his work many considerations which he had better had left in the shade.

[21]This system would divide the circumference of the earth into four quadrants, each of which is divided into 100 parts, as proposed in the original metric system.

Finally, I count Crookes among the chemists who brought Béguyer's work to the attention of other chemists late in the nineteenth century by publishing a translation of the paper of Lecoq de Boisbaudran and Lapparent [40]. In the editorial that precedes the translation [41], Crookes offered sage advice to those who delve into the history of ideas: "In going over old researches we often find in them matter which we may now regard as a forecast of subsequent discoveries." He added that "there is no sufficient evidence that the author [Béguyer] disentangled such matter from accompanying speculations." It is undeniable that Béguyer failed to disentangle periodicity of elementary properties from abundant unfounded speculations. Crookes seemed to think that failure should deprive Béguyer of credit for having reported chemical periodicity. If Crookes required a discovery to be unalloyed by unfounded speculation or error before it can be counted a discovery, he set the bar too high, in my opinion. To put it another way, Crookes and I have some different ground rules for the "Who discovered it" game. Comparing answers and comparing rules are part of what makes the game interesting and entertaining!

Acknowledgements I thank the Master and Fellows of St Catharine's College, Cambridge, the Bibliothèque nationale de France, and Bibliothèque patrimoniale numérique, MINES ParisTech for access to materials that illustrate this chapter. Thanks to my co-editors Greg Girolami and Vera Mainz for interesting discussion about the *vis tellurique*, both in person and over email, and to Greg for information about Gaudin and atomic weight order. Thanks also to Isabelle Maurin-Joffre of the Archives of the Académie des Sciences, to Amélie Dessens, Conservatrice des bibliothèques at MINES ParisTech, and to Peter Wothers, Fellow of St Catharine's College, Cambridge for responses to emailed queries. I thank an anonymous referee of the chapter for suggestions that improved its clarity and presentation. Finally thanks to the private collectors who gave me access to their copy of the 1863 edition of the *vis tellurique* booklet and chart.

References

1. Béguyer de Chancourtois AE (1862) Mémoire sur un classement naturel des corps simples ou radicaux appelé vis tellurique. C R Séances Acad Sci 54:757–761
2. Béguyer de Chancourtois AE (1862) Sur un classement des corps simples ou radicaux appelé vis tellurique: addition au mémoire présenté à la séance du 7 avril. C R Séances Acad Sci 54:840–843
3. Béguyer de Chancourtois AE (1862) Sur un classement des corps simples ou radicaux appelé vis tellurique – addition au mémoire présenté à la séance du 7 avril. C R Séances Acad Sci 54:967–971
4. Béguyer de Chancourtois AE (1862) Tableau du classement naturel des corps simples, dit vis tellurique. C R Séances Acad Sci 55:600–601
5. Béguyer de Chancourtois AE (1863) Application de la vis tellurique dans la théorie de l'acier. C R Séances Acad Sci 56:253–255
6. Béguyer de Chancourtois AE (1863) Suite du mémoire de la vis tellurique, du 7 avril 1860 [sic], adressé à propos du thallium. C R Séances Acad Sci 56:479–482
7. Béguyer de Chancourtois AE (1862) Vis tellurique. Classement naturel des corps simples ou radicaux obtenu au moyen d'un système de classification hélicoïdal et numérique. Mallet-Bachelier, Paris

8. Béguyer de Chancourtois AE (1863) Vis tellurique. Classement naturel des corps simples ou radicaux obtenu au moyen d'un système de classification hélicoïdal et numérique. Mallet-Bachelier, Paris
9. Fuchs E (1887) Notice nécrologique sur M. A.-E. Béguyer de Chancourtois, Inspecteur général des mines. Ann Mines 11:505–536
10. Kendall MB (1970–1980) Béguyer de Chancourtois, Alexandre-Émile. In: Gillispie CC (ed) Dictionary of scientific biography, vol 1. Scribner, New York, pp 572–573
11. Élie de Beaumont L, Béguyer de Chancourtois AE (1862) Études stratigraphiques sur le département de la Haute-Marne. Mallet-Bachelier, Paris
12. Béguyer de Chancourtois AE (1844) Exploration géologique d'une partie très-peu connue de la Turquie d'Asie. C R Séances Acad Sci 18:827–832
13. Hartog PJ (1889) A first foreshadowing of the periodic law. Nature 41:186–188
14. Sainte-Claire Deville CJ (1862) Essai sur la répartition des corps simples dans les substances naturelles. C R Séances Acad Sci 54:782–787
15. MINES ParisTech (2012–2020) Bibliothèque patrimoniale numérique: Béguyer de Chancourtois, Alexandre-Émile (1820–1886): Vis tellurique. https://patrimoine.mines-paristech.fr/document/Vis_tellurique. Accessed 31 Jul 2020
16. Sainte-Claire Deville CJ (1862) Essai sur la répartition des corps simples dans les substances minérales naturelles (deuxième communication). C R Séances Acad Sci 54:880–888
17. Cannizzaro S (1858) Sunto di un corso di filosofia chimica fatto nel R. Università di Genova. Nuovo Cimento 7:321–366
18. Ampère AM (1816) Essai d'une classification naturelle pour les corps simples. Ann Chim Phys 1[ser 2]:295–308, 373–394; 2:5–32, 105–125
19. Pelouze J, Fremy E (1865) Traité de chimie, générale, analytique, industrielle et agricole, vol 2. Masson, Paris, pp 4–9
20. Delafosse G (1860) Nouveau cours de minéralogie, vol 2. Roret, Paris, pp 63–65
21. Jahresbericht über die Fortschritte der Chemie und verwandter Theile anderer Wissenschaften (1862) 15:6–7
22. Nicklès J (1863) Correspondence of Jerome Nicklès, dated at Nancy, France, Nov. 2, 1862. Am J Sci Arts 85:260–270 at 270
23. Dumas JB (1858) Mémoire sur les équivalents des corps simples. Ann Chim Phys 55:129–210
24. Ihde AJ (1984; originally published 1964) The development of modern chemistry. Dover, New York, pp 238–240
25. Scerri ER (2007) The periodic table: its story and its significance. Oxford, New York, pp 52–53
26. Ihde AJ (1984; originally published 1964) The development of modern chemistry. Dover, New York, pp 226–230, 240
27. Scerri ER (2007) The periodic table: its story and its significance. Oxford, New York, pp 64–67
28. Girolami GS (2019) A book collector's view of the periodic table: key documents before Mendeleev's contributions of 1869. Substantia 3(2) Suppl. 5:109–124. https://doi.org/10.13128/substantia-592
29. Royal Society of Chemistry (2019) The story of the periodic table. https://www.rsc.org/events/detail/40485/the-story-of-the-periodic-table. Accessed 1 Aug 2020
30. Musée de Minéralogie (2019) Exposition temporaire "Avant Mendeleiev: genèse d'un tableau" http://www.musee.mines-paristech.fr/Evenements/ExpoTemp2019/avantmendeleiev/. Accessed 1 Aug 2020
31. Leach MR (2019) Internet database of periodic tables, 1862, Béguyer de Chancourtois' vis tellurique https://www.meta-synthesis.com/webbook/35_pt/pt_database.php?PT_id=7. Accessed 1 Aug 2020
32. Brush SG (1996) The reception of Mendeleev's periodic law in America and Britain. Isis 87:595–628

33. Wurtz A (1879) La théorie atomique. Baillière, Paris, p 123
34. Wurtz A (1885) Introduction à l'étude de chimie. Masson, Paris, pp v, 256
35. Berthelot M (1885) Les origines de l'alchimie. Steinheil, Paris, p 302
36. Fuchs E (1887) Notice sur M. A. E. Béguyer de Chancourtois. Bull Soc géol Fr 15:489–507
37. Mallard E (1887) Notice nécrologique sur M. B. de Chancourtois. Bull Soc Fr Minéral 73–77
38. Mendeleev D (1889) The periodic law of the chemical elements. J Chem Soc 55:634–656
39. Lecoq de Boisbaudran F, de Lapparent A (1891) Sur une réclamation de priorité en faveur de M. de Chancourtois, relativement aux relations numériques des poids atomiques. C R Séances Acad Sci 112:77–81
40. Lecoq de Boisbaudran F, de Lapparent A (1891) A reclamation of priority on behalf of M. de Chancourtois referring to the numerical relations of the atomic weights. Chem News 63:51–52
41. [Crookes W] (1891) The telluric screw. Chem News 63:51
42. Newlands JAR (1864) Relations between equivalents. Chem News 10:59–60
43. Newlands JAR (1884) On the discovery of the periodic law and on relations among the atomic weights. Spon, London, p vi
44. Seubert K (1895) Anmerkungen. In: Seubert K (ed) Das natürliche System der chemischen Elemente. Ostwalds Klassiker, vol 68. Engelmann, Leipzig, pp 119–125
45. Venable FP (1896) The development of the periodic law. Chemical Publishing, Easton, PA, pp 73–76, 81
46. Venable FP (1896) The development of the periodic law. Chemical Publishing, Easton, PA, p 39
47. Gladstone JH (1853) On the relations between the atomic weights of analogous elements. Philos Mag 5:313–320
48. Dalton J (1808) A new system of chemical philosophy, pt 1. Bickerstaff, Manchester, UK, p 219
49. Dalton J (1805) On the absorption of gases by water and other liquids. Mem Lit Philos Soc Manchester 1:271–287
50. Gaudin MA (1833) Recherches sur la structure intime des corps inorganiques définis … Ann Chim Phys 52:113–133
51. Cole TM Jr (1975) Early atomic speculations of Marc Antoine Gaudin: Avogadro's hypothesis and the periodic system. Isis 66:334–360
52. Rudorf G (1900) The periodic classification and the problem of chemical evolution. Whittaker, London, pp 20–22
53. van Spronsen JW (1969) The periodic system of chemical elements: a history of the first hundred years. Elsevier, Amsterdam, pp 98–102
54. van Spronsen JW (1951) Prioriteit bij de ontdekking van het periodiek systeem. Chem weekblad 47:807–816
55. Gordin MD (2012) The textbook case of a priority dispute: DI Mendeleev, Lothar Meyer, and the periodic system. In: Biagioli M, Riskin J (eds) Nature engaged: science in practice from the Renaissance to the present. Palgrave Macmillan, New York, pp 59–82

6 Periodicity in Britain: The Periodic Tables of Odling and Newlands

Carmen J. Giunta, Vera V. Mainz, and Julianna Poole-Sawyer

Abstract

Two British chemists, William Odling and John A. R. Newlands, published separate periodic arrangements of the elements in the middle 1860s, several years before Mendeleev's first periodic table. This chapter provides an overview of the periodic systems of Odling and Newlands, their development over time, the reception of their work, and our perspective on the often-discussed question of who deserves recognition as a discoverer of the periodic law.

6.1 Introduction

Among those who made attempts to construct a periodic system before Dmitri Mendeleev (1834–1907) were two British chemists, William Odling and John A. R. Newlands. In 1864, Odling published his periodic table in the *Quarterly Journal of Science* and a slightly different table in an encyclopedia article and in a textbook in the following year. In 1863, Newlands began classifying elements in the journal *Chemical News*, and he published several periodic arrangements over the next three years. Despite these accomplishments, Odling and Newlands remain less well

C. J. Giunta (✉)
Department of Chemistry, Le Moyne College, Syracuse, NY, USA
e-mail: giunta@lemoyne.edu

V. V. Mainz (✉)
School of Chemical Sciences, University of Illinois at Urbana-Champaign, Urbana, IL, USA
e-mail: mainz@illinois.edu

J. Poole-Sawyer
Publications Division, American Chemical Society, Columbus, OH, USA

C. J. Giunta et al. (eds.), *150 Years of the Periodic Table*, Perspectives on the History of Chemistry, https://doi.org/10.1007/978-3-030-67910-1_6

known as independent discoverers of the periodic law than Mendeleev and Lothar Meyer. In addition to giving brief overviews of the lives and careers of these two London chemists, this chapter examines the papers and books that Odling and Newlands published on periodic systems, summarizes the reception of their work, and discusses the credit they deserve as discoverers of the periodic law.

6.2 William Odling

6.2.1 Biographical Information

William Odling (1829–1921) (Fig. 6.1) [1–7] had a long and successful career as a chemist. The son of a surgeon, Odling was privately educated until he entered the medical school [8, 9] at Guy's Hospital at the age of 16. His principal chemistry teacher at Guy's was Alfred Swaine Taylor (1806–1880), who lectured on chemistry and medical jurisprudence. In 1851 Odling became one of the first students to receive an M.D. degree from London University, but he never practiced medicine, preferring instead to turn his talents to teaching and research in chemistry.

At Guy's Hospital, Odling was appointed demonstrator in chemistry in 1850, the Director of the Chemical Laboratory in 1851, and Professor of Practical Chemistry in 1856. He also was the first Medical Officer of Health for the borough of Lambeth in London, starting in 1856. He held this post until 1863, when he succeeded Edward Frankland (1825–1899) as Lecturer in Chemistry at St. Bartholomew's Hospital, where he served until 1870. In 1867, upon Faraday's death, he became

Fig. 6.1 William Odling, approximately 1867 (Reproduced courtesy of the Royal Society of Chemistry Library)

Fullerian professor of chemistry at the Royal Institution, serving until 1872, when he was appointed Waynflete Professor of Chemistry at the University of Oxford, as a successor to Benjamin Brodie (1817–1880). Odling held this position for the next forty years, retiring in 1912. As was not uncommon for that time, several of these appointments overlapped, especially early in his career.

Upon succeeding Brodie at Oxford in 1872, Odling married Elizabeth Mary Smee (1843–1919); she was the daughter of Alfred Smee, an English surgeon, chemist, metallurgist, electrical researcher, inventor, and orchid enthusiast. William and Elizabeth had three sons and a daughter: George Smee Aldersey Odling-Smee (1873–1926), Mary Elizabeth Odling (1875–1887), Maj. William Alfred Odling (1879–1943), and Marmaduke Odling (1886–1956). Marmaduke became an analytical chemist and geologist.

Odling had a long (sixty-five-year) relationship with the Chemical Society of London. Elected as a Fellow in 1848 (at age 19), he served as secretary of the Society (1856–1869), vice-president (1869–1873), and president (1873–1875). He was a Fellow of the Royal Society (1859) and a Fellow of the Royal College of Physicians (1859). As Marsh noted, with respect to the length of his service in the Chemical Society [1], "At his death he was senior Fellow but one."

Odling published 79 papers by our count (some seemed to have been published in multiple journals; we have counted them once). A complete listing can be put together from [7] and the relevant *Catalogues of Scientific Papers Compiled by the Royal Society of London*.

Odling's first published paper, on arsenical poisoning [10], appeared in 1851. His last, in 1887, concerned the decomposition of aromatic acids [11]. In between, he was interested in many topics, including the constitution of the hydrocarbons [12], analytical tests for arsenic [13], alkaline emanations from sewers and cess-pools [14], an analysis of the common contaminants in the Guy's Hospital well and their comparison to Thames water and to other deep-wells in London [15], the nomenclature of organic compounds [16], the classification of silicates [17], and the valency of aluminum [18].

A particular interest was the constitution of chemical compounds. As part of this effort, in 1855 he proposed a system of what he called dashes but which are more commonly referred to as "prime symbols" (e.g., Ag' or P''') to indicate the equivalent value or substitution value (valence) of an atom in a chemical compound [19]. Odling's 1858 paper [20], "Remarks on the Doctrine of Equivalents," contained a summary of the reasons behind this proposal:

> The idea of multi-equivalent or polyatomic radicals was, I believe, first announced by Williamson in a paper "On the Constitution of Salts," published in the year 1851. Some time afterwards he found a disciple in myself, who, in 1854 and 1855, extended his original notion, illustrated it by a variety of formulae, and moreover, rendered it more precise by applying to it the equivalent notation by dashes, to which I have before referred.

> The word equivalent was originally employed by Wollaston as a substitute for Dalton's word atom. Wollaston's notion of equivalency appears to have been derived chiefly from the phænomena of what was then considered chemical combination, as instanced more particularly in the case of neutralization. Thus, a given proportion of potash united with an

> equivalent quantity of oxalic acid to form a neutral salt, and with twice its equivalent to form an acid salt, and so forth. But this idea of equivalency, which accords much the same signification to the words atoms and equivalent, has been greatly called in question. The controversy was begun by Gerhardt and Laurent, who first clearly established the difference between the atomic weight of a body and its equivalent value. …
>
> To facilitate comparison, I proposed some few years back, that mode of indicating differences in equivalent value which has now come into very general use, namely, by the use of one or more dashes placed to the right or left of the symbol, so as to establish a difference to the eye between the atom of a body and its equivalent or substitution value, as shown in the following formulæ:— $Ag'^{3}PO^{4}$ Phospate of silver; $Bi'''PO^{4}$ Phosphate of bismuth.

This interest in chemical constitution led to Odling's papers on type theory and the classification of the elements discussed later in Sect. 6.2.2.

Note, by the way, that equivalent value as used here by Odling differs from equivalent weight or simply equivalent.[1] Equivalent value corresponds to a concept that we call valence, and that during the middle of the nineteenth century was variously called atomicity and quantivalence. Assigning definitions to terms such as "equivalent" and "atom" was discussed during the Karlsruhe Congress in 1860 [22], although the distinction between those terms that was later observed was not yet consistently in use in the years immediately following the Congress. Moreover, as has been noted by Alan Rocke, even William Hyde Wollaston, who had popularized the word "equivalent" in chemistry, used it in a way that was functionally identical to atomic weight, although his equivalents had different numerical values than the atomic weights derived by some of his contemporaries [23]. Odling, however, did not use the terms in this way. He states in his 1855 article that the "dashes" indicate how many hydrogen atoms are replaced by the atom of interest to form related compounds. His "equivalent" or "equivalent value" is a stoichiometric number and thus an integer (or a fraction), not a weight.

Odling's first book, *A Course of Practical Chemistry, Arranged for the Use of Medical Students*, appeared in 1854 [24]; it went through five editions, the last appearing in 1876. In the same year that he published the first edition of his textbook, Odling translated Auguste Laurent's *Méthode de chimie* into English (1854) [25, 26]. Laurent (1807–1853) and Charles Frédéric Gerhardt (1816–1856) were proponents of the type theory of chemical constitution. Odling, who had studied with Gerhardt in 1851, became a convert to their modified type theory. Odling's lectures and papers on this topic laid the foundation for the theory of valency, as originally defined by Frankland [4].

[1]As Trevor Levere succinctly described it [21]: "Take the combination between hydrogen and oxygen. Everyone could agree that eight parts by weight of oxygen combined with one part by weight of hydrogen. Why not simply say that eight parts by weight of oxygen were *equivalent* to one part by weight of hydrogen, and then present that empirically determined result by saying that if one took the *equivalent weight* of hydrogen as 1, then the equivalent weight of oxygen was 8. Formulas could then represent the number of equivalent weights involved in a compound: carbon dioxide would have one equivalent of carbon to two equivalents of oxygen. For practical purposes, at least in the early nineteenth century, it made no difference whether chemists used a system of atoms or of equivalents."

In 1861, Odling published *A Manual of Chemistry, Descriptive and Theoretical, Part 1* [27] in which he "adopted the plan of treating chemistry as a whole without subdividing it into organic and inorganic" [1]. In this book, Odling further developed his reasons for defining the atomic weight of oxygen as 16 (thereby making the formula for water H_2O) although he still had not accepted all the revisions in atomic weights recommended by Cannizzaro. The book was also notable because it contained the first correct structure of ozone.[2] Although there was only one edition of the *Manual*, it was translated into German (1865), French (1868), and Russian (1865). Part 2 was never published in book form. Selections from part 2 were published in the *Chemical News* but much remained unpublished in manuscript form at the time of Odling's death [1].

It is possible that translating Laurent's work led Odling to think more deeply about valency, and in writing his own textbooks he sought ways to systematize chemical knowledge for easy consumption by students. It is, perhaps, no surprise that three of the people who made significant contributions to the development of periodic systems, Odling, Meyer, and Mendeleev, all were heavily involved in writing textbooks in the 1860s [29].

Odling published very few articles after he moved to Oxford in 1872. According to Marsh [1]:

> Odling was never the slave of the laboratory. Even in early life his taste was for the philosophic and speculative rather than for the practical side of the science. Instead of directing the research work of his assistants and students, he preferred to let them work out their own ideas, and he held that the "best of all endowments for research is that with which the searcher, relying on his own energies, succeeds in endowing himself . . ."

The last book Odling wrote appeared in 1916 during his retirement [30], *The Technic of Versification: Notes and Illustrations*. In his obituary of Odling, Marsh noted [1]:

> Odling retired from the professorship at Oxford in 1912. In the first year of his retirement he seemed somewhat restless. Then he settled down to write another book, and that on a subject which rather surprised his family and friends. It was called "The Technic of Versification." To the mere chemist it appears like a kind of type theory of verse with a symbolic notation, almost chemical. It was written in his favourite style without the use of any principal verb. It contains a fine anthology of English poetry. There is no incongruity here, that poetry should appeal to a man of science. Is not Discovery the poetry of Science?

6.2.2 1857: "On the Natural Groupings of Elements"

Between 1855 and 1857, Odling wrote several articles related to Gerhardt's theory of types. Gerhardt had formulated four inorganic types—water H_2O, ammonia NH_3, hydrochloric acid HCl, and hydrogen H_2—from which, in his view, all

[2]According to his family [28], this resulted in his nickname of "Ozone Odling."

$H \cdot Cl$			
$H \cdot OH$ (*Laurent.*)	$H^2 \cdot O$		
$H \cdot NH^2$ (*Kane.*)	$H^2 \cdot NH$ (*Wurtz.*)	$H^3 \cdot N$	
$H \cdot CH^3$ (*Liebig.*)	$H^2 \cdot CH^2$ (*Dumas.*)	$H^3 \cdot CH$ (*Odling.*)	$H^4 \cdot C$

Fig. 6.2 Table from "On the Constitution of the Hydrocarbons" [12] (Private collection)

organic compounds could be derived. One of the important consequences of Gerhardt's theory was that unknown compounds could be predicted in large numbers by this scheme of classification.

In February 1855, Odling gave an important lecture at the Royal Institution, "On the Constitution of the Hydrocarbons" [12] in which he added the CH_4 type (what came to be known as the marsh-gas type) to Gerhardt's H_2, HCl, H_2O, and NH_3 types [31, 32] (see Fig. 6.2).

This addition to type theory was, however, incidental to the main thesis of the lecture, which was how a scientist should use and could abuse theory. Odling began his lecture as follows:

> Every chemical compound may be regarded in a great number of different aspects. Each of the different theories that have been propounded concerning the chemical constitution of bodies, is true in reference to one particular aspect,—untrue in reference to all others. Theories are of the highest service when they enable us to look upon a larger number of bodies from a single point of view, —of the highest detriment, when they prevent us from making use of all other points of view.

Odling offered up this new type, the CH_4 type, in order to show that it did not clarify the situation with respect to radicals. In his view, radicals did not denote anything intrinsic about the structure of molecules. As Rocke has stated [33]:

> His [Odling's] goal was to demolish the whole notion of radicals as preexisting parts of molecules. Radicals, he [Odling] averred, are nothing more than a convenient means of accounting for and notating the components of compounds, and have no further signification.

Odling ended his lecture by summing up his thesis:

> In the three best known hydro-carbons, coal-gas, olefiant-gas, and benzene, as in many other bodies ordinarily represented as containing compound radicals, the conception of self-existent constituent compound radicals, is not only unnecessary but irrational. The particular groupings of atoms, which we denominate compound radicals, do not have an existence apart from the other constituents of the bodies, into which they are said to enter.

In other words, radical theory was too confining and too artificial. What was needed was a new, more universal theory to help chemists understand the structure of molecules. Odling's call for a new way of thinking was realized through the development of the structural theory of Kekulé, Couper, and Butlerov, which began to be formulated about three years later [34]. Rocke has suggested [33] that "[Odling's] paper surely influenced contemporaries, Kekulé foremost among them. Kekulé may well have attended Odling's lecture."

Odling's two 1857 papers [35, 36] "On the natural groupings of elements" extend the ideas about the relationships of the atomic weights of the elements that had been published by John Hall Gladstone (1827–1902) in 1853 [37]. Gladstone had stated:

> The numerical relations are of three kinds. The atomic weights of analogous elements may be the same; or may be in multiple proportion; or differ by certain increments.

Gladstone had found the first rule by listing the atomic weights of the elements (e.g., H = 1, C = 6, O = 8, N = 14) according to increasing numerical values, and pointing out that the metals of the iron group, the platinum metals, and some rare earths have approximately the same atomic weight. The second rule involved elements whose atomic weights, as they were known at the time, were related (or nearly related) by integer multipliers. One example was O = 8 and S = 16, but Gladstone noted other examples, such as B = 10.9, Si = 21.3, Ti = 25, Mo = 46, Sn = 58, Y = 68.6, W = 92, Ta = 184 (multiples of 11.5). The third rule involved triads,[3] which were defined by differences in their atomic weights.

In his two 1857 communications, Odling modified Gladstone's relationships by dividing the elements into 13 groups, which he arranged in tabular form. The first of Odling's 1857 communications [35] dealt exclusively with the acid-forming ("chlorous") elements, whereas the second dealt with the metals ("basylous" elements) [36]. Odling stated his intentions at the beginning of the first paper:

> That certain elements have certain properties in common is now a time-honoured doctrine in chemical science; but the majority of chemists have been satisfied with a simple admission of the fact: they have not investigated the extent of the association, nor availed themselves of it as a means of classification. On the contrary, although the groupings of the elements are as real and certain as the natural families of plants and animals, yet we find constantly, in our systematic treatises, that bodies manifesting the strongest analogies are widely separated from one another, while bodies belonging to very different groups are conventionally associated.
>
> The existence of certain important natural families has been successively pointed out by different chemists. I propose to make some modifications in the groupings hitherto proposed, to construct a few new groups, and to point out the principal analogies by which the members of each particular group, old or new, are associated; …
>
> In attempting a classification of the elements, we must have regard, though not an equal regard, to all the properties they manifest; or in other words, we must be guided by the totality of their characters.

[3]Triads are formed of three elements, in which the atomic weight of the middle element is the average of the atomic weights of the first and third. One example of a triad is Cl, Br, and I. J. W. Döbereiner had first proposed the existence of triads in 1817 [38–40]. A general discussion of developments of periodic systems before Mendeleev can be found in Girolami [41].

Because the atomic weights Gerhardt and Laurent had used for the non-metals were already correct, i.e., equal to the modern values, Odling's first paper greatly improved Gladstone's ideas by extending the usual triads of non-metals to groups of four or five by adding elements before and sometimes afterward (F, Cl, Br, I; O, S, Se, Te; N, P, As, Sb, Bi; C, B, Si, Ti, Sn).

At the end of this first communication Odling remarked:

> The elements fluorine, oxygen, nitrogen, and carbon, which would thus stand at the head of the four groups, present a curious numerical sequence: C = 12 forms H_4C, N = 14 forms H_3N, O = 16 forms H_2O, F = 18 (?) forms HF. … In all four groups a gradual development of metallic character accompanies the increase in atomic weights, illustrating forcibly the extremely artificial character of the division of the elements into metallic and non-metallic bodies.

Odling was the first to point out this correlation between atomic weight and valence.

The second of the 1857 communications was affected by the fact that many of the atomic weights Odling used for metallic elements were not true atomic weights. As a result, Odling was led, like Gladstone before him, to classify into triads some elements that were not related in a periodic fashion. But some of Odling's triads of metallic elements corresponded to modern groups, e.g., Ca-Sr-Ba formed one triad and Mg-Zn-Cd formed another. Several elements, however, appeared in more than one triad.

6.2.3 Interlude

Dmitri Mendeleev and Lothar Meyer both attended the famous 1860 Congress in Karlsruhe where Cannizzaro gave his lecture on the revision of atomic weights. Odling was one of the secretaries at the Congress [5]. Whereas Mendeleev and Meyer were profoundly impressed with Cannizzaro's ideas, Odling was not. In an article on atomic weights for *Watts's Dictionary* in 1863, Odling rejected Cannizzaro's new atomic weights [42] and continued to use the same atomic weights he had used from 1857 to 1861. Among his reasons for objecting to Cannizzaro's ideas were the following: "Cannizzaro's proposal … would involve the dissociation of silver from lead, and that of the metals of the alkalis from those of the alkaline earths." Both of those consequences are in fact correct, as Odling soon came to appreciate.

6.2.4 1864: "On the Proportional Numbers of the Elements"

Only a few months later, Odling changed his mind and adopted Cannizzaro's numbers for the atomic weights of the elements, publishing his conclusions in the October 1864 issue of the *Quarterly Journal of Science*. In this article, "On the

H	1	Hydrogen.	Fe	56	Iron.	Cd	112	Cadmium.
L	7	Lithium.	Co	59	Cobalt.	Sn	118	Tin.
G	9	Glucinum.	Ni	59	Nickel.	U	120	Uranium.
B	11	Boron.	Cu	63·5	Copper.	Sb	122	Antimony.
C	12	Carbon.	Yt	64	Yttrium.	I	127	Iodine.
N	14	Nitrogen.	Zn	65	Zinc.	Te	129	Tellurium.
O	16	Oxygen.	As	75	Arsenic.	Cs	133	Cæsium.
F	19	Fluorine.	Se	79·5	Selenium.	Ba	137	Barium.
Na	23	Sodium.	Br	80	Bromine.	V	137	Vanadium.
Mg	24	Magnesium.	Rb	85	Rubidium.	Ta	138	Tantalum.
Al	27·5	Aluminium.	Sr	87·5	Strontium.	W	184	Tungsten.
Si	28	Silicon.	Zr	89·5	Zirconium.	Cb	195	Niobium.
P	31	Phosphorus.	Ce	92	Cerium.	Au	196·5	Gold.
S	32	Sulphur.	La	92	Lanthanum.	Pt	197	Platinum.
Cl	35·5	Chlorine.	Dy	96	Dydymium.	Ir	197	Iridium.
K	39	Potassium.	Mo	96	Molybdenum.	Os	199	Osmium.
Ca	40	Calcium.	Ro	104	Rhodium.	Hg	200	Mercury.
Ti	50	Titanium.	Ru	104	Ruthenium.	Tl	203	Thallium.
Cr	52·5	Chromium.	Pd	106·5	Palladium.	Pb	207	Lead.
Mn	55	Manganese.	Ag	108	Silver.	Bi	210	Bismuth.
						Th	231·5	Thorinum.

Fig. 6.3 Table from "On the Proportional Numbers of the Elements" [43] (Private collection)

Proportional Numbers of the Elements [43]," Odling arranged all of the elements in increasing order of their atomic weights (Fig. 6.3), just as Gladstone had in his 1853 article but with the new atomic weights.

He then stated:

> With what ease this purely arithmetical seriation may be made to accord with a horizontal arrangement of the elements according to their usually received groupings, is shown in the following table [Fig. 6.4], in the first three columns of which the numerical sequence is perfect, while in the other two the irregularities are but very few and trivial…

Odling's insight was that, by ordering the elements according to Cannizzaro's atomic weights, elements with similar properties can be made to align in "horizontal arrangements."

Through a series of eight additional tables, Odling showed that many pairs of elements with similar properties had atomic weights that differed by about 90; for other pairs, the difference was about 44, and for others, the difference was about 16. Odling drew some conclusions in connection with these relationships.

> If we compare together certain pairs of more or less analogous elements, we find in a considerable number of instances, embracing one-half the entire number of elements, a difference in atomic weight ranging from 84.5 to 97, …

> In about one-half of the above instances, the two elements associated with one another are known to be the first and third terms respectively of certain triplet families; and the discovery of intermediate elements in the case of some or all of the other pairs, is not by any means improbable.

			Ro 104	Pt 197
			Ru 104	Ir 197
			Pt 106·5	Os 199
H 1	"	"	Ag 108	Au 196·5
"	"	Zn 65	Cd 112	Hg 200
L 7	"	"	"	Tl 203
G 9	"	"	"	Pb 207
B 11	Al 27·5	"	U 120	"
C 12	Si 28	"	Sn 118	"
N 14	P 31	As 75	Sb 122	Bi 210
O 16	S 32	Se 79·5	Te 129	"
F 19	Cl 35·5	Br 80	I 127	"
Na 23	K 39	Rb 85	Cs 133	"
Mg 24	Ca 40	Sr 87·5	Ba 137	"
	Ti 50	Zr 89·5	Ta 138	Th 231·5
	"	Ce 92	"	
	Cr 52·5	Mo 96	V 137	
	Mn 55		W 184	
	Fe 56			
	Co 59			
	Ni 59			
	Cu 63·5			

Fig. 6.4 Table from "On the Proportional Numbers of the Elements" [43] (Private collection)

Although a few others (such as Lenssen) had previously predicted the existence of new elements based on completing a triad [44], here Odling's prediction is of a different nature because it is based on a periodic system involving both intragroup and intergroup relationships [45, 46]. Odling stated:

> Since many of the elements that have analogous positions in different groups have closely approximating atomic weights, it is evident that the mere determination of the atomic weight of a newly discovered element assists us but little in deciding to what group it

belongs, but only indicates its position in the group; among the members of every well-defined group the sequence of properties and sequence of atomic weights are strictly parallel to one another.

Doubtless some of the arithmetical relations exemplified in the foregoing tables and remarks are simply accidental; but taken altogether, they are too numerous and decided not to depend upon some hitherto unrecognised general law.

Odling clearly recognized three important aspects of a true periodic system. First, if the elements are ordered according to increasing atomic weight, repeating patterns in properties emerge. Second, elements within groups have a regular progression of atomic weights, but more importantly, similar progressions are seen in different groups. Third, these patterns are indicative of an underlying law of nature.

Among the most notable features of Odling's 1864 article are the following: (1) Odling classified 57 of the then known 60 elements into main groups, subgroups and transition elements, as part of a universal periodic system; (2) he interpreted the gaps in his system to be placeholders for elements that had not yet been discovered; (3) he was one of the first to recognize that the positions of iodine (at. wt. 127) and tellurium (at. wt. 129) must be inverted relative to strict atomic weight order (Newlands [47] and Meyer [48] also did this in publications in 1864); (4) he noted that the atomic weight difference between B (11) and Al (27.5) is about the same as the differences between F and Cl, O and S, N and P, C and Si, G [Be] and Mg, Li and Na, or Na and K; (5) he was the first to put boron at the top of a group, pairing it with aluminum and noting a likely vacancy below Al; and (6) he showed additional valence relationships (such as between the alkaline earths and the Zn, Cd, Hg triad) via dotted lines which presaged the arrangement of groups of elements into A and B subgroups (Mendeleev's subgrouping of the elements into "odd" and "even" numbered rows).

6.2.5 1865: *Watts's Dictionary*

Two years after writing an article on the topic of atomic weights for volume 1 of *Watts's Dictionary*, Odling wrote a second article entitled "Metals, Atomic Weights and Classification of" for volume 3 [49] in February 1865.[4] In this second article, Odling gave the reasons that led him to adopt Cannizzaro's atomic weights in 1864, one of which was the law of Dulong and Petit. At the end of the article, Odling gave a brief summary of Dumas's ideas about the atomic weights of elements within groups and then reprinted his table from the *Quarterly Journal* with a few modifications (Fig. 6.5). The pertinent text preceding the table is:

[4]Cassebaum [6] assigned this date for the article by noting that Odling sent Oppenheim a copy of the galley proofs of his *Watts's Dictionary* article on Metals and a reprint of his paper on Al-triakyls [50] from the "Royal Institution of Great Britain, Weekly Evening Meeting, Friday, February 3, 1865." In the *Dictionary* article, these vapor density determinations are referred to as "unpublished experiments." This timing is also consistent with Odling's statement that his article in the *Quarterly Journal* had appeared "a few months back."

			Mo 96	W 184
			—	Au 196·5
			Pd 106·5	Pt 197
L 7	Na 23	—	Ag 108	—
G 9	Mg 24	Zn 65	Cd 112	Hg 200
B 11	Al 27·5	—	—	Tl 203
C 12	Si 28	—	Sn 118	Pb 207
N 14	P 31	As 75	Sb 122	Bi 210
O 16	S 32	Se 79·5	Te 129	—
F 19	Cl 35·5	Br 80	I 127	—
	K 39·1	Rb 85	Cs 133	—
	Ca 40	Sr 87·5	Ba 137	—
	Ti 48	Zr 89·5	—	Th 231
	Cr 52·5	—	V 138	—
	Mn 55 &c.			

Fig. 6.5 Table from *Watts's Dictionary* [49], p 975. (In Odling's 1864 table (Fig. 6.4), Mn is linked by brackets to 4 other transition metals: Fe, Co, Ni, and Cu. Perhaps to save room, Odling wrote Mn &c in his 1865 table, thus omitting explicit mention of Fe, Co, Ni, and Cu.) Note the hand corrections of the atomic weights of K (from 3 to 39.1) and Cl (from 3.35 to 35.5). These values were correct in the galley proof in Fig. 6.6, indicating that the table was reset before the *Dictionary* was printed, perhaps because of the addition of the bold lines (Private collection)

> Allowing chromium and manganese to stand proxy for the iron metals, and palladium and platinum for their respective congeners, the following table, slightly modified from one published by the author a few months back (Quart. Journ. of Science, i, 643) [43], contains a list of all the well-known elements arranged horizontally in the order of their generally received groups, and perpendicularly in the order of their several atomic weights.

Odling then summarized many of the arithmetical relationships mentioned in his 1864 article, and ended his article with almost the exact same conclusion as in the *Quarterly Journal* article:

> Doubtless some of the arithmetical calculations exemplified in the foregoing table are merely accidental, but, taken altogether, they are too numerous and decided not to depend upon some hitherto unrecognised general law.

In the proof sheets of Odling's 1865 *Dictionary* article [6, 51], Odling's description of his table reads as follows:

> … a list of all the well-known elements arranged in the order of their generally received groups, and **almost** [emphasis added] in the order of their several atomic weights.

The text was corrected by hand to the final form quoted above (Fig. 6.6). Particularly noteworthy is the deletion of the word "almost," thus de-emphasizing the

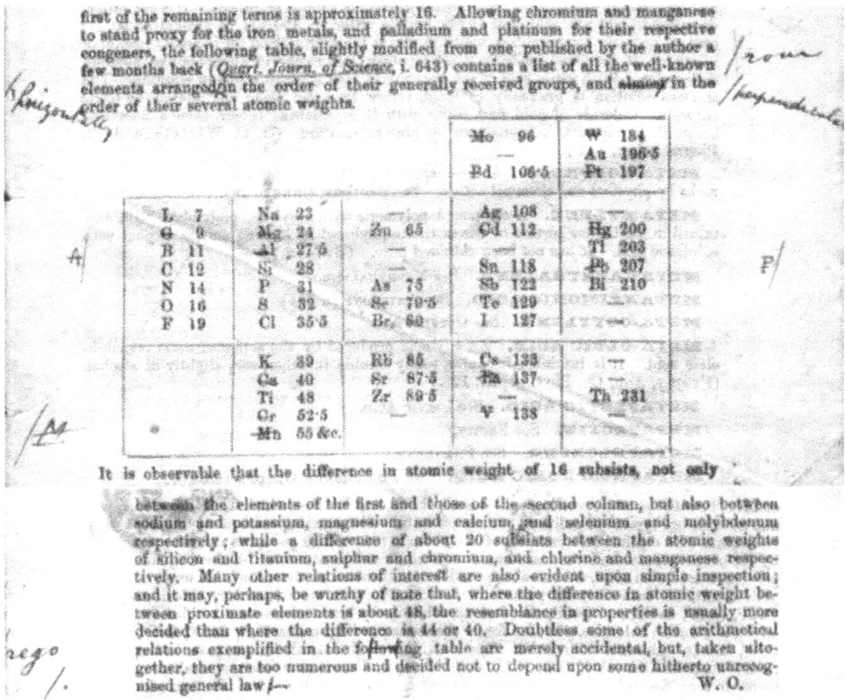
first of the remaining terms is approximately 16. Allowing chromium and manganese to stand proxy for the iron metals, and palladium and platinum for their respective congeners, the following table, slightly modified from one published by the author a few months back (*Quart. Journ. of Science*, i. 643) contains a list of all the well-known elements arranged in the order of their generally received groups, and always in the order of their several atomic weights.

			Mo 96	W 184
			—	Au 196·5
			Pd 106·5	Pt 197
L 7	Na 23		Ag 108	—
G 9	Mg 24	Zn 65	Cd 112	Hg 200
B 11	Al 27·5	—	—	Tl 203
C 12	Si 28	—	Sn 118	Pb 207
N 14	P 31	As 75	Sb 122	Bi 210
O 16	S 32	Se 79·5	Te 129	—
F 19	Cl 35·5	Br 80	I 127	—
	K 39	Rb 85	Cs 133	—
	Ca 40	Sr 87·5	Ba 137	—
	Ti 48	Zr 89·5	—	Th 231
	Cr 52·5	—	V 138	—
	Mn 55 &c.			

It is observable that the difference in atomic weight of 16 subsists, not only between the elements of the first and those of the second column, but also between sodium and potassium, magnesium and calcium, and selenium and molybdenum respectively; while a difference of about 20 subsists between the atomic weights of silicon and titanium, sulphur and chromium, and chlorine and manganese respectively. Many other relations of interest are also evident upon simple inspection; and it may, perhaps, be worthy of note that, where the difference in atomic weight between proximate elements is about 48, the resemblance in properties is usually more decided than where the difference is 44 or 40. Doubtless some of the arithmetical relations exemplified in the following table are merely accidental, but, taken altogether, they are too numerous and decided not to depend upon some hitherto unrecognised general law.

W. O.

Fig. 6.6 Galley pages, *Watts's Dictionary* [51], pp 975–976 (Courtesy of Humboldt-Universität zu Berlin, University Library, Sammlung Hofmann, 280)

Te-I inversion and instead emphasizing that groups of elements with similar properties were arranged into horizontal rows in which the atomic weight increased from left to right.[5] Another change Odling made at the galley stage, but later than the galleys shown in Fig. 6.6, was to highlight the central three columns (i.e., what we might call the original triads) by reinforcing the two vertical lines that border them (e.g., P-As-Sb; Cl-Br-I, etc.).

Two other tables in Odling's article in *Watts's Dictionary* are of interest. The table given in the "Tetrad Metals" section[6] (Fig. 6.7) comes after the following text:

[5]The reader may be interested in a recent article by Petr A. Druzhinin [52] which used an analysis of galleys to provide an accurate publication date for Mendeleev's periodic system of elements. We thank a reviewer for this reference.

[6]Monad, dyad, triad, tetrad, pentad, hexad, and heptad refer to the valence in modern nomenclature. For example, Odling's tetrad group included C, Si, Sn, Pb, Ti, Zr, Th, and Ta, and his heptad group included Cl, Br, and I.

Monads.		Dyads.		Triads.		Tetrads.		Pentads.	
Li	7	Gl	9	Bo	11	C	12	N	14
Na	23	Mg	24	Al	27·5	Si	28	P	31
,,	. .	Zn	65	,,	. .	,,	. .	As	75
Ag	108	Cd	112	,,	. .	Sn	118	Sb	122
,,	. .	Hg	200	Tl	203	Pb	207	Bi	210
K	39	Ca	40	,,	. .	Ti	48	,,	. .
Rb	85	Sr	87·5	,,	. .	Zr	89·5	,,	. .
Cs	133	Ba	137	,,	. .	,,	. .	,,	. .
						Th	231		

Fig. 6.7 Table from *Watts's Dictionary* [49], p 965 (Private collection)

$Li'Cl$	$G''Cl^2$	$B'''Cl^3$	$C''''H^4$	$N'''H^3$	$O''H^2$	$F'H$
$Na'HO$	$Mg''H^2O^2$	$Al'''H^3O^3$	$Si''''H^4O^4$	$P^vH^3O^4$	$S^{vi}H^2O^4$	$Cl^{vii}HO^4$

Fig. 6.8 Table from *Watts's Dictionary* [49], p 969 (Private collection)

> The tetrad elements agree with those of the monad and dyad groups in being divisible into a more especially reguline,[7] and a more especially basylous sub-group, the lower members of each of which possess the respective distinctive characters of the sub-group in their greatest intensity. The relations of the first five groups to one another is shown in the following table [Fig. 6.7]:

The table given in Fig. 6.8 is discussed in the surrounding text:

> The succeeding table of chlorides, hydrides, and hydrates is not without interest as exhibiting the characteristic atomicities of the different groups of elements:
>
> Having regard to the upper line of compounds, carbon may be regarded as placed at the apex of two converging series of elements of increasing atomicities; while having regard to the lower line of compounds, it will appear as the middle term of a single series. It is observable that the atomic weights of carbon and silicon correspond very closely with the arithmetical means of atomic weights of the elements placed respectively on the same lines with them, for 88/7 = 12.6 and 201/7 = 28.7.

These comments show that in early 1865 Odling had clearly grasped the division of the elements into "reguline" (metallic) groups and associated "subgroups." These are the A and B groups of later periodic tables, and Odling's description of this relationship is more explicit than that given in his 1864 article. In the *Watt's Dictionary* articles Odling has also clearly grasped the relationships between

[7]"Reguline" metals were those that formed water-stable, insoluble sulfides; as opposed to "alkaligenous" metals, which did not. "Basylous" substances were more electropositive, as opposed to "chlorous" substances, which were more electronegative. According to these classifications, Si, Sn, and Pb were reguline, whereas Ti, Zr, and Th were basylous.

valence, including maximum valence, and periodicity (although Odling did not use these terms). The elements Tl and Pb are now associated with their modern congeners (Fig. 6.6). Although uranium, which appeared in the October 1864 table, is now absent (compare Figs. 6.4 and 6.5), Pd and Pt are now correctly put together to represent all platinum metals (Fig. 6.5). In the text of the article Rh is grouped with Ir, and Ru with Os.

Odling's comments on indium, an element that had been discovered only two years before, in 1863, whose atomic weight and valence were still in dispute, are highly interesting:

> Its equivalent weight is about 37 and its atomic weight may accordingly approximate to 37, 74, 111 or 148, &c. Its properties are too imperfectly known to allow of its being referred satisfactorily to any particular family of elements, but it evidently belongs to the reguline division of the metals and seems to have much the same relationship to thallium that tin has to lead.

Odling thus proposed to place a new element into his system by choosing an atomic weight that put it in a location near to known elements with similar chemical and physical properties; for indium, however, it was too early to put this proposal into practice. Only Strecker had previously proposed revising atomic weights so as to create more regular numerical relations among groups of related elements [53].

6.2.6 1865: *A Course of Practical Chemistry*, 2nd Edition

Odling's textbook, *A Course of Practical Chemistry*, was originally published in 1854. A second edition, published in 1865 [54],[8] contained a table (Fig. 6.9) that was very similar to the table (Fig. 6.5) in the 1865 article in *Watts's Dictionary*. The bold lines emphasizing the triads in Fig. 6.5 receive additional emphasis through the use of double lines and the label "Triplet Groups."

If one compares this table to Mendeleev's 1869 periodic table (Fig. 6.10) [56], the similarities are obvious, down to the insertion of special symbols (— by Odling and ? by Mendeleev) for the gaps after Al and Si, and other places. Some differences are that Odling has Hg and Pb in the same position as do modern periodic tables whereas Mendeleev does not; that is, in Odling's table, Hg is grouped with

[8]According to Thornton and Wiles [7], the publication history of *A Course of Practical Chemistry* is somewhat muddled. It is easy to find the 1st, 2nd, 4th, and 5th editions but there seems to be no 3rd edition. Thornton and Wiles postulated, and it seems the most likely explanation, that as the 1st edition (1854) was published by Samuel Highley, when the next edition (1863) was published by Longman it again appeared to be the first (and it was, by this publisher). The 2nd edition published by Longman appeared in 1865. The next edition, published in 1869, although the third published by Longmans, was called the 4th, and it seems likely that at this point the error in edition numbers was rectified to include all editions, irrespective of the publisher. This interpretation is strengthened by the introduction to the French translation of 1869 [55], in which Odling, in 1868, stated in his Preface that this translation was taken from the third edition. Since this French translation has the same appendices as the 1865 "2nd" edition, it seems that Odling and the publishers were already trying to get the edition numbers straightened out.

Atomic Weights and Symbols.

	Triplet Groups			
			Mo 96	W 184
H 1			—	Au 196·5
			Pd 106·5	Pt 197
....	....	....	....	....
Li 7	Na 23	—	Ag 108	—
G 9	Mg 24	Zn 65	Cd 112	Hg 200
B 11	Al 27·5	—	—	Tl 203
C 12	Si 28	—	Sn 118	Pb 207
N 14	P 31	As 75	Sb 122	Bi 210
O 16	S 32	Se 79·5	Te 129	
F 19	Cl 35·5	Br 80	I 127	
....	....	....	....	....
	K 39	Rb 85	Cs 133	
	Ca 40	Sr 87·5	Ba 137	
	Ti 48	Zr 89·5	—	
	Cr 52·5	—	V 138	
	Mn 55· &c.	—		

Fig. 6.9 Table from the 2nd edition of *A Course of Practical Chemistry* [54] (Courtesy of HathiTrust)

Zn, Tl with B, Pb with C; in Mendeleev's table, Hg is grouped with Cu, Tl with Li, and Pb with Ca. Mendeleev included 63 elements vs. 45 elements in Odling's table.

The 1867 Russian translation of Odling's *A Course of Practical Chemistry* [57], which was based on the "second" English edition (1865), plays a role in the priority dispute, as we will discuss in Sect. 6.2.7. The French translation of 1869 [55] was also based on this 1865 edition. Two more English editions of Odling's *A Course of Practical Chemistry* were published, in 1869 and 1876 [7]. It is interesting that the table shown in Fig. 6.9 did not appear in either of these later editions.

6.2.7 1871: A Question of Priority

Rudolph Gerstl was the London correspondent for the Deutsche Chemische Gesellschaft, and among his duties was reporting on the activities of the Chemical Society, the Royal Institution, and other similar British organizations. At the end of his report from January 1871 [58] he referred to Odling's article on Atomic Weights

ОПЫТЪ СИСТЕМЫ ЭЛЕМЕНТОВЪ,

ОСНОВАННОЙ НА ИХЪ АТОМНОМЪ ВѢСѢ И ХИМИЧЕСКОМЪ СХОДСТВѢ.

			Ti=50	Zr=90	?=180.
			V=51	Nb=94	Ta=182.
			Cr=52	Mo=96	W=186.
			Mn=55	Rh=104,4	Pt=197,4.
			Fe=56	Ru=104,4	Ir=198.
			Ni=Co=59	Pl=106,6	Os=199.
H=1			Cu=63,4	Ag=108	Hg=200.
	Be=9,4	Mg=24	Zn=65,2	Cd=112	
	B=11	Al=27,4	?=68	Ur=116	Au=197?
	C=12	Si=28	?=70	Sn=118	
	N=14	P=31	As=75	Sb=122	Bi=210?
	O=16	S=32	Se=79,4	Te=128?	
	F=19	Cl=35,5	Br=80	I=127	
Li=7	Na=23	K=39	Rb=85,4	Cs=133	Tl=204.
		Ca=40	Sr=87,6	Ba=137	Pb=207.
		?=45	Ce=92		
		?Er=56	La=94		
		?Yt=60	Di=95		
		?In=75,6	Th=118?		

Fig. 6.10 Periodic Table, short-form, from volume 1 of *Osnovy Khimii* [56] (Private collection)

in *Watts's Dictionary* and claimed that Odling deserved priority for creating a "natural system of elements." It is known [59, 60] that Mendeleev by this time was aware of Odling's system. As van Spronsen notes:

> Mendeleev's first acquaintance with Odling dates from April 1869, i.e., after Mendeleev had read his first paper on his system to the Russian Chemical Society in March of 1869. At the April meeting Savshchenkov, the Russian translator of Odling's A Course of Practical Chemistry (1867 ed.) called his attention to Odling's system, but added the comment that the composer of the system had not grasped its meaning, upon which Mendeleev remarked that if Odling had seen its theoretical implications he would no doubt have discussed them.

Savchenkov's translation had been published in St. Petersburg, the city where Mendeleev lived, and it contained a version of the table in Fig. 6.9. The implication is that Mendeleev could have seen Odling's table at any time in the two years before he published his own periodic system in 1869.

Mendeleev replied to this challenge immediately [59, 61]:

> Perhaps the cause of these misunderstandings lies in an insufficient familiarity with my detailed papers, or in the coincidence that I gave my system the same designation (natural system) as that adopted by Odling for his system. It would be more correct to call my system "periodic," because it springs from a periodic law, which may be expressed as: "The measurable chemical and physical properties of the elements and their compounds are a periodic function of the atomic weight of the elements."

As far as we know, only once in his career did Odling make a claim for his own priority in the development of the periodic table. In a lecture to the Royal Institution on January 19, 1872 [62], he included a table of the elements (Fig. 6.11) and stated:

> This Table is based on one published by the author in 1864-5. Similar tables have been constructed by Newlands, Meyer, Mendelejeff, and others. The positions marked X^a, X^b, and X^c, are assigned by Mendelejeff to yttrium, didymium, and cerium, respectively. The recognition of the atomic weight of uranium as 240, is also due to Mendelejeff.

6.2.8 The End

Odling's fame faded after he stopped doing original research in the 1870s, and especially after his death in 1921. As Thornton and Wiles noted in 1956 [7]:

> The early life of William Odling was replete with the promise of a brilliant career, and in fact he achieved both honour and professional recognition during that period. He was industrious as a lecturer, writer and research worker, but after 1876 his writing appeared to cease, and an attempt to discover some details concerning his career was unsuccessful until it was discovered that he survived until 1921, in which year several obituary notices paid ample justice to his work as a chemist.

William A. Tilden (1842–1926) wrote one of Odling's obituaries [63], in which he summarized Odling's research contributions as follows:

> Although it is vain to look in the Royal Society Catalogue of Scientific Papers for outstanding discoveries the result of experimental work under Odling's name, it should not be

TABLE III.—ELEMENTS, IN ORDER OF ATOMIC WEIGHT.*

1.	2.	3.	4.	5.	6.	7.	8.	Type.
I. H 1	Li 7	Na 23	K 39	..	Rb 85	Ag 108	Cs 133	R Cl
II.	G 9	Mg 24	Ca 40	Zn 65	Sr 87·5	Cd 112	Ba 137	RCl_2
III.	B 11	Al 27·5	..	..	X^a	In 113	X^b	RCl_3
IV.	C 12	Si 28	Ti 50	..	Zr 89	Sn 118	X^c	RCl_4
V.	N 14	P 31	V 51	As 75	Nb 94	Sb 122	..	RCl_5
VI.	O 16	S 32	Cr 52·5	Se 79	Mo 96	Te 129	..	RCl_6
VII.	F 19	Cl 35·5	Mn 55	Br 80	..	I 127	..	RCl_7
VIII.	..	..	Fe 56 Co 59 Ni 59	..	Ru 104 Ro 104 Pd 106	..	..	RCl_8
	Na 23	..	Cu 63·5	..	*Ag* 108			

Fig. 6.11 Table from Odling's lecture "On the last new metal, indium" [62], p 398 (This image is taken from the copy Odling sent and inscribed to Cannizzaro. Private collection)

> forgotten that he contributed several very important articles on theoretical subjects to Watts's "Dictionary," and among them one on atomic weights, in which he came very near the discovery of the periodic law now always associated with the name of Mendeléeff.

Although accurate as far as it goes, the single sentence hardly captures the extent of Odling's scientific achievements.

6.3 John A. R. Newlands

In some ways, John Alexander Reina Newlands (1838–1898) (Fig. 6.12) and William Odling were similar: they were both British chemists, and they were even born in the same London district roughly a decade apart. But that is where their similarities seem to end. Newlands did not have the status of Odling within the chemistry community when he began his foray into classifying elements; in fact, Odling was an officer of the Chemical Society at a time when Newlands failed to have his classification published in its Journal. Newlands's engagement with periodicity was longer than Odling's, and he was much more adamant about staking his claim.

Fig. 6.12 John A. R. Newlands (Reproduced courtesy of the Royal Society of Chemistry)

6.3.1 Biographical Information

Newlands was born in Southwark, London, on November 26, 1837. His parents were Rev. William Newlands of Glasgow, a minister of the Established Presbyterian Church of Scotland, and Mary Sarah Reina Newlands. John was the third child of five. The youngest of the family, Benjamin Edward Reina Newlands, was also a chemist, with whom John worked for many years. John's interest in chemistry arose from some lectures at Charing Cross Hospital he attended with a friend. He entered the Royal College of Chemistry under August Hofmann in 1856. The next year he became assistant to John Thomas Way, chemist of the Royal Agricultural Society [64–66].

Newlands left Way to fight for Italian unification under Garibaldi in 1860, and he saw action outside Capua.[9] Newlands's participation in the fight for Italian unification is sometimes mentioned in discussions of his contributions to the periodic law. His action in this cause is usually attributed to his Italian descent on his mother's side, and his presence in Italy precluded his attending the Karlsruhe Congress of 1860—although, as Scerri notes, Newlands was not in a position in which he was likely to have been invited [67]. Newlands's Italian parentage may well have been a motivating factor: although his nearest Italian ancestor moved to

[9]Garibaldi's army fought the forces of the Kingdom of the Two Sicilies (capital Naples), a monarchy that ruled the southern third of mainland Italy and the island of Sicily. Garibaldi's campaign began on the island of Sicily, and it included a few British volunteers even at that early stage. More British volunteers, including Newlands, joined Garibaldi after his forces crossed to the mainland. At the siege of Capua, Garibaldi's army was joined by soldiers of the Kingdom of Sardinia (capital Turin), under whose monarchy most of Italy was unified in 1860.

England at least a century earlier, the name Reina was clearly important in the family.[10] Italian heritage was not a prerequisite for sympathy for the Garibaldian cause, though: many British men and women contributed money and about 1000 men joined his forces [68].

After the Italian campaign, Newlands returned to Way. In 1862, he married Emily Elizabeth Jane Richings, who was employed in Way's household. Over the next decade, the couple had two daughters and a son (who also went into chemistry). In 1864, Newlands set up on his own as an analytical chemist in the City. In the 1860s he taught chemistry at the Grammar School of St. Saviour's and the City of London College. During this eventful decade, he published a volume of poetry [69]. And during the middle years of the 1860s he published several notes on the classification of elements and relations among the atomic weights, to be treated in more detail below.

From 1868 to 1886, Newlands was chief chemist at James Duncan's sugar refinery at Victoria Docks, which also employed his brother Benjamin. He continued to teach, at least near the beginning of this period: into the 1870s he taught elementary chemistry at the Ladies' Medical College in London and he gave private lessons in chemical analysis. In 1886 he joined Benjamin as private consulting chemists. The brothers collaborated with Charles G. Warnford Lock to publish *Sugar: a Handbook for Planters and Refiners* in 1888 [70]. He was still attending meetings of the Chemical Society in what proved to be his last year, participating in discussions of the papers presented [71]. He died on July 29, 1898, of influenza.

6.3.2 Feb 7, 1863: "On Relations Among the Equivalents"

Newlands's first article on classifying elements was published in the journal *Chemical News*, on February 7, 1863 [72], as a letter titled, "On Relations among the Equivalents." Note that when Newlands referred to "equivalents," he was essentially referring to atomic weights. Here the values of the equivalents he employed are similar to those of Wollaston [23]. In later papers, Newlands changed the set of values he used to the atomic weights based on Cannizzaro's principles, but he still used the term equivalent.

In his article, Newlands began thus,

> Many chemists, and M. Dumas in particular, have, on several occasions, pointed out some very interesting relations between the equivalents of bodies belonging to the same natural family or group; and my present purpose is simply to endeavour to proceed a little further in the same direction.

Thus Newlands began by pointing out that elements of "natural" groups (which we can construe as those categorized together on the basis of chemical similarities)

[10]According to the Wood Family Tree website [65], that ancestor was Peter Anthony Reina, John's mother's grandfather, born in 1725 in Milan. **His** children were born in or near London. John and his three brothers were all given the name Reina as a middle name, a practice that persisted in the family for several generations.

Member of group having lowest equivalent.		One immediately above the preceding.		Difference.
Magnesium	12	Calcium	20	8
Oxygen	8	Sulphur	16	8
Carbon	6	Silicon	14·2	8·2
Lithium	7	Sodium	23	16
Fluorine	19	Chlorine	35·5	16·5
Nitrogen	14	Phosphorus	31	17

Fig. 6.13 Table comparing equivalent weights of related elements [72] (Private collection)

are related to one another through their atomic weights as well. Newlands intended to further develop this insight, which, he noted, was not original with him.

In the meat of the paper, Newlands organized 56 elements into 11 groups, based on their properties. What he then did, as others had done before him including Dumas [73] and Döbereiner [39], is look for numerical relationships among the equivalents of elements in a group. He argued, as can be seen in Fig. 6.13, that the difference between the equivalent of the lightest element in a group and that of the next heaviest is a multiple of eight. In the next table he went further and extended the same observation to the lightest and heaviest elements in a triad.

So what does this mean? It is not clear what Newlands thought the implications of these mathematical relationships were. Indeed, there is quite a bit of tentativeness in the letter. The opening sentence quoted above is followed by "I must, however, premise that many of the observations here collected together are well known already, and are only embodied in my communication for the purpose of rendering it more complete." Furthermore at the end of the article, he wrote, "I also freely admit that some of the relations above pointed out are more apparent than real; others, I trust, will prove of a more durable and satisfactory description."

Newlands eventually was to elaborate the implications of these relations, but only after being drawn out by another correspondent to the *Chemical News.*

6.3.3 July 30, 1864: Studiosus and Newlands's Rebuttal

On July 2, 1864, almost a year and a half after Newlands's letter appeared, the *Chemical News* published a short note titled, "Numerical Relations of Equivalent Numbers" [74] by someone using the pseudonym "Studiosus," which is Latin for student. Studiosus did not reference Newlands's work, but he made a claim about a mathematical relationship among the atomic weights. He wrote, "the atomic weights of the elementary bodies are, with few exceptions, either exactly or very nearly multiples of eight." As evidence, he provided a chart (Fig. 6.14), which contains 40 elements whose atomic weights are (roughly) multiples of eight. This is the paper that prompted Newlands to revisit and rework his ordering of the elements.

Li = 7	8
O = 16; N = 14	16 = 8 × 2
Mg = 24; Na = 23	24 = 8 × 3
S = 32; P = 31	32 = 8 × 4
Ca = 40; K = 39	40 = 8 × 5
Ti = 50	48 = 8 × 6
Ni = 58·5; Co = 58·5; Mn = 55; Fe = 56	56 = 8 × 7
Cu = 63·5; Zn = 65; Y = 64	64 = 8 × 8
Se = 79·5; Br = 80	80 = 8 × 10
Zr = 89·5; Sr = 87·5	88 = 8 × 11
Mo = 96; Di = 96	96 = 8 × 12
Ru = 104; Ro = 104	104 = 8 × 13
Cd = 112	112 = 8 × 14
Sn = 118; U = 120; Sb = 122	120 = 8 × 15
Te = 129; I = 127	128 = 8 × 16
Ta = 138; V = 137; Ba = 137	136 = 8 × 17
W = 184	184 = 8 × 23
Os = 199; Hg = 200	200 = 8 × 25
Pb = 207; Bi = 210	208 = 8 × 26
Th = 238	240 = 8 × 30

Fig. 6.14 Table illustrating elements having atomic weights close to multiples of eight according to Studiosus [74] (Private collection)

By the end of July, Newlands's rebuttal was published. In a long letter titled, "Relations between Equivalents" [47], Newlands made three main assertions. One, that Studiosus was wrong; two, that Newlands got there first; and three, that there are indeed patterns in the atomic weights, but not the one Studiosus found. On the first count, that Studiosus was wrong, Newlands provided a table, Fig. 6.15, in which he arranged the elements in order of increasing atomic weight, with the atomic weights listed in the second column, and the differences in atomic weights provided in the third column. About this table, Newlands wrote, "Now, it will be observed that in all the above differences the number eight occurs but once, and we never meet with a multiple of eight." Newlands argued that if all of the atomic weights were multiples of eight, then their differences would also be multiples of eight. Since that is not true, the "Law of Studiosus," as he called it, must be false. Newlands noted that the atomic weights used here were based on those reported recently by Alexander Williamson, which were in turn based on the principles set forth by Cannizzaro. Newlands did not state that many of the atomic weights he used in this letter were different from those he had used in his previous letter. Many in the 1864 letter were double those he had used the year before, and thus close to values still accepted in the twenty-first century.

H .	1		Ca .	40	1	Ce .	92	2·5	V .	137	0
Li .	7	6	Ti .	50	10	La .	92	0	Ta .	138	1
G .	9	2	Cr .	52·5	2·5	Di .	96	4	W .	184	46
B .	11	2	Mn.	55	2·5	Mo .	96	0	Nb .	195	11
C .	12	1	Fe .	56	1	Ro .	104	8	Au .	196	1
N .	14	2	Co .	58·5	2·5	Ru .	104	0	Pt .	197	1
O .	16	2	Ni .	58·5	0	Pd .	106·5	2·5	Ir .	197	0
Fl .	19	3	Cu .	63·5	5	Ag .	108	1·5	Os .	199	2
Na .	23	4	Y .	64	0·5	Cd .	112	4	Hg .	200	1
Mg .	24	1	Zn .	65	1	Sn .	118	6	Tl .	203	3
Al .	27·5	3·5	As .	75	10	U .	120	2	Pb .	207	4
Si .	28	0·5	Se .	79·5	4·5	Sb .	122	2	Bi .	210	3
P .	31	3	Br .	80	0·5	I .	127	5	Th .	238	28
S .	32	1	Rb .	85	5	Te .	129	2			
Cl .	35·5	3·5	Sr .	87·5	2·5	Cs .	133	4			
K .	39	3·5	Zr .	89·5	2	Ba .	137	4			

Fig. 6.15 Table of elements in order of atomic weight [47] (Private collection)

Newlands then segued into his own ideas, providing a table (Fig. 6.16) which highlights certain patterns in atomic weights, and seems to argue that the atomic weights follow mathematical patterns. Note that group III has a gap between Si and Sn, about which he wrote "Silicon and tin stand to each other as the extremities of a triad. Titanium is usually classed along with them, and occupies a position

			Triad.			
			Lowest term.	Mean.	Highest term.	
I.		Li 7	+17 = Mg 24	Zn 65	Cd 112	
II.		B 11				Au 196
III.		C 12	+16 = Si 28		Sn 118	
IV.		N 14	+17 = P 31	As 75	Sb 122	+88 = Bi 210
V.		O 16	+16 = S 32	Se 79·5	Te 129	+70 = Os 199
VI.		F 19	+16·5 = Cl 35·5	Br 80	I 127	
VII.	Li 7	+16 = Na 23	+16 = K 39	Rb 85	Cs 133	+70 = Tl 203
VIII.	Li 7	+17 = Mg 24	+16 = Ca 40	Sr 87·5	Ba 137	+70 = Pb 207
IX.			Mo 96	V 137	W 184	
X.			Pd 106·5		Pt 197	

Fig. 6.16 Table illustrating some groups of elements and relations between their atomic weights [47] (Private collection)

intermediate between silicon and the central term or mean of the triad, which is at present wanting." That is, the middle element of a triad containing silicon and tin ought to have an atomic weight of 73 and it was not yet known ("wanting"). More than 20 years later, after the discovery of germanium, Newlands pointed to this passage as a prediction [75]. Like Odling's predictions mentioned above, this prediction was based on a periodic system, albeit a partial one. As we will see, gaps for undiscovered elements disappeared and reappeared from Newlands's systems over the next decade or so.

6.3.4 Aug 20, 1864: Response from Studiosus and New Observations by Newlands

The August 20, 1864, issue of the *Chemical News* contained both a response from Studiosus [76] and new observations from Newlands in two short letters [77, 78]. Studiosus wrote, "Few would call chemistry a mathematical science; and, such being the case, I appeal against its being treated mathematically, as it was by Mr. Newlands in your impression of the 30th [last month]." Studiosus argued that one must "[grant] necessarily some amount of latitude (for atomic weights being merely results of experiment, and not mathematical, are therefore subject to error), …" Thus Studiosus attributed numerical irregularities to experimental error, retreating from his early assertion of rigid regularity. We will return to this point.

Newlands's letters in this issue are much more interesting. In the first [77] periodicity can be seen—not just patterns within groups, but actual periodicity of groups among a large number of elements considered together. He began thus

> In addition to the facts stated in my late communication, may I be permitted to observe that if the elements are arranged in the order of their equivalents, calling hydrogen 1, lithium 2, glucinum [beryllium] 3, boron 4, and so on … it will be observed that elements having consecutive numbers frequently either belong to the same group or occupy similar positions in different groups, …

He illustrated with a table, Fig. 6.17.

In this table, he was no longer looking at numerical relationships among atomic weights; instead he was just looking at the elements' order. He wrote "Here the difference between the number of the lowest member of a group and that

		No.		No.		No.		No.		No.
Group *a*.	N	6	P	13	As	26	Sb	40	Bi	54
,, *b*.	O	7	S	14	Se	27	Te	42	Os	50
,, *c*.	Fl	8	Cl	15	Br	28	I	41	—	—
,, *d*.	Na	9	K	16	Rb	29	Cs	43	Tl	52
,, *e*.	Mg	10	Ca	17	Sr	30	Ba	44	Pb	53

Fig. 6.17 Table illustrating periodicity among 24 elements [77] (Private collection)

immediately above it is 7; in other words, the eighth element starting from a given one is a kind of repetition of the first, like the eighth note of an octave in music." This is the first instance of what he later called "the law of octaves." As the table illustrates, phosphorus is seven elements away from nitrogen, sulfur is seven elements away from oxygen, chlorine seven from fluorine, etc. Moving to the right, arsenic is 13 elements away from phosphorus, and antimony 14 away from arsenic. Selenium is 13 elements away from sulfur, and tellurium 15 away from selenium, and so on. Thus periodicity can be seen among many of the elements, but not yet all.

In a separate short note [78], Newlands responded to another pseudonymous correspondent, "Inquirer," who had wished for an estimate of the atomic weight of the recently discovered element indium. What is noteworthy about the response is not that his guess, 182, was wrong, but rather that it was based on analogies and groups. His arrangement of elements, then, had some potential as a research tool.

These letters generated some further correspondence. On September 3, John Noble of the Royal Arsenal General School in Woolwich had a letter published in the *Chemical News* titled, "Numerical Relations of Equivalent Numbers" [79]. Noble took both Studiosus and Newlands to task. He started with Studiosus:

> I protest against the use of the term "law" when applied to a few cases of mathematical relation between atomic numbers. If there really be such a law as the "law of Studiosus" (how it sounds!)—namely, that the atomic numbers are multiples of 8—then the agreements should be many and the exceptions few.

Noble pointed out the numerous exceptions to the "law of Studiosus." Then turning to Newlands, he wrote

> It is only fair, however, to point out that what ought to be condemned, and what Mr. Newlands himself does condemn—namely, approximations and allowances—are just the means which Mr. Newlands, to a large extent, employs in his own tables of "relations" …

In his final paragraph, Noble expressed general skepticism over the project of relations among atomic weights: "The fact is, there has been a great deal of nonsense written about these 'laws' and 'relations.' (But then you know, Mr. Editor, it is much easier to sit down and make laws for these numbers than to verify the numbers themselves.)" Noble argued that most relations were "rubbish" and merely the result of random chance.

On September 24, *Chemical News* published a rebuttal from "Inquirer" entitled, "Numerical Relations of Equivalents" [80]. Inquirer agreed that Studiosus's law is indeed easily disproved, but he argued that discarding Newlands's work along with that of Studiosus would be a mistake. He placed Newlands in the company of well-known and respected chemists, including Dumas, who also made use of "approximations and allowances." Inquirer completed his article thus:

> Mr. Noble is of opinion that it is much easier to find laws for the equivalent numbers than to verify the equivalents experimentally; but really the two things have no more connection than "chalk and cheese." For my own part, if any comparison be advisable, I could mention

a hundred chemists who either have verified, or are fully capable of verifying, an ordinary equivalent; but I do not know a single chemist who can thoroughly exhibit all the relations among the atomic weights, still less one who can explain why and wherefore such relations exist.

6.3.5 August 1865: The Law of Octaves and More Relations Among the Equivalents

About a year later, on August 18, 1865, Newlands had another letter published in the *Chemical News*, this one titled "On the Law of Octaves" [81]. Here he provided a table of all the elements for which atomic weights were known (Fig. 6.18) and wrote, "This peculiar relationship I propose to provisionally term the 'Law of Octaves.'" The numbers shown in the table are not atomic weights, but ordinal numbers of the elements when placed in order of atomic weight. When the atomic weights of two elements were the same or very close, Newlands assigned both elements to the same ordinal number; six pairs of elements were thus doubled up, resulting in 62 elements occupying 56 places. The elements are arranged in order of atomic weight with a few exceptions. In the first of these, the heavier tellurium (number 43) precedes the lighter iodine (42)—an inversion that made good chemical sense, and which Newlands had made in earlier papers as well [47, 77]. The heaviest elements in Newlands's table are more out of atomic-weight order than in order.

In the next issue, published on August 25, 1865, another letter from Newlands appeared, "On the Cause of Numerical Relations among the Equivalents" [82]. He began

By way of addition to my last letter, I will, with your permission, endeavour to show that all the numerical relations among the equivalents pointed out by M. Dumas and others, including the well-known triads, are merely arithmetical results flowing from the existence of the "law of octaves" …

No.	No.	No.	No.	No.	No.	No.	No.
H 1	F 8	Cl 15	Co & Ni 22	Br 29	Pd 36	I 42	Pt & Ir 50
Li 2	Na 9	K 16	Cu 23	Rb 30	Ag 37	Cs 44	Tl 53
G 3	Mg 10	Ca 17	Zn 25	Sr 31	Bd 38	Ba & V 45	Pb 54
Bo 4	Al 11	Cr 19	Y 24	Ce & La 33	U 40	Ta 46	Th 56
C 5	Si 12	Ti 18	In 26	Zr 32	Sn 39	W 47	Hg 52
N 6	P 13	Mn 20	As 27	Di & Mo 34	Sb 41	Nb 48	Bi 55
O 7	S 14	Fe 21	Se 28	Ro & Ru 35	Te 43	Au 49	Os 51

(NOTE.—Where two elements happen to have the same equivalent, both are designated by the same number.)

Fig. 6.18 Table illustrating periodicity among 62 elements [81] (Private collection)

Symbol.	No.	Eq.	Eq.÷No.	Symbol.	No.	Eq.	Eq.÷No.
H	1	1	1	Br	29	80	2·758
Li	2	7	3·5	Rb	30	85	2·833
G	3	9	3	Sr	31	87·5	2·823
Bo	4	11	2·75	Zr	32	89·5	2·797
C	5	12	2·4	Ce	33	92	2·788
N	6	14	2·333	Di	34	96	2·824
O	7	16	2·286	Ro	35	104	2·971
F.	8	19	2·375	Pd	36	106·5	2·958
Na	9	23	2·555	Ag	37	108	2·919
Mg	10	24	2·4	Cd	38	112	2·947
Al	11	27·5	2·5	Sn	39	118	3·026
Si	12	28	2·333	U	40	120	3
P.	13	31	2·385	Sb	41	122	2·975
S.	14	32	2·286	I.	42	127	3·024
Cl	15	35·5	2·367	Te	43	129	3
K	16	39	2·437	Cs	44	133	3·023
Ca	17	40	2·353	Ba	45	137	3·044
Ti	18	50	2·778	Ta	46	138	3
Cr	19	52·5	2·763	W	47	184	3·915
Mn	20	55	2·75	Nb	48	195	4·062
Fe	21	56	2·667	Au	49	196	4
Co	22	58·5	2·659	Pt	50	197	3·94
Cu	23	63·5	2·761	Os	51	199	3·902
Yt	24	64	2·667	Hg	52	200	3·846
Zn	25	65	2·6	Tl	53	203	3·83
In	26	72	2·769	Pb	54	207	3·833
As	27	75	2·778	Bi	55	210	3·818
Se	28	79·5	2·839	Th	56	238	4·25

Fig. 6.19 Table illustrating that the ratio of atomic weight to ordinal number increases with increasing atomic weight [82] (Private collection)

He then provided a table (Fig. 6.19) that shows the element, the element's "number," that is, its numerical order, the element's atomic weight, and the element's atomic weight divided by its number.

Newlands observed, "It will be seen that the number of an element is nearly equal to its equivalent divided by a certain sum, which varies, however, as we ascend the scale, …" He then provided another table (Fig. 6.20) summarizing the gradual increase in the ratio.

Because the atomic weights increase with some average regularity, Newlands argued, "if the number of one element is the mean of those of two others … its equivalent will also be the mean of their equivalents." In effect, Newlands maintained that periodicity underlies the numerical relations among the atomic weights of related elements.

From 4 to 17, the No. = Eq. ÷ 2·5
From 18 to 34, the No. = Eq. ÷ 2·75
From 35 to 46, the No. = Eq. ÷ 3
From 47 to 56, the No. = Eq. ÷ 4

Fig. 6.20 Table illustrating ranges of the ratio of atomic weight to ordinal number [82] (Private collection)

Table II.—Elements arranged in Octaves.

	No.		No.		No.		No.		No.		No.		No.		No.
H	1	F	8	Cl	15	Co & Ni	22	Br	29	Pd	36	I	42	Pt & Ir	50
Li	2	Na	9	K	16	Cu	23	Rb	30	Ag	37	Cs	44	Os	51
G	3	Mg	10	Ca	17	Zn	24	Sr	31	Cd	38	Ba & V	45	Hg	52
Bo	4	Al	11	Cr	19	Y	25	Ce & La	33	U	40	Ta	46	Tl	53
C	5	Si	12	Ti	18	In	26	Zr	32	Sn	39	W	47	Pb	54
N	6	P	13	Mn	20	As	27	Di & Mo	34	Sb	41	Nb	48	Bi	55
O	7	S	14	Fe	21	Se	28	Ro & Ru	35	Te	43	Au	49	Th	56

Fig. 6.21 Table illustrating periodicity among 62 elements [83] (Private collection)

6.3.6 March 1866: The Law of Octaves and the Chemical Society

On March 1, 1866, Newlands presented his law of octaves to the Chemical Society, and he seems to have been met only with detractors, or at least only the detractors were recorded. A report of the presentation, which was published in the March 9 issue of the *Chemical News*, outlines Newlands's "law of octaves" as well as the critiques it received [83]. The arrangement of the elements (Fig. 6.21) differs little from the one published in August 1865 except in the last column. There the heaviest elements appear in order according to the atomic weights Newlands had, an arrangement that moved thallium and lead into the groups headed by boron and carbon, respectively.

This improvement is similar to that noted for Odling between his tables in the *Quarterly Journal* [43] and in *Watts's Dictionary* [49] (see Sect. 6.2.5). It is not known whether Newlands was aware of Odling's work in classification, although it is certain that the men knew each other from meetings at the Chemical Society [84].[11] Odling may well have been present at this March 1 meeting: the *Chemical News* reported that "the Secretary" read an abstract of a paper by C. R. Wright, and

[11]For example, comments are recorded from both Newlands and Odling about the same presentation at the Chemical Society meeting of December 15, 1864 [84].

Odling was one of the Society's two secretaries at the time. At any rate, Odling was not reported as commenting on any of the papers read at this session.

The *Chemical News* report stated, "Dr. Gladstone made objection on the score of its having been assumed that no elements remain to be discovered." Gladstone's comment makes it apparent that Newlands's thinking no longer included deliberate gaps for undiscovered elements, as it had in 1864 [47].

The report also noted "Professor G. F. Foster [*sic*, for George Carey Foster] humorously inquired of Mr. Newlands whether he had ever examined the elements according to the order of their initial letters. For he believed that any arrangement would present occasional coincidences, ..." These few words are apparently all the license needed for James Cameron to report, nearly a century later and employing more imagination than references, that the paper was received with "hilarious, uproarious laughter" [85].

Newlands published a response to these critiques in the next issue of the *Chemical News* (March 16) [86]. In response to Foster's objection that periodicity might be a coincidence, Newlands wrote that "the coincidences which I have pointed out are the rule, and not the exception."

In response to Gladstone's critique referring to the discovery of new elements, Newlands argued that new elements could certainly be discovered, but he was certain that his "law" would hold, even if, the periodicity occurred every 9 or 10 elements rather than every 8. Indeed, this explanation would be valid in case of the discovery of new families of elements, as happened near the end of the century with the discovery of the noble gases. Gladstone's objection was valid, however, in the case of piecemeal discoveries of elements, and most of the elements discovered in recorded history came in a piecemeal manner.

Even after Newlands's response, however, the law of octaves did not catch on. The Chemical Society did not publish his paper, but the other three papers read at the March 1 meeting appeared in volume 19 of the Society's *Journal*. Inquirer's endorsement of Newlands's efforts in 1864 seems to stand alone, however, at least before Mendeleev published his own table.

6.3.7 The Afterlife of the Law of Octaves

Although the basis of his claims for priority in the discovery of chemical periodicity lie in his work of the 1860s, Newlands continued to think about the classification of elements in the early 1870s, before the periodic law had gained widespread attention among chemists. The publications of Mendeleev and Meyer on chemical periodicity in 1869 and the early 1870s were neither unknown nor ignored; however, they prompted very few publications by other chemists before the discovery of gallium in 1875 [87].

Earlier in 1875, Newlands published a paper on relations among the atomic weights that included a table (Fig. 6.22) showing seven-element periods (as in his law of octaves) as well as blank spaces and only one element per position [88]. The gaps present in his 1864 arrangement [47] and absent from his 1866 law of octaves

	1	2	3	4	5	6	7	8
a.	—	Li 7·0	Na 23·0	K 39·1	—	Cu 63·4	Rb 85·4	—
b.	—	Be 9·4	Mg 24·0	Ca 40·0	—	Zn 65·2	Sr 87·6	—
c.	—	B 11·0	Al 27·4	—	Fe 56·0	—	Y 88·0(?)	—
d.	—	C 12·0	Si 28·0	Ti 50·0	—	—	Zr 89·6	Rh 104·4
e.	—	N 14·0	P 31·0	V 51·2	—	As 75·0	Nb 94·0	Ru 104·4
f.	—	O 16·0	S 32·0	Cr 52·2	Ni 58·8	Se 79·4	Mo 96·0	Pd 106·6
g.	H 1	F 19·0	Cl 35·5	Mn 55·0	Co 58·8	Br 80·0	—	—

9	10	11	12	13	14	15	16
Ag 108·0	Cs 133·0	—	—	Au 197·0	—	—	—
Cd 112·0	Ba 137·0	—	—	—	Hg 200·0	—	—
In 113·4	Di 138·0(?)	—	Er 178·0(?)	—	Tl 203·6	—	—
Sn 118·0	Ce 140·0(?)	—	La 180·0(?)	Pt 197·4	Pb 207·0	—	Th 235·0
Sb 122·0	—	—	Ta 182·0	Ir 198·0	Bi 210·0	—	—
Te 128·0	—	—	W 184·0	Os 199·2	—	—	U 240·0
I 127·0	—	—	—	—	—	—	—

Fig. 6.22 Table illustrating periodicity among 63 elements [88]. In the original, all 16 columns were arranged side by side; the table has been broken into two stacked sections here (Courtesy of HathiTrust)

[83] have returned. Newlands speculated that the blank spaces might eventually be occupied by elements not yet discovered or by known elements whose atomic weights required adjustment. Each row in this 1875 table placed together elements whose usual valence ("quantivalence" in the usage of the day) changed regularly: 1 for a, 2 for b, 3 for c, 4 for d, 3 (or 5) for e, 2 (or 6) for f, 1 (or 7) for g. Although he did not emphasize the fact, valence gave a good justification for the existence of seven groups, and it made for better groupings—at least among elements we would assign today to main groups. This paper illustrates that Newlands had not stopped thinking about systematizing the elements or exploring relationships among their atomic weights, and that his thought continued to evolve. Of course he was well aware of the work done by Mendeleev and Meyer in this area, and this paper included a reference to his earlier notes on the law of octaves and a claim of priority.

Newlands had already claimed priority for his law of octaves as early as 1873 at a meeting of the Chemical Society. There he read a short note calling attention to an abstract by Meyer that referred to the periodical arrangement of elements by Mendeleev. The abstract had been published in a recent number of the Society's *Journal*. Newlands called attention to his 1866 presentation before the Society, which was not published in its *Journal*, and he requested "as a simple matter of justice, the insertion of this brief note in the Society's *Journal*." The President of the Society, who in 1873 was none other than Odling, said that the reason the paper had not been printed was because "they had made it a rule not to publish papers of a purely theoretical nature, since it was likely to lead to correspondence of a

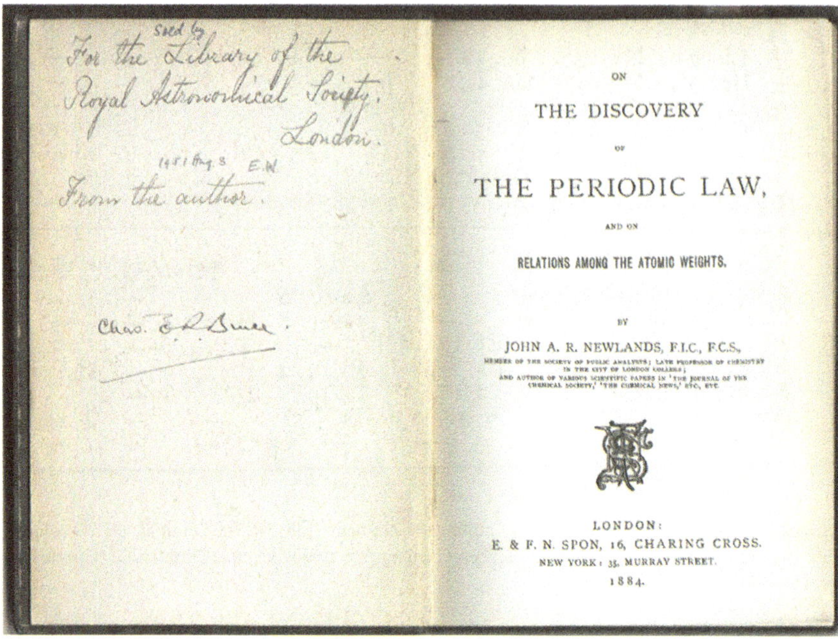

Fig. 6.23 Title page of Newlands's 1884 monograph *On the Discovery of the Periodic Law and on Relations among the Atomic Weights* with inscription on facing page (Private collection)

controversial nature" [89].[12] Newlands's 1873 note did not appear in the Society's journal. A year earlier, Newlands had noted the attention given at the time to the atomic weights of Cannizzaro in unnamed papers at home and abroad, and he drew attention to the fact that only the atomic weights of Cannizzaro worked as an ordering principle for his law of octaves [90].

The first assertion of Newlands's priority to appear in the *Journal of the Chemical Society* was a comment inserted by a young abstractor, Maurice Lichtenstein in the April 1876 issue. That issue carried several abstracts of papers on gallium, including one by Mendeleev, in which he drew attention to the likelihood that gallium was his predicted eka-aluminium. Immediately following the abstract was printed "Note by Abstractor.—The periodic law was first enunciated in 1864 by J. Newlands (*Chem. News*, x, 59, 94)" [91].

After the Royal Society recognized Mendeleev and Meyer with the Davy Medal (in 1882), Newlands collected his writings on the classification of elements and published them in a short monograph (Fig. 6.23), *On the Discovery of the Periodic*

[12]Odling was in a position to know the practices of the journal. He served on a four-person committee of publication from 1862 to 1865. He was not on the committee in 1866, the year Newlands presented to the Society on the law of octaves. He returned to the committee in 1870, and he served on an expanded committee of publication into 1877.

Law and on Relations among the Atomic Weights (1884) [92]. In 1887 the Royal Society recognized Newlands by awarding him the Davy Medal.

By the time they received the Davy Medal, both Mendeleev and Meyer were aware of priority claims made by Newlands or on his behalf [93, 94]. Signs of recognition of Newlands's priority could be seen in at least some areas of the British scientific establishment from the early 1880s. Addresses by presidents of the British Association for the Advancement of Science or its chemical section that mentioned the periodic law began to acknowledge Newlands's priority; the first that we could find was Sir John Lubbock's inaugural address at the 1881 meeting [95]. In the *Chemical News*, short reviews of books that treated the periodic law often mentioned whether or not the author acknowledged Newlands's work on the subject to the extent the reviewer considered appropriate. This practice began shortly before [96]. Newlands's own book was published and reviewed [97] in the *Chemical News*.

6.4 Conclusions

It is natural, particularly for chemists curious about the development of their discipline, to wonder who was the first to formulate the periodic law: Odling? Newlands? Someone else? After all, priority, awards, and competition are very much a part of the culture of science, both now in the early twenty-first century and in the nineteenth century when Newlands asserted his priority and received the Davy medal. Attempting to answer questions such as who discovered the periodic law leads one into historical and philosophical difficulties, including drawing arbitrary lines around the essence of the discovery. See Michael Gordin's article on the role of writing textbooks in the development of Mendeleev's and Meyer's periodic systems for more on problems of attempting to adjudicate questions of priority in scientific discovery [98].[13] Regardless of any conclusions we come to about priority, we believe that looking into the classifications of elements prepared by Newlands and Odling and other chemists of the 1860s reinforces the widely accepted view that the periodic law is a classic example of multiple independent discovery [99, 100].

The ordering of pairs of analogous elements in particular shows that Odling already held the key to the periodic table in his hands in 1864 when he wrote "On the Proportional Numbers of the Elements" for the *Quarterly Journal of Science* [43]. Exactly when in 1864 he developed his insight is difficult to pin down further: the paper appeared in the October issue and no date is given for receipt or composition of the article.

[13]Gordin's article prompts us to wonder whether pedagogical considerations were present in the minds of the subjects of this chapter. As has already been noted, one of Odling's classifications appeared in his textbook. Newlands published no textbooks, but he was engaged in teaching chemistry during the same years when he developed the law of octaves. But the role of pedagogy, if any, in the classifications of either Odling or Newlands is purely speculative.

The contributions of Newlands are easier to date, for the *Chemical News* was a weekly. Moreover, most of Newlands's contributions were in the form of letters that included a date. Using old equivalent weights, Newlands stated in 1863 [72] that the difference between the first and second members of the family of analogous elements is always about 8 or 16. Thus he saw a similar pattern of atomic weight progression in different groups, but no larger pattern and no ordering by atomic weight—which would not have yielded insights in any event from the set of atomic weights employed. In his letters in the summer of 1864, though, we see the foundations of a periodic classification (although not yet one that extended to all the elements): ordering by atomic weight and recurrence of chemical similarities at approximately regular intervals. These key letters were published on 30 July [47] and 20 August [77] 1864 and they were dated 12 July and 8 August, respectively. Was this before Odling wrote his paper? Could Odling have seen these letters before writing his paper? We do not know. We know that Newlands's 1864 letters were in print before Odling's article. And we know (see below) that in the 1870s Odling acknowledged Newlands's priority. In any event, though, Odling devised a comprehensive periodic system before Newlands did, for Newlands's system did not extend to all known elements until 1865 [81].

On the other hand, Odling and Newlands were almost certainly unaware of Meyer's book *Die modernen Theorien der Chemie* when they published their articles in 1864. Odling's article came out in October 1864[14]; Meyer had submitted his manuscript for *Modernen Theorien* to his publisher in June 1864 [101] and some time must have elapsed before it was printed. Further evidence comes from the German edition of Odling's *Manual*, which in 1864 was being translated by Alphons Oppenheim (1833–1877) [6]. On November 14, 1864, Oppenheim wrote in the foreword to the German edition [102]: "Lothar Meyer's text on modern theories … became known to me only after completing this adaptation." It seems reasonable to assume that Odling became aware of Meyer's text at the same time.

And then there is Alexandre-Émile Béguyer de Chancourtois, who published an arrangement of the elements that featured ordering by atomic weight and recurrence of similarities at approximately regular intervals in 1862 and 1863, earlier than either Newlands or Odling [103]. His arrangement, called the *vis tellurique*, was little known and certainly not useful to chemists of the time.[15] And while the same can be said of the arrangements of Newlands and Odling, theirs (unlike the *vis tellurique*) resembled in important ways the arrangements of Mendeleev and Meyer that other chemists **did** find useful.

Whether or not Newlands and Odling were acquainted with each others' classifications of elements before they published their own, they were connected at least from the 1870s and later when Newlands pressed his claims for priority. As already noted, Odling was the President of the Chemical Society in 1873 when Newlands asked the Society to publish a short note about his priority claim in its journal [89]. In 1877, in a lecture on gallium to the British Association, Odling stated,

[14]See Chapters 8 and 9 of this volume for more on Meyer and his *Modernen Theorien*.
[15]See Chapter 5 of this volume for more on Béguyer de Chancourtois and his *vis tellurique*.

"Mr. Newlands was the first chemist to arrange the elements in such a seriation that new ones might be predicted to exist where certain gaps are observed in the seriation of atomic weights" [104]. In this way, he acknowledged Newlands's priority, and on this occasion did not mention his own contemporary work in arranging elements. And Newlands in turn quoted this acknowledgment in his monograph about his work on the periodic law [105].

Neither Odling nor Newlands had an audience who developed their ideas, which were not enthusiastically received. Why did their systems not take hold? Gordin suggested that we take seriously the reason given in 1873 for the Chemical Society's not having published Newlands's paper on the law of octaves in 1866: that it was too theoretical. Gordin argued that Foster's question at Newlands's presentation was motivated by a judgment that Newlands's order numbers were too far removed from empirical chemical data [106]. Whereas a twenty-first-century chemist might look back at Newlands's use of order numbers instead of atomic weights as an insightful substitution of a regularly varying quantity for a "noisy" one, a nineteenth-century chemist might well find such a substitution unwarranted and Newlands's justification of it unconvincing. Odling's classifications were more closely tied to empirical data than the law of octaves, though, but they similarly failed to take root.

When chemical periodicity claimed the attention of chemists, Odling rarely mentioned his own work on the subject, while Newlands pressed for recognition. Today, neither Newlands nor Odling is a household name—or rather a laboratory name—among chemists. It is often a matter of controversy who should be regarded as the discoverer of a natural law, a person who publishes first and thus clearly has priority [107–109], or a person who later finds the same object independently, pronounces it more clearly, recognizes its meaning better and uses broad, effective language. Application and propagation provide recognition. For the periodic table, a case could be made that Odling and Newlands, among others, fall in the first category, whereas Mendeleev falls in the latter. But we should not forget the predecessors.

Acknowledgements We would like to thank Greg Girolami for helpful discussions leading to many of the insights presented in this chapter. We would also like to thank a reviewer for many helpful comments. VVM also thanks Ida-Marie Maeder and others at the Humboldt-Universität zu Berlin, University Library, for the help in obtaining the scan of the galley proof shown in Fig. 6.6, and David Allen, Librarian at the Royal Society of Chemistry for the scan of William Odling (Fig. 6.1).

References

1. Marsh JE (1921) William Odling. J Chem Soc Trans 119:553–564
2. Dixon HB (1922) William Odling 1829–1921. Proc Roy Soc 100A:i–vii
3. Webb KR (1957) Former Presidents of the Institute William Odling 1829–1921 Third President 1883–1888. J Roy Inst Chem 81:728–733

4. Russell C (2006, May 25) Odling William (1829–1921), chemist. In: Oxford dictionary of national biography. https://www.oxforddnb.com/view/10.1093/ref:odnb/9780198614128.001.0001/odnb-9780198614128-e-37818. Accessed 18 June 2020
5. Brock WH (2008) Odling William. In: Complete dictionary of scientific biography, vol 10. Charles Scribner's Sons, Detroit, MI, pp 177–179
6. Cassebaum H (1969) 100 Jahre Periodensystem der Elemente: Die bedeutenden Beiträge von William Odling zur Entdeckung des Periodensystems. Chemiker-Ztg 93:929–935
7. Thornton JL, Wiles A (1956) William Odling, 1829–1921. Ann Sci 12:288–295
8. Thomson A (1960) History and development of teaching hospitals in England. Brit Med J 2:749–751
9. Agha R, Agha M (2011) A history of Guy's, King's and St. Thomas hospitals from 1649 to 2009: 360 Years of innovation in science and surgery. Internat J Surgery 9:414–427
10. Odling W (1851) On the action of poisons, and on the nature and treatment of arsenical poisoning [Read at the Guy's Hospital Physical Society, 22 March 1851]. Lond Med Gaz 12:716–722; 756–760
11. Odling W, Marsh JE (1887) On some xenoene or diphenyl products and reactions. In: Report of the Fifty-seventh meeting of the British Association for the Advancement of Science held at Manchester in August and September 1887. Spottiswood & Co, London, p 646
12. Odling W (1854–1858) On the constitution of hydro-carbons. Roy Instit Proc 2:63–66
13. Odling W (1859) On some failures of Marsh's process for the detection of arsenic. Guy's Hosp Rep 5:367–374
14. Odling W (1856) On the alkaline emanations from sewers and cess-pools. Guy's Hosp Rep 2:245–248
15. Odling W (1860) An account of Guy's Hospital well. Guy's Hosp Rep 6:433–451
16. Odling W (1885) Plea for the empiric naming of organic compounds. Chem News 52:181–182; 203–205; 216–218
17. Odling W (1859) On ortho- and meta-silicates. Phil Mag 18:368–374
18. Odling W (1864) On the hexatomicity of ferricum and aluminium. Phil Mag 27:115–119
19. Odling W (1855) On the constitution of acids and salts. Quart J Chem Soc 7:1–21
20. Odling W (1858) V. Remarks on the doctrine of equivalents. Phil Mag [4th ser] 16:37–45
21. Levere TH (2001) Transforming matter: a history of chemistry from alchemy to the buckyball. Johns Hopkins University Press, London and Baltimore, p 87
22. Wurtz CA (1984) Account of the sessions of the international congress of chemists in Karlsruhe, on 3, 4, and 5 September 1860. Greenberg J, Clark W transl. In: Nye MJ (ed) The question of the atom. Tomash, Los Angeles, pp 5–28
23. Rocke AJ (1984) Chemical atomism in the nineteenth century. From Dalton to Cannizzaro. Ohio State University Press, Columbus, pp 12, 80–81
24. Odling W (1854) A course of practical chemistry, arranged for the use of medical students. Highley, London
25. Laurent A (1854) Méthode de chimie. Mallet-Bachelier, Paris
26. Laurent A (1855) Chemical method. Odling W, transl. Cavendish Society, London
27. Odling W (1861) A manual of chemistry, descriptive and theoretical. Part I. Longmans, London, p 94
28. Halton B (2019) Periodicity and approaches to periodic classification. Chem N Z 83:12–19
29. Girolami G (2020) Personal communication
30. Odling W (1916) The technic of versification: notes and illustrations. Parker, Oxford and Simpkin Marshall, London
31. Gerhardt CF (1853–1856) Traité de chimie organique, 4 vols. Chez Firmin Didot Frères, Paris
32. Rocke AJ (1984) Chemical atomism in the nineteenth century. From Dalton to Cannizzaro. Ohio State University Press, Columbus, pp 191–214
33. Rocke AJ (2010) Image and reality. Kekulé, Kopp, and the scientific imagination. University of Chicago Press, Chicago, pp 53–54

34. Rocke AJ (1981) Kekulé, Butlerov, and the historiography of the theory of chemical structure. J. Hist. Sci. 14:27–57
35. Odling W (1857) LIX. On the natural groupings of the elements. Part 1. Phil Mag 13:423–439
36. Odling W (1857) LXIV. On the natural groupings of the elements. Part 2. Phil Mag 13:480–497
37. Gladstone JH (1853) L. On the relations between the atomic weights of analogous elements. Phil Mag 5:313–320
38. Wurzer F (1817) Auszug eines Briefes vom Hofrath Wurzer, Prof. der Chemie zu Marburg. Ann Phys 56:331–334
39. Döbereiner JW (1829) Versuch zu einer Gruppirung der elementaren Stoffe nach ihrer Analogie. Ann Physik (Pogg) 15:301–307
40. van Spronsen JW (1969) The periodic system of chemical elements: A history of the first hundred years. Elsevier, Amsterdam, pp 63–68
41. Girolami G (2019) A book collector's view of the periodic table: before Mendeleev. Substantia (Suppl 5) 3:109–124
42. Odling W (1863) Atomic weights. In: Watts H (ed) A dictionary of chemistry and the allied branches of other sciences, vol 1. Spotiswoode and Co, London, pp 452–473
43. Odling W (1864) On the proportional numbers of the elements. Q J Sci 1(October):642–648
44. Lenssen E (1857) Ueber die Gruppirung der Elemente nach ihren chemisch-physikalischen Character. Ann Chem 103:121–131
45. Scerri E (2007) The periodic table: its story and its significance. Oxford University Press, New York, pp 82–85
46. van Spronsen JW (1969) The periodic system of chemical elements: A history of the first hundred years. Elsevier, Amsterdam, p 113
47. Newlands JAR (1864) Relations between equivalents. Chem News 10:59–60
48. Meyer L (1864) Die modernen Theorien der Chemie. Breslau, Germany
49. Odling W (1865) Metals, atomic weights and classification of. In: Watts H (ed) A dictionary of chemistry and the allied branches of other sciences, vol 3. Spotiswoode and Co., London, pp 957–976
50. Odling W (1865) On aluminium ethide and methide. Proc R Inst G B 4:343–346
51. Galley pages from [49], filed with the Manual, from Humboldt-Universität zu Berlin Universitätsbibliothek
52. Druzhinin PA (2020) The first publication of Mendeleev's periodic system of elements: a new chronology. Historical Studies Nat Sci 50:129–182
53. Strecker A (1859) Theorien und Experimente zur Bestimmung der Atomgewichte der Elemente. Friedrich Vieweg und Sohn, Braunschweig
54. Odling W (1865) A course of practical chemistry, arranged for the use of medical students, 2nd edn. Longmans, London, p 226
55. Odling W (1869) Cours de chimie pratique: analytique, toxicologique, animale à l'usage des étudiants en médecine. Naquet A, transl. (from the 3rd English Ed), F. Savy, Paris, France
56. Mendeleev DI (1869) Osnovy khimii [Principles of chemistry], vol. 1. Tovarishchestvo 'Obshchestvennaia Pol'za', St. Petersburg, Russia
57. Odling W (1867) Kurs prakticheskoi khimii. Savshchenkov TN, transl. (from the 2nd English Ed), St. Petersburg, Russia
58. Gerstl R (1871) Correspondenzen. Chem Ber 4:128–132
59. Mendelejeff [Mendeleev] D (1871) Zur Fräge über das System der Elemente. Chem Ber 4:348–352
60. van Spronsen JW (1969) The periodic system of chemical elements: a history of the first hundred years. Elsevier, Amsterdam, pp 349–350
61. Mendeleev D (2002) On the question concerning the system of elements (Jensen WB, transl.) In: Mendeleev on the periodic law selected writings, 1869–1905. Dover Publications, Mineola, NY, pp 115–119 (original work published in 1871)
62. Odling W (1872) On the last new metal, indium. Proc R Inst G B 6:386–401

63. Tilden WA (1921) Prof. W. Odling, F.R.S. Nature 107:19–20
64. Tilden WA (1898) John A. R. Newlands. Nature 58:395–396
65. Wood family tree, Family of John Alexander Reina NEWLANDS and Emily Elizabeth Jane RICHINGS. https://woodlloydfamilyhistory.com/fam575.html. Accessed 6 July 2020
66. (1898) John A. R. Newlands: the discoverer of the periodic law. Chem Drug 52:175–176
67. Scerri E (2007) The periodic table: its story and its significance. Oxford University Press, New York, pp 72–82
68. (1860) Garibaldian volunteers. Illustrated London News 37:371
69. Newlands J (1866) Percy Villiers, and other poems. Bennett, London
70. Lock CGW, Newlands BER, Newlands JAR (1888) Sugar: a handbook for planters and refiners. Spon, London
71. (1898) Proceedings of societies: Chemical society. Chem News 77:78–81
72. JARN [Newlands JAR] (1863) On relations among the equivalents. Chem News 7:70–72
73. Dumas J (1859) Mémoire sur les équivalents des corps simples. Ann Chim Phys 55:129–210
74. Studiosus (1864) Numerical relations of equivalent numbers. Chem News 10:11
75. (1886) Proceedings of societies: Chemical society. Chem News 53:257–261
76. Studiosus (1864) Numerical relations of equivalents. Chem News 10:95
77. Newlands JAR (1864) On relations among the equivalents. Chem News 10:94–95
78. Newlands JAR (1864) Equivalent of indium. Chem News 10:95
79. Noble J (1864) Numerical relations of equivalent numbers. Chem News 10:120
80. Inquirer (1864) Numerical relations of equivalents. Chem News 10:156
81. Newlands JAR (1865) On the law of octaves. Chem News 12:83
82. Newlands JAR (1865) On the cause of numerical relations among the equivalents. Chem News 12:94–95
83. (1866) Proceedings of societies: Chemical society. Chem News 13:113–114
84. (1864) Proceedings of societies: Chemical society. Chem News 10:302–305
85. Cameron JA (1948) J. A. R. Newlands (1837–1898): a pioneer whom the chemists ridiculed. Chem Age 59:354–356
86. Newlands JAR (1866) Proceedings of societies: Chemical society. Chem News 13:130
87. Brush SG (1996) The reception of Mendeleev's periodic law in America and Britain. Isis 87:595–628
88. Newlands JAR (1875) On the relations among the atomic weights of the elements when arranged in their natural order. Chem News 32:21–22
89. (1873) Proceedings of societies: Chemical society. Chem News 27:316–318
90. Newlands JAR (1872) Relations between the atomic weights of Cannizzaro. Chem News 25:252
91. ML [Lichtenstein M] (1876) Remarks suggested by the discovery of gallium, by D. Mendelejeff (translated abstract). J Chem Soc 520–521
92. Newlands JAR (1884) On the discovery of the periodic law and on relations among the atomic weights. Spon, London
93. Meyer L (1880) On the history of atomistic periodicity (translated extract). Chem News 41:203
94. Mendelejeff [Mendeleev] D (1881) On the history of the periodic law (translated extract). Chem. News 43:15
95. Lubbock J (1881) British Association for the Advancement of Science, York meeting, August 31, 1881, inaugural address of the president. Chem. News 44:109–121
96. (1883) Review of a manual of chemistry, Vol. I, Physical and inorganic chemistry, by H. Watts. Chem News. 48:269–270
97. (1884) Review of On the discovery of the periodic law and on relations among the atomic weights, by John A. R. Newlands. Chem News. 49:198–200
98. Gordin MD (2012) The textbook case of a priority dispute: DI Mendeleev, Lothar Meyer, and the periodic system. In: Biagioli M, Riskin J (eds) Nature engaged: science in practice from the Renaissance to the present. Palgrave Macmillan, New York, pp 59–82

99. van Spronsen JW (1969) The periodic system of chemical elements: A history of the first hundred years. Elsevier, Amsterdam, pp 102–111
100. Scerri E (2015) The discovery of the periodic table as a case of simultaneous discovery. Phil Trans R Soc A 373:20140172. https://doi.org/10.1098/rsta.2014.0172
101. Rocke AJ (2019) Lothar Meyer's pathway to periodicity. Ambix 66:265–302
102. Odling W (1865) Beschreibendes und theoretisches Handbuch der Chemie, Vol. 1. Oppenheim A, transl. Enke, Erlangen, Germany
103. Béguyer de Chancourtois AE (1863) Vis tellurique. Classement naturel des corps simples ou radicaux obtenu au moyen d'un système de classification hélicoïdal et numérique. Mallet-Bachelier, Paris
104. (1877) The month. Pharm J (3rd series) 8:141–146
105. Newlands JAR (1884) On the discovery of the periodic law and on relations among the atomic weights. Spon, London, p 37
106. Gordin MD (2018) Paper tools and periodic tables: Newlands and Mendeleev draw grids. Ambix 65(1):30–51
107. van Spronsen JW (1969) The periodic system of chemical elements: A history of the first hundred years. Elsevier, Amsterdam
108. Cassebaum H (1969) Hundert Jahre Periodensystem der Elemente. Z Chem 9:81–87
109. Cassebaum H, Kauffman GB (1971) The periodic system of the chemical elements: the search for its discoverer. Isis 62:314–327

Gustavus Hinrichs and His Charts of the Elements

7

Gregory S. Girolami

Abstract

In this paper, I analyze the efforts of the German-American chemist Gustavus Detlef Hinrichs (1836–1923) to construct a periodic system between 1867 and 1869. Included is a transcription and translation into English of major sections of his *Programme der Atomechanik* (1867), and a discussion of Hinrichs's "pantatom" theory of matter. My principal conclusions are: (1) Hinrichs's chart of 1867 is actually a double spiral that begins in a clockwise fashion but then reverses direction and continues in a counterclockwise direction, (2) the nitrogen and oxygen groups are swapped because Hinrichs felt that that order resulted in more consistent trends in the stoichiometries of the highest oxides, (3) in his chart the trigonoids and tetragonoids each subtend one-third of a circle, and the spokes are arranged so that the maximal valences of the elements increase from right to left, (4) Hinrichs devised an ingenious theory to account for isomorphism, (5) the transition elements in Hinrichs's 1869 table are listed in reverse order for the same reason that the spiral in his 1867 chart reverses direction, (6) the transition elements in the 1869 system are arranged in a slanted fashion to reflect their relative atomic weights, whereas other elements are not arranged in this way, possibly owing to a printer's omission, (7) Hinrichs was the first to point out that one advantage of the "long" form periodic tables is that the metals and non-metals can be separated by a single line, and (8) simultaneously with Meyer and Mendeleev, Hinrichs also pointed out the periodic relationship of atomic volume to atomic weight, but only in his oral presentation to the AAAS meeting of August 1869.

G. S. Girolami (✉)
School of Chemical Sciences, University of Illinois at Urbana-Champaign, Urbana, IL, USA
e-mail: ggirolam@illinois.edu

C. J. Giunta et al. (eds.), *150 Years of the Periodic Table*, Perspectives on the History of Chemistry, https://doi.org/10.1007/978-3-030-67910-1_7

7.1 Introduction

Of the pre-Mendeleev attempts to construct a periodic system, by far the most puzzling and least understood are those of the German-American chemist and polymath Gustavus Hinrichs. Hinrichs's first system, published in 1867 [1, 2], is summarized in a two-dimensional graph in which related elements (such as the halogens) are arranged on spokes radiating from a central point, elements with larger atomic weights being located farther from the center. In 1869, Hinrichs published two new charts of his system, in which the elements are arranged in tables rather than a graph [3–5].

Over the years [6–10], scholars have discussed Hinrichs's periodic systems[1] [11] and compared his achievements with those of others who proposed periodic systems in the 1860s, such as Alexandre-Émile Béguyer de Chancourtois (1820–1886), William Odling (1829–1921), John A. R. Newlands (1837–1898), Lothar Meyer (1830–1895), and of course Dmitri Mendeleev (1834–1907).[2] But many aspects of Hinrichs's periodic systems have remained puzzling even today.

In this paper, after a short biography of Hinrichs, I will offer some new insights into why Hinrichs constructed his periodic systems the way he did. Specifically, I will address the following questions:

- Are the dotted arcs in the 1867 chart circular or spiral?
- Why are the nitrogen and oxygen groups in the 1867 chart out of order?
- Why are the radial spokes in the 1867 chart located where they are?
- Why are the transition elements in the 1869 chart listed in reverse order and arranged in slanted columns?
- What are Hinrichs's ideas about how the periodic table gives insights into isomorphism, the relation between metals and non-metals, and atomic volumes?

7.2 Short Biography of Gustavus Detlef Hinrichs

Several articles [6, 8, 12], books [13, 14], and a thesis [15] give much information about Hinrichs's life and accomplishments; a list of his publications has also been compiled [16]. Many of Hinrichs's original publications, and documents about him, can be found today at the University of Iowa [17]. In addition, Hinrichs's personal papers are located at the University of Illinois, having been deposited there by one

[1] I will refer to Hinrichs's classification schemes as "periodic systems" because the elements are arranged (mostly) according to increasing atomic weight, and elements with analogous properties are arranged in the same group. I use the word "mostly" in this definition because even modern periodic systems reverse the order of Ni and Co, and Te and I. For discussions of the definition of the term "periodic system" see [10] and [11].

[2] Other chapters in this volume contain discussions of the classification schemes of Béguyer de Chancourtois (5), Odling & Newlands (6), Meyer (8 and 9), and Dmitri Mendeleev (1).

of his grandsons between 1959 and 1964 [18]. Here I will briefly summarize some of the details available in these sources.

Gustavus Detlef Hinrichs (Figs. 7.1 and 7.2) was born on 2 December 1836 in the town of Lunden in the Holstein (i.e., southern) portion of the Jutland peninsula. Lunden was then part of Denmark but today is in Germany, about 50 km south of the Denmark-Germany border. He was the third of six sons of Johann Detlef Hinrichs (b. ca. 1802), a musician; Hinrichs's mother, Carolina Cathrina Elisabeth von Andersen (b. 4 October 1809), was the daughter of an artillery captain. In 1850, at the age of 13, Gustavus ran away from home to participate in the Schleswig-Holstein War, the unsuccessful first rebellion of ethnic Germans to achieve the secession of Holstein (and the adjacent state of Schleswig) from Denmark to the German Confederation. In July of that year, he took part in the battle at Idstedt as a uniformed drummer boy. He returned to Lunden in 1853 after hostilities ended, and shortly thereafter he enrolled in the Polytechnic School of the University of Copenhagen, where he completed the regular course of studies in 1856. He continued at the University for advanced work in mathematics, physics, and chemistry.

While at the University, Hinrichs earned money as a private instructor of students. In 1856 he wrote his first book, *Die electromagnetische Telegraphie*, and in 1860 he passed the exam at the University of Copenhagen for the *Candidatus mathematicus* degree,[3] equivalent to a master's degree. At Copenhagen, he had been particularly influenced by the Danish biologist Daniel Frederik Eschricht (1798–1863) and the meteorologist and geologist Johan Georg Forchhammer (1795–1865). In April 1861 he married Auguste Margaretha Friederike Springer (1839–1865), and in May–July 1861 he immigrated to the United States with his new wife, most likely to avoid service in the Danish military.

In 1861 Hinrichs settled in Davenport, Iowa,[4] where initially he taught high school. In 1862 he was appointed Professor of Modern Languages at the University of Iowa in Iowa City (he was fluent in Danish, French, German, Italian, and English, and knew some Greek and Latin), and in the next year he was appointed Professor of Natural Philosophy and Chemistry at that same institution, giving up his former title. Hinrichs's first wife died in 1865, leaving two children, and in July 1867 Hinrichs married Anna Catharina Springer (1842–1910; Auguste's younger sister) in Iowa City; presumably, Anna had come to America to care for Gustavus's children. With Anna, Hinrichs had two more children.

In 1875 Hinrichs founded the Iowa Weather and Crop Service [19, 20], and in 1886 he was dismissed from the University of Iowa (on trumped-up charges) [13, 15, 21]. In 1889 he was appointed Professor of Chemistry at St. Louis University, and he retired in 1907. He died 14 February 1923 in St. Louis (age 86).

[3]In several papers Hinrichs published in 1860, this degree follows his name. Records at the University suggest that this certification may have been later invalidated [14].

[4]The Schleswig-Holstein Wars stimulated a large number of people to immigrate to Iowa. Gustavus's older brother, Carl Peter Hinrichs (1831–1894), a laborer, immigrated in 1868 with his wife Marie (1831–1916) to Clinton, Iowa, about 100 km from Iowa City.

Fig. 7.1 Gustavus Hinrichs in his middle years. *Left* photo courtesy of University of Iowa. *Right* photo from the *Souvenir and Annual*, 1881–1882

Fig. 7.2 Gustavus Hinrichs in his later years. Left: photo courtesy of University of Iowa. Right: photo from The Palimpsest, 1930

7.3 Hinrichs and Atomic Weights, 1866

Hinrichs published his first ideas about atomic weights in an 1866 article [22] in the *American Journal of Science* entitled, "On the Spectra and Composition of the Elements." Although much of this paper relates to finding regularities in the spectra of the elements (which has been discussed elsewhere [8]), I will instead focus on his ideas about the structure of atoms.

In this paper, Hinrichs states: "We suppose all elementary atoms to be built up of the atoms of *one single matter,* the *urstoff....*" Hinrichs proposes that the atomic weight of hydrogen, referred to this prime element, is 4, but for the rest of the article he gives atomic weights relative to H = 1. He continues,

> the laws of mechanics force them [i.e., the particles of the urstoff] to arrange themselves regularly—and the most stable form will be the prism. If quite rectangular, and *a, b, c* be the number of primary atoms, in the three directions, we shall have [where A = atomic weight] ...
>
> $A = a{\cdot}b{\cdot}c.$
>
> If the atom has a quadratic base, $a = b$, we have
>
> $A = a^2{\cdot}c.$
>
> If provided with one or several pyramidal additions, we have
>
> $A = a{\cdot}b{\cdot}c + k.$

Thus, Hinrichs clearly believed Prout's hypothesis [23–27] that all atomic weights are integer multiples of that of hydrogen (or a fraction thereof). As we will discuss below, Hinrichs makes no mention of the 1860 or 1865 publications of Jean-Servais Stas [28, 29] (1813–1891) discrediting Prout's hypothesis, or of the 1858 publications of Stanislao Cannizzaro [30, 31] (1826–1910) on atomic weights. Hinrichs mentions that he was using the atomic weights given in 1863 by Heinrich Will [32] (1812–1890), which for non-metals, the alkali metals, and the coinage metals mostly resembled the modern values, but for other elements were mostly one-half the modern values.

Examples of how Hinrichs tried to apply a common formula for the atomic weights of elements within groups are shown in Fig. 7.3. In attempting to fit the atomic weights of elements in a group to a common formula, Hinrichs was following efforts made in 1853 by the English chemist John H. Gladstone [33] (1827–1902), in 1854 by the American chemist Josiah Parsons Cooke [34][5] (1827–1894), in 1851 and 1858 by the French chemist Jean Baptiste André Dumas [35, 36] (1800–1884), and in 1860 by the American chemist Mathew Carey Lea [37] (1823–1897). All tried to fit the weights to formulas of the type $a + md$ (or to more complicated polynomial formulas), where a and d were numbers that were invariant within a group, and m was an integer that differed from element to element in that group. Only Hinrichs, however, proposed that the polynomial formulas reflected specific geometric (i.e., prismatic) arrangements of the basic building blocks.

[5]See Chap. 1 in this volume for more on Cooke's work.

Oxygen group; quadratic. Formula $A = n.4^2$.

	n	A Calc.	Obs.	Error.
Oxygen,	1	$1{\cdot}4^2 = 16$	16	0·0
Sulphur,	2	$2{\cdot}4^2 = 32$	32	0·0
Selenium,	5	$5{\cdot}4^2 = 80$	80	0·0
Tellurium,	8	$8{\cdot}4^2 = 128$	128	0·0

Chlorine group; quadratic. Formula $A = n.3^2 \pm 1$.

	n	A Calc.	Obs.	Error.
Fluorine,	2	$2{\cdot}3^2 + 1 = 19$	19	0·0
Chlorine,	4	$4{\cdot}3^2 - 1 = 35$	35·5	+·5
Bromine,	9	$9{\cdot}3^2 - 1 = 80$	80	0·0
Iodine,	14	$14{\cdot}3^2 + 1 = 127$	127	0·0

Alkaline group; quadratic with pyramid. $A = 7 + n.4^2$.

	n	A Calc.	Obs.	Error.
Lithium,	0	7	7	0·0
Sodium,	1	$7 + 1.4^2 = 23$	23	0·0
Potassium,	2	$7 + 2.4^2 = 39$	39	0·0
Rubidium,	5	$7 + 5.4^2 = 87$	85·4	−1·6
Cæsium,	8	$7 + 8.4^2 = 135$	133·0	−2·0

Alkaline-earths group; quadratic. $A = n.2^2$.

	n	A Calc.	Obs.	Error.
Magnesium,	3	$3{\cdot}2^2 = 12$	12	0·0
Calcium,	5	$5{\cdot}2^2 = 20$	20	0·0
Strontium,	11	$11{\cdot}2^2 = 44$	43·8	−·2
Barium,	17	$17{\cdot}2^2 = 68$	68·5	+·5

Fig. 7.3 Table from Hinrichs's 1866 attempt to find numerical regularities in atomic weights [22]

Hinrichs mentions two of these predecessors in this 1866 paper [22]:

> We cannot here go into any detail as to the relation of these formulae to the numerical relations discovered by Carey Lea, Dumas and others; we hope soon to be enabled to publish our labors on the constitution of the elements. Neither can we here discuss these formulae in the sense of the mechanics of atoms, deducing the physical and chemical properties of the elements from these formula; these interesting relations also we must delay till some future, but I hope not a very distant, time.

Among Hinrichs's handwritten papers at the University of Illinois are two pages summarizing the polynomial formulas of Cooke and Dumas. We do not know whether Hinrichs had seen any of the classification schemes constructed between 1862 and 1864 by Béguyer de Chancourtois [38], Odling [39], Newlands [40], or Meyer [41].

7.4 Hinrichs and Atomechanics, 1867

One year later, in 1867, Hinrichs published an expanded version of his ideas about the inner structure of atoms, which he had briefly discussed in his 1866 paper. These new ideas appear in a privately lithographed reproduction of a 44 page handwritten treatise entitled *Programme der Atomechanik oder die Chemie eine Mechanik der Panatome* (Fig. 7.4; called *Programme* from here on). It is written entirely in German, except that copies not intended for Germany also include an abstract in French on pages 45–48. At the same time Hinrichs published a four-page English abstract of his *Programme* in the American periodical *Journal of Mining* [2]. The English abstract is not a straight translation of the French abstract.

A total of 112 copies of *Programme* were printed [42]. Hinrichs sent most of these to societies and universities, with only a few going to individuals. Among the latter were the Irish physicist John Tyndall (1820–1893), the German physicist and editor Johann Christian Poggendorff (1796–1877), the German chemists August Hofmann (1818–1892), Heinrich Will (1812–1890), Justus von Liebig (1803–1873), and Carl Remigius Fresenius (1818–1897), the German dictionary editor Felix Flügel (1820–1904), the London publisher of the *Mining Journal* Edward David Hearn (1832–1909), and the biologist Charles Darwin (1809–1882). He also sent copies to several geologists and mineralogists in Austria, Germany, and Russia: Hans Bruno Geinitz (1814–1900), Wilhelm Haidinger (1795–1871), Karl

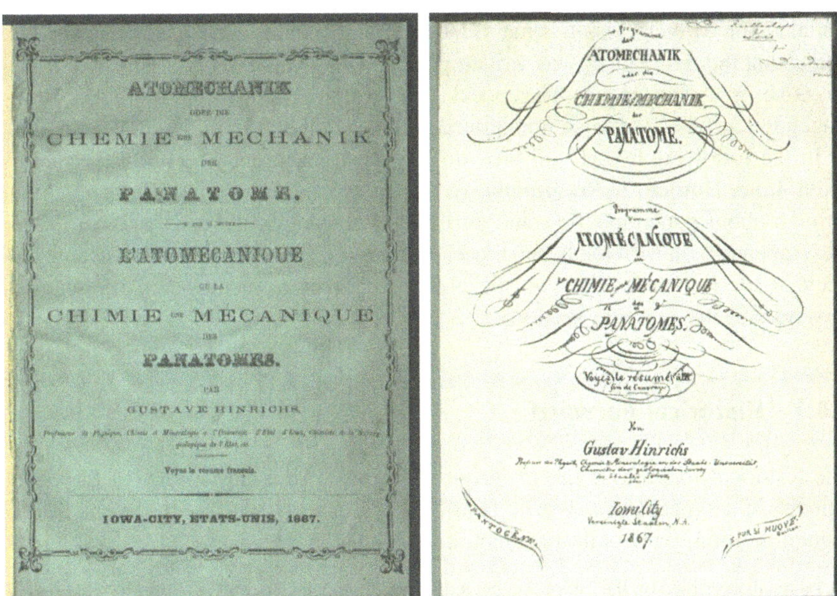

Fig. 7.4 Cover and title page of G. D. Hinrichs, *Programme der Atomechanik, oder die Chemie eine Mechanik der Panatome*, Iowa City (1867). Images courtesy of the University of Dresden

Friedrich August Rammelsberg (1813–1899), Carl Friedrich Naumann (1797–1873), Albrecht Schrauf (1837–1897), Aristides Brezina (1848–1909), and Nikolai Koksharov (1818–1893). In all, he sent 37 copies to Germany, 11 to the United States, 10 to France, 10 to England, 8 to Russia, 6 to Austria-Hungary, 6 to Scandinavia, 4 to Switzerland, 3 to Holland, 3 to Italy, 2 to Belgium, and 1 each to Greece, Portugal, and Spain [2]. His personal copy resides among the Hinrichs papers at the University of Illinois along with his hand-annotated list of recipients.

The lithographed text of the *Programme* is handwritten in English (or Latin) cursive, rather than the now-obsolete Kurrent script that was commonly used by German writers in the nineteenth century. Hinrichs's handwriting is relatively neat and mostly legible, although letters and words are sometimes sufficiently ill-formed that transcribing them involves guesswork (mit/mir/wir/wie are particularly vexing). In addition to the usual problems associated with distinguishing similarly-fashioned letters, long words at the ends of lines are often compacted and slanted downward to avoid the margin, and are frequently difficult to read as a result.

Hinrichs's *Programme* begins with a three-page historical forward (which states that the document was written between November 1866 and June 1867) and a short vocabulary list with definitions. The forward is followed by the main text of the *Programme*, which Hinrichs organizes into 400 numbered paragraphs. The main text is divided into an Introduction (paragraphs 1–5) and four main sections: Pantogen and the Elements (paragraphs 6–56), Chemical Characteristics of the Elements (paragraphs 57–120), Physical Characteristics (paragraphs 121–228), and Morphological Characteristics or Crystal Forms (paragraphs 229–399). The 400th paragraph contains some brief closing remarks. The text concludes with a colophon, which attests that the monograph was written personally by Gustavus Hinrichs and printed on stone by Augustus von Hageboeck, Lithographer in Davenport, Iowa. August Hageboeck (1836–1907) had immigrated to the United States from Germany in 1857.

In the following text, I will provide English translations of some key excerpts taken from Hinrichs's *Programme* (a transcription of original German text up through about paragraph 100, and an English translation thereof, can be found in the Appendix). I will focus on those portions of the *Programme* that are most relevant to Hinrichs's ideas on the classification of elements: i.e., the historical forward and the first two sections.

7.4.1 Historical Forward

The historical forward of the *Programme* begins with a chronology of the development of atomechanics, which Hinrichs says started with a document he had written in 1855, while still a student at Copenhagen. He states:

> I officially certified on the 7th of August a document in characters ("Zeichenschrift") entitled
>
> 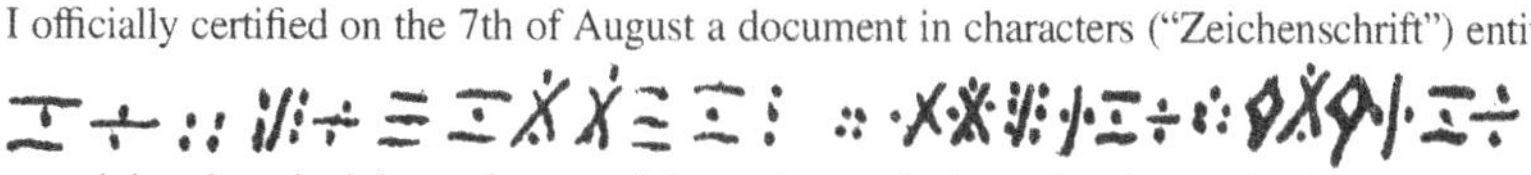
>
> containing the principles and some of the main conclusions already obtained. This document is also in my possession.

A 1969 article about Hinrichs [6] depicts this sentence in a figure bearing the caption: "The puzzling hieroglyphics in the [historical forward] of *Atomechanik*, referring to some secret 'Zeichenschrift' known only to the author."

The shortness of the encoded text makes the cryptanalysis more difficult, but after a few blind alleys I solved the puzzle of the "hieroglyphics:" they are a simple substitution cipher in which the plaintext reads HINRICHSSCHE NATURPHILOSOPHI (*Hinrichs's Nature Philosophy*), the final "E" having been omitted.

There is no document with this title in the archive of Hinrichs's personal papers at the University of Illinois, but there is a 13-page handwritten document entitled *Atom-Mechanik*. On the cover of this document is an annotation by Hinrichs written in 1920: "Old document on Atom-Mechanics. June 15, 1868. The main part, in my own shorthand, is not dated, so far as I can see." Most of the document is unreadable (at least to me): the words are written in a shorthand that resembles one for German writing introduced by Franz Xaver Gabelsberger (1789–1849) in 1834. If Hinrichs's dating from 50+ years after the fact is wrong, then this document may be the one in "Zeichenschrift" that the *Programme* says was written in 1855.

This 13-page manuscript, which mentions Pettenkofer, Dumas, and Prout (all of whom had written about atomic weights before 1855), contains some equations, such as $O = 32 = 2{\cdot}4^2 = 2{\cdot}\square_2$, and $Mg = 48 = 3{\cdot}4^2 = 3{\cdot}\square_2$, that suggest that Hinrichs is trying to rationalize (doubled) atomic weights on the basis of points arranged into geometric arrays such as squares. Loosely inserted into the document is a page showing a triangular grid, and calculations of the number of points in regular hexagons based on this grid.

Also in the Hinrichs archive at the University of Illinois (Box 3) are two longer handwritten documents, one entitled *Vorläufige Entwurf der Atomechanik* (*Preliminary Outline of Atomechanics*) that is undated but must have been written after April 1856 (the date of one of the articles Hinrichs cites), and the other entitled *Entwurf der Atomechanik* that is dated August and September 1858. The latter consists of 16 chapters, totaling 74 pages plus index.

Neither document contains any kind of periodic system. The earlier document does contain an inserted sheet, evidently added after the main text was written, summarizing an 1857 article [43] on the atomic weights of the elements by the German analytical chemist Heinrich Rose (1795–1864). Another inserted sheet, also bearing a reference to an article from 1857, gives atomic weights for 13 elements, all given on an $H = 2$ basis; these doubled atomic weights are based on a unit equal to half of a hydrogen atom.

The 1858 document contains an attempt to devise formulas for the (doubled) atomic weights of the elements, similar to the analysis he published eight years later in 1866. The logic behind the formulas, however, is far from clear: hydrogen $= 2 = 1^2{\cdot}2$, fluorine $= 38 = 3^2{\cdot}4 + 2$, chlorine $= 71 = 2^2 + 3^2{\cdot}7 + 2^2$, bromine $= 160 = 4^2{\cdot}10$, iodine $= 254 = 5^2{\cdot}10 + 2^2$, lithium $= 14 = 2^2{\cdot}3 + 2$, sodium $= 46 = 3^2{\cdot}5 + 1$; and potassium $= 78 = 3^2{\cdot}7 + 5$ [*sic*].

The earlier version of the *Entwurf* may be the one referred to in the next section of the *Programme*:

> 1856. In October of this year, I thought I was sufficiently advanced in my work to rewrite the publication. I ended up sending shorter reports to many of the existing men of science and also to some academies, expecting this to open up closer scientific communication. I had miscalculated in that. Passive resistance and the lack of help were to persuade me to further develop atomechanics.

The *Programme* then lists the scientific leaders Hinrichs had contacted in order to inform them of his ideas and to enlist their help in arranging for publication. The historical forward also describes the responses he received. Among those he contacted were Forchhammer (his former teacher), John Tyndall, J. C. Poggendorff (both mentioned above), the French editor Eugène d'Arnoult (fl. 1830s–1873; founder of *l'Institut, Journal Universel des Sciences*), the German chemist and physicist August Karl Krönig (1822–1879; known for his kinetic theory of gases, and then secretary of the Physical Society in Berlin), the Austrian mineralogist Wilhelm Haidinger (mentioned above; member of the Imperial Academy of Sciences in Vienna), the German physician and physicist Emil du Bois-Reymond (1818–1896), the Austrian chemist and mineralogist Anton Schrötter (1802–1875), the Danish-born German astronomer Peter Andreas Hansen (1795–1874), and the Prussian naturalist Alexander von Humboldt (1769–1859).

Although Hinrichs writes that he was grateful for the (mostly meager) responses he received, he accuses Krönig of plagiarizing some of his ideas in a monograph *Neues Verfahren zur Ableitung der Formel einer Verbindung aus den Gewichtmengen der Bestandtheile* (*New procedure for deriving the formula of a compound from the weights of the constituent parts*) [44]. On p. 53 of Krönig's monograph, as part of an eight-page section entitled "On the atomic weights of the elements," Krönig states: "As for the atomic weights I have assumed, I have used only Berzelius's numbers, taking into account the necessary corrections, based on half an atom of hydrogen as a unit.... So I set H = 2, O = 32, Cl = 71, and for example Ca = 80. Thus, all atomic weights appear as whole numbers." Krönig later refers to "the primordial particles, from which I think all bodies are composed, and of which 2 should form an atom of hydrogen, 32 an atom of oxygen, 71 an atom of chlorine, 80 an atom of calcium."[6]

In the historical forward to the *Programme*, Hinrichs implies that this "half-hydrogen" basis for atomic weights had been included in the reports that Hinrichs circulated privately beginning in late 1856. Hinrichs further quotes from a

[6]"Was nun die von mir angenommenen Atomgewichte betrifft, so habe ich unter Berücksichtigung der nothwendigen Correktionen nur die Berzelius'schen Zahlen auf ein halbes Atom Wasserstoff als Einheit bezogen. Ich setze also H = 2, O = 32, Cl = 71, und noch beispielsweise Ca = 80. So erscheinen denn sämmtliche Atomgewichte als ganze Zahlen.... die Urtheilchen, aus denen ich mir alle Körper zusammengesetzt denke, und von denen 2 ein Atom Wasserstoff, 32 ein Atom Sauerstoff, 71 ein Atom Chlor, 80 ein Atom Calcium bilden sollen....".

letter he received back from Krönig dated April 15, 1857, in which Krönig said that he had read Hinrichs's report carefully and offered to circulate it to members of the Physical Society in Berlin who were part of Krönig's reading circle.

Even if we accept Hinrichs's implication that his circulated report contained his half-hydrogen hypothesis, it is perhaps forgivable if by 1866 Krönig had forgotten that he had first encountered the idea in a still-unpublished report sent to him nine years earlier by an obscure correspondent. Even more to the point, however, the idea that atomic weights are multiples of some fraction of a hydrogen atom had been proposed before 1856. The first to do so was Prout himself, in a letter sent on September 12, 1831 to the English chemist Charles Daubeny (1795–1867), and printed by Daubeny as an appendix in his textbook of that same year, *Introduction to the Atomic Theory* [45]. In his letter, Prout states that: "there seems to be no reason why bodies still lower in the scale than hydrogen ... may not exist, of which other bodies may be multiples, without being actually multiples of the intermediate hydrogen."

In 1843, the Swiss chemist Jean Charles Marignac (1817–1894) was even more explicit [46]:

> In reality, so far, among all the bodies whose equivalent has been able to be determined with some precision, chlorine alone is obviously the exception. But still, by making a slight modification to Prout's law, this anomaly could be eliminated. It would suffice to admit that for some bodies, and chlorine would be in this case, the equivalent would be a multiple, not of the equivalent, but of the half-equivalent, of the hydrogen atom.[7]

In 1846, the French chemist Edme-Jules Maumené (1818–1898) promoted the same idea [47].

By 1867, however, almost all of the former advocates of Prout's hypothesis, even in such a modified form, had abandoned it. For example, Marignac, who had been the foremost skeptic of Stas's 1860 work [28] on atomic weights, was convinced by Stas's follow-up 1865 monograph [29]. That same year, in a review of the latter work, Marignac stated [48]:

> I can now no longer raise any doubts about the accuracy of the above numerical results, and I perfectly recognize with Mr. Stas that the atomic weights of bodies do not strictly offer among them the simple relationships that Prout's hypothesis would require.[8]

[7]"En réalité, jusqu'ici, parmi tous les corps dont l'équivalent a pu être déterminé avec quelque précision, le chlore seul fait évidemment exception. Mais encore, en apportant à la loi de Prout une légère modification, on pourrait faire disparaître cette anomalie. Il suffirait d'admettre que pour quelques corps, et le chlore serait dans ce cas, l'équivalent serait un multiple, non plus de l'équivalent, mais du demi-équivalent, de l'atome d'hydrogène."

[8]"Je ne saurais plus maintenant élever aucun doute sur l'exactitude des résultats numériques qui précédent, et je reconnais parfaitement avec M. Stas que les poids atomiques des corps n'offrent point rigoureusement entre eux les rapports simples qu'exigerait l'hypothèse de Prout."

Hinrichs concludes the historical forward by mentioning his 1866 article in the *American Journal of Science*, and adds that in the Fall of 1866 he had come to an agreement with the editor-in-chief, James Dwight Dana (1813–1895), that Hinrichs's further work would be published in the same journal. But he adds, "I wanted to start with the crystallographic part, [and] work inductively to the pantogen.... Circumstances have caused the current publication."

The "circumstances" were a dispute with Dana; evidently, Hinrichs took offense at suggested changes in the manuscript that Dana and the referees suggested [12]. Hinrichs was later [49] to accuse Dana of plagiarizing his ideas in an 1867 article Dana wrote [50] on the relationship of crystalline form to chemical composition.

7.4.2 Introduction (§ 1–5) and Pantogen and the Elements (§ 6–35)

Beginning in Sect. 7.3, Hinrichs proposes that "everything of a material nature has arisen from a former substance. We may call this original element pantogen." He proposes that free pantogen "probably occurs in the outermost solar atmosphere (appears light-producing) and in the planetary nebulae. Hydrogen is closest to it. The relationship to the luminiferous ether remains undecided." But later in the *Programme* Hinrichs seems to use the word pantogen not only to mean this original form of matter, but also as matter as it exists today on the atomic scale.

Hinrichs continues by making the a priori assumption that atoms are composed of fundamental units which he calls pantatoms,[9] which mutually attract one another. The atomic weight of an atom is then proportional to the number of pantatoms it consists of. To account for the non-integral atomic weight of chlorine (35.5) when scaled to the atomic weight of a hydrogen atom, Hinrichs further assumes (unlike the assumption made in his 1866 paper) that the weight of a pantatom must be 0.5 —i.e., half the atomic weight of hydrogen—so that, for example, there are 2 pantatoms in hydrogen, and 71 pantatoms in chlorine.

In paragraph 10, Hinrichs introduces his additional assumption that atoms consist of layers of pantatoms that are stacked to form prisms. He begins this proposal with the statement:

> There are only two possible compound arrangements in a plane for equal material points: at the corners of an equilateral triangle or a square. Accordingly, there are two kinds of pantogen compounds or elements; namely trigonoids and tetragonoids.

Hinrichs then discusses various ways to arrange pantatoms into plates based on triangular and square grids. He begins by depicting and counting the number of pantatoms included in plates based on triangular grids having equal-length sides (Fig. 7.5, parts a–i): for six-sided plates (i.e., regular hexagons) the numbers are 1, 7, 19, ... and for four-sided plates (i.e., diamond shapes) the numbers are 4, 9, 16,

[9]Although Hinrichs uses the spelling "panatom" in the title of his monograph, he uses this spelling throughout the main text.

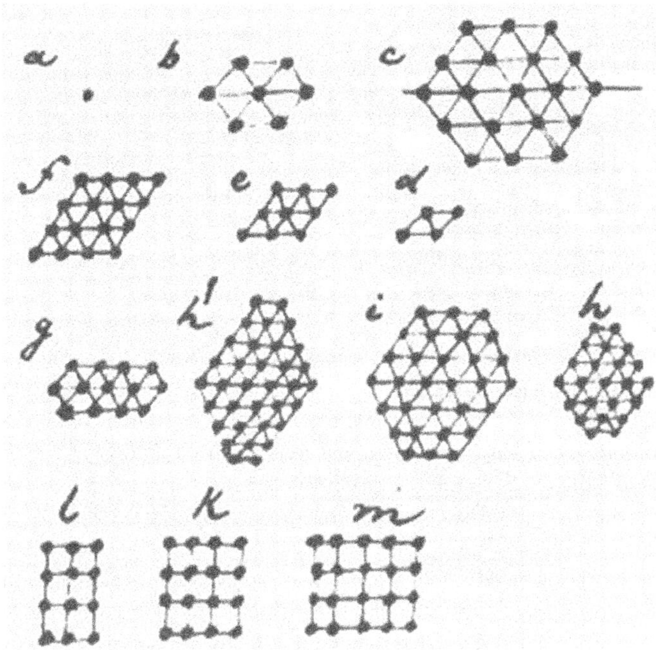

Fig. 7.5 Various rafts or layers of pantatoms from paragraph 11 of *Programme der Atomechanik*

.... But he also analyzes certain irregular hexagons with unequal-length sides (examples include those with 13, 23, 30, and 34 pantatoms).

For the plates based on square grids (Fig. 7.5, parts k–m), Hinrichs mentions both plates with equal-length sides (i.e., squares), as well as rectangular plates having unequal sides (examples include those with 12, 16, and 20 pantatoms). Hinrichs does not mention the fact that the four-sided plates based on triangular grids (Fig. 7.5, parts d–f) are slanted versions of (with the same number of total points as) square plates with the same length side: e.g., a plate with 4, 9, or 16 points can be represented either way. This is just one of the many degrees of freedom in Hinrichs's system.

Hinrichs then suggests that "by placing these pantatom plates vertically above one another, prisms emerge as the atoms of the elements." The number of pantatoms in the base of the prism he calls the atomare (symbol a), and the number of layers—i.e., the height of the prism—he calls the atometer (symbol m). Therefore, "if there is only one prism, then the sum total of the pantatoms in the element atom or the atomic weight = atogram $g = m \cdot a$".

Hinrichs proposes a Linnean nomenclature (order, genus, species, variety) for classifying elements:

> [the elements] divide into 2 orders (the trigonoids and the tetragonoids) according to the mutual association of the pantatoms on average. This order divides into genera according to the external form of the figure.... The species (the element) is determined by the atometer *m*. Varieties (very closely related elements [having] nearly the same atomic weight) are created by merging the subsequently determined caps onto the prism.

Hinrichs's orders divide all the elements into two large categories (see below); his genera are the equivalent of our groups (alkali metals, halogens, etc.); his species are the elements within a group. In allowing for varieties, Hinrichs acknowledges that pure prismatic structures cannot be assigned for some atoms; it is sometimes necessary to add caps ("Aufsätze") of additional pantatoms.

Hinrichs says shortly thereafter that trigonoids are metalloids (what we would call non-metals) and tetragonoids are metals. Thus, fluorine, oxygen, and nitrogen (along with their heavier congeners) are trigonoids, whereas all other elements are tetragonoids.[10] Hinrichs evidently made this choice (rather than the reverse) because he believed it led to a more consistent set of pantatomic structures for the various atoms, and thus a more consistent classification scheme. Of course, Hinrichs's proposal that there was any such correlation was another of his a priori assumptions.

A key assumption in Hinrichs's system is that atoms of elements that belong together in a group have similarly shaped prismatic structures with the same base: "Thus the general character of the element will be determined by *a*, while the particular determination of this character will be expressed by *m*, the height".

Based on this assumption, Hinrichs fits the atograms (i.e., twice the atomic weights) of the elements within a group to a common formula, much like he had done in his 1866 paper, except now the formulas correspond to stacks of plates based on triangular or square grids (rather than square grids only), and the basic unit has an atomic weight of ½ rather than ¼ or 1 on a H = 1 scale. Examples of some of his proposed arrangements of pantatoms are as follows (see Fig. 7.6):

- H (2) = 2 layers of 1 pantatom
- F (36) = 5 layers of 7-pantatom hexagons +1
- N (28) = 4 layers of 7-pantatom hexagons
- O (32) = 8 layers of 4-pantatom double triangles (diamonds)
- K (78) = 2 layers of 7-pantatom hexagons + 4 layers of 4 × 4 squares
- Ca (80) = 2 layers of 4 × 4 squares + 2 layers of 4 × 6 rectangles.

From the fact that Hinrichs proposes that calcium is composed of 80 pantatoms, it is apparent that in the year between 1866 and 1867 he has switched over entirely to the system of atomic weights advocated by Cannizzaro and others.

Note that the most straightforward proposal for the structure of calcium would be five layers of 4 × 4 squares, but instead Hinrichs proposes a more complicated

[10]Hinrichs's categorization of metals and non-metals is different from ours, and from that of most of his contemporaries. For example, he placed carbon and boron among the metals, and antimony and tellurium among the nonmetals.

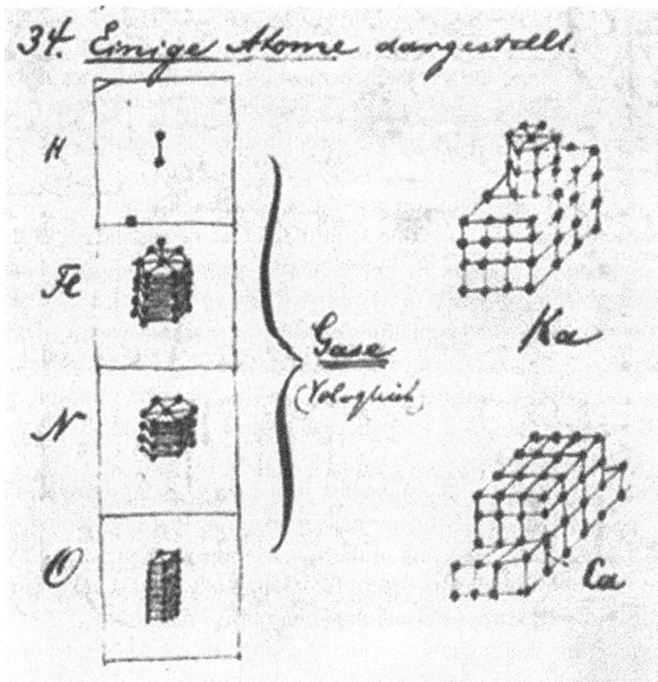

Fig. 7.6 Atoms as stacks of pantatom layers, from paragraph 34 of *Programme der Atomechanik*

structure involving the stacking of both square and rectangular plates (Fig. 7.6). He does this in order to devise a formula that also applies to the congeners of calcium: strontium and barium. Also, despite Hinrichs's proposal that metals are tetragonoids, he proposes that potassium (Ka), along with the other alkali metals, consists of hexagon-shaped triangular plates on top of square plates. We will return to this point later.

This section of Hinrichs's monograph contains a series of 13 tables, one for each of his different groups of elements: the pantoids (containing only hydrogen), the chloroids (halogens), phosphoids (pnictogens[11]), sulphoids (chalcogens), kaloids (alkali metals through Rb and perhaps also including In, Cs, and Tl), calcoids (heavier alkaline earths starting with Ca), kadmoids (Mg, Zn, Cd, and Pb), ferroids (Al, Fe, Rh, Ir), molybdoids (Cr, Mo, V, W), cuproids (Cu, Ag, Au),[12] titanoids (C, Si, Ti, Pd, Pt, and perhaps also including Zr, Sn, Ta, and Th), sideroids (Cr, Mn, Fe Ni, Co, and U), and "undetermined" (B, Be, Hg). Chromium and iron each appear in two of these groups. Hinrichs also mentions but does not give formulas

[11]For convenience, I use here a modern term for the nitrogen group [51].

[12]Hinrichs was the second person to place copper, silver, and gold into the same group within the context of a periodic system; this had first been done in 1864 by Lothar Meyer.

for two additional groups, rhodoids (Rh, Ru) and iridoids (Ir, Os), claiming that too little is known about their atograms (i.e., atomic weights). In fact, this is not correct: accurate atomic weights for these elements had been reported for Ru in 1845 by the Russian chemist (and discoverer of Ru) Carl Ernst Claus [52] (1796–1864) and for the other three elements in 1828 by the Swedish chemist Jöns Jacob Berzelius [53] (1779–1848).

Each of the 13 tables contains a general parameterized formula for the atogram (i.e., twice the atomic weight), along with the values of parameters in the formula that correspond to each element. For example, the table for the chloroids (i.e., halogens), the general formula is $(1) + m{\cdot}(p)$, where (p) denotes a six-sided raft (not necessarily with equal sides) containing p pantatoms arranged in a triangular grid. Fluorine has $m = 5$ and $p = 7$, chlorine has $m = 10$ and $p = 7$, bromine has $m = 12$ and $p = 13$, and iodine has $m = 11$ and $p = 23$. The "predicted" atomic weights for fluorine and bromine show a discrepancy of 1.0 and 1.5, respectively, with respect to the then-accepted values.

Although Hinrichs states that elements in the same group have similar shapes with the same base, the table for the halogens shows that his "atomic structures" conform to a looser rule: elements in the same group are represented by the same formula, but the base can change from, for example, a regular hexagon of 7 or 13 pantatoms to a six-sided (but not regular) plate of 23 pantatoms.

The formulas for the atograms of all 13 groups of elements according to Hinrichs's classification scheme can be found in Table 7.1.

Table 7.1 Hinrichs's formulas for the atograms (i.e., doubled atomic weights) of the elements[a]

Pantoids (H)	$m{\cdot}(1)$
Chloroids (F, Cl, Br, I)	$(1) + m{\cdot}(p)$
Phosphoids (N, P, As, Sb, Bi)	$m{\cdot}(p)$
Sulphoids (O, S, Se, Te)	$m{\cdot}/p^2/ = \mu{\cdot}[2p^2]$
Kaloids (Li, Na, K, In, Rb, Cs, Tl)	$2(7) + 2{\cdot}m{\cdot}4^2$
Calcoids (Ca, Sr, Ba)	$2{\cdot}4^2 + m{\cdot}2{\cdot}[4{\cdot}6]$
Kadmoids (Mg, Zn, Cd, Pb)	$2{\cdot}4^2 + m{\cdot}2{\cdot}[4.6]$ or $2{\cdot}4^2 + \mu{\cdot}[8{\cdot}12]$
Ferroids (Al, Fe, Rh, Ir)	$K + m{\cdot}2{\cdot}[3{\cdot}4]$ m even
Molybdoids (Mo, V, W)	$K + m{\cdot}2{\cdot}[3{\cdot}4]$ m odd
Cuproids (Cu, Ag, Au)	$K + m{\cdot}[5{\cdot}8]$
Titanoids (C, Si, Ti, Pd, Pt)	$2{\cdot}[3{\cdot}4] + m{\cdot}[4{\cdot}5]$
Sideroids (Cr, Mn, Fe, Ni, Co, U)	$K + 4{\cdot}2{\cdot}[3{\cdot}4]$
Undetermined (B, Be, Hg)	

[a]In this table, (p) denotes a six-sided triangular raft containing p pantatoms, $/p^2/$ denotes a square raft with sides of length p, and $[i{\cdot}j]$ denotes a rectangular raft with sides of length i and j. The variables m and K represent integers, and $\mu = m/2$; although Hinrichs does not say so, K seems to be restricted to either 7 or a multiple of 4

7.4.3 Hinrichs's Chart of 1867 (§ 36–56)

In paragraph 37 of the *Programme*, the chart shown in Fig. 7.7 appears. Hinrichs introduces his chart as follows:

> To illustrate the mechanical or rational classification of the elements contained in the foregoing [tables], I present them in the following drawing. The pantogen forms the midpoint, the genera are represented by rays from this point, and the species are recorded in these rays where the distance from the center equals the atogram *g* ….

Because the distance from the center is the atogram (=twice the atomic weight), the dotted lines in Hinrichs's diagram have usually been referred to as a spiral [6, 8, 9, 54]. But such a description is actually only half-true. In (unnumbered) paragraph 54, Hinrichs states explicitly that the dotted lines in his diagram form a spiral, but with a twist:

> …the known elements in our chart follow one another in spiral lines. See in 37 [i.e., his chart] the lines H-Li-C-O-N-Fl-Na–Mg-Al–Si-S-P-Cl-Ka-Ca–Ti and then in the opposite direction Ti–Fe-Zn-In-Br-As-Se-Pd-Rh-Cd-Cs-Jo-Sb-Te.

The words *and then in the opposite direction* ("in entgegengesetzte Richtung") indicate that Hinrichs considered that the elements form *two* spirals, not one (Fig. 7.8). Elements from H to Ti lie on a *clockwise* spiral, but from Ti through Te

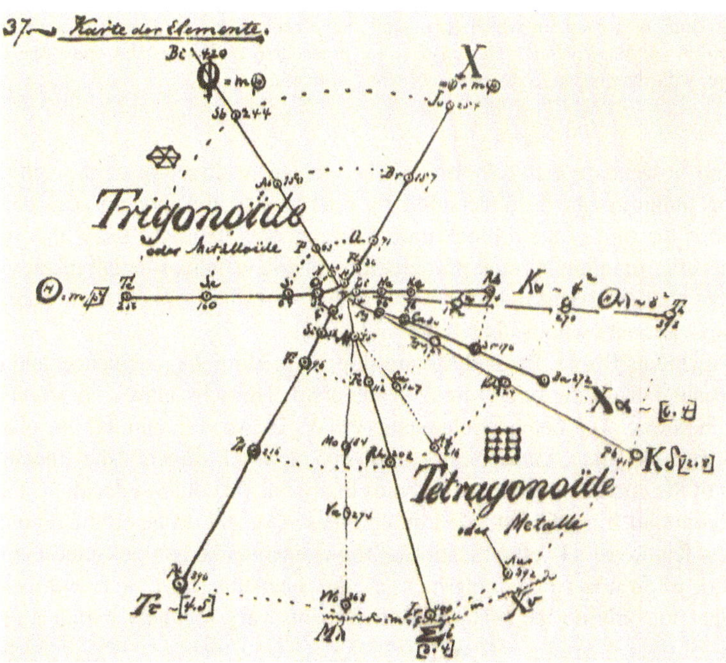

Fig. 7.7 Hinrichs's 1867 chart of the elements from paragraph 37 of his *Programme der Atomechanik*. Image courtesy of the University of Dresden

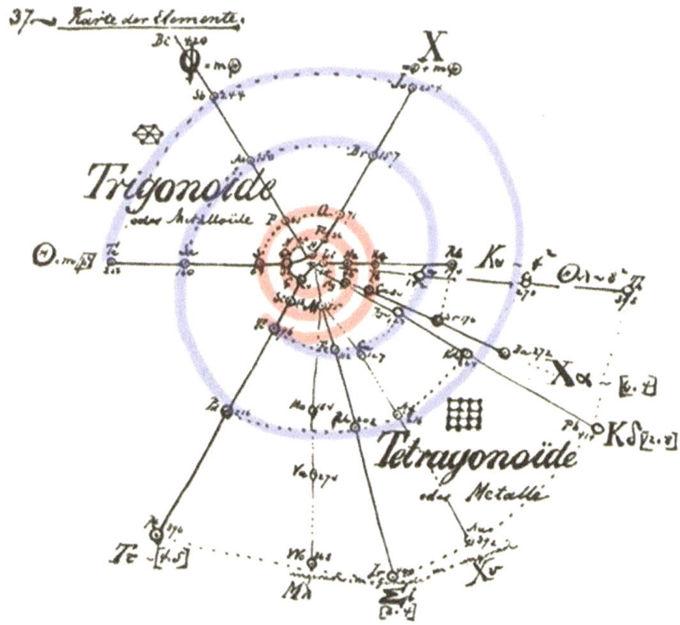

Fig. 7.8 Hinrichs's chart, in which two spirals have been added as described in unnumbered paragraph 54 of *Atomechanics*, the inner spiral proceeding clockwise, the outer one counterclockwise, with the change in direction occurring at titanium

the spiral proceeds *counterclockwise*. The change in direction of the spiral upon reaching titanium can be understood by reference to the modern periodic table: titanium is the first of the d-block transition elements. Hinrichs knew that the next elements after titanium in order of increasing atomic weights were all metals, and he concludes that the spiral must reverse direction in order to place these succeeding transition elements within his tetragonoid sector.

The elements Rb, Sr, Ba, and Mo are skipped in Hinrichs's sequence, and indeed they do not fall on the dotted lines in the chart; Hinrichs makes no comment on these omissions. The heaviest elements, Pt, W, Ir, Au, Pb, and Tl, are placed on their own dotted line; Hinrichs does not say in which direction he intended this portion of the curve to spiral, and it seems likely that he couldn't decide. Cu and Ag are also omitted from the list, even though these element do lie on the dotted lines.

A key feature of any classification of the elements that can be called periodic is the recognition that the similarities of the elements have a two-dimensional character: not only do the properties of elements vary regularly *within* a group,[13]

[13]As first pointed out for atomic weights by Johann Döbereiner (1780–1849) some 50 years previously. But non-quantitative chemical similarities (such as found, for example, in the compounds of Sr and Ba, and in the hydrohalic acids of F and Cl) had been recognized in the late 1700s.

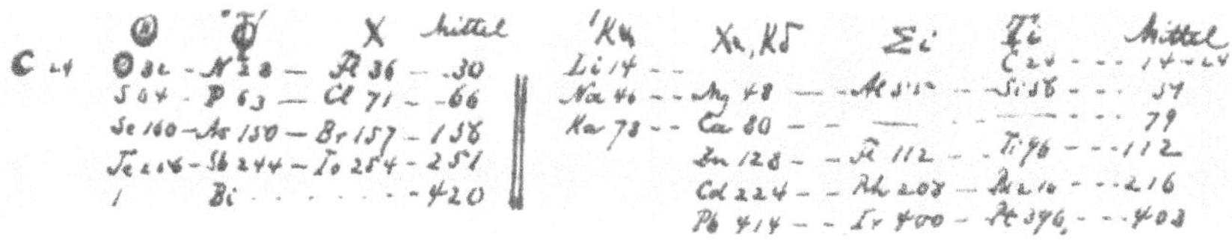

	Θ	Φ	X	Mittel	Kα	Xα, Kδ	Σι	Tτ	Mittel
C 24	O 32	N 28	Fl 36	30	Li 14			C 24	14-24
	S 64	P 63	Cl 71	66	Na 46	Mg 48	Al 55	Si 56	51
	Se 160	As 150	Br 157	156	Ka 78	Ca 80			79
	Te 256	Sb 244	Io 254	251		Zn 128	Fe 112	Ti 96	112
		Bi		420		Cd 224	Rh 208	Pd 216	216
						Pb 414	Ir 400	Pt 396	403

Fig. 7.9 Table showing inter-group relationships of Hinrichs's doubled atomic weights.[14] From paragraph 53 in Hinrichs's *Programme der Atomechanik*

there are regularities that relate *different* groups (which Hinrichs called genera). Hinrichs recognizes this two-dimensional character in paragraph 53: "…nearly equal values of *g* [i.e., Hinrichs's atogram] can correspond to different forms; that is, near equal *g* give species of different genera. Our chart § 37 shows this quite clearly." This sentence is followed by a table (Fig. 7.9) that emphasizes recurrent similarities in the atomic weights of elements from different genera.

Thus, the elements O–N–F all have atograms near 30, the heavier congeners S–P–Cl all have atograms near 66, and so on for the subsequent members of those three groups. Similarly, Hinrichs points out that the elements Na–Mg–Al–Si all have atograms near 51, although the heavier congeners show quite a bit more scatter, in part because some of the elements Hinrichs includes are today recognized as belonging to other groups.

The table in Fig. 7.9 is significant because, as a (partial) tabular representation of Hinrichs's periodic system, it more closely resembles other periodic systems devised in the 1860s, and thus makes comparisons a little easier. This table includes all three of the trigonoid groups, but only four of the tetragonoid groups. The latter include the alkali metal groups and the three tetragonoid groups which Hinrichs calls in paragraph 49 "the 3 main genera with a rectangular base, namely the ferroids, … the titanoids …, and the kadmoids …." We will return to this idea of there being a smaller number of "main" tetragonoids below.

Two further aspects of Hinrichs's chart deserve comment: the first is why the halogens, chalcogens, and pnictogens (Hinrichs's X, Θ, and Φ groups, respectively)

[14]Hinrichs sometimes uses Tτ and sometimes (as here) Tι to denote the titanoid group; for clarity, I use only Tτ.

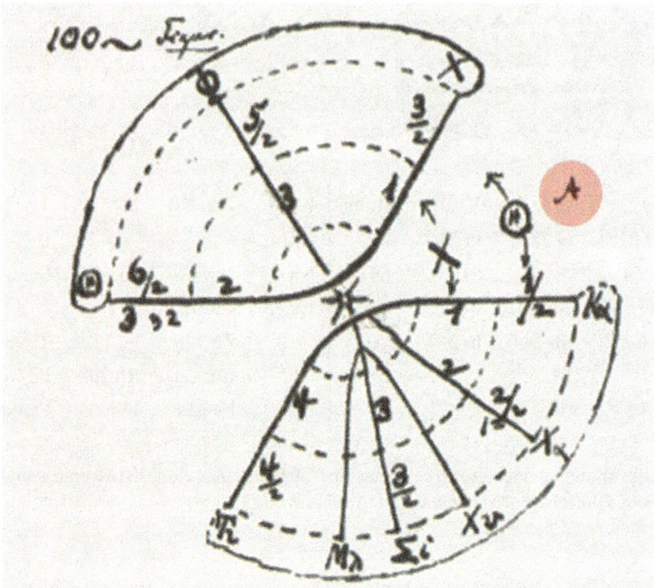

Fig. 7.10 Diagram showing the combining ratios of elements with hydrogen and the halogens (*inner circle*) and with the chalcogens (*outer circle*). A circle has been added to emphasize what Hinrichs calls "point A". From paragraph 100 in *Programme der Atomechanik*

are out of order when compared with the modern table, and the second is why the spokes appear where they do (i.e., what is the basis for their angular positions). The first of these aspects has long been noted [8] but never explained.

The reason Hinrichs chose the order chalcogens–pnictogens–halogens (i.e., O–N–F) rather than pnictogens–chalcogens–halogens (N–O–F) can be deduced from a diagram he gives in paragraph 100 (Fig. 7.10). This diagram, which appears in the next section on the chemical properties of the elements, immediately follows a summary of combining ratios with hydrogen and the halogens.

In reference to this diagram, Hinrichs states,

> In the adjacent figure, the main result of the compound ratios is compiled. The number indicates how many atoms of X (inner circle) or Θ (outer) unite with one atom of the different genera (their symbol in the outermost circle) to [give] the main compound.[15] One can see that the ratio in both the X [i.e., halide] and the Θ [i.e., chalcogenide] compounds grows regularly from [point] A in every direction, reaching its maximum at Θ [O-S-Se-Te group] among the trigonoids [and] at Tτ [C-Si–Ti-Pd-Pt group] among the tetragonoids.

Point A in Hinrichs's diagram occupies a place that today would be filled with the noble gases, which have a characteristic combining ratio of zero. Starting from this location, Hinrichs places the groups of elements in order of increasing

[15]"Hauptverbindung".

combining ratios. Indeed, if one moves column by column away from the lighter noble gases in the modern periodic table, the valence of E in the EH_n hydrides increases in the order 1, 2, 3, 4… in both directions (i.e., increasing and decreasing atomic number).

Hinrichs had previously stated in paragraphs 92 and 95–98 that the various groups have combining ratios with the chloroids (X) as follows: XX for F and its congeners, ΘX_2 for O and its congeners, and ΦX_3 for N and its congeners (these are the trigonoid groups), and $T\tau X_4$ for C and its congeners, $\Sigma\iota X_2$, $\Sigma\iota X_3$, or $\Sigma\iota X_4$ for Al and its congeners, $X\alpha X_2$ for Mg and its congeners, and $K\alpha X$ for Li and its congeners (these are the tetragonoid groups listed in his table in paragraph 53; Fig. 7.9). In the diagram in paragraph 100, Hinrichs chose $\Sigma\iota X_3$ as the "main" composition of the group of elements headed by Al. Had Hinrichs based his chart on these trends in combining ratios, he would have placed the F, O, and N groups in the correct order.

Instead, Hinrichs evidently places greater weight on the combining ratios with chalcogens (oxygen in particular). In the 1860s [55], the highest known chlorine oxide was Cl_2O_3; although I_2O_5 and I_2O_7 had been claimed, the evidence in support of these higher oxides was ambiguous (and even today the existence of I_2O_7 is doubtful).[16] Thus, Hinrichs mentions that F and its congeners form binary compounds whose maximum chalcogen content is embodied in the formula $X_2\Theta_3$, whereas N and its congeners form binary compounds up to $\Phi_2\Theta_5$ (such as P_2O_5 and As_2O_5), and among O and its congeners the compounds SO_3 and SeO_3 exist. Correspondingly, in his diagram, Hinrichs arranges the F–N–O groups in order of increasing (maximum) combining ratio with oxygen: 3/2, 5/2, and 6/2.

Thus, it is clear that, to Hinrichs, atomic weights play an important role in determining what elements belong together in the same groups, but they play a lesser role in determining how the groups are ordered with respect to one another. Hinrichs focuses more on the *similarity* of the atomic weights of elements from different groups, and less on arranging them in strict order of increasing atomic weight. Instead, Hinrichs places greater importance on combining ratios when deciding on the relative arrangement of the groups in his periodic system.

Almost all of Hinrichs's atomic weights are close to the modern values but, like everyone else in the 1860s, he used wildly incorrect atomic weights for vanadium, tantalum, and uranium (137, 137.6, and 120, respectively, vs. modern values of 51, 181, and 238). Of these, only vanadium appears in his chart, and not surprisingly this element is located farthest from any of the arcs of his spiral. It is interesting that this deviation did not prompt Hinrichs to consider whether his assigned atomic weight of vanadium might be wrong.

[16] In paragraph 67 of the *Programme*, Hinrichs mentions Cl_2O_5 and Cl_2O_7 as example molecules, but in 1867 these two substances had never been prepared, although analogous chlorate and perchlorate salts (and the corresponding acids) were known.

Fig. 7.11 Possible structure of Hinrichs's chart. Compare with Fig. 7.10

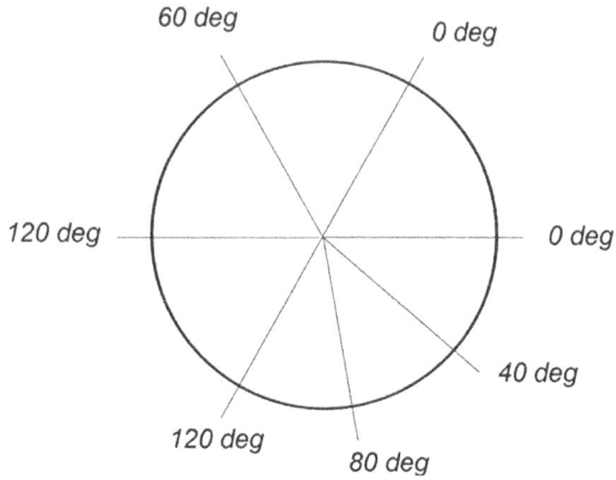

As to the second aspect of Hinrichs's chart, i.e., the basis for the angular positions of the various spokes, Hinrichs says nothing in the *Programme*.[17] But the diagram in paragraph 100 also helps explain why the spokes are where they are.

Evidently, Hinrichs intended the trigonoids collectively to subtend approximately 120° of arc, and the same for the tetragonoids; he places these two arcs symmetrically within a circle, separated by 60° (Fig. 7.11). Hinrichs chose the alkali metal and chalcogen spokes to define the horizontal axis of his chart. He does not give reason for this arrangement, but one possibility is that Hinrichs orients the spokes in this way so that the maximum valence (as given by the outermost values in Fig. 7.10) increases from right to left.

Hinrichs considers the molybdoids Mλ (Cr, Mo, V, W) and the cuproids Xυ (coinage metals) as subordinate to the ferroids Σι (Al, Fe, Rh, Ir): in paragraph 47 he states "the molybdenoids and cuproids are … only side branches of the ferroids." This subordinate role is reflected in his omission of the molybdenoids and cuproids from his list of the "main" tetragonoids, as given in the table in paragraph 53 (Fig. 7.9), and also in his construction of the chart. Thus, the spokes for the three trigonoid groups and the four "main" tetragonoids are spaced equally, by 60° of arc for the trigonoids, and 40° for the tetragonoids. The subordinate molybdoids and cuproids lie on partial spokes that flank the spoke for the ferroids.

Hinrichs justifies his designation of the molybdenoids and cuproids as side branches because their atograms can be represented by formulas of the type $a + md$ where m is odd, whereas the ferroids have formulas $a + md$ where m is even. The flanking relationship of the molyboids and cuproids is evident in Hinrichs's chart in

[17]In his 1867 abstract published in the *Journal of Mining*, Hinrichs says, "The most convenient representation of this classification is obtained by representing each genus by a line radiating from the centre … the trigonoids upwards and to the left, the others downward and to the right." But this sentence is merely descriptive, not explanatory. Interestingly, in 1894 Hinrichs added, "My original conception was the arrangement on a right cone with π at the vertex. This brings the genera less far apart for the high atomic weight, but cannot be readily reproduced" [56, p. 241]. Here, π is Hinrichs's symbol for pantogen, which is placed at the very center of the diagram.

paragraph 37: these flanking spokes are labeled "m = ungerade (odd)" whereas the ferroid spoke is labeled "m = gerade (even)".

We can speculate that, because vanadium was on one of these flanking partial spokes, it was of less concern to Hinrichs that this element did not lie on any of the dotted spiral lines. But Hinrichs is silent as to the reasons that Rb, Sr, and Ba also do not lie on his dotted lines.

7.4.4 Chemical Characteristics (§ 57–110)

The next section of the *Programme* deals with the chemical characteristics of the elements. He begins with a statement about chemical bonding: "The chemical bonding between two atoms A and B, viewed as a mechanical phenomenon, can only consist of the side-by-side arrangement (juxtaposition) of atoms, AB." He then devotes a lengthy discussion to the geometries of molecules (three-atom molecules are triangles, etc.) and to the contraction in volume that occurs when, for example, hydrogen and oxygen gases react to form water.

Hinrichs then tries to explain the atomicities (i.e., valences) of the atoms in terms the pantatomic structures he has assigned. The following is taken from his English abstract [2]:

I. One atom of any *chloroid* combines with one of hydrogen; for (1)+$m(p)$ shows one prominent centre of attraction.
II. One atom of a *sulphoid* is saturated by two atoms of H: for the atomare[18] 2 2^2 of O shows two equal centres of attraction.
III. One atom of any *phosphoid* requires three atoms of H for saturation; for the regular hexagon gives as foci the centre of gravity of the three rhombs into which it is divisible.

Thus, Hinrichs proposes that the halogens have pantatomic structures in which the single capping pantatom is responsible for the valence of one, the divalent chalcogens have structures that consist of two identical halves, and the trivalent pnictogens have structures based on hexagons, which can be considered as consisting of three joined rhombuses. Here is another example of Hinrichs formulating ad hoc hypotheses, which in this case are not even internally consistent.

Hinrichs's "explanations" of gas volume changes in chemical reactions and interatomic bonding are outside of the scope of the present discussion. Those interested in these aspects of Hinrichs's theories are encouraged to consult the appendix to the present article.

I will, however, mention one aspect of Hinrichs's theory of atomic structure, which as far as I am aware has not been pointed out previously. This is Hinrichs's explanation of the cause of isomorphism.

---

[18]Hinrichs has evidently made a mistake here; presumably he intended to say that the atomare is 2^2.

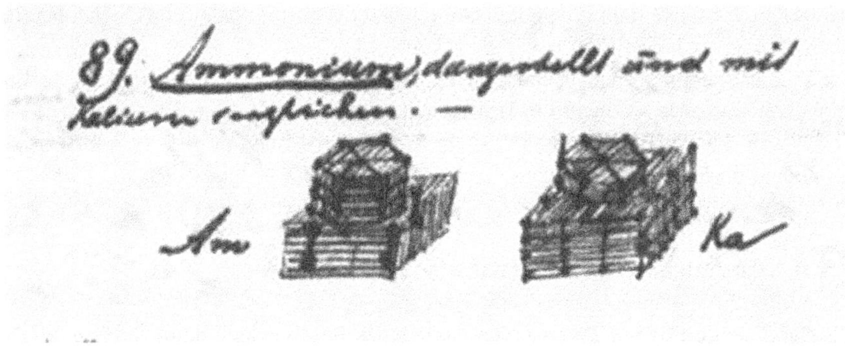

Fig. 7.12 Hinrichs's explanation of the isomorphism of ammonium and potassium salts. From paragraph 89 of *Programme der Atomechanik*

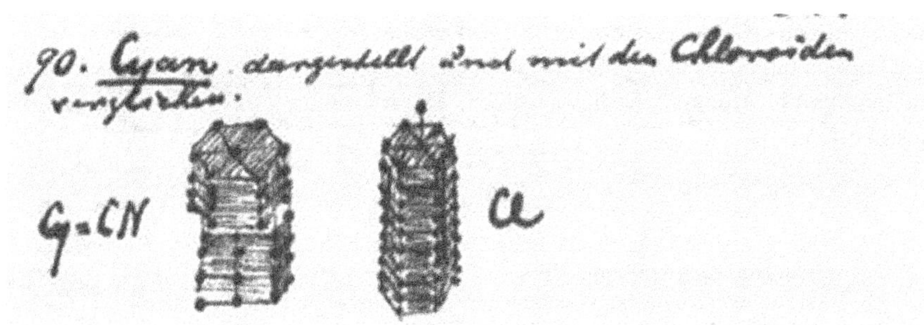

Fig. 7.13 Hinrichs's explanation of the isomorphism of cyanide and chloride salts. From paragraph 90 of *Programme der Atomechanik*

Hinrichs proposes that, in the ammonium ion (NH_4^+), the N atom (4 layers of 7-pantatom hexagons) is attached to four H atoms (each consisting of 2 pantatoms), which he places below the N atom as four "legs" that define a square-based parallelepiped (the overall shape is reminiscent of NASA's lunar module). His model for potassium (shown in his paragraph 34; Fig. 7.6) consists of 2 layers of 7-pantatom hexagons on top of 4 layers of 4 × 4 squares. Therefore, according to Hinrichs, ammonium and potassium commonly form isomorphous compounds because these chemical species have about the same three-dimensional shape (Fig. 7.12).

Similarly, cyanide and chlorine also form many isomorphous compounds, and Hinrichs accounts for this phenomenon in a similar way (Fig. 7.13). The cyanide ion he depicts as a nitrogen atom on top of a carbon atom; the height and cross section of the resulting shape are similar to those of his proposed structure of chlorine (10 layers of 7-pantatom hexagons plus 1).

Hinrichs evidently kept the phenomenon of isomorphism in mind as he made his choices for the arrangements of pantatoms within atoms. Although he would have disagreed strongly with the comparison,[19] this part of Hinrichs's *Programme* is similar in intent to the work of the French chemist and crystallographer Marc Antoine Gaudin (1804–1880), who devoted his life (unsuccessfully) to working out the arrangements of atoms within molecules and molecules within crystals [57, 58].

7.4.5 Physical Characteristics (§ 121–228), and Morphological Characteristics or Crystal Forms (§ 229–399)

The last two sections of the *Programme* contain detailed discussions of trends and mathematical relationships between chemical formulas and various properties of the elements, such as specific weights (i.e., densities), specific heats, melting points, boiling points, refraction equivalents (i.e., the refractive index minus unity, divided by the density and multiplied by the atomic weight), spectral lines, and the axial ratios, external forms, and optical properties of crystals. In a few cases, Hinrichs notes regular trends in these properties as a function of increasing atogram (atomic weight): for example, he mentions that the melting points of Li, Na, and K decrease in that order. But other than a few scattered examples, Hinrichs is silent on the relationship of these properties to his periodic system.

Hinrichs also published his ideas on these topics in two contemporary summaries [2, 59].

NAME.	GROUPS.	SYMBOL OF THE ELEMENTS.							
Pantoïds........	Γ	H	—	—		—		—	
Kaloïds........	Kχ	Li	Na	Ka		Rb		—	
Chalcoïds.......	λχ	—	—	Ca		Sr		Ba	
Cadmoïds.......	Kδ	(Be ?)	Mg		Zn		Cd		Pb
Hydrargoïds...	Γγ	—	—		—		—		Hg
Cuproïds.	Kν	—	—		Cu		Ag		Au
Ferroïds........	Σν	(Bc ?)	Al		Co-Ur Ni Fe		Rh		Ir
Molybdoïds.....	Mλ	Bo	—		Mn Cr		Mo		Wo
Titanoïds.......	Tτ	C	Si		Ti		Pd		Pt
?							Sn		
Nioboïds.......	Nβ	—	—		Va		Nb		Ta
Phosphoïds.....	Φ	N	P	P	As	As		Sb	Bi
Sulphoïds.......	θ	O		S		Se		Te	—
Chloroïds.......	X	Fl		Cl		Br		Io	—
Pantoïds........	Γ	H		—		—		—	—

Fig. 7.14 Hinrichs's chart from *The Pharmacist* (1869) [3]

[19]In the English abstract of his *Programme* [2], Hinrichs states in reference to his mechanical theory of the forms of crystals, "The investigations of Gaudin are simply geometrical." So Hinrichs was aware of Gaudin's work.

Genera.		Species.							
x =		1	2	3		4		5	
Υ		*H*							
Κα		*Li*	*Na*	*Ka*		*Rb*			
Χα		—	—	*Ca*		*Sr*		*Ba*	
Κδ		—	*Mg*		*Zn*		*Cd*		*Pb*
Υγ		—	—		—		—		*Hg*
Κν		—	—		*Cu*		*Ag*		*Au*
Σι			*Al*		*Co* *Ni* *Fe* *Mn* *Cr*		*Rh*		*Ir*
Ττ		*C*	*Si*	*Ti*		*Pd*		*Pt*	
						Sn			
Φ		*N*	*T*	*As*		*Sb*		*Bi*	
Θ		*O*	*S*	*Se*		*Te*		—	
Χ		*Fl*	*Cl*	*Br*		*Io*		—	
Υ	*H*								

Fig. 7.15 Hinrichs's 1869 chart from *Proceedings of the AAAS Meeting* (1869) [4]

7.5 Hinrichs's Charts of 1869

Two years later, in 1869, Hinrichs published modified versions of his periodic system; one of them (Fig. 7.14) appeared in the July 1869 issue of a rather obscure Chicago-based journal, *The Pharmacist* [3], whereas the other (Fig. 7.15) was presented at the August 1869 meeting of the American Association for the Advancement of Science and published in their proceedings [4]. Hinrichs also issued his AAAS paper separately as a privately printed offprint [5]. The two new versions are very similar, the one presented at the AAAS meeting including fewer of the transition elements.

The tables also resemble Mendeleev's first table of 1869, in that the elements within a group are arranged in rows rather than columns. Mendeleev's table had appeared in the *Zeitschrift für Chemie* and the *Journal für praktische Chemie* a few months before Hinrichs's tables appeared, but it is unclear whether Hinrichs had seen it.

Several features of Hinrichs's 1869 tables have been discussed in some detail [6–9] and here I will just summarize some of the main conclusions.

(1) Hinrichs chose fusibility and volatility, i.e., "the deportment of the elements in increasing temperature, as the basis of classification. For heat is merely motion of the particles, in fact this classification is a mechanical one, expressing the relative mobility of the atoms of the elements …" [3] In a textbook Hinrichs published in 1871 [60], he elaborates:

> The order of the genera … is determined [as follows]. The least fusible and volatile is placed in the middle. The most fusible and volatile are at the top and at the bottom; the metals standing above, the metalloids below. The upper elements in this table are decidedly electropositive; the lower equally electronegative.

(2) Unlike his 1867 system, in which combining ratios played a dominant role, Hinrichs now uses his new organizational principles, fusibility, volatility, and electronegativity to reverse the relative ordering of the nitrogen and oxygen groups, so that in his 1869 system these groups are in the modern order, C–N–O–F. Although Hinrichs does not emphasize the point, this rearrangement also gives an order in which the atomic weights increase monotonically.

(3) Hinrichs points out [4] that

> in this table the elements of like properties, or their compounds of like properties, form groups bounded by simple lines. Thus a line drawn through C, As, Te, separates the elements having metallic lustre from those not having such lustre.

With regard to the last feature, it is relevant to point out that, in terms of overall structure, Hinrichs's 1869 system can be regarded as a long-form periodic table. Two aspects of the table support this conclusion. First, except for Pd and Pt, the transition elements are not co-mingled with pnictogen, chalcogen, or halogen groups: for example, V is not grouped with the pnictogens, Cr is not grouped with the chalcogens, and Mn is not grouped with the halogens, as they are in short-form tables. Second, the first-, second-, and third-row transition elements (except for Pd, Pt, and the elements in the Zn and Cu groups) collectively occupy the three places corresponding to the next three heavier congeners of aluminum. Hinrichs thus treats the transition elements the same way as the lanthanide and actinide elements are handled in many forms of today's periodic tables (i.e., as collectively occupying the two places after yttrium in group 3). This way of handling the transition elements is not like that in the short form of the periodic table; instead, Hinrichs's 1869 system is best regarded as a compacted version of a long-form periodic table.

Hinrichs points out that, in his 1869 table, a line can be drawn that separates the metals from the non-metals. This is the first recognition by anyone of this advantage of the long-form periodic table (although Gmelin had come close in 1843 [61]). Interestingly, Mendeleev emphasized that his system placed similar elements in adjacent locations, and he occasionally published long-form periodic tables (most notably in 1872 [62]), but as far as I know he never explicitly pointed out that the long form makes it possible to divide the metals and the non-metals with a single line. Hinrichs's idea was not resurrected until the Scottish chemist James Walker (1863–1935) independently recognized this advantage of the long-form table in 1891 [63].

In the longer of his two 1869 articles [4], Hinrichs proposes new formulas for the atomic weights of elements within a group. Instead of polynomials based on integers, Hinrichs now proposes formulas that include exponential quantities, but of course even these formulas do not match the experimental values exactly. He comments

> We do not mean to have the observed values corrected, for what here appears us "corrections" may in fact represent the links which hold together the various portions of the resulting atom. A negative correction would thus indicate that some projecting point had been removed before combination was effected....

So here Hinrichs is proposing that his formulas give a sort of idealized atomic weight, which is modified when the atom engages in chemical combinations. He further says:

> Most chemists seem to think that the chief importance of the painstaking work of Stas is to disprove and forever reject the so-called hypothesis of Prout; and with the destruction of this hypothesis they seem to think all the palpable harmonies of the atomic weights, and particularly all relating to "pantogen" is annihilated. We are inclined to think that just such careful determinations will demonstrate the correctness of the law of a common divisor (equal one-half the atomic weight of hydrogen?) for all elements, and prove some essential features of the structure of the element atoms.

Hinrichs was nothing if not steadfast in his views.

Several additional aspects of the 1869 charts are worthy of comment. One is why the transition metals in Hinrichs's 1869 tables are listed in reverse order and in slanted columns. Van Spronsen has speculated [7, 8], I think correctly, that the reverse order "might also be explained by the fact that Hinrichs wanted in any case to see the elements Zn, Cd, and Pb classified as a group next to Mg." My sense is that the reverse order can also be viewed a holdover from his 1867 system, in which he proposed a reversal in the direction of the spiral beginning with titanium.

In reference to the arrangement in slanted columns, Hinrichs says in his 1869 AAAS article: "By printing their symbols at distances from that of the genus, nearly proportional to the atomic weight, we obtain the following chart ...."[20] This sentence explains the slanting of the block of elements formed by the transition metals: the atomic weights of Zn, Cd, and Hg/Pb are the largest in their respective d-block rows, and thus they are placed farthest to the right.

But this explanation raises the question why the main group elements are not arranged with distances from the leftmost column "nearly proportional to the atomic weight". I found a possible answer in a copy of the handwritten original of his 1869 AAAS paper, which is present among the Hinrichs papers at the University of Illinois (Fig. 7.16).

[20] By the word "genus," Hinrichs is referring to the leftmost column in the table, which lists what we would call groups.

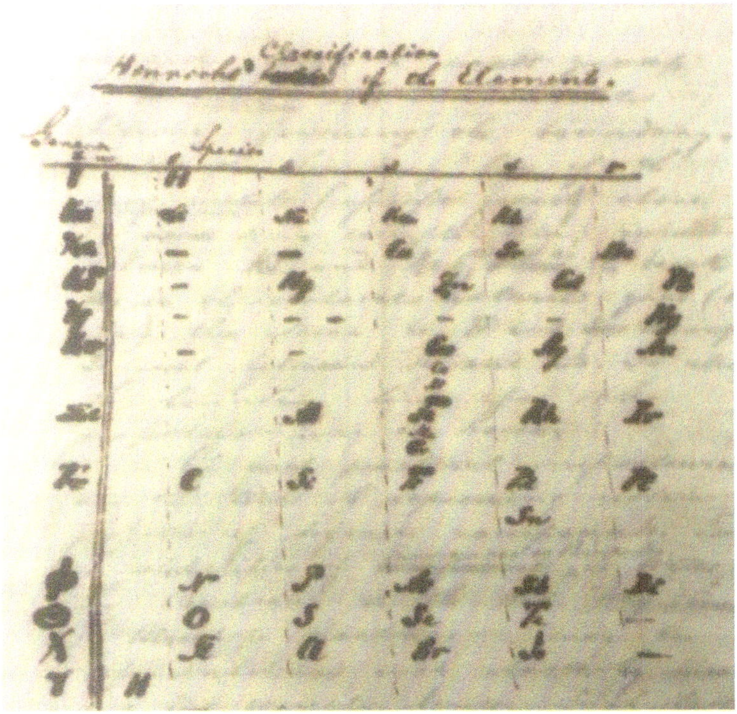

Fig. 7.16 Handwritten original of Hinrichs's 1869 paper given to the American Association for the Advancement of Science. Courtesy of the University of Illinois at Urbana-Champaign

The dotted lines that precede each column of elements are slightly slanted with respect to the double line in the "genera" column, as they should, because the atomic weights increase in a period. I think it is possible that Hinrichs meant to have all the columns slanted in the printed table (i.e., with the bottoms of the columns set further right than the tops), but the typesetters didn't notice this subtlety and Hinrichs didn't insist on correcting it.

Another aspect of Hinrichs's 1869 tables is that they contain numerous gaps, but in neither of his 1869 publications does Hinrichs comment on the gaps in any way. Even if he had proposed that these gaps represented elements still to be discovered, however, his predictions would not have been borne out. For example, his placement of Ti, Pd, and Pt in the same group with C and Si made it impossible for him to predict the existence of germanium, and his grouping of aluminum with a number of transition metals (but not with boron) made it impossible for him to predict the existence of either scandium or gallium.

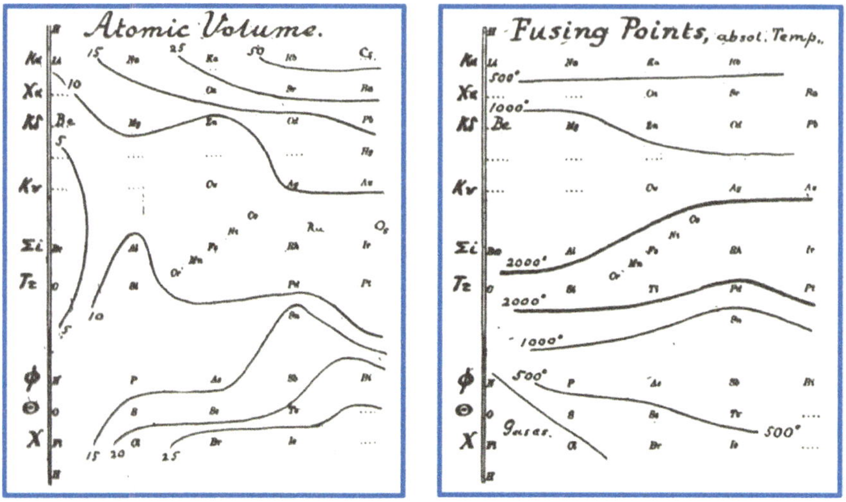

Fig. 7.17 Diagrams showing atomic volumes and melting points, mapped onto Hinrichs's 1869 table of the elements. From G. D. Hinrichs, *The Elements of Atom-Mechanics*, St. Louis, 1894

I will end with one aspect of Hinrichs's work that has largely been overlooked, which is his use of his system to illustrate periodic trends (Fig. 7.17). In a book he wrote in 1894 [56, pp. 231–239], Hinrichs gives examples of "some of the charts which were exhibited before the Salem Meeting of the American Association in August, 1869." In these charts, Hinrichs maps various chemical properties onto his 1869 table: among the properties he plots are atomic volume, fusing (i.e., melting) points, and specific gravity (i.e., density), but also the method and date of discovery of the elements, and reactions in the wet and dry way.

Hinrichs's atomic volume plot is especially notable for its anticipation of the important role that atomic volumes played in the formulation of the periodic systems of Lothar Meyer and Dmitri Mendeleev [64]. Of course, Hinrichs's plot was not published in 1869 and became available to the scientific public only after a lapse of 25 years.

7.6 Conclusions

Hinrichs used a number of chemical properties to construct his 1867 and 1869 periodic systems, including atomic weights, electronegativities, volatilities, valence, and specific gravities. In his 1869 article in *The Pharmacist*, Hinrichs states his approach as follows: "all the previous attempts [to classify the chemical elements] were founded upon only some one property of the elements, and hence necessarily led to an *artificial* classification." He was thus led "to propose a classification which we believe to be *natural*, because it does rest upon fundamental and essential properties, and because all other properties of the elements not directly involved in this classification nevertheless harmonize therewith" [3]. In a real way, he anticipated the holistic approach Mendeleev took when constructing his periodic system.

Both Hinrichs and Mendeleev used atomic weights as well as combining ratios to create their periodic systems, but Mendeleev placed greater importance on atomic weights as an organizing principle. As a result, Mendeleev's system not only more closely resembled today's periodic table, it facilitated his ability to make predictions about undiscovered elements, to correct incorrect atomic weights, and more generally to convince others that the system was useful.[21] Hinrichs's placing of greater importance on combining ratios led him initially to invert the relative locations of the nitrogen and oxygen groups, and his system was sufficiently flawed that it made it difficult for him to duplicate Mendeleev's achievements.

Hinrichs had a very ingenious theory for isomorphism, in which he proposed that chemical units such as ammonium and potassium, or cyanide and chloride, had similar shapes and sizes and thus formed crystals with similar shapes. He was entirely correct in general, but entirely wrong in his particular explanation.

The finding that Hinrichs's chart of 1867 is actually a double spiral—which begins in a clockwise fashion but then reverses direction and continues in a counterclockwise direction—stems simply from reading what Hinrichs wrote. The additional proposal that the transition elements in his 1869 system are arranged in a slanted fashion to reflect their relative atomic weights—whereas other elements are not arranged in this way owing to a printer's oversight—comes from a consultation of his original manuscripts.

Hinrichs was not well connected to other chemists, and was evidently unaware of (or uninterested in) many of the considerable developments that had taken place in chemistry in the decade or so leading up to 1867. There is, for example, no hint of the "new" structural organic chemistry in any of his publications before 1870. It seems that, at the time he constructed his *Atomechanics*, Hinrichs's knowledge of chemistry was based almost entirely on what he learned as a student in the early 1850s, augmented by some selected reading of more recent chemical papers (Dumas's 1858 paper for example).

Hinrichs never carried out chemical research of his own and, after he moved to the United States, he was scientifically isolated and kept busy with his instructional and organizational duties. Very likely, he had little time or opportunity to keep up with the newest ideas. Thus, in the field of chemistry Hinrichs was, to a great degree, intellectually frozen in the 1850s. From this perspective, it is even more remarkable that he was able to devise a periodic system at all.

To call Hinrichs a crank, as some have, is in my opinion an oversimplification. To be sure, he invented novel comprehensive systems in vastly divergent areas of science, often proposing new vocabulary that proved more of a barrier than an aid to understanding. He was often guilty of forcing data to fit a preconceived idea, and finding more meaning in correlations and trends than actually exist. He resorted to

[21]For a detailed discussion of this topic, see Ann Robinson's Chap. 1 in this volume.

private publication when he could not get his ideas into the mainstream scientific press; furthermore, he distributed his *Atomechanics*, his magnum opus, as a lithographed handwritten manuscript rather than a printed monograph. He sent his work to influential people, such as Darwin and Humboldt, even when it was far outside their area of expertise. He accused established scientists of either ignoring his ideas or trying to steal them.

Such behavior is certainly displayed by true cranks, but many of these tendencies are also displayed by those who, like Hinrichs, are outside the scientific mainstream. For example, not long before, in the 1840s, both Julius Robert Mayer (1814–1878) and Hermann Helmholtz (1821–1894) resorted to private distribution of their philosophical (and rather hand-wavy) ideas on the conservation of energy when they could not get them published; also like Hinrichs, their ideas were largely ignored at the time, and both later became involved in priority disputes [65].

In my view, Hinrichs was a polymath with an encyclopedic (if not completely up-to-date) knowledge of the natural world. His unpublished work, as it survives in the University of Illinois archives, is massive in quantity (multiple large handwritten volumes) and impressive both in scope and content. Although ultimately his penchant for uncritical systematizing prevented him from devising a more compelling periodic system, Hinrichs deserves credit for being one of the first to organize the chemical elements into a useful two-dimensional arrangement having both groups and periods based on increasing atomic weights, and to point out that many properties of the elements change in systematic ways not only within a group but also between them.

Acknowledgements I thank the William and Janet Lycan Fund of the University of Illinois for support, and Alan Rocke, Carmen Giunta, and Vera Mainz for numerous helpful suggestions.

Appendix: Transcription and Translation of Hinrichs's 1867 Monograph

Note: Underlining (which Hinrichs uses liberally) and figures are generally omitted from this transcript.

Atomechanik oder die Chemie eine Mechanik der Panatome	**Atomechanics or Chemistry a Mechanics of Panatoms**
(French title)	**(French title)**
Von Gustav Hinrichs	**By Gustav Hinrichs**
Professor der Physik, Chemie & Mineralogie an der Staats-Universitat, Chemiker der geologischen Survey des Staates Iowa, etc.	**Professor of Physics, Chemistry & Mineralogy at the State University, Chemist of the Geological Survey of the State of Iowa, etc.**
Iowa City, Vereinigte Staaten, N[ord] A[merika] 1867	**Iowa City, United State, North America 1867**
Geschichtliche Vorbemerkung	**Historical Foreword**
1855. Durch Untersuchungen über das Verhältniss der Wärme zum Licht und namentlich über die Erzeugung beider, wurde ich Anfang Februar 1855 auf das Princip geführt, worauf gegenwärtige Atomechanik fusst, nämlich das von der Existenz eines einzigen Elementes.	1855. By investigations of the relation of heat to light, and especially of the production of both, I was led to the principle at the beginning of February, 1855, on which the present atomechanics is based, namely, that of the existence of a single element.
Ich begann sogleich dieses Princip zu beweisen durch Ableitung von Folgerungen und Vergleich derselben mit den Resultaten der Erfahrung. Mehrere Personen in Kopenhagen, wo ich der Zeit studirte, machte ich mehr oder weniger ausführliche Mitteilungen darüber. So auch meinem betrauerten Lehrer, Professor G. Forchhammer.	I immediately began to test this principle by deriving conclusions and comparing them with the empirical results. Several persons in Copenhagen, where I studied at the time, made more or less detailed communications about it. So did my lamented teacher, Professor G. Forchhammer.
Eine Zeichenschrift des Titels „Hinrichssche Naturphilosophi[e]" enthaltend die Principien und einige bereits erlangter Hauptfolgerungen liess ich unteren 7te August obrigkeitlich beglaubigen. Dieses Schriftstück ist auch in meinem Besitz.	I officially certified on the 7th of August a document in characters entitled "Hinrichs's Naturphilosophie" containing the principles and some of the main conclusions already obtained. This document is also in my possession.
1856. Im October dieses Jahres glaubte ich hinreichend weit fortgeschritten in meiner Arbeit um an die Veröffentlichung zu schreiben. Ich sandte zu dem Ende kürzerer Anzeigen an viele der bestanden Männer der Wissenschaft und auch an einige Akademien, erwartend dass mir dadurch die wissenschaftlichen Zuschriften für das Nähere eröffnet werden wurden. Darin hatte ich mich aber sehr verrechnet. Ich sollte durch passiven Widerstand und den Mangel an Hielfe [sic] dazu gebracht werden, die Atomechanik noch merher [sic for mehrer] auszubilden.	1856. In October of this year, I thought I was sufficiently advanced in my work to rewrite the publication. I ended up sending shorter reports to many of the existing men of science and also to some academies, expecting this to open up closer scientific communication. I had miscalculated in that. Passive resistance and the lack of help were to persuade me to further develop atomechanics.

1856/7. Folgend Herren und Institute erhielten kürzere oder längerer Mittheilungen von mir: G. Forchhammer, Kopenhagen; John Tyndall und ausserdem die Editors of the Philosph[ical] Magazine, London; l'académie des sciences, sowie M. E. Arnoult, Redakteur des „Institut" u[nd] a[ndere] in Paris; Poggendorff, Krönig, A. Dubois Reymond, die physikalische Gesellschaft, in Berlin; W. Haidinger, A. Schrötter, K[önigliche] Akad[emie] der Wissenschaften, in Wien; Hansen in Gotha, v[on] Humboldt, Berlin, etc. etc. Alle diese Mitteilungen wurden Anfang 1857 oder in den letzten quartal 1856 abgegeben.

1856/7. The following gentlemen and institutes received shorter or longer reports from me: G. Forchhammer, Copenhagen; John Tyndall and also the Editors of the Philosphical Magazine, London; l'académie des sciences, as well as Mr. E. Arnoult, editor of "L'Institut" and others in Paris; Poggendorff, Krönig, A. Dubois Reymond, the Physical Society, in Berlin; W. Haidinger, A. Schröther, Royal Academy of Sciences, in Vienna; Hansen in Gotha, von Humboldt, Berlin, etc. etc. All of these communications were made in early 1857 or in the last quarter of 1856.

Wenn vielleicht Keine dieser Schriften gedruckt wurde, so darf ich dieselben doch als Veröffentlichung meiner Idee und der Auffassung sowie der Theilweisen Hörung des Problems anzusehen berechtigt sein. Denn obige Namen zeigen dass die ihnen gewordene Mitteilung dem durch Druck fast gleiche kommt; jedenfalls aber hatte ich zur Veröffentlichung alles mir mögliche gethan.

Even if none of these documents was printed, I may be entitled to regard them as a publication of my idea and of my opinion as well as the partial hearing of the problem. Because the above names show that the communication they received is almost the same as the printed one; at least I had done everything I could to publish it.

Trotzdem dass meine damaligen Mitteilungen sehr schwach waren, so war doch das ganze Problem der Atomechanik deutlich und klar aufgefasst, und deren Lösung bereits vielfach hervor geführt.

In spite of the fact that my communications at the time were very weak, the whole problem of atomechanics was distinctly and clearly understood, and whose solution had already been given multiple times.

Die ausführlichsten Mitteilungen gingen an Forchhammer in Kopenhagen, an A. Krönig in Berlin für die physikalischen Gesellschaft, deren Schriftführer er damals war, und an Dr. W. Haidinger in Wien für die wesentlich von ihm ins Leben gerufener K[onigliche] Akademie der Wissenschaften.

The most detailed communications went to Forchhammer in Copenhagen, to A. Krönig in Berlin for the physical society of which he was the secretary at the time, and to Dr. W. Haidinger in Vienna for the Royal Academy of Sciences, which he essentially founded.

Der jetzt schon entschlafene Forchhammer hat sich mehrfach für meine ferneren Arbeiten verwendet durch seinen grossen Einfluss in der dänischen Residenz. Seinem Andenken alle Ehre! Er führte mich sowohl in die Chemie als in der Krystallographie ein.

The now departed Forchhammer has several times applied for my further work due to his great influence in the Danish royal capital. All honor to his memory! He introduced me to both chemistry and crystallography.

Der humane W. Haidinger schrieb mir unverzüglich in einer Weise, die dem Manne eben so sehr zum Ehre geneigt, wie die grossen wissenschaftlichen Arbeiten in ihm den ausgezeichneten Gelehrten zur Geltung gebracht haben. Dass er dem jugendlichen Ungestüm Rechnung trug und Nachsicht übte danke ich dem hochwürdigen Greise nochmals.

The humane W. Haidinger immediately wrote to me in a way just as much to the man's honor as the great scientific work brought out the excellent scholar in him. I thank the venerable old man once again that he took account of youthful impetuosity and practiced forbearance.

Herr Dr. A. Krönig antwortet mir auf meine Zuschrift vom 26 Januar 1857 erst unteren 15 April 1857, und zwar nachdem ich mich bei dem Secretär der Gesellschaft geschickter Arbeit erkundigt hatte. Er sagt: „Ich habe Ihre Abhandlung mit Aufmerksamkeit durchgelesen ..." sie aber nicht imstande dieselbe der Gesellschaft vorzutragen, da er als dann dieselbe zu vertreten haben würde; u[nd] fährt fort: Um übrigens, so viel in meinen Kräften steht dafür zu thun, dass Ihre Atomechanik den Mitgliedern der physikalischen Gesellschaft bekannt wird, will ich dieselbe morgen unseren Lesezirkel übergehen, wo sie in die Hände jeden einzelner Mitglieder gelangt: Hatte Krönig die Abhandlung in dem Vierteljahr seit dem Empfang dieselbe der Gesellschaft noch gar nicht angezeigt.

Dr. A. Krönig replied to my letter of January 26, 1857 only at the end of April 15, 1857, after I had inquired about the work sent to the secretary of the Society. He said: "I read your paper carefully..." but was not in a position to present it to the Society, because then he would be responsible for it; and he continued: In order to ensure to the best of my power that your atomechanics becomes known to the members of the Physical Society, I want to pass it on to our reading circle tomorrow, where it will get into the hands of each individual member: Krönig had not reported the paper to the Society even in the quarter since it was received.

Kürzlich habe ich die mit so vielen und so humanem (!) Eifer von Herrn Krönig an Liebig gerichtete „Werthlosigkeit etc." sowie das „Neues Verfahren etc." Berlin, 1866 gelegen. Mit mir wird jeder Wissenschaftsmann die ernste und würdige Vertretung der freien Forschung und des ungehinderten Fortschritte darin gebührend würdigen. Am Schlusse des „Neuen Verfahrens" dem die „Werthlosigkeit" vorangestellt ist, scheint in der That ein wirklich neues Verfahren offenbar, zu sein!

Recently I have received the [monographs] "Worthlessness etc." as well as "New Procedure etc." Berlin, 1866, which Mr. Krönig addressed to Liebig with so much and so humane (!) zeal. Every scientist will duly honor with me the serious and worthy representation of free research and the unhindered progress in it. At the end of the "New Procedure", which is preceded by "Worthlessness", a really new process seems indeed to be evident.

[Translator's note: Krönig's monograph *Neues Verfahren zur Ableitung der Formel einer Verbindung aus den Gewichtmengen der Bestandtheile* (*New Procedure for Deriving the Formula of a Compound from the Weights of the Constituent Parts*), was issued with another by Krönig, entitled *Die werthlosigkeit einer grossen Anzahl von chemischen Formeln* (*The Worthlessness of a Large Number of Chemical Formulas*). In this latter monograph, Krönig argued that assigning a definite formula to many chemical substances was often unjustified. For example, he showed that the then-current analytical data for linoleic acid (Leinölsäure) were consistent with formulas having between 16 and 18 carbon atoms – and between 29 and 31 hydrogen atoms – for every 2 oxygen atoms (in terms of modern atomic weights; Krönig used H = 1, C = 6, O = 8). The modern formula for linolenic acid is $C_{18}H_{32}O_2$.]

Er heisst nämlich s[iehe] ob[en] im „Neuen Verfahren" dass der Herr Krönig sich zu den Verehrern der in neuester Zeit auch von Graham befürworteten Hypothese von der Einheit der Materie zähle! Meine ihm geschickte Abhandlung gehört ja schon nicht mehr der neuesten Zeit an, sondern 1856 und 1857. Seite 53, immer in „Neuen Verfahren": „Ich habe ... Ende 1863 ... ein eigenes (!) System von Atomgewichten aufgestellt. Dieselben unterscheiden sich von den Berzelius'schen fast nur hinsichtlich der Einheit auf welche sie be-

Namely, see above in the "New Procedure" that Mr. Krönig is one of the admirers of the hypothesis of the unity of matter, which Graham has recently also advocated! My treatise sent to him is no longer up to date, but [is from] 1856 and 1857. Page 53, again in "New Procedure" [Here Hinrichs quotes from Krönig's monograph, adding an exclamation point]: "At the end of 1863 ... I set up my own (!) system of atomic weights. They differ from Berzelius's almost exclusively in terms of the unit to which they

zogen sind." Aber Herr Krönig, in unterscheiden sich in nichts von denen ihnen in meiner Abhandlung vom Januar 1857 als dem Sekretär der physikalischen Gesellschaft anvertrauten Atomzahlen (für welcher Vertrauen Sie beruflich in 1857 dankten; doch das war vor dem Neuen Verfahren!) Ferner (S[eite] 56) Ich setze also H = 2, O = 32, Cl = 71, Ca = 80, etc u[nd] sagt auch noch: chlor betrachte Er als aus 71 Molekülen bestehend u[nd] s[o] w[eiter].

Es wird nicht nöthig sein, diesem ein Mehreres hinzuzufügen. Ich gehe es den Vertretern der Wissenschaft anheim, und namentlich meinen Landsleuten unter diesen, u[nd] als solche betrachte ich selbstverständlich alle Deutsche. Ich beklage mich nicht darüber dass ich damals kein grösserer Entgegenkommen fand; denn ich blicke jetzt als mehrfahrener u[nd] durch Schicksalsschlage gereifter kann auf den jugendlichen vertrauensvollen und selbst überschätzenden Enthusiasmus zurück. Ich danke es Männern wie Haidinger, du Bois Raymond [sic], Forchhammer u[nd] a[ndere] wenn dieselben in dem jugendlichen Enthusiasmus einen zu achtenden Kern entdecken; aber wer im Stande ist, erst eine anvertraute Entdeckung zurückzuhalten dann zu verwerfen – u[nd] nachdem der Kämpfend, sein Leben für die Idee, all seine Thatkraft der Entwickelung derselben widmend Jüngling der Ungunst politischer Umstande weichend (ich bin Schleswig-Holsteiner u[nd] Deutscher ! war aber in Kopenhagen) den fernsten Westen aufsuchen muss dort trotz aller Schwierigkeiten fortfährt an der Lebensaufgabe zu arbeiten – je nachdem 6 Jahren des Kampfes und Schweigens vielleicht ein „Verschollen" wahrscheinlich machen – wir dann nach einem Gott sei Dank neuen oder doch seltenen Verfahren dieselben Ideen, die ihm als Officier einer gelehrten Gesellschaft anvertraut wurden in kaum verändertet Gestalt – u[nd] mir es scheint Kaum weiter entwickelt! – als Eigene Arbeit veröffentlicht – ein[?]bringt man einfach vor das Forum der strebenden Fachmänner. Damit übergebe ich diese Sache der Männern der Wissenschaft, besonders in Deutschland.

Ferner habe ich dieses Fundamental-Princip (Pantogen oder Urstoff) ausgesprochen bei folgenden Gelegenheiten:

relate." But Mr. Krönig, there are no differences between yours and the atomic numbers in my treatise from January 1857 that I entrusted to the secretary of the Physical Society (for which trust you thanked [me] professionally in 1857; but that was before the New Procedure!) Furthermore (page 56) [Krönig says] I put H = 2, O = 32, Cl = 71, Ca = 80, etc., and also [he] says: He regards chlorine as consisting of 71 molecules etc.

It will not be necessary to add more to it. I leave that to the representatives of science, and especially my compatriots among them, and as such I naturally consider all Germans. I do not complain about the fact that I did not find a greater accommodation at the time; because I can now look back as a more experienced and more mature man on the youthful, trusting and overestimating enthusiasm. I thank men like Haidinger, du Bois Reymond, Forchhammer and others when they discover a kernel to be respected in youthful enthusiasm. But who is able first to quash a confided discovery and then to reject it – after struggling, devoting his life to the idea [and] all his energy to developing it, giving way to the unfavorability of political circumstances (I'm from Schleswig-Holstein and German! but was in Copenhagen) and having to go to the distant west despite difficulties to continue working there on the life's work. Depending on 6 years of struggle and silence, perhaps making "lost without a trace" probable – we then have, according to (praise God!) [Here Hinrichs is being sarcastic, and in the next sentence alludes to Krönig's "New Procedure"] a new or at least unusual procedure, the same ideas that were entrusted to him as an officer of a learned society, in hardly changed form – and as it seems to me hardly developed further! – published as his own work – [and] simply put in front of the forum of aspiring experts. Thereby I hand this matter over to men of science, especially in Germany.

I have also expressed this fundamental principle (pantogen or primordial substance) on the following occasions:

1860 – Versammlung die skandinavischer Naturforscher, Juli 1860, Kopenhagen. Auch meine Schrift „Die Erdmagnetismus als Folge der Bewegung der Erde in Aether.“ Kopenhagen 1860, S. 8, §15.

1865 – Introduction to the mathematical principles of the nebular theory or Planetology, Silliman's Journal, 1865, Bd. 39 p. 65 im §6. Hier ist wiederum dasselbe Princip ausgesprochen. In dieser Schrift ist unter anderer beweisen, dass die Planeten nach gleichen Zeitintervallen euch vom Centralkörper ablösten, so dass die Bewegung eine Pulsation, Schwingung war.

Im Sommer und Herbst desselben Jahres war ich schon nicht weit mit meine Krystallographischen Arbeit, basirt auf die Constitution der Salze wie gegenwärtig angenommen in dieser Schrift. Der Tod meines Lebensgefährten setzte aber meinem Wissenschaftlichen Werke einstweilen ein Ziel. 12 October 1865.

1866 Im Novemberhefte von Silliman's Journal ist meine Abhandlung „On the Spectra and Composition of the Elements“ vol 42, p. 364 unter §30 gebe ich einige Formeln für Genera, dort aber alle auf den rectangulären Base bezeigen. Hier mag es gesagt sein, dass ich viele hundert Gesetze versucht habe, ehe ich mit dem jetzigen Zufrieden war (siehe 21 etc. dieses Schrift).

Schon letzten Herbst kam ich mit Dana, einem der Herausgeber des American Journal, zu einen Verständnis über die Veröffentlichung meine Arbeit in diesem Journal; 15 Seiten in jeder Nummer. Einen frühern Verständnisse gemäss wollte ich mit dem Krystallographischen Theile anfangen, inductiv bis zum Pantogen hinarbeiten. Es meiner mehr als ein Jahr darüber vergangen sein, ehe ich auch nur die Hauptsachen würd angeregt haben. Umstände haben die gegenwärtige Veröffentlichung hervorgerufen.

Hiermit übergebe ich diese Arbeit dem wissenschaftlichen Publikum. Es ist die Frucht ernsten Strebens, und unverdrossener Ausdauer. Sie ist unvollkommen – unendlich unvollkommen vergleichen mit dem Ideale das mir vorschwebte, und dem ich mich stets zu nahern suchte. Aber ich glaube dass man doch darin einen Fortschritt der Wissenschaft finden wird.

1860 – Meeting of the Scandinavian Naturalists, July 1860, Copenhagen. Also my paper "The Earth's Magnetism as a Result of the Movement of the Earth in Aether." Copenhagen 1860, p. 8, §15.

1865 – Introduction to the mathematical principles of the nebular theory or Planetology, Silliman's Journal, 1865, vol. 39 p. 65 in §6. Here again the same principle is expressed. This document proves, among other things, that the planets detached themselves at equal time intervals from the central body, so that the movement was a pulsation, vibration.

In summer and autumn of the same year I was not far with my crystallographic work, based on the constitution of the salts as currently assumed in this document. The death of my life's companion [wife], however, set a goal for my scientific work for the time being. 12 October 1865.

1866 In the November issue of Silliman's Journal is my treatise "On the Spectra and Composition of the Elements" vol. 42, p. 364 under §30 I give some formulas for genera, but all point to the rectangular base there. It may be said here that I tried many hundreds of laws before I was satisfied with the present one (see 21 etc. of this paper).

Last fall I came to an understanding with Dana, one of the editors of the American Journal, about the publication of my work in this journal; 15 pages in each number. According to an earlier understanding, I wanted to start with the crystallographic part, [and] work inductively to the pantogen. It must have been more than a year since I even suggested the main points. Circumstances have caused the current publication.

I hereby hand over this work to the scientific audience. It is the fruit of serious striving and undaunted perseverance. It is imperfect - infinitely imperfect compared to the ideal that I had in mind and to which I always tried to approach. But I think you will find progress in science in it.

So mögen denn andre Kräfte, unter günstigern Verhältnissen mit grösseren Fähigkeiten und bedeutenderen Mitteln diesen Weg fruchtbringend finden.

So other forces may find this path fruitful under more favorable conditions with greater abilities and more important means.

Möge dieses Arbeit der Wissenschaft als ein Beitrag zum wahren Fortschritt erscheinen! Das war mein Wunsch und mein ernstes Streben während der langen, frühen Jahre die seit der jugendlichen Sturm und Drang periode in 1856 u[nd] 1857 verflossen!

May this work of science appear as a contribution to true progress! That was my wish and my serious pursuit during the long, early years that had passed since the juvenile Sturm und Drang period in 1856 and 1857!

Gustavus Hinrichs, Iowa City, im Staat Iowa; Redaktion von Novbr 1866 – Juni 1867

Gustavus Hinrichs, Iowa City, in the state of Iowa; Edition of November 1866 – June 1867

Bei der wissenschaftlich fast isolirten Stellung werden Briefe von Fachmännern wir sehr wollkommen sein; jede Meinung ab für oder wider, ist mir gleich werth. Man addressiere:

In the scientifically almost isolated location, letters from experts will be very welcome; any opinion for or against is of equal value to me. One addresses:

Prof. Gustavus Hinrichs
Iowa City
Im Staat Iowa
U. St. Nord Amerika

Prof. Gustavus Hinrichs
Iowa City
Iowa
U. S., North America

Nur unter diese Addresse sind Briefs der Bestellung sicher. Wenn Briefe, wie neulich, bloss – Iowa US – als Localbezeichnung habend, sich ankommen, ist es ein reiner Zufall.

Only at this address are letters of order certain [to arrive]. If letters, like the other day, only have – Iowa US – as the local name, their arrival is a pure coincidence.

Vocabulaire

Atomechanik – für Mechanik der Atome

Ato – für Atom in Zusammensetzungen, wie

Atometer (*m*) Maass, Hohe der Atom, §.14.
Atobar, der schwere Theil des Atoms, §.123
Atostere, der Gesammtraum des Atoms, §.20
Atogramme, das Gewicht des Atoms, §.14
Atomare, *a*, das Areal oder der Durchschnitt des At[oms], §.14

Pantogen oder Panogen, §.3
Panatom oder Pantatom, §.7
Trigonoid, etc. §.10 u[nd] s[o] w[eiter].

Vocabulary

Atomechanik – for mechanics of atoms

Ato – for atom in combinations, like

Atometer (*m*) measure, height of the atom, §.14.
Atobar, the heavy part of the atom, §.123
Atostere, the total space of the atom, §.20
Atogramme, the weight of the atom, §.14
Atomare, *a*, the area or the base of the atom, §.14

Pantogen or Panogen, §.3
Panatom or Pantatom, §.7
Trigonoid, etc. §.10 and so on.

Einleitung

1. Es giebt überhaupt keinen absoluten Unterschied zwischen den chemischen Elementen. Dieselben physikalischen und chemischen Eigenschaften finden sich bei allen

Introduction

1. There is no absolute difference between the chemical elements. The same physical and chemical properties are found in all elements, only in different magnitude. Therefore,

Elementen wieder, nur in verschiedener Grösse. Daher dürften dieselben füglich als quantitative Modificationen eines Stoffes zur betrachten sein. Diese Annahme wäre auch an sich philosophischer, als die der Ursprünglichkeit der vielen chemischen Elemente.

they may be considered as quantitative modifications of a [common] substance. This assumption would also be more than [the assumption] of the originality [unrelatedness?] of the many chemical elements.

2. Es ist daher eine wichtige und legitime Aufgabe der Wissenschaft zu erforschen ob diese Annahme in ihren mathematischen Folgerungen nicht nur genügt, die chemischen Elemente in ihren Eigenschaften darzustellen, wie die Gravitation die Bewegungen der Planeten ausdrückt; sondern auch die bekannten Gesetze der chemischen Verbindungen erklärt als rein mechanische Folgen neue Gesetze entdeckt und vor Allem auch den lange gesuchten Schlüssel zu den Krystallformen und den mit derselben so wesentlich modificirten physikalischen Eigenschaften an die Hand giebt – etwa wie die Gravitation auch die kleinern Abweichungen als Perturbationen nachwies, selbst Form und Dichte bestimmte und gar Planeten (Neptun) und Sonnen (Companion des Sirius) errechnete. Wir werden manche bis jetzt unbekannte Elemente andeuten.

2. It is therefore an important and legitimate task of science to investigate whether this assumption in its mathematical conclusions is not only sufficient to show the properties of the chemical elements, [and] how gravity expresses the movements of the planets; but also the known laws of chemical compounds are explained as purely mechanical consequences of newly discovered laws and, above all, the long-sought-after key to crystal forms and the physical properties that have been so significantly modified with them - such as how gravitation also calculated the smaller deviations as perturbations, even determined shape and density and even calculated planets (Neptune) and suns (companion of Sirius). We will hint at some previously unknown elements.

3. Wir wollen daher annehmen, dass Alles der materiellen Natur aus einem einsigen [sic for einstigen] Stoffe hervorgegangen ist. Dieses ursprüngliche Element nennen wir füglich Pantogen.

3. We therefore want to assume that everything of a material nature has arisen from a former substance. We may call this original element pantogen.

4. In den vorliegenden Blättern gebe ich nur eine vorläufige Anzeige, einen Ueberblick meiner Arbeit, doch hinreichend dass der Fachmann ein bestimmtes Urtheil über dieselbe erlangen kann. In den der Physik und Chemie gewidmeten Zeitschriften hoffe ich zur vollständigern Darlegung der Beweise Gelegenheit zu bekommen. Wenn meine Theorie als auf gesunder Basis ruhend und den Weg zeigend zur Ergründung der Constitution der Materie, erkannt wird, so ist meine Aufgabe gelöst. Den vollständingen Aufbau muss grössern und mannigfältigern Kräften vorbehalten bleiben.

4. In the present pages I give only a preliminary announcement, an overview of my work, but sufficient that the expert can get a certain judgment about it. In the journals devoted to physics and chemistry, I hope to have the opportunity to fully present the evidence. If my theory is recognized as resting on a healthy base and pointing the way to the exploration of the constitution of matter, then my task is solved. The complete structure must be reserved for large and varied forces.

5. In dem felsenfesten Vertrauen auf die Universalität und Einheit der Gesetze der Mechanik und in dem durch ernstes und unbefangener Studium der exacten

5. This work is rooted in the rock-solid trust in the universality and unity of the laws of mechanics and in the fact that through serious and impartial study of the exact

Wissenschaften erlanget Bewusst sein, dass die chemischen Processe ebenso sehr wie die Planeten Bewegungen einfache mechanische Bewegungs-Erscheinungen sind, wurzelt diese Arbeit; die mich in beiden Hemisphären, unter drei verschiedenen Nationen und mannigfachen Wechselfällen schon reichlich zwölf Jahre beschäftigt hat. Als Knabe las ich dass die Physik in mechanische und chemische Physik eingetheilt werde; Wärme und Electricität gehörten zur chemischen (!) Physik. Als ich Student war zeigten die Lehrbücher die ganze Physik schon als mechanische Wissenschaft. Jetzt giebt es schon eine schöne physikalische Chemie – welch ein Fortschritt seit den Tagen der chemischen Physik. Ich versuche jetzt eine Chemie als Mechanik, die Mechanik der Atome, wie Astronomie die Mechanik der Himmelskörper – die auch nur kosmische Atome sind.

sciences, that chemical processes are phenomena involving simple mechanical movement just like planetary movements; this has occupied me in both hemispheres, among three different nations and manifold vicissitudes for a good twelve years. As a boy I read that physics is divided into mechanical and chemical physics; Heat and electricity were part of chemical (!) Physics. When I was a student, textbooks showed all of physics as a mechanical science. Now there is beautiful physical chemistry – what progress since the days of chemical physics. I am now trying chemistry as mechanics, the mechanics of atoms, like astronomy is the mechanics of celestial bodies – which are also just cosmic atoms.

Erster Abschnitt. Das Pantogen und die Elemente.

6. Die Hypothese. Das Pantogen besteht aus gegenseitig gleichen Atomen, die sich gegenseitig anziehen und durch Wärme oder Bewegung von einander gehalten werden.

Diese Atome sind hier selbstverständlich als materielle Puncte der Mechanik zu betrachten; Keinerlei occulte Eigenschaften dürfen ihnen beigelegt.

7. Pantogen Atom kürzen wir ab als Pantatom; das Gewicht eines solchen setzen wir als Einheit. Das Gewicht irgend einer Anzahl *n* Pantatome ist also = *n*.

8. Freier Pantogen kommt wahrscheinlich vor in der äussersten Sonnenatmosphäre (wirkt lichterregend) und in den planetarischen Nebeln. Wasserstoff steht demselben am nächsten. Verhältniss zum Lichtäther bleibe dahingestellt.

9. Zur Darstellung erscheinen im Folgenden als am Geeignetsten: die schwereren Alkali-Metalle (Ka, Rb), ferner Cl, P, Sb, S; Ca, Sr, Ba, u[nd] a[ndere] Dissociations Versuche müssen mit Hülfe hoher Hitzegrade oder electrischer Spannung angestellt werden.

10. Verbindungsweisen in einer Ebene sind für gleiche materielle Punkte nur zwei mög-

First section. Pantogen and the Elements.

6. The hypothesis. Pantogen consists of mutually identical atoms, which attract each other and are held to one another by heat or movement.

These atoms are here of course to be regarded as material points of mechanics; no occult features may be attached to them.

7. Pantogen atom we abbreviate as pantatom; the weight of such we set as unity. The weight in a number *n* of pantatoms is therefore = *n*.

8. Free pantogen probably occurs in the outermost solar atmosphere (appears light-producing) and in the planetary nebulae. Hydrogen is closest to it. The relationship to the luminiferous ether remains undecided.

9. For the purposes of illustration, the following are the most significant: the heavier alkali metals (Ka, Rb), furthermore Cl, P, Sb, S; Ca, Sr, Ba, among others. Dissociation experiments must be carried out with a high degree of heat or electrical tension.

10. There are only two possible compound arrangements in a plane for equal material

lich: als Ecken eines gleichseitigen Dreiecks oder quadrats. Demgemäss ergeben sich zweierlei Pantogen-Verbindungen oder Elemente; nämlich Trigonoide und Tetragonoide.

points: at the corners of an equilateral triangle or a square. Accordingly, there are two kinds of pantogen compounds or elements; namely trigonoids and tetragonoids.

11. Regelmässige Trigonoid-Formen sind die beistehenden. Jeder Punkt bezeichnet einem Pantatom. Als Kurze Bezeichnung dieser Formen gebrauchen wir im Folgenden eine der beiden nebenstehenden, je nachdem Seitenzahl oder Pantatomzahl der Figur erfordert wird.

11. Regular trigonoid forms are the [ones shown] adjacent. Each point denotes a pantatom. As a short description of these forms, we have described below one of the two ones [shown] adjacent, depending on the number of sides or the number of the number of atoms in the Figure.

[Translator's note: Calculations are given here, related to Figs. 11a-f, on the number of points in six- and four-sided rafts of having sides of equal length. For hexagons, the series is 1, 7, 19, ... and for squares is 4, 9, 16, ...]

Uebergangsformen die sehr einfach sind und wirklich vorkommen sind.

Transitional forms that are very simple and that really occur.

[Translator's note: Calculations are given here, related to Figs 11g-i, on certain six-sided rafts having sides of unequal length; examples include those with 13, 23, and 23 [sic; should be 30] points.]

Figur [11]h' ist 33 [sic for 34]. Kommt nicht vor.

Figure 11h' [a hexagonal raft with unequal sides] is 34 [points]. Does not occur.

12. Tetragonoid-Formen [Figs 11k-m] sind, als Vorkommend quadrat $4^2 = 16$, Rechteck $[4 \cdot 3] = 12$, $[4 \cdot 5] = 20$ und deren Multipla.

12. Tetragonoid forms [Figs 11k-m] are, as the previous square $4^2 = 16$, rectangle $[4 \cdot 3] = 12$, $[4 \cdot 5] = 20$ and their multiples.

13. Indem diese Pantatom-Platten sich vertikal übereinander lagern entstehen Prismen als die Atome der Elemente. Diese Lagerung ist durch die Mechanik absolute geboten.

13. By placing these pantatom plates vertically above one another, prisms emerge as the atoms of the elements. This arrangement is absolutely required by the mechanics.

14. Die Basis oder der Durschnitt [sic] dieser Prismen (obige Figuren) nennen wir Atomare (areal) und bezeichnen es durch a. Die Anzahl solcher Platten giebt die Höhe der Prismen; wir nennen es Atometer, da es das Maass [sic] der verschiedenen Glieder die zur gleichen Basis gehören ist. Wenn nur ein Prisma, so ist die Gesammtsumme der Pantatome im Element-Atom oder das Atomgewicht = Atogramme $g = m \cdot a$ (siehe §.7.)

14. We call the base or section of these prisms (above figures) atomare (areal) and designate it by a. The number of such plates gives the height of the prisms; we call it the atometer, because it is the measure of the various members that belong to the same base. If there is only one prism, then the sum total of the pantatoms in the element atom or the atomic weight = atogram $g = m \cdot a$ (see §.7.)

15. Es ergiebt sich nur leicht mathematisch, dass

1° nicht jeder Zahl n entspricht ein Element g (Zahl des Elemente ist beschränkt).

2° Gleicher Atomzahl können verschiedene Formen entsprechen.

15. Now it is easy to find mathematically that

1st. Not every number n corresponds to an element g (the number of elements is limited).

2nd. The same number of atoms can correspond to different forms.

16. Die Eigenschaften der Elemente sind hier als mathematische Functionen des Gewichts *g*, der Höhe *m* und des Durchschnitts *a* zu betrachten, indem diese 3 Grössen die Zahl und Lage der Pantatome völlig Bestimmen. Das Allgemeinere von diesen Grössen ist aber *a*, die Form der Basis des Atoms; also wird der allgemeine Charakter des Elementes bestimmt sein durch *a*, während die besondere Bestimmung dieses Charakters durch *m*, die Höhe, ausgedruckt wird.

17. Dadurch ist die natürlich Klassification der Elemente gegeben. Sie zerfallen in 2 Ordnungen (die Trigonoide und die Tetragonoide) nach der gegenseitigen Verbindung der Pantatome im Durchschnitt. Diese Ordnung zerfällt in Genera nach der äussern Form der Gestalt (Fig. a–m in 11 u[nd] 12). Die Species (das Element) wird bestimmt durch das Atometer *m*. Varietäten (äusserst nahe verwandte Elemente nahe gleichen Atomgewichts) entstehen durch Anwachsen der im Folgenden bestimmten Aufsätze auf dem Prisma.

18. Die genaue Bestimmung von *m* und *a* (da *g* durch Erfahrung bestimmt ist) ist nicht leicht. Die nun folgende Bestimmung ist aber jedenfalls im Ganzen richtig, da dieselben allen Anforderungen der Erfahrung genügen. In der Tafel ist immer: *A* das jetzt gebräuchliche Atomgewicht; *g* unser Atogramm und da H aus 2 Pantatomen bestehend angenommen wird, $g = 2A$. Wir führen keine besonderer Zeichen für diese Werthe ein, sondern benutzen die Anfangsbuchstaben die lateinischen Namen. Wenn nöthig unterscheidet man unsre Werthe leicht von den alten Equivalenten durch die Bezeichnung: Atogramm oder Hinrichs' Atomzahlen.

19. Die Genera bezeichne ich durch dem Anfangsbuchstaben ihres griechischen Namens. Dem entspricht die Formel wie der lateinische Anfangsbuchstaben und der Zahl die Species bezeichnet. Dadurch wird der Chemie eben so sehr gedient als durch die gebräuchlichen Zahlen und Symbole. Beispiel:

	Symbol		Werth]		
Genus:	Phosphoide = Φ	=	$m{\cdot}(p)$		
Species:	Phosphor = P	=	$9{\cdot}(7)$	=	63

16. The properties of the elements are to be regarded here as mathematical functions of the weight *g*, the height *m*, and the base *a*, in which these 3 measures completely determine the number and position of the pantatoms. But the most general of these quantities is *a*, the form of the base of the atom; Thus the general character of the element will be determined by *a*, while the particular determination of this character will be expressed by *m*, the height.

17. This gives the natural classification of the elements. They divide into 2 orders (the trigonoids and the tetragonoids) according to the mutual association of the pantatoms on average. This order divides into genera according to the external form of the figure (Figs. a-m in 11 and 12). The species (the element) is determined by the atometer *m*. Varieties (very closely related elements of almost the same atomic weight) are created by merging the subsequently determined caps onto the prism.

18. The exact determination of *m* and *a* (as *g* is determined by experience) is not easy. In any case, the following determination is correct on the whole, because it satisfies all the requirements of experience. In the table is always: *A* is the atomic weight now used; *g* our atogram, and because H [i.e., hydrogen] is assumed to consist of 2 pantatoms, $g = 2A$. We introduce no special symbols for these values, but use the initials of the Latin name. If necessary, our values are easily distinguished from the old equivalents by the term atogram or Hinrichs's atomic numbers.

19. I designate the genera by the initial letter of its Greek name. This corresponds to the formula as the Latin initial letter and the number denoting the species. This serves chemistry as much as the ordinary numbers and symbols. Example:

	Symbol		Value		
Genus:	Phosphoid = Φ	=	$m{\cdot}(p)$		
Species:	Phosphorus = P	=	$9{\cdot}(7)$	=	63

Verbindungen: Kα X stellt alle Verbindungen irgend eines Kaloids (Li, Na, Ka, etc) mit irgend einem Choroid (Fl, Cl, Br, I) dar.

Compounds: Kα X represents all compounds of any kaloid (Li, Na, Ka, etc.) with any choroid (Fl, Cl, Br, Jo).

20. Die bis jetzt betrachteten Atome der Elemente sind natürlich nicht in Contact; der Raum den sie beherrschen ist also grösser als das Prisma $m{\cdot}a$. Den ganzen Raum den ein Atome erfüllt nennen wir Kurz das Atostere (atom-Volum ist unbequem und barbarisch); bezeichnen wir es durch S, so ist offenbar $g = s{\cdot}S$ wo s das specifische Gewicht ist.

20. The atoms of the elements considered so far are of course not in contact; the space they dominate is thus larger than the prism $m{\cdot}a$. The whole space that an atom fills is called, for short, the atostere (atomic volume is inconvenient and barbaric); if we denote it by S, then obviously $g = s{\cdot}S$ where s is the specific gravity.

I Ordnung. Trigonoide (Metalloide).

1st Order. Trigonoids (Metalloids)

[Translator's note: in the following tables, (p) denotes a six-sided triangular raft containing p pantatoms, $/p^2/$ denotes a square raft with sides of length p, and $[i{\cdot}j]$ denotes a rectangular raft with sides of length i and j.]

21. Genus I. Pantoide [Pantoids]. $\Upsilon = m{\cdot}(1)$

	A	α	g	m	(p)	s	S
H	1	0.0	2	2	(1)	--	--

α ist die Abweichung meiner theor[etische] Werth von der Erfahrung, A.

α is the deviation of my theoretical value from the empirical value, A.

22. Genus II. Choroide [Chloroids]. $X = (1) + m{\cdot}(p)$

	A	α	g	m	(p)	s	S
F	19	−1.0	36	5	(7)	--	--
Cl	35.5	0.0	71	10	(7)	1.33	53.4
Br	80	−1.5	157	12	(13)	2.97	53.8
Jo	127	0.0	254	11	(23)	4.95	57.2

23. Gen. III. Phosphoide [Phosphoids]. $\Phi = m{\cdot}(p)$

	A	α	g	m	(p)	s	S
N	14	0.0	28	4	(7)	--	--
P	31	0.5	63	9	(7)	1.83	33.8
As	75	0.0	150	8	(19)	5.75	26.0
Sb	122	0.0	244	13	(19)	6.70	36.0
Bi	210	0.0	420	12	(30)	9.9	41.0

[Translator's note: In Hinrichs's 1867 paper in the *Journal of Mining*, he says "In the table of the phosphoids in my "Atomechanics" the product for As and Sb, and the atomare for Bi are mis-stated, having been written in the bustle of a printing office." Thus g should be 152 for As and 247 for Sb, and (p) for Bi should be 35.]

24. Gen. IV. Sulphoide [Sulphoids]. $\Theta = m{\cdot}/p^2/ = \mu{\cdot}[2p^2]$ (auch: Thionoide [also: Thionoids])

	A	α	g	m	$/p^2/$	s	S
O	16	0.0	32	8	2^2	--	--
S	32	0.0	64	8	$2{\cdot}2^2$	2.08	30.8
Se	79.4	+0.6	160	10	4^2	4.8	33.4
Te	128	0.0	256	8	$2{\cdot}4^2$	6.26	37.6

25. Genus V. Kaloide [Kaloids]. $K\alpha = 2(7) + 2 \cdot m \cdot 4^2$

		A	α	*g*	*m*	*s*	*S*
	Li	7	0.0	14	0	.6	23.2
	Na	23	0.0	46	1	.97	47.6
	Ka	39.1	–0.1	78	2	.865	90.0
?	In	71	0.0	142	4	7.362	19.3
	Rb	85.4	+1.6	174	5	1.52	112.2
?	Cs	183	+2.0	160	8		
?	Tl	204	–5.0	256	12	11.57	34.0

26. Genus VI. Calcoide [Calcoids]. $X\alpha = 2 \cdot 4^2 + m \cdot 2[4 \cdot 6]$

	A	α	*g*	*m*	*s*	*S*
Ca	40	0.0	80	1	1.6	50.0
Sr	87.6	+0.4	176	3	2.5	70.4
Ba	137	–1.0	272	5	4–5	54–68

27. Genus VII. Kadmoide [Kadmoids]. $K\delta = 2 \cdot 4^2 + m \cdot 2 \cdot [4.6]$ oder [or] $2 \cdot 4^2 + \mu \cdot [8 \cdot 12]$

da *m* gerade! [where *m* is even!; i.e., $m = 2\mu$]

	A	α	*g*	*m*	μ	*s*	*S*
Mg	24	0.0	48	2·[4.6]		1.74	27.6
– –							
Zn	65.2	–1.2	128	2	1	6.8–7.2	18.0–18.4
Cd	112	0.0	225	4	2	8.7	25.6
Pb	207	0.0	414	8	4	11.5	36.0

28. Genus VIII. Ferroide [Ferroids]. $\Sigma\iota = K + m \cdot 2[3 \cdot 4]$

m gerade [*m* even] $m = 2\mu$

	A	α	*g*	*m*	*K*	*s*	*S*
Al	27.4	+0.1	55	2	(7)	2.5–2.6	21–22
Fe	56	0.0	112	4	16	7.0–7.6	15–16
Rh	104.4	–0.4	208	8	16	12	17.3
Ir	198	+2.0?	400	16	16?	21	19

29. Genus IX. Molybdoide [Molybdoids]. $M\lambda = K + m \cdot 2[3 \cdot 4]$ *m* odd (except Cr)

m ungerade [*m* odd] $m = 2\mu + 1$

	A	α	*g*	*m*	*K*	*s*	*S*
[Cr	52.2	–0.2	104	4	8	6.0	17]
Mo	92	0.0	184	7	16		
Va	137.2	–0.2	274	11	16		
Wo	184	0.0	368	15	8	17.5	21

30. Genus X. Cuproide [Cuproids]. $X\upsilon = K + m[5 \cdot 8]$

	A	α	*g*	*m*	*K*	*s*	*S*
Cu	63.4	+0.1	127	3	(7)?	8.85	14
Ag	108	0.0	216	5	16	10.5	25.7
Au	197	–1.0	392	9	32	19.5	20.0

31. Genus XI. Titanoide [Titanoids]. Tτ = 2[3·4] + *m*[4·5]

	A	α	*g*	*m*	*s*	*S*
C	12	0.0	24 = 2[3·4]		2.2	11.0
--						
Si	28	0.0	56	1	2.34	23.7
Ti	50	−2.0	96	3		
Pd	106.6	+1.4	216	9	11–12	18
Pt	198	0.0	396	18	21.5	18

vielleicht auch nach [perhaps also after]

Zr	89.6	−1.6	176	7	4.15	41.1
Sn	118	0.0	236	10	7.3	32.3
Ta	137.6	+0.4	276	12		
Th	231	−3.0	456	21		

32. Unbestimmt [Undetermined]

	A	α	*g*	*m*	*s*	*S*
Bo	11					
Be	9.3					
Hg	200					

Doch besteht Hg höchst wahrscheinlich aus Hexagonen mit folgenden Seitenlängen	However, Hg most likely consists of hexagons with the following side lengths
1 + 2 + 4 + 6 + 6 + 5 + 4 + 2 + 2	1 + 2 + 4 + 6 + 6 + 5 + 4 + 2 + 2
Also aus	That is
1 + 7 + 37 + 61 + 91 + 91 + 37 + 7 + 7 = 400	1 + 7 + 37 + 61 + 91 + 91 + 37 + 7 + 7 = 400
Pantatomen. Demnach fiele Hg zwischen Kα und den Metalloiden; entspräche H unter den Metallen.	pantatoms. Accordingly, Hg would fall between Kα and the metalloids; would correspond to H among the metals
33. Varietaten der Ferroide sind	33. Varieties of ferroids are

Sideroide [Sideroids], Σδ = *K* + 4·2[3·4] = *K* + 96

	A	α	*g*	*K*	*s*	*S*
[Cr	52.2	−0.2	104	8	6.0	17]
Mn	55	−1.0	108	12	8.07	13.5
Fe	56	0.0	112	16	7.0–7.6	15–16
Ni	58	0.0	116	20	8.6	13.5
Co	60	0.0	120	24	8.5	14.1
--						
U	120	0.0	240			

Rhodoide, Rh, Ru	Rhodoids, Rh, Ru
Iridoide, Ir, Os	Iridoids, Ir, Os
denen Atogramme aber noch zu wenig bekannt; werden den Σδ entsprechend gefunden werden. Für Co, Ni Schneider's Werthe bestätigt von v. Sommaruga; für Mn, Cr sind neue Bestimmungen geboten. Cr wahrscheinlicher Sideroid als Molybdoid.	However, too little is known about their atograms; the Σδ will be found accordingly. For Co, Ni Schneider's values confirmed by von Sommaruga; for Mn, Cr new determinations are required. Cr probable sideroid as molybdoid.

[Translator's note: Erwin von Sommaruga wrote a paper on the equivalents of Co and Ni, "Ueber die Aequivalente von Kobalt und Nickel," *Wien. Akad. Sitzungsber.* **1866**, *54* (Abth. 2), 50-60; *J. Prakt. Chem.* **1867**, *100*, 106-117.]

34. Einige Atome dargestellt.	34. Some atoms illustrated.
H, Fl, N, O } Gase (Vol[um] gleich)	H, Fl, N, O } Gases (Volume equal)
Ca, Ka	Ca, Ka

---- diagrams -----

35. Die vorstehenden Zahlen enthalten alles was der Erfahrung zu entnehmen ist und eigentlich schon mehr, da die Form theoretisch bestimmt ist. Es wird eigentlich von der Erfahrung nur der Name des betreffenden Elements entlehnt, welches den verschiedenen Atom kern m entspricht. Wie die analytische Bestimmung auch unsre Atogramme und Atometer nicht als entgültig dahingestellt, sondern mögen auch einselne Modificationen erfahren. Jedenfalls aber sind sie im Wesentlichen richtig. Diese Werthe von *m* und *a* sind gewissermassen die Fundamental Constanten der materiellen Nature, die fortschreitenden Wissenschaft wird alle Erscheinungen aus diesen Zahlen errechnen, da aller Stoff Pantogen und alle Vorgänge nur mechanisch sind. Die Grundlage dieser Atomechanik geben wir hier – den Ausbau geben wir den Männern der Wissenschaft anheim.	35. The preceding numbers contain everything that can be inferred from experience, and actually more, because the form is theoretically determined. It has actually borrowed empirically only the name of the element concerned, which corresponds to the different atomic nucleus m. Because the analytic determinations of our atogram and atometer are not final, they may also undergo some modifications. In any case, they are essentially correct. These values of *m* and *a* are, so to speak, the fundamental constants of material nature; advancing science will calculate all phenomena from these numbers, because all matter is pantogen and all processes are only mechanical. The basis of this atomechanics we give here – the expansion we submit to men of science.
36. Zur Veranschaulichung der im Vorstehenden enthaltenen mechanischen oder rationellen Klassification der Elemente stelle ich sie in folgender Zeichnung dar. Das Pantogen bildet den Mittelpunkt, di Genera sind nach ihrem Generschnitt in Strahlen dieses Punktes dargestellt und die Species werden in diesen Strahlen dort verzeichnet wo der Abstand vom Mittelpunkte dem Atogramm *g* gleich kommt (Maassstab *S* einheiten in $g = 1$ mm). Die Symbole und Formeln der Genera, sowie die Symbole und Zahlen *g* der Species sind für Bezeichnung beigesetzt.	36. To illustrate the mechanical or rational classification of the elements contained in the foregoing, I present them in the following drawing. The pantogen forms the midpoint, the genera are represented by rays from this point, and the species are recorded in these rays where the distance from the center equals the atogram *g* (*S* scale units in 1 mm). The symbols and formulas of the genera, as well as the symbols and numbers *g* of the species are appended for designation.

37. [Hinrichs's chart]

38. Die Atometer *m* folgen ganz einfachen Gesetzten, die wir jetzt näher betrachten wollen um zu sehen wie die Atogramme an sich indem sie obenstehende Klassification der Elemente gestalten schon ein wichtiger Beweisgrund des Pantogens sind.	38. The atometer *m* follows very simple laws, which we now want to examine closer to see how the atograms in themselves by the above classification of the elements are already important evidence of pantogen.

39. Die Chloroide geben $m = 5$ (Fl) und $= 10$ (Cl) für gleiches $a = (7)$; also wie 1:2, Br, Jo representiren einfache Modificationen von a; denn Cl ist 2-2-2, Br ist 2-4-2 und Jo ist umgekehrt 4-2-4.

40. Die Phosphoide geben $a = (7) = 2\text{-}2\text{-}2, (19) = 3\text{-}3\text{-}3$ und $30 = 4\text{-}3\text{-}4$, also consecutiv. $m = 4$ (P), 8 (As) und 12 (Bi) also = 1:2:3. Für gleiche a noch P, $4 + 5 = 9$, Sb $= 8 + 5 = 13$, constant 5.

41. Sulphoide haben alle m gleich ausgenommen Se. Die Aren [sic for Areal?] varieren regelmässig, O ~ 2^2, Se ~ 4^2 und S und Te das Doppelte $2 \cdot 2^2$ und $2 \cdot 4^2$.

42. Kaloide; m für Li : Na : Ka : Rb sind 0 : 1 : 2 : 5. Für die noch zweifelhaften In : Cs : Tl = 4 : 8 : 12 = 1 : 2 : 3 so dass dieselben füglich als $4 \cdot 4^2 = 8^2$ Basis inzwischen wären mit den Atometern $\mu = 1, 2, 3$. Wir nennen sie Thalloide, $\Theta\lambda = 2\,(7) + 2\,\mu\,8^2$. Dem entsprechen ihre Eigenschaften.

43. Calcoide bilden $m = 1, 3, 5$, mit constanter Differenz 2.

44. Kadmoide haben die geraden $m = 2, 4, 8$, so dass wahrscheinlich $K\delta = 2 \cdot 4^2 + \mu[8 \cdot 12]$ wo $\mu = 1, 2, 4$.

45. Die Ferroide geben die Meter Al : Fe : Rh : Ir = 2 : 4 : 8 : 16 = 1 : 2 : 4 : 8 = $2^0 : 2^1 : 2^2 : 2^3$ oder successiv Verdopplung; wie μ der Kadmoide, wie Cl [und] Fl, As und N.

46. Die ungeraden Werthe m der Molybdoide sind 7 : 11 : 15 zeigen also die constante Differenz 4 welche auch Glied der Ferroide ist.

47. Die Cuproide geben ähnliche 3 : 5 : 9 die Differenzen 2 u[nd] $4 = 2 \cdot 2$. Die Molybdenoide und Cuproide sind also, wie es auch das gleiche Atomare [3.4] erheischt, nur Seitenzweige der Ferroide.

48. Die Titanoide haben die Atometer 1 : 3 : 9 : 18 = $3^0 : 3^1 : 3^2 : 2 \cdot 3^2$ oder eine änliche Beziehung wie die der Ferroide.

49. Ganz besonders wichtig ist noch die gegenseitige Beziehung der 3 Haupt Genera mit rectangulärer Basis, nämlich die Ferroide [3·4] die Titanoide [5·4] und die Kadmoid [6·4] oder [12·8]. Dies ist ausgedrückt durch den fast gleicher Werth des Atometers Atogramms g für die successivter Glieder dieser Genera, nämlich

39. The chloroids give $m = 5$ (Fl) and $= 10$ (Cl) for the same $a = (7)$, that is 1: 2. Br, Jo represent simple modifications of a; because Cl is 2-2-2, Br is 2-4-2 and Jo is conversely 4-2-4.

40. The phosphoids give $a = (7) = 2\text{-}2\text{-}2, (19) = 3\text{-}3\text{-}3$ and $(30) = 4\text{-}3\text{-}4$, i.e. consecutive. $m = 4$ (P), 8 (As) and 12 (Bi) = 1: 2: 3. For the same a, P $= 4 + 5 = 9$, Sb $= 8 + 5 = 13$, constant 5.

41. Sulphoids all have the same m except Se. The areas vary regularly, O ~ 2^2, Se ~ 4^2 and S and Te doubled $2 \cdot 2^2$ and $2 \cdot 4^2$.

42. Kaloids; m for Li : Na : Ka : Rb are 0 : 1 : 2 : 5. For the still doubtful In : Cs : Tl = 4 : 8 : 12 = 1 : 2 : 3 so that the same would be acceptable as a $4 \cdot 4^2 = 8^2$ base with the atometers $\mu = 1, 2, 3$. We call them thalloids, $\Theta\lambda = 2\,(7) + 2\,\mu\,8^2$. This corresponds to their properties.

43. Calcoids form $m = 1, 3, 5$, with constant difference 2.

44. Kadmoids have even $m = 2, 4, 8$, so that probably $K\delta = 2 \cdot 4^2 + \mu\,[8 \cdot 12]$ where $\mu = 1, 2, 4$.

45. The ferroids give the measures Al : Fe : Rh : Ir = 2 : 4 : 8 : 16 = 1 : 2 : 4 : 8 = $2^0 : 2^1 : 2^2 : 2^3$ or successive doubling; like μ for the Kadmoids, like Cl [and] Fl, As and N.

46. The odd values of the molybdoids are 7: 11: 15, so they show the constant difference 4, which is also a [mathematical] term for the ferroids.

47. Similarly The cuproids give 3 : 5 : 9 the differences 2 and $4 = 2 \cdot 2$. The molybdenoids and cuproids are thus, as the same atomic area [3.4] requires, only side branches of the ferroids.

48. The titanoids have the atometers 1 : 3 : 9 : 18 = $3^0 : 3^1 : 3^2 : 2 \cdot 3^2$ or a similar relationship as that of the ferroids.

49. Especially important is the mutual relation of the 3 main genera with a rectangular base, namely the ferroids [3·4] the titanoids [5·4] and the kadmoids [6·4] or [12·8]. This is expressed by the almost equal value of the atometer [and] atogram g for the successive members of these genera, viz.

						Sauerstoffverbindung [Oxygen compound]	
Kadmoide		Mg 48	Zn 128	Cd 224	Pb 414	$K\delta\,\Theta$	1 : 1
Ferroide		Al 55	Fe 112	Rh 208	Ir 400	$\Sigma\iota_2\,\Theta_3$	1 : 1½
Titanoide	C 24	Si 56	Ti 96	Pd 216	Pt 396	$T\iota\,\Theta_2$	1 : 2

Jeder Chemiker wird die vollkommene Harmonie dieser Verhältnisse leicht werden ausführen können. Bo entspricht wahrscheinlich dem C unter den Mλ. Be auch die Σι.

Any chemist will easily be able to explain the perfect harmony of these relationships. Bo probably corresponds to C [i.e., carbon] under the Mλ. Be also [probably with] the Σι.

50. Diese einfachen Atometer-Beziehungen wären an sich wichtige Stützen für die Richtigkeit unserer Atom formeln; Sie sind aber ein positiver Beweis, wenn man bedenkt dass alle die in einem Genus vereinten Elemente die durch diese Atometer-Beziehungen verknüpft sind, höchst natürliche Gruppen darstellen, Gruppen der die besten bisher versuchten Klassificationen in sich schliessen.

50. These simple atometer relationships would in themselves be useful supports for the correctness of our atomic formulas; but they are positive proof, considering that all the elements united in one genus are linked by these atometer relationships, representing the highest natural groups, groups that contain the best classifications tried so far.

51. Schon hier wird eine hoch bedeutsame Analogie sich aufdrängen: nämlich die zwischen den bekannten Reihen organischer Verbindungen und unsern Genera. So ist die Formel der monatomischen Alkohole = $C^nH^{2n+2}O$ = H_2O + $n[CH_2]$ ganz von die Form unserer Chloroide X = (1) + m (p) oder Kaloide Kα = 2(7) + $m[2{\cdot}4^2]$. Die Eigenschaften sind in beiderlei Reihen ähnlich bestimmbar.

51. Already here a highly significant analogy will arise: namely, that between the known series of organic compounds and our genera. Thus, the formula of monatomic alcohols = $C^nH^{2n+2}O$ = C^nH^{2n} = H_2O + $n[CH_2]$ is entirely of the form of our chloroids X = (1) + m (p) or kaloids Kα = 2 (7) + m $[2{\cdot}4^2]$. The properties are similarly determinable in both series.

52. Allgemein mathematische Untersuchungen zeigen dass nicht aus jeder Zahl n Pantatome eine geschlossene prismatische Atomgestalt sich bilden lässt (§.15). Denn in $g = m{\cdot}a$ muss erstens das Atomare a eine geschlossene Figur sein, denen Pantatomare also schrittweise wächst und m kann auch nur regelmässig nach bestimmten Gesetzen fortschritten. Siehe Beispiele §.11 u. §§.38-50.

52. General mathematical investigations show that not every number n of pantatoms allows a closed prismatic atomic form to be built (§15). For in $g = m{\cdot}a$, first of all, the atomare a must be a closed figure, to which pantatomares thus grow gradually, and m can only progress regularly according to certain laws. See examples §.11 and §§.38-50.

53. Daraus folgt aber ferner dass nahezu gleiche Werthe von g verschiedenen Formen entsprechen können, d[as] h[eisst] nahe gleiche g geben Species verschiedene Genera. Dies zeigt unsre Karte §.37 recht deutlich. Beispiele:

53. It follows, however, that nearly equal values of g can correspond to different forms; that is, near equal g give species of different genera. Our chart § 37 shows this quite clearly. Example:

	Θ	Φ	X	Mittel [Average]	Κα	Χα, Κδ	Σι	Ττ	Mittel [Average]
C 24	O 32	N 28	Fl 36	30	Li 14			C 24	14-24
	S 64	P 63	Cl 71	66	Na 46	Mg 48	Al 55	Si 56	51
	Se 160	As 150	Br 157	156	Ka 78	Ca 80			79
	Te 256	Sb 244	Io 254	251		Zn 128	Fe 112	Ti 96	112
		Bi		420		Cd 224	Rh 208	Pd 216	216
						Pb 414	Ir 400	Pt 396	403

[54.] Man ersieht auch dass die Gruppen abwechselnd Metalle und Metalloide umfassen; so dass die bekannten Elemente in unsrer Karte in Spiralgeraden auf einander folgen. Siehe in 37 der Linien H-Li-C-O-N-Fl-Na-Mg-Al-Si-S-P-Cl-Ka-Ca-Ti und dann in entgegengesetzte Richtung Ti-Fe-Zn-In-Br-As-Se-Pd-Rh-Cd-Cs-Jo-Sb-Te. Da unsrere Spire [sic] umfasst nahezu alle Elemente die allgemeinere Bedeutung haben.

[54.] One sees also that the groups alternately include metals and metalloids; so that the known elements in our chart follow one another in spiral lines. See in 37 the lines H-Li-C-O-N-Fl-Na-Mg-Al-Si-S-P-Cl-Ka-Ca-Ti and then in the opposite direction Ti-Fe-Zn-In-Br-As-Se-Pd-Rh-Cd-Cs-Jo-Sb-Te. So our spiral includes almost all elements of more general significance.

55. Je kleiner überhaupt *g*, je näher das Element dem Pantogen steht, um so bildungsfähiger muss es sein. Daher sind C mit H, N und O aus dem innersten Kreise die Hauptbestandtheile aller organischen Körper (organogene) So auch die Elemente bis *g* gegen 100 (Fe = 112) die hauptsächlichsten Körper der unorganischen Welt – nicht nur auf Erden, sondern auch in den Meteoriten und Sternen, wie die Spectralanalyse zeigt.

55. The smaller the *g*, the closer the element is to pantogen, [and] the more capable of development it must be. Therefore, C with H, N and O from the innermost circle [of the chart] are the main constituents of all organic bodies (organogenic). So the elements up to *g* around 100 (Fe = 112) are the most important bodies of the inorganic world – not only on Earth, but also in the meteorites and stars, as spectral analysis shows.

56. Nach diesen allgemeinen Betrachtungen schreiben wir zur speciellen Uebersichtung der Eigenschaften der Elemente um endgültig zu entscheiden, ob unsre Annahme des Pantogens zur Erklärung der Elemente ausreicht und daher die Frage nach der Constitution der Materie löst. Wir betrachten am zweckmässigsten zuerst die chemische Eigenschaften oder eigentlich die Verbindungsweisen an sich. Hiernach folge die Untersuchung der physikalischen Eigenschaften und schliesslich die durch die Constitution gegebenen Morphologischen Eigenschaften (die Krystallformen). Die strenge Sonderung der Elemente und ihre Verbindungen wäre unzweckmässig; wir werden daher beide miteinander betrachten.

56. After these general considerations, we set down a special overview of the properties of the elements in order to finally decide whether our assumption of the pantogen is sufficient to explain the elements and therefore solve the question of the constitution of matter. We best consider first the chemical properties or actually the bonds per se. This is followed by the examination of the physical properties and finally the morphological properties (the crystal forms) given by the constitution. The strict separation of the elements and their compounds would be inappropriate; we will therefore consider both of them together.

Zweiter Abschnitt. Chemische Eigenschaften der Elemente.

57. Die chemische Verbindung zweier Atome A und B als mechanische Erscheinung betrachtet kann nur in einer nebeneinander Lagerung der Atome (juxtaposition) AB bestehen. So auch wären *m* Atome A sich mit *n* Atomen B verbinden. Um das Nähere zu erforschen wollen wir hier das im nächsten Abschnitt erörterte Gesetz der Gleichheit der Atostere der Element in Gasform annehmen, worauf wir auch in §.34 schon Bezug nahmen.

58. Wenn wir nun mit den heutigen Chemikern annehmen, dass je 2 Atome in Gasform ein Molekül bilden, so wäre nebenstehende Zeichnung wohl eine Darstellung eines solchen Moleküls. Die Atomen A haben ihrer Schwerpunkt in C; die Ebene durch C senkrecht zu der sei Verbindung Achse theilt das Molekül-Volumen in 2 gleich Theile. Dies wäre der Wasserstoff-Typus der modernen Chemie.

59. Für das eine A liesse sich ein Atom B eines demselben Genus angehörenden andern Elementes substitutiren, ohne irgend etwas zu andern. Die chemische Verbindung AB bestünde also aus 2 Volumen, ohne Condensation. Dies ist der Chlorwasserstofftypus der modernen Chemie.

60. Wenn 2 B sich mit A verbinden sind 2 Fälle denkbar. Entweder bilden sie 3 Atome eine gerade Linie (1 [und] 1‘) oder nicht (2); der letztere Fall giebt aber immer ein Dreieck.

Wären die Atome in einer geraden Linie so müssten sie durch die geringste Veranlassung (Wärmeschwingung Licht, Electricität etc.) für einen Augenblick aus derselben gerissen werden. Dann würde aber die gegenseitig Anziehen die drei Atome ein Dreieck BAB‘; und zwar ein gleichschenklicher da beide B mechanisch völlig gleich sind, also die Bezug auf A gleiche Lage haben müssen. A ist daher der Scheitel, B B‘ der Grundlinie des Dreiecks.

61. Sei nun C der Schwerpunkt des Dreiecks (also AC = 2 CO); dann ist das Gewicht BB‘ in Bezug auf die Axe EF, gleich dem Gewichte B in A‘ wenn CA = 2 CO also CA‘ = CA. Die Ebene EF theilt den Raum des zusammengesetzten Moleküls in 2 Theile. Da nun das System A,2B equivalent ist dem A,

Second Section. Chemical Properties of the Elements.

57. The chemical bonding between two atoms A and B, viewed as a mechanical phenomenon, can only consist of the side-by-side arrangement (juxtaposition) of atoms, AB. So also would *m* atoms of A combine with *n* atoms of B. In order to explore the details, we want to adopt the law of equality of the atostere of the elements in the gaseous form discussed in the next section, which we also referred to in §.34.

58. If we now assume with today’s chemists that every 2 atoms form a molecule in the gaseous form, the drawing opposite would be a representation of such a molecule. The atoms A have their center of gravity at C; the plane through C perpendicular to the connection axis divides the molecular volume into 2 equal parts. This would be the hydrogen type of modern chemistry.

59. An atom B of a different element belonging to the same genus could be substituted for A without changing anything. The chemical compound AB would therefore consist of 2 volumes, without condensation. This is the hydrogen chloride type of modern chemistry.

60. If 2 B connect to A, 2 cases are conceivable. Either they form 3 atoms in a straight line (1 [and] 1') or not (2); the latter case always gives a triangle.

If the atoms were in a straight line, they would have to be torn out of it for a moment by the slightest reason (heat vibration, light, electricity, etc.). Then, however, the mutual attraction of the three atoms would form a triangle BAB'; namely an isosceles because both B are mechanically completely the same, that is, they must have the same position with respect to A. A is therefore the vertex, B B' the baseline of the triangle.

61. Let C be the center of gravity of the triangle (so AC = 2 CO); then the weight BB‘ with respect to the axis EF is equal to the weight B in A‘ if CA = 2 CO thus CA' = CA. The level EF divides the space of the compound molecule into two parts. Thus the system A,2B is now equivalent to A,A‘, but

A', dies aber nach §.59, aus gleichen Volumen besteht, es folgt. In Verbindungen A,2B verdichten sich die 2 Volumina B zu einem, welcher ohne weitere Verdichtung mit diese Volum A zu 2 Volumen zusammentritt. Die Verdichtung ist also 2/3; die graphische Darstellung das Dreieck ABB. Dies ist der Gerhardt'schen Typus Wasser HHO.

according to §.59, it follows that it consists of the same volume. In compounds A,2B, the 2 volumes of B compress to one, which combine these volumes A to 2 volumes without further compression. The compression is therefore 2/3; the graphical representation of the triangle ABB. This is the Gerhardt water type HHO.

62. Für 2 A und 1 B würde ganz dasselbe gelten; dies sehen wir beispielweise am Stickstoffsoxidul N_2O und der Untersalpetersäure NO_2. Im ersten verdichten sich die 2N zu 1 Volum, im letztern die 2O. Dieser wohl bewiesene Fall zeigt deutlich, dass das Gesetz ein universeller ist, unabhängig von den „besondern" Eigenschaften den sich verbindenden Element. So auch in Wasser verdichten sich 2 Vol[umen] H zu einem Volum das mit dem O zusammentritt.

62. The same would apply to 2 A and 1 B; we can see this, for example, in nitrogen oxide N_2O and sub-nitric acid NO_2. In the first the 2N compress to 1 volume, in the latter the 2O. This well-proven case clearly shows that the law is universal, regardless of the "special" properties of the connecting element. So too in water, 2 volumes of H condense to a volume that combines with O.

63. Diese Condensation hat durchaus nichts Auffälliger. Wissen wir doch, dass in 1 Vol[um] Cl 35 ½ mal so viele Pantatome sind, als in dem gleichen Volum Wasserstoff u[nd] s[o] w[eiter]. So können dann auch leicht 2 H auf den Raum (Atostere) eines H zusammengedrängt werden. Ebenso 3-4 etc. und allgemein *n* H, so wohl als *n* A und m B.

63. This condensation is nothing particularly striking. We know that there are 35½ times as many pantatoms in 1 volume of Cl as there are hydrogen in the same volume, and so on. In this way, 2 H can easily be pushed together into the space (atostere) of one H. Likewise 3-4 etc. and generally *n* H, as well as *n* A and *m* B.

64. Diese ganze Schlussweise, sowohl was Stellung als auch was der Raumfüllung betrifft, kann leicht auf 1 A und *n* B ausgedehnt werden, und noch ferne auf *n* A und *m* B. In A mit 3 B müssen die 3 B, da sie gegenseitig gleich sind, auch zu A gegenseitig ganz gleiche Stellung haben; sie bilden also ein gleichseitiger Dreieck, und A steht gerade über dem Schwerpunkt dieses Dreiecks. Dieser Schwerpunkt representiert gewissermaassen einen neuen Atom B' und dieser ist mit A ohne Verdichtung verbunden. Die 3 Volumina B auf 1 reducirt sind zu dem 1 Volum A gefugt, so dass 2 Volum der Verbindung resultieren. Dies wäre der Ammoniak Typus mit Bestimmung der gegenseitigen Lage der Atome, welche Bestimmung der Typenlehre fehlt. Ebenso wie auch der Nachweis der Wie der Condensation erst durch dieser Theorie gegeben werden konnte. Im Allgemeinen:

64. This whole procedure, both in position and in terms of space filling, can easily be extended to 1 A and *n* B, and still further to *n* A and *m* B. In A with 3 B the 3 B, because they are equal to one another, also have mutually identical positions with respect to A; so they form an equilateral triangle, and A is just above the center of gravity of this triangle. This center represents, so to speak, a new atom B' and this is connected to A without compression. The 3 volumes of B reduced to 1 are added to the 1 volume of A, so that 2 volumes of the compound result. This would be the ammonia type with determination of the mutual position of the atoms, a determination that the type theory lacks. Likewise, the proof of how condensation [occurs] could only be given by this theory. In general:

65. Wenn *m* Atome A sich mit *n* Atomen B zu dem A^mB^n verbinden, so geschieht es, indem die m A sich auf 1 Vol[um] zusammenziehen, ebenso die n B auf 1 Volum, welche sich nun

65. If *m* atoms A combine with *n* atoms B to form the A^mB^n, so it happens, that the *m* A contract to 1 volume, as well as the *n* B to 1 volume, which are now next to each other, as

neben einander, lagern wie im Molekül eines Gases, so dass die Ebene der Atome A derjenigen von B parallel ist und die Verbindungslinie der Schwerpunkt auf beiden Ebenen sinkrecht steht. Im resultierende Condensation $c = s/S = 2/(n + m)$ wo s das Volum der Verbindung, S die Summe der noch unverbundenen Atostere ist.

in the molecule of a gas, so that the level of atoms A is parallel to that of B and that the connecting line of gravity is perpendicular to both levels. In the resulting condensation $c = s/S = 2/(n + m)$ where s is the volume of the compound, [and] S is the sum of the still unconnected atosteres.

66. Wenn $m = 1$, und $n = 1, 2, 3, 4$, so erhalten wir die gewöhnlichen Typen der modernen Chemie. Ich gebe nebenstehend eine graphische Darstellung derselben, welche sowohl Anordnung (Constitution) als auch Volum-Verhältnis und Condensationsgrad zeigt; ferner die rationellen Symbole AB^n und endlich noch die gebräuchlichen Typen-Symbole.

66. If $m = 1$, and $n = 1, 2, 3, 4$, we get the common types of modern chemistry. I give a graphical representation of the same on the left, which shows the arrangement as well as the volume ratio and degree of condensation; also the rational symbols AB^n and finally the common type symbols.

Wir werden Gelegenheit haben diese Gesetze durch reichlichen Erfahrungs dahin zu belegen.

We will have the opportunity to prove these laws through extensive experience.

67. Es ist jedoch einseitig, diese als die einzigen Typen zu betrachten. 2 A mit 3 B, 2 A mit 5 B etc. sind ebenfalls Typen. Als Beispiel führen wir nur an N_2O_3, N_2O_5, Cl_2O_5, Cl_2O_7 etc., sowie die vielen Sesquiverbindungen der Ferroide.

67. However, it is one-sided to consider these as the only types. 2 A with 3 B, 2 A with 5 B etc. are also types. For example, we list only N_2O_3, N_2O_5, Cl_2O_5, Cl_2O_7 etc., as well as the many sesqui-compounds of the ferroids.

68. In seiner Allgemeinheit gilt dieses Gesetz nur für den gasförmigen Zustand, da ist auf die Gleichheit der Atostere in gasform beruht. Im Folgenden machen wir diese Gleichheit als eine Folge der gegenseitig grossen Abständen der Atome in Bezug auf ihr Atometer m und Atomare a kennenlernen. Für den fest-flüssigen Zustand sind diese Grösser sehr merklich im Vergleich zum Atostere; also werden für diese Zustände obige Gesetze nur in Bezug auf die allgemein-gültige Anlagerung der Theile gelten, nicht aber in Übersicht auf Volumen oder Condensation. In Bezug auf letztere ergiebt sich leicht, dass die Condensation um so mehr dem obigen Gesetz für Gase entsprecht, je grösser das Prisma A im Vergleich zu den damit verbundenen n Prismen B ist.

68. In general, this law applies only to the gaseous state, because it is based on the equality of the atostere in gaseous form. In the following we get to know this equality as a consequence of the large distances between the atoms in relation to their atometer m and atomare a. For the solid-liquid state, these sizes are very noticeable compared to the atostere; so the above laws will apply only to these conditions with regard to the generally applicable attachment of the parts, but not with regard to volume or condensation. With regard to the latter, it is easy to see that the larger the prism A is in comparison to the associated n prisms B, the more the condensation conforms to the above law for gases.

Zum Beispiel ΦX_3 sollte geben $c = ½ = 0.50$. Man findet aber für

For example ΦX_3 should give $c = ½ = 0.50$. But one finds for

$PCl_3 = 0.88$, $AsCl_3 = 0.88$, $SbCl_3 = 0.87$, $BiCl_3 = 0.68$, eine Mittel 0.83

$PCl_3 = 0.88$, $AsCl_3 = 0.88$, $SbCl_3 = 0.87$, $BiCl_3 = 0.68$, an average [of] 0.83

$PBr_3 = 0.94$, $AsBr_3 = 0.91$, ---, $BiBr_3 = 0.80$, eine Mittel 0.91

$PBr_3 = 0.94$, $AsBr_3 = 0.91$, ---, $BiBr_3 = 0.80$, an average [of] 0.91

69. Die Substitutions-Erscheinungen bestehen

69. The phenomena of substitution exist in

zu der Typenlehre da diese aber in unsrer Theorie inbegriffen ist, so folgt dass auch die Substitutionserscheinungen unsre Theorie bestätigen. Wir werden auf Volum und Krystall Verhältnisse ausführlich zurückkommen, und darin ganz besondere Bestätigungen der Verdichtung und Constitution finden. So auch in der jetzt folgenden Abteilung der Gerhardt'schen Typen selbst in Bezug auf Atomicität.

relation to type theory, but since this is included in our theory, it follows that the phenomena of substitution also confirm our theory. We will come back to volume and crystal conditions in detail, and find very special confirmations of the compression and constitution. So also in the following section of Gerhardt's types themselves in terms of atomicity.

Pantoid-Verbindungen

Pantoid [Hydrogen] Compounds

70. Chloropantoide XY, oder zu gleichen Atomen. Da X = (1) + $m(p)$ ein einzelner Pantatom als Scheitel hat, ist dieses der Punkt, grösster Anziehung, oder Kurzweg ein chemischer Pol. Daferner die Atome mit der Längsrichtung parallel stehen, so kann jedes X also, nur ein Y = 2(1) aufnehmen, wie obige Formel ausdrückt. Die einfache Anlegung von einem Atom X an einem Atom Y = H kann offenbar keine Volumänderung hervorbringen; also 1 Vol[um] X + 1 Vol[um] Y = 2 Vol[um] XY. Dies der Chlorwasserstoff-Typens der modernen Chemie. Beispiele, FlH, ClH, BrH, JoH.

70. Chloropantoids [hydrogen chloroids] XY, or equivalent atoms. Because X = (1) + $m(p)$ has a single pantatom as the apex, this is the point, greatest attraction, or short path, of a chemical pole. Furthermore, because the atoms are parallel with the longitudinal direction, each X can take up only one Y = 2(1), as the above formula espresses. The simple application of an atom X to an atom Y = H obviously cannot produce a volume change; thus 1 volume X + 1 volume Y = 2 volumes XY. This is the hydrogen chloride type of modern chemistry. Examples, FlH, ClH, BrH, JoH.

71. Phosphopantoide ΦY_3. Die Endfläche ist ein regelmässiges Sechseck (Bi sehr nahe). Also 3 gleiche Anziehungs-maxima, die Punkte X in der Figur. Die 3 H werden aber keine grössre Breite (Atomare) einnehmen, als der eine Atom Φ. Folglich 3 H-Volume verdichten sich auf 1 Vol[um], und verbinden sich dann ohne weitere Verdichtung mit Φ. Dies ist der Ammoniak-Typus der modernen Chemie. NH_3, PH_3.

71. Phosphopantoids [hydrogen phosphoids] ΦY_3. The end face is a regular hexagon (Bi very close). So [there are] 3 equal maxima of attraction, the points X in the figure. However, the 3 H will not have a greater breadth (atomare) than the one atom Φ. Consequently, 3 H volumes compress to 1 volume, and then combine with Φ without further compression. This is the ammonia type of modern chemistry. NH_3, PH_3.

Verbindungen in andern Verhältnissen sind natürlich nicht ausgeschlossen, aber doch nicht es beständig wie der typisch ΦY_3.

Compounds in other ratios are of course not excluded, but [are] not as stable as the typical ΦY_3.

72. Sulphopantoide, ΘY_2. Da $\Theta = m/p^2/$ oder $= \mu[2 \cdot p^2]$ so ist $O = 8 \cdot 2^2$ oder auch $= 4 \cdot [2 \cdot 2^2]$. Die Entscheidung lässt sich aus dem Gewicht natürlich nicht ableiten. Beide Annahmen geben 2 chemische Pole, B aber deutlicher, als A. Die Formel deutet an, dass jedes 2^2 eine ganz gesonderte Stellung hat, indem also der Abstand beider 2^2 grösser oder kleiner ist, als die Dimension jeder 2^2. Wäre A wichtig, so müsste ΘY eben so häufig sein wie ΘY_2. Da aber letztere fast ausschliesslich beobachtet ist – der Wasser Typus der modernen Chemie -

72. Sulphopantoids [hydrogen sulfoids], ΘY_2. Because $\Theta = m\ /p_2/$ or $= \mu\ [2 \cdot p^2]$ then $O = 8 \cdot 2^2$ or also $= 4 \cdot [2 \cdot 2^2]$. The decision cannot, of course, be derived from the weight. Both assumptions give 2 chemical poles, but B more clearly than A. The formula indicates that each 2^2 has a very special position, i.e. the distance between the 2^2 is larger or smaller than the dimension of each 2^2. If A were important, ΘY should be just as common as ΘY_2. But because the latter is observed almost exclusively – the water type of

so muss also B richtig sein; dies wird sich im Folgenden in den physikalischen Eigenschaften vollkommen bestätigen. So stimmt z[um] B[eispiel] das Refractions-equivalent genau mit B, durchaus nicht mit A. Die Allotropie des O ist ebenfalls nur in B erklärlich. Hier sind denn wie diesen 2 Υ auf den Querschnitt des Θ zusammengedrängt, ohne das ihre Länge irgendwie hätte geändert werden können.

73. Titanopantoide, $Ti\Upsilon_4$. C = 2[3·4]. Basis [3·4] giebt gewissermaassen ein Dreieck von 3 Pantatomen 4 mal, wie B anzeigt. Daher die Verbindung CH_4 und wie vorhin der Condensation 2/5 = 2/(1+4). Der Kohlenwasserstoff Typus.

74. Veranschaulichung des Typen, sowohl in ihrer Zusammensetzung als in den Condensations-Verhältnissen ist nebenstehend versucht. Im gleichen Quadrate bezeichnen gleiche Volumen.

Zum Zweck der kurzen Darstellung von Verbindungen sowohl an sich, als namentlich in Bezug auf Krystallform, hat die Anwendung der verticalen Stellung grosse Vorzüge. Wir werden dieselbe daher im Folgenden fast ausschliesslich benutzen. Der Kurze wegen werden wir H einfach durch ein verticales Comma ′ bezeichnen. Demnach erscheinen der Typen, perspectivisch

′ ′ ′ ′ ′ ′ ′ ′
X Θ ′Φ′ ′Ti′ oder ′C′.

75. Es scheint mir durch Vorstehendes (§.§.70-74) in Verbindung mit den Früher entwickelten Gesetzen sei nicht nur die Typenlehre auf eine rationelle, mechanischen Basis zurück geführt und erklärt, sondern auch wesentlich näher bestimmt; statt blos einer Ausdruck der Reactionen zu geben, ist jetzt durch meine Symbole eine perspectivisch Darstellung der wirklichen Constitution gegeben. Darauf fusst meine Krystall-lehre und darauf können wir auch Einsicht in der wirkliche Constitution der organischen Reihen erlangen. Dies ist jetzt zu thun.

76. Der typische Kohlenwasserstoff C'''' ist gesättigt; *n* Atome können sich eben so wenig verbinden mit von irgend einer andern Verbindung.

modern chemistry – B must be correct; this will be fully confirmed in the physical properties below. For example, the refraction equivalent matches B, not at all A. The allotropy of O is also understandable only by B. Here, 2 Υ are squeezed like this onto the profile of the Θ, without which their length could have been changed in any way.

73. Titanopantoids [hydrogen titanoids], $Ti\Upsilon_4$. C = 2[3·4]. The base [3·4] gives to a certain extent 4 times a triangle of 3 pantatoms, as B indicates. Hence the compound CH_4 and as before the condensation 2/5 = 2/(1 + 4). The hydrocarbon type.

74. An illustration of the type, both in its composition and in the condensation ratios, is attempted below. Equal squares denote equal volumes.

For the purpose of the short description of compounds, both as such and especially in relation to crystalline form, the application of the vertical position has great advantages. Therefore, we will use it almost exclusively in the following. For the sake of brevity, we will simply denote H by a vertical comma ′. Accordingly, the types appear, in perspective

′ ′ ′ ′ ′ ′ ′ ′
X Θ ′Φ′ ′Ti′ or ′C′.

75. It seems to me from the above (§.§.70-74) in connection with the earlier developed laws that not only the type theory can be traced and explained on a rational, mechanical basis, but also much more precisely determined; instead of just expressing the reactions, my symbols now give a perspective representation of the real constitution. My crystal theory is based on this and we can also gain insight into the real constitution of the organic series. This is to be done now.

76. The typical hydrocarbon C'''' is saturated; *n* atoms can cannot bind with any other compound.

77. Mehrerer C-Atome können sich aber parallel nebeneinander zu einer Riehe zusammenstellen in der beigezeichneten Weise. Dem Atom a) = CH_4 fügt sich noch 1 C bei, der 2 H rechts dienen als Verbindungsglied, als Kette der beiden; siehe a[1]). Das letztere C_2H_4 kann noch offenbar 2 H aufnehmen; daraus entsteht b = 2C6H = C_2H_6. Hieraus wieder c = C_3H_8 und wie leicht ersichtlich allgemeine, C_nH_{2n+2} welches ich Y^o_n (Hydrocarbon, o -atomisch) bezeichne.

77. Several C atoms can, however, assemble parallel to each other to form a row in the manner shown. 1 C can be added to the atom a) = CH_4, the 2 H on the right serve as a link, as a chain of the two; see a[1]). The latter C_2H_4 can apparently still take up 2 H; this results in b = 2C6H = C_2H_6. From this again c = C_3H_8 and as can be seen in general, C_nH_{2n+2} which I call Y^o_n (hydrocarbon, o - atomic).

[Translator's note: paragraphs 78-87 give Hinrichs's theory of organic chemistry; mainly based on the idea that the hydrogen atoms are directly involved in the formation of C-C (and also C-X) bonds]

88. Es könnten zwar auch 6 Atome auf einem N Platz finden, besser aber schön 4, die mit *o* bezeichneten. Dies letzten giebt Ammonium, Am = NH_4 und seine Derivative. Die Verbindungen derselben, namentlich die Haloide, ergeben sich leicht (siehe bei gefügte) perspectivische Ansicht. NX bildet die Axe, 4 H eine Equator-Ebene. Es ist der Prototyp der Spiralle, u[nd] wie diese tessular.

88. There could also be 6 atoms on one N site, but better 4, which are marked with *o*. The latter gives ammonium, Am = NH_4 and its derivatives. The compounds of the same, namely the haloids, arising from a simple perspective view (see attached). NX forms the axis, 4 H an equatorial plane. It is the prototype of the spiral and like this cubic.

89. Ammonium dargestellt und mit Kalium vergleichen.

89. Ammonium shown and compared with potassium.

90. Cyan dargestellt und mit den Choroiden vergleichen.

90. Cyanide shown and compared with chloride.

91. Nachdem der Hydrogen oder Pantoid-Verbindungen mit einiger Ausführlichkeit betrachtet sind, dürfen wir im Folgenden uns einer kürzer fassen in Bezug auf die X, Φ, Θ-Verbindungen.

91. Having considered the hydrogen or pantoid compounds in some detail, in the following let us take a brief look at the X, Φ, [and] Θ compounds.

Chloroide Verbindungen, etc.

Chloroid Compounds, etc.

92. Da der X = (1) + *m* (*p*) mit (1) enden, so werden ihrer Verbindungen sich XX, ΘX_2, ΦX_3, TiX_4 schreiben. Die Volumina in Gasform folgen den oft erwarteten Gesetzen. Im fest-flüssigen Zustande müssen wir aber erinnern, dass mit dem Atomare X im Vergleich zum Radical R in RX_n das resultirende Atostere wachsen muss (§. 68).

92. Because the X = (1) + *m* (*p*) ends with (1), [we] will write their compounds thus XX, ΘX_2, ΦX_3, TiX_4. The volumes in gaseous form follow the often expected laws. In the solid-liquid state, however, we must remember that with [an increase in] the atomare X, just like the radical R in RX_n, the resulting atostere must grow (§ 68).

93. $X_2\Theta$, $X_2\Theta_3$ mit (a) als Ort für O der Basis; (b) für die beiden Θ in $X_2\Theta_5$.

93. $X_2\Theta$, $X_2\Theta_3$ with (a) as the location for O of the base; (b) for the two Θ in $X_2\Theta_5$.

94. ΦX_3 dargestellt das mittleren Sechseck ist Φ.

94. ΦX_3 depicted the middle hexagon is Φ.

95.Titanochloroide, TiX_4.

95.Titanochloroids, TiX_4.

96. Ferrochloroide, $\Sigma\iota X_2$, $\Sigma\iota X_3$, $\Sigma\iota X_4$.	96.Ferrochloroids, $\Sigma\iota X_2$, $\Sigma\iota X_3$, $\Sigma\iota X_4$.
97. Calcochloroide, $X\alpha X_2$.	97.Calcochloroids, $X\alpha X_2$.
98. Kalochlorode, $K\alpha X$.	98.Kalochloroids, $K\alpha X$.
99. Wie in den organischen Reihen, so ist hier die Endung der Atome von Einfluss. So ist $K\alpha$ durch die Endung monatomisch in Bezug auf X.	99. As in the organic series, the terminus of the atoms is of influence here. Thus $K\alpha$ is monatomic with respect to X.
100. Figur [Figure; A schematic diagram of Hinrichs's chart]	
101. In der nebenstehenden Figur ist das Hauptresultat der Verbindungsverhaltnissen zusammengestellt. Die Zahl giebt an, wie viele Atome X (innerer Kreis) oder Θ (äusserer) sich mit eineres Atom der verschiedenen Genera (deren Symbol in äussersten Kreis) zur Hauptverbindung vereinigen.	101. In the adjacent figure, the main result of the compound ratios is compiled. The number indicates how many atoms of X (inner circle) or Θ (outer) unite with one atom of the different genera (their symbol in the outermost circle) to [give] the main compound.
102. Man sieht was sowohl die Verhältniss der X als der Θ Verbindungen von A an regelmässig nach jeder Richtung wachsen, bei Θ unter der Trigonoiden bei $T\iota$ unter der Tetragonoiden ihr Maxima erreichen. Die Ordnung des Elements ist dieselbe wie in § 37.	102. One can see that the ratio in both the X and the Θ compounds grows regularly from [point] A in every direction, reaching their maxima at Θ under the trigonoid [and] at $T\iota$ under the tetragonoids. The order of the elements is the same as in § 37.
103. Die Ursache, warum allgemein Υ_2 oder X_2 durch Θ ersetz werden kann ist in §.72, §.74 anzugeben.	103. The reason why Υ_2 or X_2 in general can be replaced by Θ is given in §.72, §.74.
104. Affinität, wie gewöhnlich verstanden (?) ist nur ein Scheinwort, Mysticismus. Dagegen erklären Berthollet's Gesetze sehr viele chemische Reactionen indem die dieselben auf schon vorher bekannte physikalische Eigenschaften (Flüchtigkeit und Löslichkeit) beziehen. Von diesen ist die Flüchtigkeit in gegenwärtige Schrift auf ihre Grundursache zurückgeführt; damit sind also alle die äusserst zahlreichen dadurch bestimmenden chemischen Reactionen erklärt. In Bezug auf Löslichkeit ist auch einige Klarheit gewonnen, wenn auch bei weitem noch nicht genug.	104. Affinity, as usually understood, is only a pseudoword, a mysticism. In contrast, Berthollet's laws explain many chemical reactions by referring to well-known physical properties (volatility and solubility). Of these, the volatility, in current work, is traced back to its root cause; this explains all the numerous chemical reactions that determine it. Some clarity has also been gained with regard to solubility, if also by far not enough.
105. Die Verbindungswärme, d[as] h[eisst] die durch chemisch Verbindung hervorgebunden Wärme wird wohl in folgende Weise aufzufassen sein. Ein Körper von der Masse m entwickelt auf der Erde indem er sich der Oberfläche um h nährt eine lebendige Krafte $= 2\ g\ m\ h$; um derselben er muss zu entfernen, wäre ein ebenes grosse Aufwand an lebendiger Kraft nöthig.	105. The heat of bonding, that is, the heat which is produced by chemical bonding, may well be interpreted in the following way. By approaching to the surface of the earth by [distance] h, a body of mass m develops a vis viva $= 2\ g\ m\ h$; in order to remove it, a great deal of vital force would be required.

Wärme ist aber lebendige Kraft. Jeder Atom ist ein Mikrokosmischen Himmelskörper, ebenso wie jeder Himmelskoerper ein Kosmischer Atome ist. Daher wird die in Verbindungen stattfindende annäherung der Atome eine der obigen Formel entsprechende Wärmeentwicklung hervorbringen wenn eine Annährung stattfindet, und eine wärmeabgabe ein entgegengesetzten Fälle. Es handelt sich, also um die Bestimmung von *h*; die Mechanik wird diese Bestimmung aus unsern Atomformel ableiten können, sobald die Frage in Bezug auf Aetherhüllen oder Schwingungen genauer erördert ist, d[as] h[eisst] sobald die abstossende Kraft (wenn es ein solche giebt, oder ihr Aequivalenz ein Bewegung) wodurch die Atome auseinander gehalten werden bestimmt ist.	But heat is living force. Every atom is a microcosmic celestial body, just like every celestial body is a cosmic atom. Therefore, the assumption of atoms approaching in compounds will produce a heat generation corresponding to the above formula when a convergence takes place, and a heat release in opposite cases. It is about the determination of h; mechanics will be able to derive this determination from our atomic formula, as soon as the question regarding ether envelopes or vibrations has been more precisely discussed, that means as soon as the repulsive force (if there is any, or its equivalency a movement) by which the atoms are kept apart is determined.
106. Jedenfalls aber sind die bei verschiedenen Verbindungen derselben Elemente fanwendenden Wärmemengen durch Obiges erklärt. In von Farve gefundenen Resultaten geben:	106. In any case, the amounts of heat used in different compounds of the same elements are explained by the above. Results found by Favre give:

As_2O_3:As_2O_5 = 3:4.64 statt [instead of] 3:5 SO:SO_2:SO_3 = 1:1.95:3.32 statt [instead of] 1:2:3

P_2O_3:P_2O_5 = 3:4.64 statt [instead of] 3:5 SeO:SeO_2:SeO_3 = ?:2:3.12 statt [instead of] 1:2:3

P_2O_3:P_2O = 3:1.05 statt [instead of] 3:1 CO:CO_2 = 1:3.27 statt [instead of] 1:2

PCl_3:PCl_5 = 1:1.06 statt [instead of] 3:5

Hier ist ganz offenbar, dass die 5 Cl nicht als dicht an P treten können, wie 3 Cl (siehe §.94), daher die geringe Wärmeentwicklung für PCl_5.	Here it is quite clear that the 5 Cl cannot come as close to P as 3 Cl (see §.94), hence the low heat development for PCl_5.
107. Es bedarf kaum eines Beweises dass die stattfindende Condensation in den Verbindungen RH_n für $n > 1$ durch aus unzureichend ist für die Erklärung der Wärmeerscheinungen. Fanden doch Favre [und] Silberman nur eine Erhöhung der Temperatur des H um 13.2 °C wenn es auf das halben Volume zusammengedrückt wurde.	107. It is hardly necessary to prove that the condensation taking place in the RH_n compounds for $n > 1$ is inadequate for the explanation of the heat phenomena. Favre [and] Silberman found an increase in the temperature of H of only 13.2 °C when it was squeezed to half the volume.

[Translator's note: paragraphs 108-120 deal with ternary compounds.]

Dritter Abschnitt. Physikalische Eigenschaften.	**Third section. Physical Properties.**

[Translator's note: Paragraphs 121-228 omitted]

Vierter Abschnitt. Morphologische Eigenschaften oder die Krystallformen.	**Fourth section. Morphological Properties or Crystal Forms.**

[Translator's note: Paragraphs 229-399 omitted]

Schluss	Closing
400. Nachdem wir nun an dem uns einstweilen gesteckten Ziele angelangt sind, und auf das Geleistete zurückblicken gemahnen wir nur zu wohl der vielen Unvollkommenheiten, dir manchen Unvollständigkeiten. Aber diese Arbeit ist nur das Programm der Atomechanik, nur die erste Anzeige dessen, was der Forschung auf diesem neuen Gebiete möglich sein wird. Wenn man berücksichtigt, dass der Raum jede Ausführlichkeit von vorne herein verbot (daher auch seitens der Leser der grösste Aufmerksamkeit nöthig sein wird) so wird man doch wohl zugeben, dass redlich Alles geleistet worden, was unter den obwaltenden Verhältnissen möglich war.	400. Now that we have reached the goals we have sent for the time being, and looking back on what we have achieved, we are reminded only too well of the many imperfections, the many omissions. But this work is just the program of atomechanics, just the first indication of what research will be possible in this new area. If one takes into account that space forbids any elaborateness from the outset (therefore the reader will also need the greatest attention), one will surely admit that everything that was possible under the prevailing conditions was honestly achieved.
Eigenhändig geschreiben von Gustavus Hinrichs	Written personally by Gustavus Hinrichs
Auf Stein überdruckt	Printed on stone
Augustus von Hageboeck	Augustus von Hageboeck
Lithograph in Davenport, Iowa	Lithographer in Davenport, Iowa
Zur Nachricht.	Note.
Den für Deutschland bestimmten Exemplaren werden das französische Résumé (S. 45-48) im Allgemeinen nicht beigegeben werden.	The copies intended for Germany will in general not include the French résumé (pp. 45-48).

References

1. Hinrichs GD (1867) Programme der Atomechanik oder die Chemie eine Mechanik der Panatome. Privately printed, Iowa City, IA
2. Hinrichs G (1867) A programme of atomechanics, or chemistry as a mechanics of the panatoms. Am J Min 4:66, 82, 98, and 114. This paper was also issued as an offprint, of which very few copies survive
3. Hinrichs G (1869) Natural classification of the elements. Pharmacist (Chicago) 2:10–12
4. Hinrichs G (1870) On the classification and the atomic weights of the so-called chemical elements, with reference to Stas' determinations. In: Proceedings of the American Association for the Advancement of Science: eighteenth meeting, held in Salem, Massachusetts in August, 1869, Joseph Lovering, Cambridge, MA, pp 112–124
5. Hinrichs G (1870) Contributions to molecular science, or atomechanics, Nos. 3, 4. Essex Institute Press, Salem, MA
6. Zapffe CA (1969) Gustavus Hinrichs, precursor of Mendeleev. Isis 60:461–476
7. van Spronsen JW (1969) The periodic system of chemical elements. Elsevier, Amsterdam

8. van Spronsen JW (1969) Gustavus Detlef Hinrichs discovered, one century ago, the periodic system of chemical elements. Janus 56:46–62
9. Kauffman GB (1969) American forerunners of the periodic law. J Chem Educ 46:128–135
10. Giunta CJ (1999) J. A. R. Newlands' Classification of the Elements: Periodicity, But No System. Bull Hist Chem 24:24–31.
11. Scerri ER (2001) A philosophical commentary on Giunta's critique of Newlands' classification of the elements. Bull Hist Chem 26:124–129
12. Wylie CC (1930) Gustavus Detlef Hinrichs. Palimpsest 11:193–201
13. Berg CP (1980) The University of Iowa and biochemistry from their beginnings. The University of Iowa, Iowa City
14. Stolz G (1998) Gustav Dethlef Hinrichs. Ein Naturforscher von Weltruf aus Dithmarschen 1836–1923. Verein für Heimatgeschichte des Kirchspiels Lunden, Lunden, Germany
15. Lang WC (1941) A history of the State University of Iowa: the collegiate department from 1879 to 1900. Thesis, State University of Iowa
16. Keyes C (1924) Contributions to knowledge by Gustavus Detlef Hinrichs, M.D., LL.D. Proc Iowa Acad Sci 31:79–94
17. Gustavus Hinrichs Papers. University of Iowa Archives. RG99.0039
18. Gustavus D. Hinrichs Papers, 1837–1917. Illinois History and Lincoln Collections, University of Illinois at Urbana-Champaign. IHLC MS 712
19. Wolf RA (2020) Brief history of Gustavus Hinrichs, discoverer of the derecho. https://www.spc.noaa.gov/misc/AbtDerechos/hinrichs/hinrichs.htm. Accessed 13 Feb 2020
20. Giacomelli J (2017) Unsettling gilded-age science: vernacular climatology and meteorology in the "Middle Border". Hist Meteorol 8:15–34
21. Palmer WP (2007) Dissent at the University of Iowa: Gustavus Detlef Hinrichs—chemist and polymath. Chemistry: Bulg J Sci Educ 16:534–553
22. Hinrichs G (1866) On the spectra and composition of the elements. Am J Sci 42:350–368
23. Prout W (1815) On the relation between the specific gravities of bodies in their gaseous state and the weights of their atoms. Ann Phil 6:321–330
24. Prout W (1816) Correction of a mistake in the essay on the relation between the specific gravities of bodies in their gaseous state and the weights of their atoms. Ann Phil 7:111–113
25. Thomson T (1816) Some observations on the relations between the specific gravity of gaseous bodies and the weights of their atoms. Ann Phil 7:343–346
26. Farrar WV (1965) Nineteenth-century speculations on the complexity of the chemical elements. Brit J Hist Sci 2:297–323
27. Brock WH (1985) From protyle to proton: William Prout and the nature of matter, 1785–1985. Adam Hilger Ltd., Bristol and Boston
28. Stas JS (1860) Recherches sur les rapports réciproques des poids atomiques. Bull Acad R Belg 10:108–336
29. Stas JS (1865) Nouvelles recherches sur les lois des proportions chimiques, sur les poids atomiques et leurs rapports mutuels. Mem Acad R Sci Lett Beaux-Arts Belg 35:3–311
30. Cannizzaro S (1858) Lezioni sulla teoria atomica fatte nella R. Università di Genova. La Liguria Medica, G Sci Med Nat 3:113–142
31. Cannizzaro S (1858) Sunto di un corso di filosofia chimica fatto nella R. Università di Genova dal Prof. S. Cannizzaro. Nuovo Cimento 7:321–366
32. Will H (1863) [published in 1864] Jahresber Fortschr Chem Verw Theile Anderer Wiss, p xxiv
33. Gladstone JH (1853) On the relations between the atomic weights of analogous elements. Phil Mag 5:313–320
34. Cooke JP (1855) The numerical relation between the atomic weights, with some thoughts on the classification of the chemical elements. Am J Sci 17:387–407. Cooke's paper was also printed in (1855) Mem Am Acad Arts Sci 5:235–257 and 412

35. Dumas JBA (1851) Observations on atomic volumes and atomic weights, with considerations of the probability that certain bodies now considered elementary may be decomposed. The Athenaeum 750; L'Inst, J Univ Sci Soc Sav Fr. L'Etrang 19:303–304
36. Dumas JBA (1857–58) Mémoire sur les équivalents des corps simples. C R Séances Acad Sci 45:709–731; 46:951–953; 47:1026–1034
37. Lea MC (1860) On numerical relations existing between the equivalent numbers of elementary bodies. Am J Sci Arts 29:98–111
38. Béguyer de Chancourtois A-E (1863) Vis tellurique. Classement naturel des corps simples ou radicaux obtenu au moyen d'un système de classification hélicoïdal et numérique. Mallet-Bachelier, Paris
39. Odling W (1864) On the proportional numbers of the elements. Quart J Sci 1:642–648
40. Newlands JAR (1864) Chem News 10:59–60 and 94–95; (1865) Chem News 12:83; (1866) Chem News 13:113
41. Meyer L (1864) Die modernen Theorien der Chemie und ihre Bedeutung für die chemische Statik. Maruschke & Berendt, Breslau
42. Darwin Correspondence Project. 2020. Letter of G. D. Hinrichs to Charles Darwin, 31 Aug 1868. "Letter no. 6337," https://www.darwinproject.ac.uk/letter/DCP-LETT-6337.xml. Accessed 25 Feb 2020
43. Rose H (1857) Ueber die Atomgewichte der einfachen Körper. Ann Phys Chem 176:270–291
44. Krönig A (1866) Neues Verfahren zur Ableitung der Formel einer Verbindung aus den Gewichtmengen der Bestandtheile. Springer, Berlin
45. Daubeny C (1831) An introduction to the atomic theory. John Murray, Oxford
46. Marignac C (1843) Analyses diverses destinées à la vérification de quelques équivalents chimiques. Biblioth Univ Genève 26:350–373
47. Maumené E (1846) Sur les équivalents chimiques du chlore, de l'argent et du potassium. Ann Chim Phys 18:41–79
48. Marignac C (1865) Remarques sur le mémoire de M. Stas, intitulé "Nouvelles recherches sur les lois des proportions chimiques, &c". Arch Sci Phys Nat 24:371–376
49. Hinrichs GD (1874) The principles of chemistry and molecular mechanics. Day, Egbert, & Fidlar, Davenport, IA; B. Westermann & Co., New York, NY, p 166
50. Dana JD (1867) On a connection between crystalline form and chemical constitution, with some inferences therefrom. Am J Sci 44:89–95 and 252–263
51. Girolami GS (2009) Origin of the terms pnictogen and pnictide. J Chem Educ 86:1200–1201
52. Claus CE (1845) Fortsetzung der Untersuchung des Platinrückstandes, nebst vorläufiger Ankündigung eines neuen Metalles. Bull Cl phys-math Acad imp sci St-Pétersbg 3:col. 353–371
53. Berzelius JJ (1828) Försök öfver de Metaller som åtfölja Platinan, samt öfver sättet att analysera Platinans nativa legeringar eller Malmer. K Sven Vetenskapsakad Handl 25–116
54. Venable FP (1896) The development of the periodic law. Chemical Publishing Co., Easton, PA
55. Watts H (1863–1868) A dictionary of chemistry and the allied branches of other sciences. Longmans, Green, and Co, London, vol 1, p 907 and vol 3, p 308
56. Hinrichs GD (1894) The elements of atom-mechanics. B. Westermann & Co, St. Louis, MO
57. Mauskopf SH (1969) The atomic structural theories of Ampère and Gaudin: molecular speculation and Avogadro's hypothesis. Isis 60:61–74
58. Cole TM (1975) Early atomic speculations of Marc Antoine Gaudin: Avogadro's hypothesis and the periodic system. Isis 66:334–360
59. Hinrichs G (1868) Synopsis of new memoirs on atomechanics. Privately printed, Iowa City, IA
60. Hinrichs GD (1871) The elements of chemistry and mineralogy, demonstrated by the student's own experiments and observations. Griggs, Watson, and Day, Davenport, IA, p 58
61. Gmelin L (1843) Handbuch der Chemie, K. Winter, Heidelberg, vol I, p. 457

62. Mendeleev DI (1872) Osnovy Khimii, 2nd ed., Tovarishchestva 'Obshchestvennaia Pol'za' for the author, St. Petersburg, vol 2, frontispiece
63. Walker J (1891) On the periodic tabulation of the elements. Chem News 63:251–253
64. Girolami GS, Mainz VV (2019) Mendeleev, Meyer, and atomic volumes: an introduction to an English translation of Mendeleev's 1869 article. Bull Hist Chem 44:100–115
65. Kipnis N (2014) Thermodynamics and mechanical equivalent of heat. Sci Educ 23:2007–2044

8 The Periodic Table of the Elements and Lothar Meyer

Gisela Boeck

Abstract

The General Assembly of the United Nations and UNESCO proclaimed 2019 as the International Year of the Periodic Table (IYPT). The occasion was the 150th anniversary of the creation of the periodic system, if the point of reference is February 17, 1869 (today the date is given by the Gregorian calendar as March 1, 1869), on which Dmitri I. Mendeleev (1834–1907) formulated an overview of the natural elements. The logo of the IYPT refers only to Mendeleev. However, the physical chemist Lothar Meyer also published several papers about the classification of elements.

8.1 Introduction

This article [1] is dedicated to the life and work of Meyer and also tries to find an answer to the questions as to why he was not shown on the International Year of the Periodic Table (IYPT) logo (Fig. 8.1) and why he is often not mentioned as a founder of the periodic system.

This chapter has been translated from the original German and published with permission from Ref. [1]. https://doi.org/10.1002/ciuz.201900001.

G. Boeck (✉)
Universität Rostock, 18051 Rostock, Germany
e-mail: gisela.boeck@uni-rostock.de

C. J. Giunta et al. (eds.), *150 Years of the Periodic Table*, Perspectives on the History of Chemistry, https://doi.org/10.1007/978-3-030-67910-1_8

Fig. 8.1 Logo of the International Year of the Periodic Table (www.iypt2019.org)

8.2 Lothar Meyer—His Life

Julius Lothar Meyer (Fig. 8.2) was born on August 19, 1830, in the little town of Varel near the Jade river and Jade Bay (references [2–6] were used as sources for the biographical summary). His father Heinrich Friedrich August Jacob Meyer (1783–1850) was the official physician there. His marriage to Anna Sophie Wilhelmine Biermann (1800–1853) in Varel in 1822 resulted in at least eight children, of whom, however, only four reached adulthood: Lothar, the later professor of physics Oskar August Emil Meyer (1834–1909), the future farmer Eugen Theodor Meyer (1836–1890), and Selma Corinna Helmine Meyer (1839–1928).

Lothar first received private lessons. From 1841 until his confirmation he attended a citizen's school (Bürgerschule). Because of his poor health, especially because of severe headaches, he stopped going to school. Meyer was sent as an assistant to the chief gardener of the grand duke of Oldenburg's summer palace in Rastede, recovered there, and from 1847 was able to attend school again at the Gymnasium in Oldenburg. In 1851 he successfully passed his graduation examination and made the decision to become a doctor.

On May 8, 1851, Meyer enrolled at the University of Zürich. He stayed there through the winter semester of 1852/1853 and studied medical subjects, but also chemistry, physics, mineralogy, geology, botany, and zoology. He was particularly interested in the lectures in physiology by Carl Ludwig (1816–1895), which were perhaps the source of Meyer's later interest in gas exchange in blood.

Meyer went to Würzburg at Easter 1853. With the work "On the dependence of the vessels and the pigment cells in the frog on the influence of the nerves"[1] Meyer gained his degree as Dr. med. on February 25, 1854, as Karl Seubert (1851–1942) wrote in his obituary [2]. The work is about the color changes in the pigment cells of a common or dewy frog. Unfortunately, the doctoral file has not survived in Würzburg. We can only learn from a doctorate advertisement that the doctorate was announced for February 22, 1854. Perhaps the difference in the dates is that one is

[1] "Über die Abhängigkeit der Gefäße und der Pigmentzellen beim Frosch von dem Nerveneinfluß".

Fig. 8.2 Lothar Meyer [2]

the date of the colloquium and the other is the date on which the doctoral certificate was issued [7].

At Easter 1854 Lothar Meyer moved again, having been attracted to Heidelberg to work for Robert Bunsen (1811–1899). There he again enrolled in medicine—enrollment was a prerequisite for attending lectures and working in the laboratory—but he was even more interested in chemistry. He investigated gases, specifically how much oxygen, nitrogen, and carbon dioxide are present in the arterial blood and what laws govern the exchange.

He summarized the results in 1857 in the publication "The gases of the blood,"[2] which he is said to have submitted as a subsequent medical doctoral dissertation in Würzburg. In fact, the results of the work on the frog were very meager, but since a certificate was issued, it seems unlikely that Meyer was obligated to submit another scientific paper. Due to the lack of sources [7] it cannot be conclusively clarified what significance this paper had for Meyer's graduation. Perhaps he just felt obliged to prove his scientific abilities.

In Heidelberg, Lothar Meyer had made the acquaintance of Friedrich Beilstein (1838–1906), Henry Roscoe (1833–1915), Hans Landolt (1831–1810), and August Kekulé (1829–1896). Together with his brother he went to Königsberg in the winter semester of 1856/1857 to listen to the lectures of the physicist Franz Ernst Neumann (1798–1895) on electromagnetism and the wave theory of light. He also

[2]"Die Gase des Blutes".

worked in the field of physiology again, in the laboratory of Gustav Werther (1815–1869). He summarized the results of the investigation in the work "Effect of the gas carbon monoxide on the blood,"[3] submitted them as a doctoral dissertation in Breslau, where he had relocated at Easter 1858, and on July 5, 1858, he was promoted to Dr. phil. Meyer was able to show that carbon monoxide is retained in the blood by so-called chemical forces, thereby preventing oxygen uptake, which suffocates living beings. At the time, he was unable to clarify which substance, namely hemoglobin, was involved.

Finally, Meyer completed his habilitation in Breslau in 1859 with the paper "On the chemical theories of Berthollet and Berzelius"[4] and the trial lecture "The so-called volumetric methods of chemistry."[5] From Easter 1859 Meyer took over the management of the chemical laboratory of the physiological institute at the University of Breslau. He held lectures on plant and animal chemistry, photochemistry, gas analysis, and volumetric analysis and refresher courses on organic and inorganic chemistry.

In September 1860 the first international chemists' conference took place in Karlsruhe, which had been convened to clarify some theoretical questions and terms. One of the questions raised was whether it is justified to use different relative atomic weights in inorganic chemistry than in organic chemistry. Since at that time the notion "atomic masses" was not common but rather atomic weights, the outdated term is used here in the historical context.

Stanislao Cannizzaro (1826–1910) called for a consistent approach based on the theory of Amedeo Avogadro (1776–1856) in his conference contribution. His fiery speech and the "Outline of a Course in Theoretical Chemistry," his *Sunto* that he had distributed, met with an enthusiastic reception from Meyer and Mendeleev, who had also participated in the conference.

Immediately after the conference Meyer began writing a treatise on theoretical chemistry in order to resolve contradictions regarding atomic weights, but also in the use of the terms atom, molecule, and equivalent. The first indications of the existence of the corresponding manuscript were given in various letters in the summer of 1862, including those to Kekulé and Hermann Kolbe (1818–1884) [8]. But Meyer seems to have been very prudent with his publications. It was not until July 1864 that the work *Die modernen Theorien der Chemie und ihre Bedeutung für die chemische Statik* was published by Maruschke & Berendt in Breslau (Fig. 8.3). He began working on the second edition in 1868, but it was not published until 1872; further, increasingly extensive editions appeared in 1876, 1880 (republished in 1883), and 1884. The fifth edition had 626 pages!

In 1866, Lothar Meyer accepted a call to the Neustadt-Eberswalde Forest Academy because he saw no further opportunities for an academic career in Breslau. Heavy teaching duties in the fields of mineralogy, chemistry, and physics awaited him there, and if necessary also botany.

[3]"Einwirkung des Gases Kohlenmonoxyd auf das Blut".
[4]"Über die chemischen Lehren von Berthollet und Berzelius".
[5]"Die sogenannten volumetrischen Methoden der Chemie".

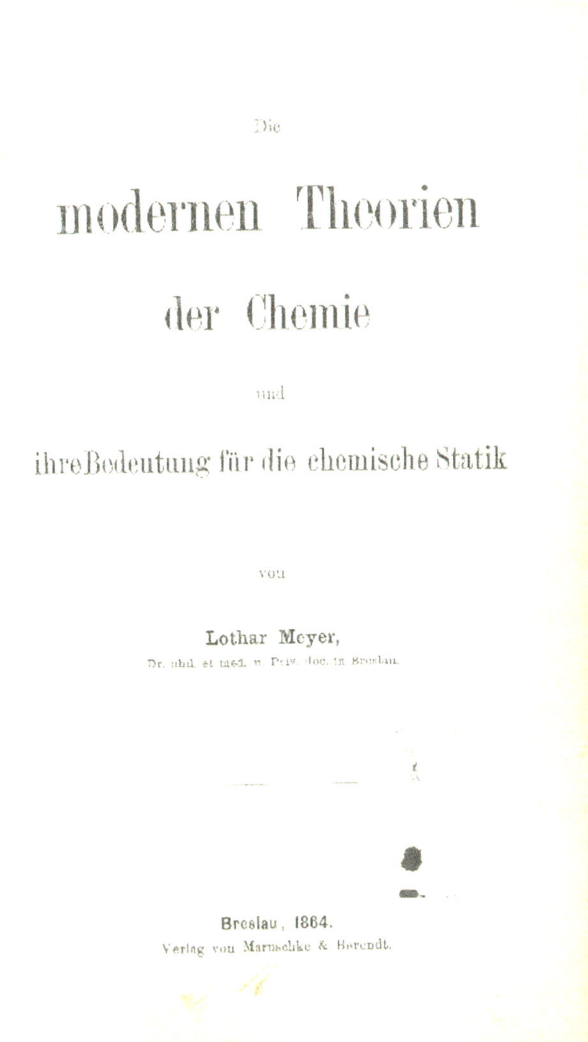
Die

modernen Theorien

der Chemie

und

ihre Bedeutung für die chemische Statik

von

Lothar Meyer,

Dr. phil. et med. u. Priv. doc. in Breslau.

Breslau, 1864.

Verlag von Maruschke & Berendt.

Fig. 8.3 Title page of the 1st edition (*Source* Berlin State Library—Prussian Cultural Heritage)

Meyer soon complained that he had too little time for scientific work. He reported that it was a feast day for him whenever an alluring hunt was announced for the students and lecturers at the Forestry Academy which required them to spend the day in the woods and fields. He then stayed at home so that he could work undisturbed in his laboratory. In 1867 he was appointed professor of inorganic natural sciences at the Forest Academy. Not only was there a lack of time for scientific work, Meyer had practically no students of his own among the future foresters.

In 1868 Lothar Meyer was appointed full professor of chemistry and head of the chemical laboratory at the Karlsruhe Polytechnic. There were more favorable working conditions there: he gave courses in his own subject and not in subsidiary ones, and found students for scientific work.

But he also paid attention to the security of his financial situation because he had married Johanna Volkmann (1842–1922) in Halle on August 16, 1866, before moving to Eberswalde. From this marriage there were four children, Roderich Lothar (1867–1944), Hildegard (1870–1954), Waldemar Lothar (1872–1948), and Ilse (1874–1943).

Meyer worked in Karlsruhe from 1868 to 1876. He turned down a call to Königsberg in 1870. During the Franco-Prussian war of 1870/1871 he used his medical training and took part in the management of an auxiliary hospital that was housed in the rooms of the Polytechnic.

In Karlsruhe, Meyer's health was not stable; he often complained of being overworked, and in the winter semester of 1874/1875 he even had to be released from teaching duties, which were then taken over by August Michaelis (1847–1916).

In 1875 Rudolph Fittig (1835–1910) had announced that he would leave Tübingen for Strasbourg. Meyer received the call to this professorship for theoretical chemistry. As a result, he not only improved financially, he finally came to a university that, unlike the Polytechnic, had the right to award doctorates. He found in the building in Wilhelmstraße 9 a generously equipped laboratory. Parts of this building exist still today. The official residence was on the upper floor. Meyer rededicated a number of rooms in the house according to his research interests and took care of the technical modernization of the laboratory [9].

The name Lothar Meyer brought a large number of students to Tübingen. His teaching activity was assessed by his closest colleague Karl Seubert as follows [2]:

> His lecture, well thought out and arranged according to a carefully considered plan, was spoken exceptionally clearly on the basis of leading notes, free of rhetorical embellishment and sensationalism. The main lecture—in winter about inorganic, in summer about organic experimental chemistry, both explained by means of numerous experiments—was continuously enriched by new ideas and enlivened by historical and personal anecdotes, which showed him as unusually well-read and with many extremely faithful memories. But the latter was particularly true for his smaller lectures, which focused on the various areas of physical chemistry, gas analysis, etc.; here he was able to show his astonishing in-depth knowledge of the relevant literature, most of which he was able to present in original form from his own extensive library which he had assembled.[6]

[6]"Sein Vortrag, wohl durchdacht und nach sorgfältig überlegtem Plane geordnet, wurde an Hand leitender Notizen klar, frei von rhetorischen Verzierungen und Effekthascherei, und ausnehmend deutlich gesprochen. Die Hauptvorlesung, im Winter über anorganische, im Sommer über organische Experimentalchemie, durch zahlreiche Versuche erläutert, wurde fortlaufend durch Einfügen von Neuem bereichert und durch historische und persönliche Erinnerungen, die ihm eine ungewöhnliche Belesenheit und ein ungemein treues Gedächtnis in großer Zahl zu Gebote stellten, belebt. Besonders galt aber letzteres von seinen kleineren Vorlesungen, die sich auf die verschiedenen Gebiete der physikalischen Chemie, Gasanalyse u.a.m. erstreckten; hier setzte seine eingehende Kenntnis der einschlägigen Literatur, die er überdies meist aus seiner eigenen, mit

During his time in Tübingen, Meyer worked intensively with Seubert on determining atomic weights and in 1883 published a corresponding work with newly calculated values. He became a member of the "Tuesday Society."[7] In the "Graeca," where he cultivated his great interest in ancient Greek, Greek classics were read in the original. At the end of each semester, he organized "chemists' pubs" that united teaching staff and students and bore witness that they had a good relationship. Meyer had established himself in Tübingen and he turned down subsequent calls to Leipzig (1887) and Breslau (1889).

Meyer received numerous awards. On November 2, 1882, he and Mendeleev received the Davy Medal from the Royal Society in London for their research on the classification of the elements. In 1883 he became an honorary member of the Chemical Society in London, in 1887 of the Physical Association of Frankfurt/Main, and in 1889 of the Manchester Literary and Philosophical Society. In 1888 he was appointed a corresponding member of the physical-mathematical class of the Prussian Academy of Sciences and in 1891 a corresponding member of the Academy of Sciences in St. Petersburg. In 1892 Lothar Meyer received the Knight of Honor Cross of the Order of the Württemberg Crown, which was associated with personal nobility.

In 1894/1895 Lothar von Meyer was elected rector of the University of Tübingen. He died suddenly on April 11, 1895. His grave is in the city cemetery in Tübingen (Fig. 8.4).

8.3 Lothar Meyer and the Order of the Elements

Lothar Meyer has left his mark on the entire field of chemistry. Here, however, only his work in connection with the creation of a system of classification of the chemical elements is highlighted.

With the revival of atomic theory and the associated possibility of determining relative atomic weights as well as the discovery of new elements, efforts to classify the elements increased. Quantitative relations based on the atomic weights were sought. Two examples that played a major role for Lothar Meyer are mentioned below.

The physician William Prout (1785–1850) had stated that the atomic weights were apparently multiples of a single quantity and—later—that hydrogen was the primordial matter, the πρωτη ὑλη. These thoughts challenged the ever

vielem Fleiße zusammengebrachten, reichhaltigen Bibliothek im Originale vorlegen konnte, geradezu in Erstaunen."

[7]This society was founded in 1862. The members were obliged to give lectures about their subject in either the sciences or the arts. Meyer was a member from 1877 until his death and gave 6 lectures during this time. The lectures had to be instructive. And finally, anyone who was late 15 min or more had to pay a 40 Pfennig penalty. From: Laupp Jr. H (1897) Die Dienstagsgesellschaft zu Tübingen 1862–1897, Tübingen.

Fig. 8.4 The grave of Lothar Meyer and his wife in the Tübingen city cemetery [personal photo]

improving experimental determination of atomic weights, which refuted the assumption that atomic weights were always multiples of the mass of hydrogen. Nevertheless, Prout's theory played a role in the considerations of many scholars throughout the nineteenth century. The building blocks known as πρωτη ὑλη were later found in the protons. Ernest Rutherford (1871–1937) probably gave this component of the atom the name proton (Greek: "the first") in 1920 not only for etymological reasons, but also to honor William Prout [10]. Lothar Meyer later also pursued the idea of building atoms from smaller aggregates in his presentations.

Johann Wolfgang Döbereiner (1780–1849), chemistry professor at University of Jena and well-known for the gas lighter, the Döbereiner Feuerzeug, was interested in numerical relations between reacting substances. As a result of his stoichiometric investigations, he published in 1816 an "Account of the numerical ratios of earthy elements in chemical compounds."[8] He also examined such relations between

[8]"Darstellung der Verhältnisszahlen der irdischen Elemente zu chemischen Verbindungen" (the triads are not mentioned in these tables).

equivalent numbers of the oxides in some minerals.[9] He established groups of three chemically similar elements with increasing atomic weights and found that the atomic weight of the middle element of each of these groups is roughly the mean value of the other two [11]:

> Berzelius' experiments to determine the atomic weights of iodine and bromine were very interesting for me, because they confirm the assumption expressed earlier in my lectures that perhaps the atomic weight of bromine is the arithmetic mean of the atomic weights of chlorine and iodine. This mean value is namely [(35.470 + 126.470)/2] = 80.470. A number that is somewhat larger than the one found by Berzelius (78.383), but comes so close to it that one can almost hope that the difference will disappear completely with (future) repeated precise determinations of the atomic weights of the three salt formers.[10]

Döbereiner collected lithium, sodium, and potassium as well as calcium, strontium and barium, but also sulfur, selenium, and tellurium into groups that were later called triads. Lothar Meyer was explicitly interested in these mathematical relationships. He expressed these thoughts editing Döbereiner's work on this subject.

In his habilitation lecture, Lothar Meyer had already dealt with the different theoretical concepts that predominate in chemistry, following on from Claude Louis Berthollet (1748–1822) and his work on chemical statics. This interest was deepened by his participation in the chemists' congress in Karlsruhe. He began with the writing of the already mentioned book *Die modernen Theorien der Chemie*. In his

---

[9]Döbereiner wrote, for the first time, about the arithmetical relations between the oxides of three elements in a letter to Johann Wolfgang von Goethe (1749–1832) on September 30, 1816 (published in Döbling H (1928) Die Chemie in Jena zur Goethezeit. Jena: Fischer): "Der bei Dornburg gefundene Coelestin zeigt merkwürdige Verhältnisse: sein spec. Gewicht ist das mittlere von dem des Anhydrits und Schwerspaths, nehmlich (2.95 + 4.47 / 2 =) 3.71 und die Aequivalentzahl seiner Base, der Strontia, ist ebenfalls die mittlere von der der Grundlage des Anhydrits, des Kalks, und der des Schwerspaths, der Baria, nehmlich (27.5 + 72.5 / 2 =) 50. Man sollte fast glauben, der Coelestin sey eine chemische Verbindung von 1 stoechiometrischen Antheile wasserleeren Gypses mit 1 solchen Antheile Schwerspath und die Strontia das Resultat einer Vereinigung des Kalks und der Baria. Es ist mir jedoch nicht gelungen, dieses Resultat (künstlich) auf synthetischem Wege zu erhalten." "The celestine* found at Dornburg shows strange proportions: its specific gravity is the average of that of anhydrite** and barite***, namely (2.95 + 4.47 / 2 =) 3.71 and the equivalent number of its base, strontia, is also the mean of that of the base of anhydrite, lime, and that of barite, baria, notably (27.5 + 72.5 / 2 =) 50. One would almost believe that celestine is a chemical compound of 1 stoichiometric part of anhydrous gypsum with 1 such part of barite, and strontia the result of a union of lime and baria. However, I have not succeeded in obtaining this result (artificially) by synthetic means." *anhydrous strontium sulfate; **anhydrous calcium sulfate; ***anhydrous barium sulfate; Lime = CaO, strontia = SrO, baria = BaO (Translation Gregory S. Girolami).

Later Ferdinand Wurzer (1765–1844) reported about this discovery (Wurzer F [1817] Auszug eines Briefes vom Hofrath Wurzer, Prof. der Chemie zu Marburg. Ann Phys 56:331–334).

[10]"Sehr interessant waren für mich Berzelius' Versuche zur Bestimmung der Atomgewichte des Jods und Broms, denn sie bestätigen die früher in meinen Vorlesungen ausgesprochene Vermuthung, dass vielleicht das Atomengewicht des Bromes das arithmetische Mittel der Atomengewichte des Chlors und Jods sey. Dieses Mittel ist nämlich [(35.470 + 126.470)/2] = 80.470. Eine Zahl, welche zwar etwas grösser ist, als die von Berzelius gefundene (78.383), aber doch dieser so nahe kommt, dass man fast hoffen darf, die Differenz werde bei (künftigen) wiederholten scharfen Bestimmungen der Atomgewichte der drei Salzbildner ganz verschwinden."

Atome:	Molekeln:	Radicale:
Li = 7,03	Holzgeist = CH_4O = 32	Methyl = CH_3 = 15
Diff. . 16,02	. . . . . CH_2 = 14	. . . . CH_2 = 14
Na = 23,05	Weingeist = C_2H_6O = 46	Aethyl = C_2H_5 = 29
Diff. . 16,08	. . . . . CH_2 = 14	. . . . CH_2 = 14
K = 39,13	Propylgeist = C_3H_8O = 60	Propyl = C_3H_7 = 43

Fig. 8.5 Comparison of the weight differences of atoms, molecules and radicals [12] (*Source* Universitätsbibliothek Braunschweig)

statements Meyer assumed the existence of atoms and even discussed their divisibility, as the following sentence in paragraph §91 (p. 135) shows [12]:

> But it is not only the nature of the interaction of the chemical atoms that has become the subject of speculation, but also the peculiar nature of these atoms themselves. The specific and regular relationships between the atomic weights of different elements, which have been found over a long period, have repeatedly prompted (particularly in recent years) treatment of the question whether our atoms are themselves again associations of higher-order atoms, making them "groups of atoms" or "molecules."[11]

Meyer pointed to the fact that the atomic weights of some groups of elements have a striking numerical relationship. For a time he endorsed Prout's hypothesis, because he assumed that atoms are composed of smaller building blocks. He also hoped to find an analogy to the homologous series in organic chemistry and therefore compared the masses of atoms, molecules, and radicals on page 136 [12] (Fig. 8.5).

In the book *Die modernen Theorien* Lothar Meyer made his first attempt to order the elements, taking into account the atomic weights and valences. He pointed out that these are relations "for six groups of elements that are well characterized as belonging together."[12]

Meyer tried to arrange 50 elements. His first compilation on page 137 [12] (Fig. 8.6) comprised 28 elements. The atomic weights of the elements increase from left to right in the individual rows. This is consistent with a regular change in valence, although Meyer did not speak of periodicity.

Exceptions to the increase of atomic weight are found for tellurium and thallium. For reasons based on chemical properties, he did not classify tellurium among the halogens, but rather in the preceding group, despite its atomic weight being higher

[11]"Aber nicht bloss die Art der Wechselwirkung der chemischen Atome ist Gegenstand der Speculation geworden, sondern auch die eigenste Natur dieser Atome selbst. Die eigenthümlichen regelmässigen Beziehungen, welche seit lange zwischen den Atomgewichten der verschiedenen Elemente aufgefunden wurden, haben, namentlich in den letzten Jahren, wiederholt die Behandlung der Frage veranlasst, ob nicht unsere Atome selbst wieder Vereinigungen von Atomen höherer Ordnung, also Atomgruppen oder Molekeln seien" (I thank Vera Mainz for the translation which is published in Chapter 9 of this volume.).

[12]"für sechs als zusammengehörig wohl charakterisierte Gruppen von Elementen".

	4 werthig	3 werthig	2 werthig	1 werthig	1 werthig	2 werthig
	—	—	—	—	Li = 7,03	(Be = 9,3?)
Differenz =	—	—	—	—	16,02	(14,7)
	C = 12,0	N = 14,04	O = 16,00	Fl = 19,0	Na = 23,05	Mg = 24,0
Differenz =	16,5	16.96	16,07	16,46	16,08	16,0
	Si = 28,5	P = 31,0	S = 32.07	Cl = 35,46	K = 39,13	Ca = 40,0
Differenz =	$\frac{89,1}{2}$ = 44,55	44,0	46,7	44,51	46,3	47,6
	—	As = 75,0	Se = 78,8	Br = 79,97	Rb = 85,4	Sr = 87,6
Differenz =	$\frac{89,1}{2}$ = 44,55	45,6	49,5	46,8	47,6	49,5
	Sn = 117,6	Sb = 120,6	Te = 128,3	J = 126,8	Cs = 133,0	Ba = 137,1
Differenz =	89,4 = 2.44,7	87,4 = 2.43,7	—	—	(71 = 2.35,5)	—
	Pb = 207,0	Bi = 208,0	—	—	(Tl = 204?)	—

Fig. 8.6 Meyer's arrangement of the elements according to their atomic weights in the year 1864 [12] (*Source* Universitätsbibliothek Braunschweig)

Fig. 8.7 Table of a further six elements [12] (*Source* Universitätsbibliothek Braunschweig)

	4 werthig	6 werthig
	Ti = 48	Mo = 92
Differenz =	42	45
	Zr = 90	Vd = 137
Differenz =	47,6	47
	Ta = 137,6	W = 184.

than that of iodine. With regard to thallium, he seems to have been uncertain about the atomic weight, as the question mark suggests.

It is also noticeable that Meyer left gaps in his system, but did not discuss them further. Rather, he was interested in the regularity in the differences in atomic weights, which are around 16 in the first and second rows, then around 46. This difference of 46 also plays a role in the following tables from page 138 [12] (Figs. 8.7 and 8.8).

Here, too, Meyer was guided by atomic weight and valence. In the last group on page 138 [12] (Fig. 8.8), he grouped together elements (Cu, Ag, Au) despite the fact that they often exhibit different valencies.

	4 werthig	4 werthig	4 werthig	2 werthig	
	Mn=55,1 Fe=56,0	Ni =58,7	Co=58,7	Zn=65,0	Cu=63,5
Differ. =	49,2 48,3	45,6	47,3	46,9	44,4
	Ru=104,3	Rh=104,3	Pd=106,0	Cd=111,9	Ag=107,94
Differ. =	92,8=2.46,4	92,8=2.46,4	93,0=2.46,5	88,3=2.44,2	88,8=2.44,4
	Pt=197,1	Jr –197,1	Os =199,0	Hg=200,2	Au=196,7

Fig. 8.8 Table of a further 16 elements [12] (*Source* Universitätsbibliothek Braunschweig)

Meyer concluded his remarks in this section by pointing out that there can be no doubt that a certain regularity prevails in the numerical values of atomic weights. He ascribed some of the deviations that occur to incorrectly determined values of the atomic weights. He wrote on page 139 [12]:

> In part, however, these deviations can be viewed with justification, as brought about by incorrectly determined values of the atomic weights. However, this should hardly be the case for all of them; and certainly one is not entitled, as has happened all too often, for the sake of a supposed regularity to arbitrarily correct and change the empirically found atomic weights before the experiment has put more exactly determined values in their place.[13]

In order to further examine the question of the accuracy of the atomic weights, Meyer began to determine and to calculate them not later than 1866. He could no longer devote himself to this task in Karlsruhe; only in Tübingen did he resume the investigations together with Seubert [2].

In Eberswalde Meyer was already working on the 2nd edition of his book *Die modernen Theorien*. Presumably in this context he designed a more extensive overview with 52 elements in 1868, since aluminum and chromium had now also been included (Fig. 8.9). The system consisted of 16 columns, the last (16th) being empty. The element hydrogen was not taken into account.

If you compare the individual groups with today's representation of the periodic system, there is already a lot of agreement with regard to the main group elements. The assignment of the subgroup elements, however, is still problematic. It is noticeable in the table that Meyer deliberately left a space free under silicon.

[13]"Zum Theil allerdings können diese Abweichungen mit Fug und Recht angesehen werden als hervorgebracht durch unrichtig bestimmte Werthe der Atomgewichte. Bei allen dürfte indess dies kaum der Fall sein; und ganz sicherlich ist man nicht berechtigt, wie das nur zu oft geschehen ist, um eine vermeintlichen Gesetzmäßigkeit willen die empirisch gefundenen Atomgewichte willkürlich zu corrigieren und zu verändern, ehe das Experiment genauer bestimmte Werthe an ihre Stelle gesetzt hat" (Translation Vera Mainz).

Anhang. 6

Entwurf eines Systems der Elemente von Lothar Meyer. 1868.

§ 91 Nicht gedruckt. Wiedergabe nach dem Manuscript.

Lothar Meyer

1	2	3	4	5	6	7	8
		Al — 27,3 *)	Al = 27,3 *)				C = 12,00
		$\frac{28,7}{2}$ = 14,3					16,5
							Si = 28,5
							$\frac{89,1}{2}$ = 44,55
Cr = 52,6	Mn = 55,1	Fe = 56,0	Co = 58,7	Ni = 58,7	Cu = 63,5	Zn = 65,0	—
	49,2	48,3	47,3		44,4	46,9	$\frac{89,1}{2}$ = 44,55
	Ru = 104,3	Rh = 104,3	Pd = 106,0		Ag = 107,94	Cd = 111,9	Sn = 117,6
	92,8 = 2·46,4	92,8 = 2·46,4	93 = 2·46,5		88,8 = 2·44,4	88,3 = 2·44,15	89,4 = 2·44,7
	Pt = 197,1	Ir = 197,1	Os = 199,0		Au = 196,7	Hg = 200,2	Pb = 207,0

*) Im Original durchstrichen und durch daruntergesetzte Punkte wieder gültig gemacht. K. S.

Natur der Atome: Gründe gegen ihre Einfachheit.

9	10	11	12	13	14	15	16
			Li = 7,03	Be = 9,3			
			16,02	14,7			
N = 14,04	O = 16,00	Fl = 19,0	Na = 23,05	Mg = 24,0			
16,96	16,07	16,46	16,08	16,0			
P = 31,0	S = 32,07	Cl = 35,46	K = 39,13	Ca = 40,0	Ti = 48	Mo = 92	
44,0	46,7	44,51	46,3	47,6	42	45	
As = 75,0	Se = 78,8	Br = 79,97	Rb = 85,4	Sr = 87,6	Zr = 90	Vd = 137	
45,6	49,5	46,8	47,6	49,5	47,6	47	
Sb = 120,6	Te = 128,3	J = 126,8	Cs = 133,0	Ba = 137,1	Ta = 137,6	W = 184	
87,4 = 2·43,7			71 = 2·35,5				
Bi = 208,0			? Tl = 204 ?				

S. *L. Gmelin*, Hdb. 5. Aufl. I, 47 ff.; Münch. gel. Anz. 1850 Bd. **30**, S. 261, 272, abgedr. Ann. Chem. Pharm. 1858. **105**, 187; *J. Dumas*, C. r. 1857, t. **45**, p. 709; auch Ann. Chem. Pharm. **105**, S. 74 u. a.

Seite 7 hat man sich in der Weise seitlich an Seite 6 angereiht zu denken, dass N = 14,04 in Spalte 9 neben C = 12,00 in Spalte 8 zu stehen kommt, P neben Si, Sb neben Sn, Bi neben Pb. K. S.

7

Fig. 8.9 Meyer's unpublished periodic system [14]

There are also other irregularities in the assumed regular increase of atomic weights. For example: if molybdenum had been placed next to zirconium and vanadium next to tantalum, the sequence of increase would have been less disturbed. Whether this is a transmission error by Seubert cannot be determined, as the original version has not yet been found.

The fate of Meyer's representation bears some question marks. As Meyer's successor in Eberswalde, the mineralogist and geologist Adolf Remelé (1839–1915), himself reported, Meyer gave him the handwritten draft. In a necrology he wrote [13]:

> In July 1868, when I took over his inventories as his successor in the teaching post of chemistry, physics and mineralogy, he gave me a handwritten arrangement of the elements according to the atomic sizes, with the remark that he was thinking of publishing the matter soon as it represents an essential addition and improvement of the above-mentioned scheme of 1864.[14]

Why Remelé did not send this scheme to Meyer during the priority disputes that soon flared up or whether Meyer explicitly asked for it cannot be established. Remelé did not show the draft to Meyer again until 1893. A copy was given to Seubert in 1895 [15].

There is no doubt Meyer had anticipated much of Mendeleev's thoughts with this unpublished table. Until 1870 Meyer spoke exclusively of the regular relationships found in atomic weights and did not use the term "periodically recurring relationship." Meyer regarded the constant differences in atomic weights as proof of the complex nature of the atom.

In 1870, a paper by Meyer appeared, dated Karlsruhe, December, 1869. Meyer discussed the relationship between the atomic weights and the properties of 55 elements, although he spoke of 56 in the text [16]. Had he counted hydrogen but not included it in the presented table?

Meyer had already announced this work in a letter to Franz Ernst Neumann (1798–1895) in February 1870 [17]:

> In the course of next month I hope to send you a small paper that attempts to treat the properties of the chemical elements as a function of their atomic weight. The first attempt was a bit poor.[15]

Even more astonishing is a postscript of a letter that Meyer had sent on March 27, 1870, to his brother-in-law Adolf Ferdinand Weinhold (1841–1917), professor of physics at the Royal Trade School in Chemnitz: "You will have received my little atomic swindle."[16] [18] Did Meyer have doubts about his results?[17]

---

[14]"Als ich im Juli 1868 als sein Nachfolger im Lehramt der Chemie, Physik und Mineralogie die ihm unterstellten Inventarien übernahm, übergab er mir mit dem Bemerken, er denke die Sache doch bald zu veröffentlichen, eine eigenhändig geschriebene Anordnung der Elemente nach den Atomgrößen, welche eine wesentliche Ergänzung und Vervollkommnung seines vorerwähnten Schemas von 1864 darstellt."

[15]"Im Laufe des nächsten Monats hoffe ich Ihnen eine kleine Abhandlung zusenden zu können, die den Versuch macht, die Eigenschaften der chemischen Elemente als Function ihres Atomgewichts zu behandeln. Der erste Versuch ist noch etwas dürftig ausgefallen."

[16]"Meinen kleinen Atomschwindel werden Sie erhalten haben."

[17]In the original publication of this paper I had no explanation for this note. Today I know that the phrase of a swindle was often used in private correspondence in that time and was not connected with doubts about the accuracy.

In this 1870 publication, Meyer used updated values for atomic weights, e.g., from determinations of the gas density of the compounds or from the heat capacity, and referred to the fact that now clear regularities in the atomic weights come to light, and he mentioned for the first time that there was a periodicity in properties depending on atomic weight [14]: "The same or similar properties return when the atomic weight has increased by a certain magnitude, at first 16, then about 46 and finally 88 to 92 units."[18]

As an example of periodicity in the properties, he established a relationship between atomic volume and atomic weight (Fig. 8.10).

His table of elements was designed differently in this publication: the columns became rows, each representing a so-called natural family. Later, however, he returned to the representation of similar elements in a column. Meyer ordered the elements strictly according to atomic weight and marked the places of a deviation, which he explained as an imprecise determination of the atomic weights. Again, he did not include hydrogen in the system. There are also some places filled with dashes. It remains unclear whether the dashes symbolize the gaps he mentioned.[19] Meyer wrote on page 12 [14][20]:

> These elements will probably later (at least partly) fill the gaps which are still to be found in the table. Other gaps will probably be filled by elements to be discovered later. Perhaps future discoveries will displace one or the other element from its place and replace it by one that fits better.

Here Meyer, like Mendeleev, predicted the existence of other elements, but made no statements about their properties. Meyer was impressed by the periodicity in the properties, but made it clear on page 362 that it remains completely unclear what the causes are [14].[21]

> If these and similar regularities cannot possibly be the result of pure chance, then we have to admit that by empirically determining them we have by no means found the key to the knowledge of their inner causal connection. But at least a starting point seems to have been gained for the investigation of the constitution of the atoms that have not yet been decomposed, a guideline for further comparative investigation of the elements.[22]

[18]"Dieselben oder ähnliche Eigenschaften kehren wieder, wenn das Atomgewicht um eine gewisse Grösse, die zunächst 16, dann etwa 46 und schließlich 88 bis 92 Einheiten beträgt, gewachsen ist." Quote on page 13 in [14] and on page 358 in [16].

[19]In the original publication of this paper I stated the intent of the lines was unclear. Today I am sure that these mark the places for new elements.

[20]"Diese Elemente werden voraussichtlich später, z. Th. wenigstens, die Lücken ausfüllen, welche sich in der Tabelle jetzt noch finden. Andere Lücken werden möglicherweise durch später zu entdeckende Elemente ausgefüllt werden; vielleicht auch wird durch künftige Entdeckungen das eine oder andere Element aus seiner Stelle verdrängt und durch ein besser hinein passendes ersetzt werden." Quote on page 357 in [16].

[21]Quote on page 362 in [16].

[22]"Wenn diese und ähnliche Regelmässigkeiten unmöglich reines Spiel des Zufalls sein können, so müssen wir uns andererseits gestehen, dass wir mit der empirischen Ermittlung derselben noch keineswegs den Schlüssel zur Erkenntnis ihres inneren ursächlichen Zusammenhangs gefunden haben. Aber es scheint wenigstens ein Ausgangspunkt gewonnen zu sein für die Erforschung der

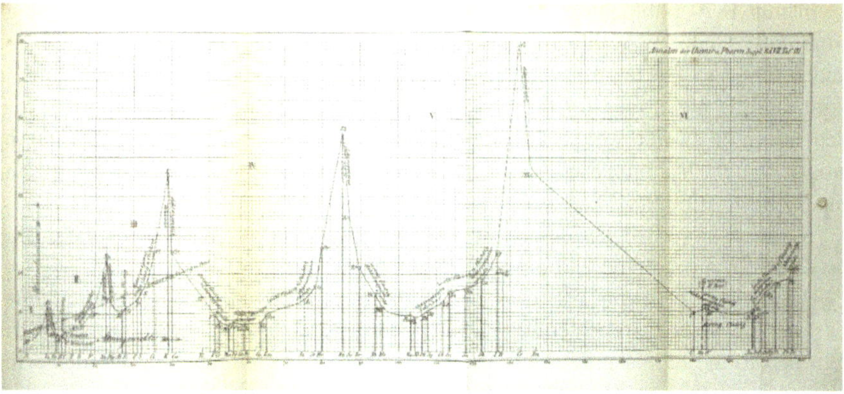

Fig. 8.10 The atomic volume curve (scan from [16])

Finally, he pointed out that the identified relationships now also allow a verification of the atomic weights.

But in the meantime, Mendeleev had also published reports on a natural system of elements, for example in 1869 in the *Journal of the Russian Chemical Society* [19]. Viktor von Richter (1841–1891) reported on this in the same year in the *Berichte der deutschen chemischen Gesellschaft* [20] and mentioned a paper published in the *Zeitschrift für Chemie* [21].

Meyer was familiar with these reports and wrote in his publication on the nature of chemical elements that his system is largely comparable to that of Mendeleev. This led some readers to believe that Meyer had not even published his own ideas. Mendeleev probably saw it that way too. He responded with two publications in 1871 and raised the question of priority.

After 1871, Mendeleev and Meyer turned to other scientific problems. But then the priority dispute flared up again. In 1879, Adolphe Wurtz (1817–1884) complained to the German Chemical Society about alleged changes in the content of the translation of his book *La théorie atomique*. He believed that the translation overvalued Meyer's contribution to establishing the periodic table. Wurtz noted that only Mendeleev had the idea of arranging the elements in two rows according to their atomic weight. Meyer would only then have completed this idea [22].

In response to this letter, Meyer published his version of the history of the periodic system [23]. He compared the tables he published with those of Mendeleev. First, he noted that the three tables of elements (see Figs. 8.6–8.8) he published in 1864 did not represent a simple arrangement of similar elements. At that time, it was not possible to draw up only one table based on atomic weights because they were not precise for all elements. However, once the exact data were

Constitution der bis jetzt unzerlegten Atome, eine Richtschnur für fernere vergleichende Untersuchung der Elemente."

available to him, he was able to construct such a system, but it was not published in time.

Second, Meyer stated that Mendeleev was actually the first to notice the periodic change in the properties related to atomic weights and to recognize the possibility of predicting elements from this system. Third, Meyer complained that Mendeleev was not actually able to formulate one and only one series with increasing atomic weight.

A few months later, Mendeleev also responded very vigorously with an article in the *Berichte der deutschen chemischen Gesellschaft*. Mendeleev's biting tone led to a new brief statement from Meyer. In 1882, the Royal Society had acted wisely to award Meyer and Mendeleev the gold Davy Medal.

But haven't there been discussions to this day about Lothar Meyer's contribution to the periodic system of the elements? If you ask students about its creator, they usually only mention Mendeleev. Lothar Meyer is not known to many. And the logo of the International Year of the Periodic Table also shows only one scholar, namely Mendeleev.

Apparently the fact of successful predictions of new elements by Mendeleev, which he dared especially in his extensive work from 1871 [24], had an enormous influence on the recognition of his scientific achievement. However, it does not take into account that some of the predictions did not come true or came true in a completely different way. Mendeleev also failed to find the predicted elements himself. There was no mathematical basis for Mendeleev's predictions; he was wrong about the presumably incorrectly determined atomic weights of iodine and tellurium [25]. As indicated above, Meyer also mentioned empty spaces, but did not predict elements explicitly. Aren't these gaps a step on the way to discovering a new element?

When Hans Landolt recommended electing Meyer as a member of the Berlin Academy in 1888, it was precisely these predictions that were at stake. The letter of recommendation contained a comparison of the work of both scholars. Mendeleev would have been interested in the gaps and predicting new elements. These elements—in fact three of the 16 predicted during Mendeleev's lifetime—were discovered, a fact that would be understood by non-chemists as well. Meyer's results were less "popular." But Landolt expressed the opinion that they were more important from a scientific point of view [26].

Two years before his death, Meyer published an article about the usefulness of the periodic system in studying inorganic chemistry, which once again made his goals clear. He saw his contribution as a modification of Döbereiner's triad system. He emphasized that the system was very well suited to give the students an overview. However, he himself said that the periodic system was given little attention in textbooks, or it was not explained or not mentioned. Only a very small number of textbooks used it as a basis for the arrangement of matter; usually it was described only very briefly and incompletely.

Meyer even tried to explain this situation: first, the leading chemists were only interested in organic chemistry and second—which is perhaps even more decisive—the system was not self-explanatory. He described his system as a didactic tool rather than a law. He also mentioned a large table chart to demonstrate his system in the classroom. Occasionally, he also used a representation of the periodic system on a rotatable cylinder [27]. Unfortunately, neither the chart nor the cylinder survived.

Today the periodic system has found its atomic theoretical explanation. Until the beginning of the twentieth century, the contributions of Meyer and Mendeleev to the periodic system were treated with about equal value. But the opinion became increasingly accepted that Mendeleev, who had rejected Prout's theory and the existence of atoms, had the much greater share in the establishment of the periodic system owing to his predictions, which he had emphasized again and again during his lifetime [28].

8.4 Conclusions

This paper presents the biography and the most important achievements of Lothar Meyer in the field of the classification of elements. Lothar Meyer spent almost his entire scientific life looking for an explanation of the Döbereiner triads and for the smallest building blocks of matter. He did not completely reject Prout's hypothesis. Concerning the predictions of new elements Meyer was very cautious—but this is an unjustified reason to undervalue his contribution to the periodic system of elements. He recognized the need to determine better atomic weights. He saw the special value of the periodic system in its role as a teaching tool, although he himself had recognized that it seemed to find its way into textbooks only very slowly. Lothar Meyer did not live to see the confirmation of the periodic system through atomic theory and the finding and classification of isotopes.[23]

A Small Selection of Books Relating to the Periodic System and the Elements

Aldersey-Williams H (2011) Periodic tales. the curious lives of the elements. Penguin Books, London

Atkins PW (1995) The periodic kingdom. a journey into the land of the chemical elements. Orion Publishing Group Ltd./Basic Books, New York

Gordin M (2004) A well-ordered thing: Dmitrii Mendeleev and the shadow of the periodic table. Perseus Books Group/Basic Books, New York.

Gray T (2010) The elements. a visual exploration of every known atom in the universe. Black Dog & Leventhal Publishers, New York

Jackson T (2012) The elements. an illustrated history of the periodic table (Ponderables: 100 breakthroughs that changed history. Who did what when). Shelter Harbor Press Ltd, New York

[23]I thank the editors for helpful hints and support in translating my paper.

Kean S (2010) The disappearing spoon and other true tales of madness, love, and the history of the world from the periodic table of the elements. Little, Brown and Company, New York

Levi P (1984) The periodic table. Schocken Books, New York

Quadbeck-Seeger H-J (2006) Die Welt der Elemente – Die Elemente der Welt. Wiley-VCH, Weinheim

Sacks O (2001) Uncle Tungsten. memories of a chemical boyhood. A. Knopf, New York/Toronto

Scerri ER (2007) The periodic table: its story and its significance. Oxford University Press, New York

Van Spronsen JW (1969) The periodic system of chemical elements. Elsevier, Amsterdam (et al.)

Strathern P (2000) Mendeleyev's dream: the quest for elements. Hamish Hamilton, London

Dingle A, Gillespie J (2018) Periodic table. Dorling Kindersley, London

A Small Selection of Further Publications by Mendeleev and Meyer on the Question of Priority

Mendelejeff DI (1871) Die periodische Gesetzmäßigkeit der chemischen Elemente. Annalen Chem Pharm Supplement VIII:133–229

Mendelejeff DI (1871) Zur Frage über das System der Elemente. Ber Dtsch Chem Ges 4:348–352

Meyer L (1873) Zur Systematik der anorganischen Chemie. Ber Dtsch Chem Ges 6:101–106

Meyer L (1880) Zur Geschichte der periodischen Atomistik. Ber Dtsch Chem Ges 13:259–265

Mendelejeff DI (1880) Zur Geschichte des periodischen Gesetzes. Ber Dtsch Chem Ges 13:1796–1804

Meyer L (1880) Zur Geschichte der periodischen Atomistik. Ber Dtsch Chem Ges 13:2043–2044

References

1. Boeck G (2019) Das Periodensystem der Elemente und Lothar Meyer. Chem unserer Zeit 53:372–381
2. Seubert K (1896) Nekrolog Lothar Meyer. Ber Dtsch Chem Ges 28:1109–1146
3. Danzer K (1974) Dmitri I. Mendelejew und Lothar Meyer: die Schöpfer des Periodensystems der chemischen Elemente. Teubner, Leipzig
4. Schwanicke G (1995) Aus dem Leben des Chemikers Julius Lothar Meyer und über seine epochemachenden Arbeiten auf dem Gebiet der Chemie im 19. Jahrhundert. Vareler Heimathefte, H. 8. Heimatverein, Varel
5. Stutte B (1997) Lothar Meyer in Tübingen. Bausteine zur Tübinger Universitätsgeschichte 8:79–88
6. Kluge H, Kästner I (2014) Ein Wegbereiter der Physikalischen Chemie im 19. Jahrhundert – Julius Lothar Meyer (1830–1895). Shaker, Aachen

7. Private communication, Würzburg University Archives, 1 and 10 December 2014
8. Deutsches Museum München, Archive, Munich, letters from Lothar Meyer to Hermann Kolbe, HS-Nr. 03535 and letters from Lothar Meyer to August Kekulé, NL Kekulé, NL 228 / 0509
9. Nawa C (2017) Das Tübinger Chemische Laboratorium von 1846. Mitteilungen, Gesellschaft Deutscher Chemiker/Fachgruppe Geschichte der Chemie 25:125–163
10. Brock WH (1985) From protyle to proton: William Prout and the nature of matter, 1785–1985. Accord, Bristol
11. Döbereiner JW (1829) Versuch zu einer Gruppirung der elementaren Stoffe nach ihrer Analogie. Annalen Phys Chem 15:301–307. Reprinted in: Meyer L (ed) (1895) Die Anfänge des natürlichen Systems der Elemente. Ostwalds Klassiker der exakten Wissenschaften, vol 66. Engelmann, Leipzig
12. Meyer L (1864) Die modernen Theorien der Chemie und ihre Bedeutung für die chemische Statik. Maruschke & Berendt, Breslau
13. Remelé A (1895) Nekrolog Lothar Meyer. Zeitschrift für Forst- und Jagdwesen 27:721–724
14. Seubert K (ed) (1913) Das natürliche System der chemischen Elemente, 2nd edn. Ostwalds Klassiker der exakten Wissenschaften, vol 68. Engelmann, Leipzig
15. Berlin-Brandenburgische Akademie Berlin, Archive, Collection of autographs, chemists' letters, Seubert-Remelé correspondence, No. 94
16. Meyer L (1870) Die Natur der chemischen Elemente als Function ihrer Atomgewichte. Ann Chem Pharm Supplement 7:354–364
17. Lower Saxony State and University Library Göttingen, letters from Lothar Meyer to F. E. Neumann 61: 9, pp 30–31, letter dated February 8, 1870
18. Chemnitz University of Technology, Archive, A. F. Weinhold estate, inventory 301, letter from Lothar Meyer dated March 27, 1870
19. Mendeleev DI (1869) Sootnošenie svojstv s atomnym vesom elementov. Žurnal Russkogo chimičeskogo obšččestva 1:60–77
20. Von Richter V (1869) Bericht aus St. Petersburg vom 17. Oktober 1869. Ber Dtsch Chem Ges 2:552–554
21. Mendelejeff DI (1869) Über die Beziehungen der Eigenschaften zu den Atomgewichten der Elemente. Z Chem 12:405–406
22. (1880) Protokoll der Vorstandssitzung vom 11. Januar 1880. Ber Dtsch Chem Ges 13:6–8
23. Meyer L (1880) Zur Geschichte der periodischen Atomistik. Ber Dtsch Chem Ges 13:259–265
24. Mendelejeff DI (1871) Die periodische Gesetzmäßigkeit der chemischen Elemente. Ann Chem Pharm Supplement 8:133–229
25. Cahn RM (2002) Historische und philosophische Aspekte des Periodensystems der chemischen Elemente. HYLE Studies in History and Philosophy of Chemistry 1, HYLE Publications, Karlsruhe
26. Greiner A, Welsch F, Klare H (1986) Chemiker über Chemiker Wahlvorschläge zur Aufnahme von Chemikern in die Berliner Akademie 1822–1925 von Eilhard Mitscherlich bis Max Bodenstein (Studien zur Geschichte der Akademie der Wissenschaften der DDR, vol 12). Akademie-Verlag, Berlin
27. Meyer L (1893) Über den Vortrag der anorganischen Chemie nach dem natürlichen Systeme der Elemente. Ber Dtsch Chem Ges 26:1230–1250
28. Gordin M (2012) The textbook case of a priority dispute. In: Biagoioli M, Riskin J (eds) Nature engaged: science in practice from the renaissance to the present. Palgrave Macmillan, New York, pp 59–82

Translation of §§ 91–94 of Lothar Meyer's *Modernen Theorien* (1864) 9

Vera V. Mainz

Abstract

An English translation of §§ 91–94 (pp 135–147) of Lothar Meyer's *Die modernen Theorien der Chemie und ihre Bedeutung für die chemische Statik* is given. In this book Meyer describes his first classification of the elements into a periodic system and presents his thoughts on theory and its application to chemistry.

9.1 Introduction: Biographical Information

Among the predecessors of Mendeleev was the German chemist Lothar Meyer. In 1864, Meyer published the earliest version of his periodic table in his book *Die Modernen Theorien der Chemie und Ihre Bedeutung für die Chemische Statik* [1]. A short biographical sketch of Meyer is presented in Sect. 9.1, up through the publication of the first edition of *Modernen Theorien*. See Chapter 8 of this volume for a fuller discussion of Meyer's life and work.

(Julius) Lothar Meyer [2] (1830–1895) was the son of a physician, Heinrich Friedrich August Jacob Meyer, and initially intended to become a physician, too. He began his studies of medicine in the summer of 1851 at the University of Zürich, moving to the University of Würzburg in 1853 to attend lectures by Rudolf Virchow (1821–1902), the eminent pathologist. He received his M.D. in 1854. At this point, encouraged by his former physiology professor at Zürich, Carl Ludwig (1816–1895), he switched from medicine to physiological chemistry and went to Heidelberg to study under Robert Bunsen (1811–1899). Meyer was particularly

V. V. Mainz (✉)
School of Chemical Sciences, University of Illinois at Urbana-Champaign, Urbana, IL, USA
e-mail: mainz@illinois.edu

C. J. Giunta et al. (eds.), *150 Years of the Periodic Table*, Perspectives on the History of Chemistry, https://doi.org/10.1007/978-3-030-67910-1_9

interested in Bunsen's studies of gas analysis, and completed his doctoral dissertation, *Ueber die Gase des Blutes*, in 1856, at Heidelberg, although his degree was awarded by Würzburg. While he attended classes in Heidelberg, Friedrich Beilstein (1838–1906), Hans H. Landolt (1831–1910), Henry E. Roscoe (1833–1915), Adolf von Baeyer (1835–1917) and Friedrich August Kekulé (1829–1896) were also at the University (see Fig. 9.1). Meyer developed a close friendship with Kekulé, in particular. The physicist Gustav Robert Kirchhoff (1824–1887) was also in Heidelberg and his lectures likely moved Meyer's interests toward physical chemistry.

Meyer next moved to Königsberg in the fall of 1856 and attended the lectures on mathematical physics given by Franz Neumann (1798–1895). He also pursued studies of the effect of carbon monoxide on the blood, resulting in a Ph.D. from the University of Breslau in 1859. He was appointed as Privatdozent in physics and chemistry in 1859 and was responsible for the chemistry laboratory in the physiological institute.

Meyer attended the Karlsruhe Conference in 1860, heard Cannizzaro's plea for the establishment of uniform atomic weights and formulas, and became a whole-hearted supporter of Cannizzaro's ideas. His *Die modernen Theorien der Chemie und ihre Bedeutung für die chemische Statik*, published in 1864, was a direct outcome of this change [3]. Figure 9.2 shows Meyer in the mid-1860s.

In an article on Meyer and his contributions to the periodic law [3], Alan Rocke concluded:

> One of the themes of this study has been the importance for Meyer of his multifaceted education, and his continuing high-level engagement with all aspects of chemistry—organic, inorganic, physical, and analytical—along with his mastery of mathematical physics as well. This range of competency was highly unusual, in his day as in ours, and we have seen how it was a central key to his success.

Meyer's 1864 periodic tables from his *Modernen Theorien* are often reproduced, but little attention has been paid to his accompanying text, which has never been translated in its entirety into English. Here we provide a translation of pages 136–147 of *Modernen Theorien*, in which Meyer describes his first classification of the elements and his thoughts on theory and its application to chemistry.

9.2 Translation, *Die Modernen Theorien*, 1864: §§ 91–94, pp 135–147

§ 91

But it is not only the nature of the interaction of the chemical atoms that has become the subject of speculation, but also the peculiar nature of these atoms themselves. The specific and regular relationships between the atomic weights of different elements, which have been found over a long period, have repeatedly prompted (particularly in recent years) treatment of the question whether our atoms are themselves again associations of higher-order atoms, making them "groups of

Fig. 9.1 The lunch table community in Heidelberg in 1857. Standing, from left to right: Victor Ernest Gaupillat (1831–1889); Agostino Frapolli (1824–1903); Adolph Wagner (1835–1917); Henry Roscoe (1833–1915); Lothar Meyer (1830–1895); Angelo Pavesi (1830–1896); Friedrich Beilstein (1838–1906). Sitting, from left to right: Prof. Johann Friedrich Bahr (1815–1875); Hans Landolt (1831–1910); Ludwig Carius (1829–1875); August Kekulé (1829–1896); Leopold von Pebal (1826–1887). (Deutsches Museum, Munich, Archive, PT11681)

Fig. 9.2 Lothar Meyer (1830–1895). (Deutsches Museum, Munich, Archive, PT11625-01)

atoms" or "molecules."[1] Indeed, the latter view in itself has an extraordinarily high probability because the atomic weights of certain groups of closely related elements offer very similar relationships among themselves, in analogy to the molecular weights of certain series of organic compounds of analogous constitution. So one has, e.g.:

Atoms:	Molecules:	Radicals
Li = 7.03	Wood Spirit [methanol]: CH_4O = 32	Methyl = CH_3 = 15
Diff. 16.02	CH_2 = 14	CH_2 = 14
Na = 23.05	Wine Spirit [ethanol] = C_2H_6O = 46	Ethyl = C_2H_5 = 29
Diff. 16.08	CH_2 = 14	CH_2 = 14
K = 39.13	Propyl Spirit [propanol] = C_3H_8O = 60	Propyl = C_3H_7 = 43

It is reasonable to assume that the difference in the atomic weights of these metals, as in the above-mentioned analogous organic compounds or radicals, also is due to a difference in the composition of their so-called "atoms." The latter would therefore not be indivisible quantities, but rather compounds of higher order atoms, that is, composite radicals. The analogy in their behavior with that of the radicals already recognized as compounds would, in this view, find a very natural explanation.

The given similar numerical relations between the atomic weights can be found in many cases. The various authors who deal with the subject have presented them in the most varied of ways. The table below gives such relations for six groups of elements that are well characterized as belonging together.

	4 valent	3 valent	2 valent	1 valent	1 valent	2 valent
	–	–	–	–	Li = 7.03	(Be = 9.3?)
Difference=	–	–	–	–	16.02	(14.7)
	C = 12.0	N = 14.04	O = 16.00	Fl = 19.0	Na = 23.05	Mg = 24.0
Difference=	16.5	16.07	16.07	16.46	16.08	16.0
	Si = 28.5	P = 31.0	S = 32.07	Cl = 35.46	K = 39.13	Ca = 40.0
Difference=	89.1/2 = 44.55	44.0	46.7	44.51	46.3	47.6
	–	As = 75.0	Se = 78.8	Br = 79.97	Rb = 85.4	Sr = 87.6
Difference=	89.1/2 = 44.55	45.51	49.5	46.8	47.6	49.5
	Sn = 117.6	Sb = 120.6	Te = 128.3	J = 126.8	Cs = 133.0	Ba = 137.1
Difference=	89.4 = 2*44.7	87.4 = 2*43.7	–	–	(71 = 2*35.5)	–
	Pb = 207.0	Bi = 208.0	–	–	(Tl = 204?)	–

[1]See especially L. Gmelin's Handbuch, 5th ed. vol. 1 p. 47 ff; M. Pettenkofer, Anz Münch Acad 1850 vol. 30 pp. 261–272; later reprinted: Ann Chem Pharm 1858 vol. 105 p. 187; J. Dumas, Compt. rend. 1857 vol. 45 p. 709; also Ann Chem Pharm Vol. 105, p. 74; etc.

It can be seen that the first (respectively the first and second) difference in each vertical row is approximately 16 everywhere, except between the still very uncertain atomic weights of beryllium and magnesium. The following two differences fluctuate around 46; the last difference is almost twice as large, namely 87–90, if we again disregard thallium's atomic weight, which is not yet sufficiently certain, and which may also (as until recently, according to a provisional determination carried out with the recently discovered substance, cesium) have been determined a little too low.

The differences near 46 also show the groups:

	4 valent	6 valent
	Ti = 48	Mo = 92
Difference =	42	45
	Zr = 90	Vd = 137
Difference =	47.6	47
	Ta = 137.6	W = 184

The penultimate and the last difference in the first table can still be found in the groups below, of which only the last group contains unequal saturation capacities.

	4 valent	4 valent	4 valent	2 valent	
	Mn = 55.1	Ni = 58.7	Co = 58.7	Zn = 65.0	Cu = 63.5
	Fe = 56.0				
Differ. =	49.2	45.6	47.3	46.9	44.1
	48.3				
	Ru = 104.3	Rh = 104.3	Pd = 106.0	Cd = 111.9	Ag = 107.94
Differ. =	92.8 = 2*46.4	92.8 = 2*46.4	93.0 = 2*46.5	88.3 = 2*44.2	88.8 = 2*44.4
	Pt = 197.1	Jr = 197.1	Os = 199.0	Hg = 200.2	Au = 196.7

From the metals of the so-called iron group (Fe, Co, Ni, etc.) aluminum deviates in its atomic weight (Al = 27.3) by approximately the same amount as Li from K, namely by slightly less than 32 = 2 • 16.

There is likely no doubt that a certain law governs the numerical values of the atomic weights. However, it is rather unlikely that it will be as simple as it appears, apart from the relatively small deviations in the values of the differences that occur. In part, however, these deviations can be viewed with justification, as brought about by incorrectly determined values of the atomic weights. However, this should hardly be the case for all of them; and certainly one is not entitled, as has happened all too often, for the sake of a supposed regularity to arbitrarily correct and change the empirically found atomic weights before the experiment has put more exactly determined values in their place.

§ 92

There is a great danger in the natural endeavor, within certain limits, to interpret and correct the results of the observations from a theoretical point of view. This danger is almost without exception connected with the establishment of every hypothesis, indeed with every attempt at theory. Even the two hypotheses on which the latest developments in chemical statics described in this document are based, Avogadro's hypothesis about the molecular size of the gases and Dulong and Petit's hypothesis about the thermal capacity of the atoms, also have raised this danger, or at least the concern of the same. This is certainly true at least of the last-mentioned hypothesis, whose alleged or too far-reaching consequences, in conjunction with uncertain and incorrect observations, threatened more than once to do violence to the conclusions drawn from other, better observations and thus to distort the recognized facts and to obscure proper insight into them.

Now, however, there should be no doubt that the validity of this hypothesis has been restricted to the correct degree, and that all the conclusions drawn from it are in complete agreement with the results of other research.

The latter also applies to Avogadro's hypothesis, against whose justification, apart from that discussed in §§ 68–71, no significant objection was ever made.

Nevertheless, neither of the two hypotheses can yet be regarded as very general, at least not expressly recognized as such. They are neither contested nor recognized by many chemists, but rather simply ignored. As a rule, only a small part of their consequences comes into play in textbooks and lectures.

This appears all the more striking when one looks at the elegant theories of atomistic statics developed from these hypotheses and thus universally confirmed by observation. One would be very wrong, however, if one wanted to draw the conclusion from this apparent neglect and disregard for these hypotheses and theories that they are in themselves worthless and deserve no special consideration.

On the contrary, there is no mistaking the fact that in today's chemistry the tendency to give great weight to theoretical considerations is remarkably low. One tries, often with a certain anxiety, to keep even well-founded theories out of the consideration of empirical facts for as long as one can do so, and as long as one can keep the rich empirical material organized without their help.

On the other hand, this contrasts most vividly with the tenacity and toughness with which hypotheses and theories once introduced into science, even after they have become untenable, have been upheld by the most careful and experienced researchers. Thus, the belief in the existence of the phlogiston has—in a way that is hardly understandable—long prevented worthy chemists from recognizing the correctness and justification of Lavoisier's conclusions. For love of his theory of mass action, Berthollet raised the most lively opposition against the recognition of stoichiometric laws. Berzelius long and stubbornly resisted Davy's theory of the atomic (as opposed to molecular) nature of chlorine because it seemed incompatible with his electrochemical theory[2] and defaced the orderliness of chemistry.[3] For the same or similar reasons, he refused to acknowledge the doctrine of substitution defended by Dumas, Laurent and others, even after almost all other chemists accepted it, and, as did many other authorities, justified the so-called dualistic view of the constitution of chemical compounds[4] as opposed to the so-called unitary or now type view of Laurent, Gerhardt, and others. The history of chemistry shows many examples of this kind.

But it will be difficult for a sensible critic to try to find the reason for the apparent or real contradiction between the sometimes too great, sometimes too little appreciation of the hypotheses and theoretical considerations in an uncertainty of judgment. In a science that has such a large number of such excellent researchers, there could (and can) be no mistake about the worth or worthlessness of hypotheses and theories. The position and validity of the same in chemistry is merely a necessary consequence of the previous and the current state of chemistry itself.

The value of hypotheses[5] is essentially of two kinds. It is based foremost on a purely practical benefit that they bring; for the progress of science is based on the hypotheses being put forward, justified, or refuted by experiment. From the beginning, this benefit has been shown to be extremely rich in chemistry; so much so that wherever the value of hypotheses is mentioned in chemical writings, only this advantage tends to be thought of. In fact, it is large enough to justify the proposition of a hypothesis on its own.

But the value and usefulness of hypotheses does not lie solely in the suggestion for new research to undertake to justify, test, or refute them. The simple knowledge of things as they are or they seem is not enough for the researching spirit of mankind, which also strives to find out the causal connection between things, all that forms and happens. This goal can never be achieved; our ideas will never coincide with the essence of things; but they can approach it more and more as the shadow is able to reproduce the shape of the object more and more sharply. However, in order to adapt our ideas more and more to the nature of things, we must first hypothesize, logically develop the consequences thereof as theory with or without the help of calculation, and compare the results of this development with the phenomena accessible to our observation. The greater the agreement between theory and observation, the greater the likelihood that our hypothesis of the essence of things will give us an approximately correct idea, at least parallel to reality, if not coincident with it. This possibility is the highest goal that the natural sciences can achieve; the increasing probability can come closer and closer to certainty without ever going into absolute certainty itself.

In this meaning of hypotheses and theories lies precisely the danger which they can pose; this is due to the difficulty in judging the degree of probability and in strictly distinguishing probability and certainty everywhere. Especially in such a

[2]Gilb. Ann. 1815 vol. 20 (i.e., F. 50) pp. 367, 410, 445 et loc. al.
[3]ib. 1812 vol. 12 (i.e., 42) s. 288.
[4]s. § 63.
[5]Cf. Especially the excellent criticism in Berthollet's statics (Statique chimique I p. 4–10).

young science as exact, measuring chemistry still is, it is often difficult to distinguish those theories that are strictly the abstract expression of observations, that is to say which have the highest degree of certainty that our senses are capable of perceiving, from the truly hypothetical assumptions that were introduced to understand observations.

However, if this distinction is not made, facts and hypotheses grow together so much that it becomes difficult to give up hypotheses that have become very unlikely. They are therefore easily kept longer than they should; and if such a hypothesis that has been incorporated into the doctrine is finally abandoned as completely unsustainable, then science is easily shaken to its foundations. Chemistry has already experienced a not negligible number of such more or less violent tremors, from the fall of the phlogistic teaching to the implementation of classification according to series and types.

But the more often the upheavals were repeated, the greater the treasure trove of verified results that has remained, and the more this core of science has become independent of prevailing theories and subjective opinions.

The current state of chemistry is particularly suitable for showing how great this independence has already become. From various points of view, it is possible to survey the rich material and present it in an orderly manner. The order is established by the facts themselves, it no longer needs to be brought in by theory.

The more science advances, the more it will become possible to avoid the harmful influence of hypotheses and theories. In chemistry too, more and more, as is now the case in physics, one will increasingly be able to see clearly the connection between each hypothesis and the theoretical conclusions derived from it and the results of observation. One will be able to make the necessary improvements, restrictions, or expansions to an underlying hypothesis immediately, as soon as the consequences drawn from it are no longer consistent with the results of observation.

In chemistry, too, it will only be possible to maintain any hypothesis as long as it seems suitable to explain the facts, but to abandon it as soon as this is no longer the case. One will then be able to carry out the most fundamental change in the premises just as smoothly and safely as in physics, e.g., the transition from emanation to undulation theory was accomplished.

The more the systematic order of chemistry is fixed, the more it will be allowed to put speculation on an equal footing with empiricism. In order to expand the theoretical system, more hypotheses, apart from those that have already been incorporated, will then likely have to be introduced into the science, and can be introduced without causing damage. The method of chemistry will again approach that of physics considerably, although each of the two disciplines will retain its individual character.

§ 93

The introduction of a few new hypotheses is likely to be necessary now or in the not too distant future. In particular, it seems that many of the fields of molecular physics that are closely related to chemistry and therefore have passed into the almost exclusive possession of chemistry cannot be successfully worked on without the theoretical considerations and hypotheses, which are currently particularly represented and developed by Clausius, which explain the different states and manifestations of matter by assuming different forms of movement of the physical molecules.

It is only from these views, derived from the basic principles of mechanics, and especially mechanical heat theory, that it seems possible to penetrate with research into the nature of the influence that the chemical nature of substances, the atomistic constitution of molecules, has on the changes in the state of matter, melting and solidification, evaporation and compression, on the tension of vapors, on the phenomena of diffusion, absorption, solution, crystallization, imbibition, endosmosis and all similar processes. Also electrolysis, and thus the whole field of electrochemistry, seems to be accessible to a successful theoretical investigation only from this approach.[6] Perhaps, even when considering all purely chemical processes—chemical decomposition and bonding—one will soon no longer do without those views; after all, from them originated independently very similar conclusions as from observation of purely chemical processes.[7]

In fact, it is not unlikely that looking at the movements of the smallest particles of matter will be allowed to have a much wider influence on chemical theories. One has already been able to explain the so-called gravitation of the celestial bodies without the difficult assumption of action at a distance, by means of only one assumption, a medium that fills space, the aether, whose particles are endowed with a very lively movement.[8] One may perhaps come to dispense also with the presupposition of other now generally accepted attractions, affinity, cohesion, etc., and to derive the phenomena ascribed to these attractive forces simply as necessary consequences from the movements of the smallest particles of matter caused by heat, light, etc.

But even if we should not get there, the observation of these movements will be a necessary and useful aid for any deeper theory of chemical processes.

§ 94

Through the introduction of these considerations, as well as the introduction of the atomistic hypothesis and the teachings of Avogadro and Dulong and Petit based on it, then the current so-called mechanical views on the nature of light and heat and probably later also of electricity and magnetism will cause Berthollet's chemical statics to receive a manifold change in design. But although almost all of the views and assumptions from which Berthollet started have undergone very profound changes, the goal he sought has not been affected and has remained unchanged: the application of the general laws of statics and mechanics to chemical phenomena.

This immutability of the goal is the best proof of justification for Berthollet's pursuit. If one day it will be possible to start again the long interrupted construction of his chemical statics, the work will be relatively easy. The unchanged frame will be filled with the new material that has been delivered by the flourishing

[6]Cf. "On Electrical Conduction", by R. Clausius. Pogg. Ann. 1857 vol. 101 p. 338 ff.

[7]Cf. "About the theory of ether formation; from Al. Williamson. Ann Chem Pharm 1851 Vol. 77 pp. 37 ff.

[8]Le Sage, Deux Traités de Physique mécanique, publiés par Pierre Prevost, Genève et Paris 1818; P. Prevost, De l'origine des forces magnétiques, 1788, chap. 2: "principes physiques."

development of science since then, and is still increasing every day. The expansion of the work will take a lot of time and effort; but it will be well worth the effort. May it build a worthy monument to the sublime spirit of its founder.

9.3 Conclusions

When I discussed this project with a native German-speaking colleague, he noted that a near-verbatim translation of Meyer's prose into English sounded stilted. This was disturbing to him because he found Meyer's German as quite elegant. This dilemma is common—is it best to capture the intended meaning (leading to a possible danger of interpretation based on our modern knowledge), or adhere very closely to the original phrase order and wording (less elegant, but less danger of putting one's own words in)?

It is notable that Meyer engages rather explicitly with questions in philosophy of science on the roles that theory and experiment ought to have and actually do have in chemistry. It is evident that he looks to physics as an example of a science which has an appropriate relationship between them, one in which theory and hypothesis are useful only when actual data are used to fit the theoretical model.

As Rocke noted [3]:

> In another friendly dispute with Kekulé five years after Karlsruhe, this time in public, Meyer expressed what he thought was a crucial difference [4]:
>
> Perhaps my honored friend will now decide to concede what he has so often contested in our private conversations and correspondence: that the fundamental hypotheses of chemistry should be derived not just from purely chemical data in a narrow sense, but rather, as in all investigations of the most intimate nature of matter, all scientific aids must be applied … The recognition of this fundamental tenet is a necessary condition for theoretical chemistry to thrive.

Acknowledgements Thanks to Gregory Girolami and Martin Gruebele for their comments on the translation.

References

1. Meyer L (1864) Die modernen Theorien der Chemie. Breslau, Germany
2. Benfey OT (1974) Lothar Meyer. Dictionary of scientific biography, vol 9. Scribners, New York, pp 347–353
3. Rocke AJ (2019) Lothar Meyer's pathway to periodicity. Ambix 66:265–302
4. Meyer L (1865) Kritik einer Abhandlung von A. Kekulé. Zeitschrift für Chemie 8:250–254

Part II
Discoveries of Elements: Successes and Challenges

Discovery of Three Elements Predicted by Mendeleev's Table: Gallium, Scandium, and Germanium

10

Mary Virginia Orna and Marco Fontani

Abstract

The 1869 publication of Dmitri Mendeleev's systematic table contained lacunae left by the author to signal the presumed existence of not-yet-discovered simple bodies. Three of the missing elements were discovered within a span of time from 1875 to 1886: gallium, scandium, and germanium. Aside from the great psychological impact, they served to decisively change the attitude of the scientific world with respect to the validity of the periodic system of the elements. Although Paul-Émile Lecoq de Boisbaudran, Lars Fredrik Nilson, and Clemens Alexander Winkler are the acknowledged discoverers of these three elements, respectively, questions arise about the nature of discovery itself. Can their discovery be attributed to Mendeleev, who *speculated* on and predicted their existence? Must discovery be attributed to those who only *detected* the elements, or to those who *isolated* the oxides of the elements, or to those who *isolated* the free elements themselves? Arguments can be made for recognizing all of these cases.

10.1 Introduction

Recently, the periodic system (1869–2019), has been hailed as indisputably "one of the most significant achievements in science, capturing the essence not only of chemistry, but also of physics and biology" [1]. We recognize this system today as

M. V. Orna (✉)
ChemSource, Inc., 39 Willow Drive, New Rochelle, NY 10805, USA

M. Fontani
Department of Chemistry "U. Schiff", University of Florence, via della Lastruccia, Sesto Fiorentino, Italy

C. J. Giunta et al. (eds.), *150 Years of the Periodic Table*, Perspectives on the History of Chemistry, https://doi.org/10.1007/978-3-030-67910-1_10

the single-page entity that contains on it all of the known primary substances in the universe—118 elements from which all other substances are formed.

Though we do not wish to overlook the giant steps along the way to the concept of the periodic system, suffice it to say that beginning in the early nineteenth century many thinkers were beginning to understand that there was a hidden order in the properties of the chemical elements. The real problem was trying to discern a relationship without a sufficient number of data points and perhaps only a fuzzy notion of an organizing principle. It is no accident that the great breakthroughs to understanding the periodic system occurred in and around 1869 when there were 63 known elements to consider—the critical mass, if you will, toward being able to see a repetitive, and more importantly a periodic, pattern.

Both Dmitri Mendeleev (1834–1907) and Julius Lothar Meyer (1830–1895) published their own ideas on the organizing principle of the elements virtually simultaneously. Meyer's first table, published in 1864 [2] contained only 28 elements arranged according to their combining power, or valence, plus separate tables of 22 transition elements arranged into a periodic grid. Four years later, he updated his table to include a total of 52 elements and hinted at the possibility of yet-undiscovered elements, but this hand-written table, containing many of the advanced ideas also espoused by Mendeleev, was only published posthumously in 1895 [3]. Meyer's great breakthrough was his plot of the molar volumes of the then-known elements against atomic weight (Fig. 10.1) published in Liebig's *Annalen* [4].[1]

Although this plot clearly displays a periodic trend in the elements, it, sadly for Meyer, appeared in 1870. One year earlier, in 1869, Dmitri Mendeleev had published his table in the *Journal of the Russian Physical-Chemical Society* (Russkoe fizikokhimicheskoe obshchestvo. Zhurnal) [5] and in the *Zeitschrift für Chemie*. Within a matter of months, both thinkers had discerned an apparent organizing principle, atomic weight, that would only be toppled almost 50 years later by Henry Moseley's (1887–1915) confirmation of atomic number [6, 7] first hypothesized by Antonius van den Broek (1870–1926) [8]. A protracted priority dispute ensued, fueled by both authors' conviction that they had uncovered one of nature's fundamental secrets. The Royal Society of London decided to hedge its bets: it honored both scientists (Figs. 10.2 and 10.3) with the prestigious Davy Medal in 1882 [9].

In 1869, when he laid out his systematic grid of the hitherto-discovered elements, Dmitri Mendeleev clearly understood that he had created a potent tool for bringing order out of the apparent chaos of a random set of elemental substances. But he never predicted the enormous and all-embracing effect that his idea would have on scientific thinking for the next 150 years: he based his own legacy on his role as a Russian public servant [12]. He formed his table [13] by arranging a set of cards containing the names and properties of the 63 elements then known in order of increasing atomic weight. Even though some of these weights were erroneous, he began to see a pattern nevertheless [14]. His genius became evident when not only did he leave open spaces for supposed missing elements (actually 16 in all), but also

[1]For more on Meyer, see Gisela Boeck's Chap. 8 in this volume.

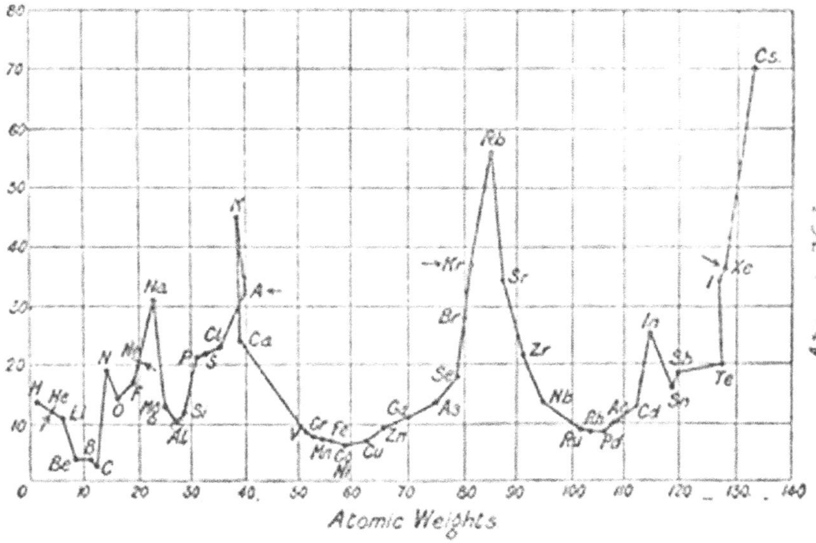

Fig. 10.1 Julius Lothar Meyer's 1870 plot of atomic volume as a function of atomic number (redrawn)

Fig. 10.2 International Year of the Periodic Table stamp issued by Magyar Posta on 3 June 2019[2] (Image courtesy of Daniel Rabinovich [10])

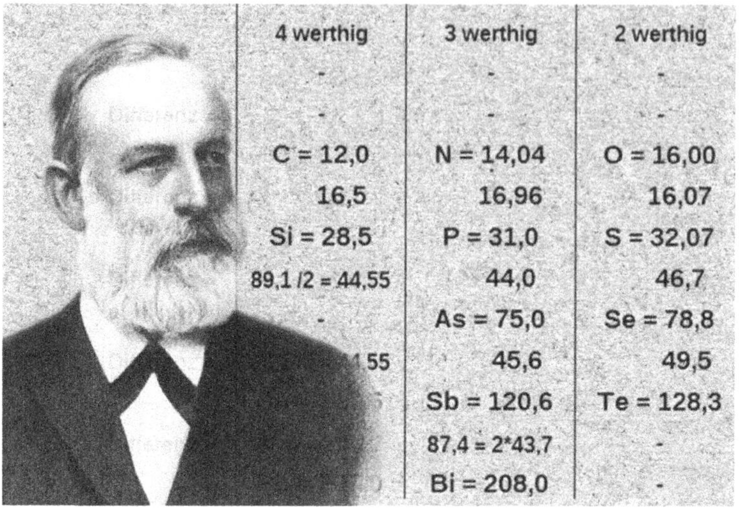

Fig. 10.3 Julius Lothar Meyer and his Table. Reproduced by permission from the Royal Society of Chemistry [11]

predicted atomic weights and chemical and physical properties for some of them. By recognizing an implied design hinted at by nature, he bestowed a prophetic attribute on his table that flowered in the discovery of three of these missing elements (see Fig. 10.4) within the following 20 years. Furthermore, by ordering of some of the elements according to properties rather than atomic weights, he left some leeway for attempts to determine more accurate atomic weights by future experimentation.

Though some historians of chemistry tend to give almost equal credit to both Mendeleev and Meyer, Ida Freund (1863–1914), an astute critic and historian of chemistry at the beginning of the last century, had this definitive statement to offer [15]:

> Comparison of [Mendeleev's] and Lothar Meyer's treatment of the inductive part of the subject shows that whilst the latter pays greater attention to the consideration of the physical properties, the former devotes himself more to a comparative study of the chemical relations...But it is especially in the deductive application of the *system*,[3] that we find the Russian scientist much in advance of the German; the scope of the phenomena encompassed, the definiteness and lucidity of the reasons adduced for the conclusions arrived at, the number and importance of the predictions made together with the marvelous way in which these have been verified, have combined to make this part of [Mendeleev's] work

[2]Behind Mendeleev's right shoulder is shown the final version of his periodic table dated 17 February 1869. The inscription in Hungarian directly under his name reads "the creator of the periodic table."

[3]Present authors' italics. Freund's use of the word "system" is entirely consonant with Mendeleev's own idea. As Van Tiggelen, et al. point out, there are thousands of periodic tables, but only one periodic system from which Mendeleev inferred his law [16].

ОПЫТЪ СИСТЕМЫ ЭЛЕМЕНТОВЪ,

ОСНОВАННОЙ НА ИХЪ АТОМНОМЪ ВѢСѢ И ХИМИЧЕСКОМЪ СХОДСТВѢ.

			Ti = 50	Zr = 90	? = 180.
			V = 51	Nb = 94	Ta = 182.
			Cr = 52	Mo = 96	W = 186.
			Mn = 55	Rh = 104,4	Pt = 197,1.
			Fe = 56	Rn = 104,4	Ir = 198.
			Ni = Co = 59	Pl = 106,6	Os = 199.
H = 1			Cu = 63,4	Ag = 108	Hg = 200.
	Be = 9,4	Mg = 24	Zn = 65,2	Cd = 112	
	B = 11	Al = 27,4	? = 68	Ur = 116	Au = 197?
	C = 12	Si = 28	? = 70	Sn = 118	
	N = 14	P = 31	As = 75	Sb = 122	Bi = 210?
	O = 16	S = 32	Se = 79,4	Te = 128?	
	F = 19	Cl = 35,5	Br = 80	I = 127	
Li = 7	Na = 23	K = 39	Rb = 85,4	Cs = 133	Tl = 204.
		Ca = 40	Sr = 87,6	Ba = 137	Pb = 207.
		? = 45	Ce = 92		
		?Er = 56	La = 94		
		?Yt = 60	Di = 95		
		?In = 75,6	Th = 118?		

Д. Менделѣевъ

Fig. 10.4 Mendeleev's 1869 Periodic Table. The circled entries eventually became gallium, scandium, and germanium in order of their discoveries [13]

> one of the greatest scientific achievements of the century, one of the most striking confirmations of the modern method.

Mendeleev based his table on the idea that all the elements are unique but that they are somehow related. He also placed his faith in gaining knowledge by experiment, not by speculation, by stating that his periodic law resulted solely through an analysis of experimental data [17]. As such, Mendeleev's accomplishment ushered in the modern chemical world.

10.2 Discovery of the "Missing Three:" Eka-Aluminum, Eka-Boron, and Eka-Silicon

Dmitri Mendeleev, in organizing his periodic table, realized that there were missing elements, among them three obviously missing elements provisionally designated eka-boron, eka-aluminum, and eka-silicon.[4] These theoretical "discoveries" eventually led to the detection and isolation, respectively, of scandium, gallium, and germanium—achievements with great scientific and psychological impact [18]. Their discoveries, over the course of a little more than ten years, electrified the scientific community and brought pleasure to Mendeleev [19] since he never expected any of them to be found within his lifetime. In fact, he confessed while delivering his famous Faraday Lecture in 1889 [17]:

> When in 1871, I described to the Russian Chemical Society the properties, clearly defined by the periodic law, which such elements ought to possess, I never hoped that I should live to mention their discovery to the Chemical Society of Great Britain as a confirmation of the exactitude and the generality of the periodic law.

They decisively enshrined the periodic system as an undisputed triumph of what has been dubbed an "interpretative theory" which became "explanatory" after the periodic table was based on atomic number [20]. The story of their appearance and incorporation into the periodic system is told in detail by Mary Elvira Weeks in her *Journal of Chemical Education* "Discovery of the Elements" series [21] eventually collected into one volume [22].[5] The "mini-biographies" that follow summarize and update this material.

10.2.1 Mini-Biography of Gallium (Eka-Aluminum)

Paul-Émile François Lecoq (1838–1912) discovered gallium (eka-aluminum) in 1875 (Fig. 10.5). He was born on 18 April 1838 in Cognac, France, into an ancient Protestant noble family whose watchwords were "justice, kindness, and a sense of personal responsibility." His parents were Paul-Aimé Lecoq (1799–1870) and Anne-Louise Alexandrine Joubart (1814–1891). On 4 May 1859, 21 years after his birth, on an addendum to his original birth certificate, his name was "rectified in the sense that his true name was Lecoq de Boisbaudran and not simply Lecoq."

About 150 years before Lecoq de Boisbaudran was born, Louis XIV (1638–1715) revoked the Edict of Nantes which, in effect for about 85 years after its enactment by Henry IV of Navarre (1553–1610), gave rare and important privileges to the Protestant minority in France. The revocation plunged the Lecoq family into dire straits. Poor for several generations, Paul-Émile's father and his father's brother, Scaevola, started a wine business in the town of Cognac into which young

[4]Mendeleev explained the use of the prefix eka thus: 'In order not to introduce new names for the expected elements, I shall call them by the name of the nearest lowest analogue from among odd- or even-numbered elements of the same group, adding Sanskrit numerals to the name of the element.

[5]See Vera Mainz's Chap. 14 in this volume, "Mary Elvira Weeks and *Discovery of the Elements*."

Fig. 10.5 **Left** Gallium, element number 31, showing its atomic number, symbol, name and atomic mass; **Right** Gallium's discoverer: Paul-Émile Lecoq de Boisbaudran

Paul-Émile was initiated at the age of 15 since the family's modest circumstances precluded further education. However, his mother was a very cultured woman who taught him the classics, history, and foreign languages. His uncles, Scaevola and Émile Albert, instilled in him a love for science, and especially for chemistry; he began to devour every chemistry text he could get his hands on. The uncles fostered their nephew's scientific vocation further by persuading Paul-Émile's father to grant him time to pursue his scientific studies; they even helped to outfit a small laboratory for him to carry out experiments that he read about in the syllabi of the École Polytechnique in Paris. By chance one day, he passed by a display case of rare earth element compounds in the École's chemistry department and immediately fell in love with their colors—an event that set the direction for all of his future scientific work: separation of rare earth ores and development of the spectroscopic expertise necessary to interpret his results [23]. Although poor, he lived very simply and spent most of what he had on chemicals and laboratory equipment [24].

Lecoq's acquisition of an initial 52 kg of ore would have cost him a pretty penny had he not received it as a gift from one M. Malgor, an engineer at the Pierrefitte mine—but the gift more than gave back in his being able to discover a new element, gallium. This discovery, one of the earliest by a spectroscopic method, occurred when Lecoq was trying to complete, to his own satisfaction, a spectroscopic principle of his own that he had derived in 1863 [25, 26]: that families of elements display a similar spectroscopic pattern. This principle derived from the similarities in the characteristic spectra of the alkali and alkaline earth metals demonstrated by the work of Gustav Kirchhoff (1824–1887) and Robert Bunsen (1811–1899) [27]. In this case, he was looking for the pattern of an element that he suspected existed between aluminum and indium. He remarked [28]:

> Amongst the conclusions that might be drawn from my attempts at chemical classification was the probability of the existence of unknown elements coming to fill up the gaps left vacant in the natural series. It is clear that the position thus occupied in a chemical family by a hypothetical body indicates approximately the properties of that body.

A bit later, he added:

> The present perfection of chemical analysis, and the care with which nearly all the known minerals have been examined, leave little hope of finding new elements; we can only expect to meet with them as minute traces disseminated in considerable masses of foreign substances.

Lecoq was to confute the first part of this statement years later when he discovered two more elements, samarium in 1880 and dysprosium in 1886, and succeeded in isolating gadolinium for the first time in that same year.

Whether he realized it or not, Lecoq was well positioned to discover at least one of the missing elements. He possessed his own private laboratory which, considering descriptions of his habits, he seemed to occupy at almost every waking moment. His unmitigated passion for chemistry, his assiduous and never-ending laboratory work, his great expertise in spectroscopy,[6] and his intense interest in whatever the various ores he collected contained, converged on the fateful evening of 27 August 1875. Working with a sample of zinc blende, he produced spectra using both a Bunsen burner and electric sparks; the latter yielded two lines in the violet never seen before, one at 417 and another at 404 (nm). Here are his own words [30][7]:

> I found indications of the probable existence of a new simple body, in the products of the chemical examination of a blende[8] from the Pierrefitte mine, Argelès valley (Pyrénées). Here are the data I have been able to collect so far...

During the following year, Lecoq went on to describe his treatment of this new body in 17 points, including attempts to separate it out from the main ore body, which consisted chiefly of zinc [31]:

> The oxide, or perhaps a sub-salt, is precipitated out by metallic zinc in a solution containing chlorides and sulfates. It does not appear to be the metal itself which is reduced by the zinc. The extremely small quantity of the substance at my disposal did not permit me to isolate the new body from the excess of the zinc accompaniment. The few drops of zinc chloride in which I concentrated the new substance gave under the impact of the electric spark a spectrum composed chiefly of a violet ray, narrow, readily visible, and situated at about 417 (nm) on the wavelength scale. I also perceived a very faint ray at 404 (nm).

He estimated that the quantity of gallium he had in hand at his first observation did not exceed 0.01 mg [22], and he attributed his ability to detect it to instrumental improvements of his own invention.

As to the rationale for the name of the element, Lecoq gives no clue in his August 1875 statement [30]:

[6]Lecoq de Boisbaudran had some general theories about the relationship between spectra (both luminescence and spark) and the atomic weights of the elements. To test them, he made a detailed study of the spectra of 35 elements and published them in a single volume in 1874 [29].
[7]All translations are the work of the authors.
[8]"Blende" can mean any of several minerals, mainly metallic sulfides. In the context of this experiment, sphalerite, or zinc sulfide, often called zinc blende, is meant.

> The experiments which I have carried out since August 29 make me realize that the observed body must be considered a new element, to which I propose to give the name of Gallium.

However, in his 1877 paper in the *Chemical News*, he adds, "…I called [it] 'Gallium' in honour of France (Gallia)" [22]. Some critics have suggested that Lecoq, as a practicing amateur chemist, played a joke on the chemical world by slyly naming the element after himself [24, p. 169]: the Latin word *gallus* translates into French as *le coq*!

A few months later, in November 1875, in an arrangement with Charles Adolphe Wurtz (1817–1884), Professor of Chemistry at the École de Médecine, Lecoq worked up about 300 kg of zinc blende ore given to him by two technical zinc-mining societies in order to prove that he had actually discovered a new element. To isolate the gallium, found in a very small amount in the ore, he had to remove interfering metals from the mixture: copper, arsenic, lead, cadmium, indium, thallium, mercury, selenium, bismuth, silver, tin, antimony, and gold. Only then was he able to precipitate the basic salts of zinc, and the hydroxides of aluminum, gallium, iron, cobalt, and chromium [22, p. 646].

He found that, although gallium sulfide did not precipitate easily on its own, it would be carried down by co-precipitation with zinc sulfide. A key point in the isolation of gallium was tracking its precipitation spectrally: as he added zinc and hydrogen sulfide to the solution, he continued to take spark spectra until the lines attributed to gallium had completely disappeared, signifying complete precipitation. He was then able to dissolve the precipitate in strong base and produce a gram of metallic gallium by electrolysis [32]. From then on, production of gram quantities of gallium became almost routine (at least for Lecoq) and he and his colleagues, by crushing several hundred kilograms of ore and treating it with 1500 L of aqua regia, succeeded in producing as much as 75 g of the element by 1878 [33]. Lecoq remarked that gallium constituted a very small percentage of its native ores, usually sphalerite and bauxite; however, it ranks number 35 in average crustal abundance, or 18.7 ppm, which makes it more abundant than two other common elements, lead and boron.

Lecoq's discovery of gallium fulfilled all the prerequisites for being proclaimed undisputed discoverer: he observed it spectroscopically, produced gallium compounds, and isolated pure gallium metal in fairly large quantities. His discovery, as stated earlier, was guided by his own spectroscopic ideas and goals. However, later in the same year, 1875, Dmitri Mendeleev, alerted by Lecoq's discovery, commented at the 6 November meeting of the Russian Chemical Society that he believed that his eka-aluminum and gallium were identical. This view was later borne out by comparison of the measured properties of the element vs. those predicted by Mendeleev [34], thus drawing substantial attention to his 1871 periodic system. This was the only instance in which Mendeleev combed through the foreign literature himself in search of possible confirmations of his predictions, and ended up making the connection himself. Mendeleev also sent a letter to Lecoq to call his attention to the reported density of gallium, 4.7 [35] that did not match his own prediction of 5.9 [36]. He informed Lecoq that his material was impure and that he should repeat his determination with a larger sample [37]. Lecoq responded at first in a rather roundabout way to this criticism [35]:

> At the time when I was preparing the mailing which I have the honor to make to the Academy, I received the reports of November 22, with a very interesting note from M. Mendeleeff relating to the classification of simple bodies and of Ga in particular…I did not want to publish my hypotheses without having subjected them to the control of experiment, and without having made some efforts to make them produce positive results which confirm them, at the same time as they allow them to be improved. I will come back to the communication from M. Mendeleeff; For the present, I will confine myself to saying that the analogy of the spectra of Al, Ga and In struck me from my first observation; I immediately calculated the equivalent of gallium by applying my first spectral law. I have communicated these remarks to M. Wurtz and other scholars; but, … considering the impurity of the matter in my possession, I judged it more prudent to simply submit to the Academy the facts which I had discovered, postponing theoretical considerations until later. I should also say that I was unaware of the description made by M. Mendeleeff of the supposed properties of his hypothetical metal: I would even add that this ignorance was probably favorable to me; because, despite the undeniable merit of the theoretical ideas of M. Mendeleeff (ideas with which I am very willing to be in accord), and assuming the predictions of this scientist verified as a whole, I would have been ready to seek gallium in the precipitates formed by ammonia, and not, as I did, in ammonia solutions. Indeed, the properties of the hypothetical metal had to display the average between those of aluminum and those of indium… I therefore consider it very probable that, without the particular method followed in the present research, neither M. Mendeleeff's theories nor mine would have led to the discovery of gallium for a long time.

After further work, Lecoq did indeed find that gallium's density was 5.935, thus corroborating Mendeleev's claim. In the note of correction, he first reviewed his initial finding of a density of 4.7 and then in two separate determinations, one at 23 °C and the other at 24.45 °C, he reported densities (or rather, specific gravities because he specifically states "relative to water") of 5.935 and 5.956. Then, without any apology, he closed his note by remarking [38]:

> There is no need to insist on the extreme importance, I believe, that is attached to the confirmation of M. Mendeleef's theoretical ideas about the density of the new element.

Despite Lecoq's dismissive words, history has given its judgment [39]:

> After Lecoq de Boisbaudran had discovered gallium in 1875, Mendeleev rightly concluded that the validity of the periodic system of elements could no longer be questioned. The confirmation of this prediction may certainly be called the culminating point in the history of the periodic system.

Lecoq de Boisbaudran went on to great fame by discovering two additional elements, samarium in 1869 and dysprosium in 1886. Samarium took its name from the mineral, samarskite, in which it was found; dysprosium became famous for its unusual etymology, "difficult to obtain." Lecoq was also the first person to isolate the element gadolinium. His scientific work garnered great honors: the Cross of the Legion of Honor, the Davy Medal, and the Prix Lacaze. After 1895, his scientific work began to slow down due to increased family responsibilities and, later on, painful illness. At the age of 74, on 28 May 1912, he passed away [40].

Gallium has 31 known isotopes, only two of which are stable. Gallium metal displays some unusual properties, one of which is its enormous liquid range, about 2370 °C (from 30 to 2400 °C), which makes it an ideal filler for thermometers needed for such a range. It tends to form low-melting alloys with many metals, precluding its shipment in metal containers. It has found limited applications in semiconductors and biomedical alloys. Since 2009, liquid gallium has found a use as a molten streaming anode target in X-ray tubes that support higher electron beam power than does solid-anode technology [41, 42]. Table 10.1 illustrates how the properties of this element predicted by Mendeleev compared to the actual element.

In its compounds, gallium occurs mainly in the +3 oxidation state, and rarely in the +1 oxidation state. Besides the normal salts formed with halogens, chalcogens, etc., gallium easily forms organogallium compounds. Alkylgalliums are liquids at room temperature and quite flammable. Pure gallium metal costs about US$0.40 per gram.

10.2.2 Mini-Biography of Scandium (Eka-Boron)

Lars Fredrik Nilson (1840–1899), during his tenure as Professor of Analytical Chemistry at Uppsala University, Sweden, discovered scandium (eka-boron) in 1879. Nilson was born at Skönberga, Östergöthland County, Sweden, about 140 miles southwest of Stockholm, but as a child his family moved to Gothland Island, about 130 miles due south of Stockholm, where the elder Nilson owned a large farm, Rosendal. It was this rural setting dotted with medieval ruins that formed young Nilson's consciousness—he never lost sight of his agrarian roots and, toward the latter part of his life, was destined to direct his scientific expertise to enhancing his country's agricultural resources (Fig. 10.6).

At the end of his high school education at Visby (Gothland Island) in 1859, Nilson, aged 19, began his higher education at Uppsala, gravitating initially to biology and geology. During pursuit of the latter, he came under the influence of the Professor of Chemistry, Lars Fredrik Svanberg (1805–1878). Svanberg, said to be passionate about chemistry, had an equal devotion to his students whom he encouraged to use unreservedly his library of chemical literature, the largest ever privately amassed in Sweden. As the academic successor to the great Jöns Jacob

Table 10.1 Selected properties of gallium (eka-aluminum, Ea)

Property	Predicted	Measured
Atomic mass	68	69.723
Melting point (°C)	Low	29.76
Specific gravity	5.5	5.9
Formula of oxide	Ea_2O_3	Ga_2O_3
Solubility of oxide	Soluble in acid and alkali	Soluble in acid and alkali
Formula of chloride	Ea_2Cl_6	Ga_2Cl_6
Volatility of chloride	Volatile	Volatile

Fig. 10.6 **Left** Scandium, element number 21, showing its atomic number, symbol, name, and atomic mass; **Right** Scandium's discoverer: Lars Fredrik Nilson

Berzelius (1779–1848), Svanberg gave up his own experimental work on Berzelius' death in order to continue the chemical education, mostly based in the laboratory, of the next generation of students.

In 1865, Nilson was ready to present himself for his final examinations in chemistry and a host of other required subjects. At this point, fate intervened by way of notification of his father's dire illness and Nilson, dropping everything, went home to the family farm to take over its management. This was not an easy decision since the Swedish system of education required students to pass all of their examinations within a small window of time or else be constrained to start all over again. Fortunately for Nilson, his self-sacrifice to family duty worked out well. Not only did his father eventually recover his health, but so did the son, who had been suffering from pulmonary hemorrhaging all the time he had been a student at Uppsala. Both were healed by fresh country air and hard work: Nilson took up his studies once more, passed his examinations and, within a year had been appointed laboratory head and demonstrator in chemistry.

Nilson's first paper in inorganic chemistry was published in 1871 on the topic of the sulfides of arsenic, taken up to complete work begun by Berzelius a half-century previously. He also did some work on selenious acid, again as a complement to selenium's discoverer. Continuing to follow in the footsteps of the great Scandinavian chemists, like Berzelius, who had discovered many elements in minerals mined in their home territories, Nilson, in collaboration with Sven Otto Pettersson (1848–1941), began to study the elements contained in the mineral euxenite since he had received a large quantity of this mineral as a gift from Peter Waage (1833–1900), Professor of Chemistry at Kristiania (later Oslo) [43].[9] Euxenite's mineralogical properties had been described for the first time by Theodor Scheerer (1813–1875) who found that it contained numerous rare constituents, including yttric and ceric earths [44].

[9]The details of Nilson's life and work are drawn mainly from this source.

Using a fractionating method pioneered by Nils Johan Berlin (1812–1891) [45] and perfected by Johan Fredrik Bahr (1805–1875) and Robert Bunsen [46], Nilson and Pettersson pursued two parallel tracks: preparation of pure beryllium metal and separation of the purest possible rare-earth oxides.[10] Their work on beryllium was carried out in order to determine its valence and its real atomic weight, in which they succeeded in 1880. However, new discoveries in rare earth chemistry were proceeding rapidly as chemists began to discover that their "pure" samples contained hidden additional elements extractable with greater and greater difficulty. These discoveries were advanced by the work of the Swiss chemist, Jean Charles Galissard de Marignac (1817–1894) whose work on erbium earths culminated in the discovery of a new element which he called ytterbium [47]. Marignac noted:

> Following my research on gadolinite earths…I obtained a few grams of an earth showing all the characteristics of erbia according to the classic work of MM Bahr and Bunsen…[but] the process by which I had separated it from other gadolinite earths was not absolutely identical with that used by these chemists…More recently, I took another look at my results…and observed a fact that greatly surprised me….that the earth which I had extracted from gadolinite, and which I had considered as erbia, was still only a mixture of two distinct oxides.

He came to this conclusion because in carrying on further fractionations, his erbia disappeared entirely, but at the same time, the equivalent weight of the remaining oxide increased. He went on to remark [47]:

> One, with a pure pink color and having a very characteristic absorption spectrum, must retain the name of erbia, since these are the characteristics which have been considered as the most distinctive of this base. The other is a new base, belonging to the same group, and for which I propose the name of ytterbia, which evokes its presence in the mineral from Ytterby…

However, since Marignac could not isolate and purify this substance for lack of sufficient material, he sent out a clarion call to other chemists who might have in their possession larger quantities of the rare earths to continue with this research [47]:

> The difficulty of obtaining gadolinite in sufficient quantity and the time-consuming work necessary to remove the erbia, besides which it is often present only in very small amounts, hardly allows me to hope to resolve these issues myself and more fully establish the properties of ytterbia. This is why I wanted to draw the attention of these facts to chemists, and in particular to those who may have a significant amount of erbia at their disposal, in the hope that they may direct their research on this subject, and make sure that this earth did not contain the new base, whose existence seems incontestable to me.

[10]Rare earth elements are often defined as elements 57 through 71, the lanthanides, but strictly speaking, yttrium and scandium should also be included due to their similar chemical behavior and placement in group 3 of the periodic table. See Chap. 11 of this volume for more on the rare earth elements. The definition of an "earth" evolved over the course of the eighteenth century from a generalized mineral lacking taste and solubility in water to, by the end of the century, a difficultly reducible metal oxide. These oxides often take the suffix -ia; for example, erbium oxide is called erbia, thorium oxide is thoria, etc.

Nilson was in a perfect position to rise to the occasion. He had in his possession a much greater amount of rare earth ores, complex oxide minerals, namely euxenite, than any other chemist of his time. He had the analytical expertise honed after long years in the laboratory employing state-of-the-art fractionation techniques. He was very organized and worked in a very methodical manner. And last, but not least, the euxenite was extraordinarily rich in the higher atomic weight rare earths, those of the gadolinite group. Nilson began to repeat Marignac's experiment, but he was able to fractionate the material much more easily because of the greater quantity, 63 grams of erbia from euxenite, he had to hand. He performed 13 series of nitrogen decompositions and while observing the separation of insoluble basic nitrates of gradually increasing atomic weight of metal, he was astonished to find another, less basic salt of lower atomic weight, separate out among them. He immediately recognized, in addition to the ytterbia found by Marignac [48], the existence of a hitherto-unknown oxide [49].

For confirmation, Nilson deferred to the expert spectrographer, Tobias Robert Thalén (1827–1905), who confirmed the observation of spectral lines that had never been seen in known elements. Thalén also reported some lines proper to ytterbium and erbium, indicating that Nilson's new earth was impure, a fact immediately recognized by Nilson himself. Nevertheless, he went on to state: "I suggest the name scandium for the element so characterized, considering that it is found in gadolinite or euxenite, minerals that have so far only been found on the Scandinavian peninsula." He concluded his paper by enumerating some chemical properties of the new earth already determined with a work plan for future investigations. On the basis of its supposed tetravalence and its supposed atomic weight (between 160 and 180), he placed the new element between tin and thorium in the scheme of the elements.

In a note presented by Marcellin Berthelot (1827–1907) [50], Nilson reported his discovery of scandium via the *Comptes rendus* session of 12 March 1879. The great Swedish chemical savant, Per Teodor Cleve (1840–1905), himself a discoverer of holmium and thulium, wasted no time in carrying out his own investigation of the new earth. He reported his results in a note read by Charles Adolphe Wurtz (1817–1884) on 18 August 1879; he remarked that Nilson's scandium was obviously a mixture with other rare earths, so he carried out his own purification process and was able report an atomic weight of 45.12 (which is remarkably close to the currently accepted value of 44.96). Later in the same paper, he observed [51, 52]:

> What makes scandium's discovery so interesting is that its existence was announced in advance. In his Memoir on the periodic law, M. Mendeleev predicted the existence of a metal with an atomic weight of 44. He called it eka-boron. Eka-boron's characteristics match those of scandium quite well.

The announcement was accompanied by a double table: in the left-hand column were listed the predicted characteristics of eka-boron, and in the right-hand column the observed characteristics of scandium. This type of presentation would become standard practice in the presentation of an eka-element's discovery. Then, on 19 August 1879, Cleve wrote to Mendeleev [19, pp. 38–39]: "I have the honor to inform you that your element eka-boron has been isolated. It is scandium, discovered by Nilson this spring."

Thus was recognized the fulfillment of a second of Mendeleev's prophecies. This second instance of a successful prediction revealed what might be deep regularities reflected in the periodic system [12]. The scientific community began to take notice.

Nilson continued his rare earth research for the next few years and published several papers with Gerhard Krüss (1859–1895), but in 1883 he was appointed Professor of agricultural chemistry at the Royal Academy of Agriculture in Stockholm. From then on, his entire career was taken up with issues of agronomy, always backing up his recommendations in this field with solid experimental work. He was largely responsible for changing large areas of infertile land in Sweden into flourishing farmland—through the transforming power of chemistry. He was a member of many of European learned societies. His motivation was a sense of duty, not a desire for success. "He was not a brilliant lecturer, but no one was listened to with greater attention" [43]. Following a brief illness, Lars Fredrik Nilson passed away on 14 May 1899.

In 1922, Victor Goldschmidt (1888–1947) reported on the crystal structures of several groups of metals [53]. Listed among those he designated as having a face-centered cubic crystal structure was scandium.[11] Goldschmidt based his prediction on scandium's position in the periodic table but, up until this point, scandium had never been prepared in a pure, metallic state. It was isolated for the first time by a group working in the inorganic chemistry department of the Freiberg School of Mines (presently the Freiberg University of Mining and Technology) [54]. The group's avowed purpose lay in the fact that scandium's physical and chemical properties had always been necessarily estimated but not measured. Although the characteristics of other elements could be extrapolated easily, scandium was at the juncture of discontinuity in the periodic table: it was the first of the so-called transition elements. The team succeeded, but with great difficulty for several reasons: the separation had to be carried out at very high temperatures and finding a suitable reaction vessel that could withstand the heat and also not react with the metal was not easy; the great volatility of scandium halides, necessary as a starting point; the search for a suitable metal that would dissolve scandium prior to electrolysis at the cathode. They selected zinc as the metal of choice and succeeded in preparing a fairly pure zinc-scandium alloy and, after distilling off the last of the zinc at very high temperatures, they obtained what they described as a heavily sintered preparation of metallic scandium with a sponge-like structure. They estimated that their sample contained about 94–98% free scandium metal. Because of the impurities, their estimated melting point of 1450 °C was off by about 100 degrees and their specific gravity of 3.1 was off by 0.76, as we can see in Table 10.2, which compares the properties predicted by Mendeleev with today's accepted values.

[11]Scandium has since been shown to have a hexagonal close packed structure.

Table 10.2 Selected properties of scandium (eka-boron, Eb)

Property	Predicted	Measured
Atomic mass	45	44.96
Melting point (°C)		1541
Specific gravity	3.5	3.86
Formula of oxide	Eb_2O_3	Sc_2O_3
Acidity of oxide	More acidic than MgO	More acidic than MgO
Formula of chloride	$EbCl_3$	$ScCl_3$
Color of salts	Colorless	Colorless

All of this information causes us to invoke the afore-mentioned discovery criteria once again: who discovered scandium? Mendeleev foresaw it; Nilson isolated an impure mixture; its spark spectrum was described by Thalén; Cleve purified it so that its properties came into conformity with those predicted; its isolation was accomplished by Werner Fischer, Karl Bürger and Hans Grieneisen. Shall we vote for all of the above?

Scandium has only one stable isotope, but 25 radioisotopes have been identified, most of which are very short-lived. Its relative crustal abundance is 22 ppm, ranking it as number 36 among the elements. Its dominant oxidation number is 3+ although some lower valencies have been observed in organoscandium compounds. Because of its high cost, almost US$5 per gram, it finds limited use, mainly in a low-density alloy with aluminum that is used in sporting equipment and as a brightening additive to mercury vapor lamps.

10.2.3 Mini-Biography of Germanium (Eka-Silicon, Es)

Clemens Alexander Winkler (1838–1904) discovered germanium, the third of the "big three" elements that conferred credibility on Mendeleev's periodic system, on 6 February 1886. Winkler was born in Freiberg, a city in Lower Saxony on the Elbe River, on 26 December 1838, the third son of Kurt Alexander Winkler (1794–1862) and Elmonde Antonie Schramm (1810–1897). Freiberg is home to the famous Freiberg School of Mines (Bergakademie).

Following his own father's early death in 1807, Kurt Winkler had to go to work in the mines at the age of 13, but two of his father's influential friends made it possible for him to attend the Freiberg School of Mines. He excelled to such a degree that the State of Saxony awarded him a "Studienreise" that allowed him to study in Sweden and Norway. He became an excellent chemist and metallurgist under the tutelage of such luminaries as Jöns Jacob Berzelius, Nils Gabriel Sefström (1787–1845) and Johan Gottlieb Gahn (1745–1818).[12] On his return to Freiberg in 1828, he joined the Saxon civil service as a member of the superior mining council, where he remained

[12]Sefström and Gahn discovered vanadium and manganese, respectively.

for 12 years. In 1840, when Clemens Alexander was two years old, he left the civil service and brought his little family to the village of Zschopenthal in order to direct the smalt works there. He outfitted an excellent laboratory at the plant, and it was there that young Clemens Alexander spent many happy hours during his school vacations. Through the entire period of his education, first at the Freiberg Gymnasium where he studied mineralogy under Johann Friedrich August Breithaupt (1791–1873), then at the Realschule in Dresden and the Gewerbeschule in Chemnitz, the young Winkler developed a growing interest in the natural world, especially in classifying plant, animal, and mineral specimens. He followed in his father's footsteps, matriculating at the Freiberg School of Mines in 1857, but because of his father's mentorship, he already knew more analytical chemistry than could be taught at the school [55]. Following his graduation as Doctor of Engineering in 1861, he accepted the position of director of the Pfannenstiel Prussian Blue Works. In 1864, he was awarded the degree of Doctor of Philosophy.

In the course of the following decade, Winkler (Fig. 10.7) concentrated on the many problems that arose from transforming raw ores and minerals into useful products. He solved some of the difficulties associated with gas analysis and the separation of elements one from the other, particularly nickel from cobalt and lanthanum from didymium.[13] He also made an exhaustive study of the element indium at the request of his two colleagues, Hieronymus Richter (1824–1898) and Ferdinand Reich (1799–1882), who had discovered this element at the Bergakademie in 1863 [56].

In 1873, Winkler was appointed Professor and Chair of the Department of Technological Chemistry and Analytical Chemistry at the Bergakademie, his *alma mater*. He would remain in this position until his retirement in 1902. During his tenure as professor, not only did he carry out very fruitful research on a variety of pertinent topics, but he also endeared himself to his students. One in particular, an American named LeRoy Wiley McCay (1857–1937), described him as a man of moderate height, thick-set, with a short dark beard, an attractive face, an engaging smile, and a low, but powerful, voice. McCay said that Winkler's lectures on chemistry were extraordinarily gripping because through them he was able to communicate his own enthusiasm for science. He also did all of his experiments himself, rarely calling upon an assistant. McCay concluded his accolade as follows [57]:

> His good nature, his winning manner, his generosity, his bonhomie, and his keen social instinct captured the love of all, students, colleagues, and fellow citizens...In my day he was the most popular man in Freiberg.

Louis Pasteur (1822–1895) is famously said to have remarked that "chance favors the prepared mind" [58]. Clemens Winkler was thus prepared—not only his mind, but his laboratory, his expertise, his chemical sense—for what would become the defining climax of his career in 1886. The train of events leading to his discovery reads thus [59]:

[13]Didymium was still considered an element in the 1860s.

Fig. 10.7 **Left** Germanium, element number 32, showing its atomic number, electron configuration, symbol, name, and atomic mass; **Right** Germanium's discoverer: Clemens Alexander Winkler

In the middle of September 1885, on the geotechnical line number 10.5, 460 m underground at a crossing of the shaft with an unknown spur in the silver mine… there was a fissure in which one of the ores caught the attention of Operations Director Eduard Wilhelm Neubert, who sent it to Rudolph Benno Wappler (1828–1888), a board member of the Mining Academy mineral dealership, for assessment. Neubert said that the ore had some similarity to silver glance, but still seemed to differ from it.

The material finally came to Albin Weisbach (1833–1901), Professor of Mineralogy at the Bergakademie, who recognized it as a new mineral and described it in 1886 [60]:

"Mr. Foreman Wappler was also convinced of the differences [between the two minerals] and therefore gave a sample to Mr. Superintendent Th. Richter for chemical analysis [by blowpipe]. He found silver and sulfur as the main constituents, but also found, with great determination, some mercury.[14] The presence of the latter seemed quite remarkable, since he had never noticed a trace of it in any of the Freiberg ores."

"Mr. Wappler was kind enough to make me aware of the new find… and to hand over a large number of pieces of the ore from the fissure in the Himmelsfürst mine…. I was put in the position to give a brief description of the new mineral, which was given by me the name argyrodite, and to circulate a few specimens, at the meeting of our local mining association on 1 October [1885] and on 15 October to be able to show the club members a wooden model representing the crystal form of argyrodite."

Given Clemens Winkler's reputation as an excellent analytical chemist, he was entrusted with the task of clarifying argyrodite's composition. Initially, Winkler corroborated Richter's results, finding that the mineral contained 75% silver and 18% sulfur, by weight. But what was the missing 7%? From comments made by Brunck [55] and McCay [57], Winkler was convinced that he had a new element on

[14]The chemical formula of argyrodite is Ag_8GeS_6. A first, incomplete, description of this mineral was given in 1831 by CGA von Weissenbach (1797–1846) and in 1832 by JFA Breithaupt. See ref. [59] for the original citations.

his hands, but he could not isolate it. Working ceaselessly night and day for over four months, he finally succeeded. Here is how he did it [61]:

> After several weeks of painstaking search, I can now state definitely: argyrodite contains a new element, very similar to antimony and yet sharply distinguished from it, to which the name "germanium" shall be given. Its discovery was connected with much difficulty and painful doubt because the minerals that accompanied argyrodite contained arsenic and antimony; their great resemblance to germanium and the absence of methods for separation were extremely disturbing.

He methodically fused the mineral with sodium carbonate and sulfur, treated the liquid with water, filtered off the residue, and precipitated first the antimony and then the arsenic as sulfides by slowly adding dilute HCl dropwise to the filtrate, accomplishing what he called fractional precipitation. What was left in solution, he realized, was his missing element. At first, he tried to boil off the solvent to retrieve the dissolved elemental substance, to no avail: he ended up with only a sodium chloride residue. (This failure was due to the high volatility of some germanium compounds—he had actually boiled off the sought-for species as the highly volatile germanium tetrachloride.) Moreover, no change occurred on addition of more acid. He repeated this procedure again and again—and each time he met with failure. Finally, the breakthrough came on 6 February 1886, the result of Winkler's utter frustration. Since he got no results on the addition of a moderate amount of acid, he impulsively threw in an excessive amount of acid and, to his great surprise, a flocculent white precipitate formed. He eventually determined that it was the sulfide of the long sought-for unknown element. He calcined the sulfide to form the oxide, which he heated strongly while running a current of hydrogen over it, producing a gray, metallic-looking powder which, when further melted with borax, gave shiny gray globules. His great stumbling block in all of this was encountering a sulfide that was soluble in dilute acid or in water, but insoluble in more concentrated acid —a veritable chemical anomaly [62]![15]

Thus, the third of the three "nationalistic" elements made its appearance. However, at first, Winkler assumed that his new element should chemically be similar to antimony because argyrodite and antimony minerals behaved similarly. So, germanium lodged in the periodic system for a time as eka-antimony, a dubious space between antimony and bismuth. He realized his error when first Victor von Richter (1841–1891), and then Lothar Meyer, pointed out[16] to both Mendeleev and Winkler (Fig. 10.8) that, according to his plot of atomic volumes (Fig. 10.1), germanium should be low-melting and volatile and identified with eka-silicon.[17] Winkler concludes [61]:

[15]For a detailed description of the exact chemical reactions involved in germanium's discovery, please see ref. [62].

[16]Meyer, Mendeleev, and Winkler corresponded voluminously on this point before they all realized germanium's true identity.

[17]In German, silicon is "silicium," so Winkler's choice of a name ending in -ium for the new nonmetal was not out of line with some other nonmetals like helium, selenium and tellurium.

Fig. 10.8 Visit of Dmitri Mendeleev and Clemens Winkler in Berlin on 19 March 1900 at the meeting celebrating the 100th anniversary of the Prussian Academy of Science. Old scientific stock photograph from the Freiberg School of Mines

> It was definitely premature when I expressed…an assumption in my first notice concerning germanium; at least there was no basis for its proof. Nor would I have ventured at first to assume argyrodite to be a sulpho salt with a quadrivalent acid radical, because there were no analogies at all for such an assumption. Thus the present case shows very clearly how treacherous it can be to build upon analogies; the quadrivalency of germanium has by now become an incontrovertible fact, and there can be no longer any doubt that the new element is no other than the eka-silicium prognosticated fifteen years ago by Mendeleev.

Shortly after the discovery, Winkler sent some of his precious sample to Lecoq de Boisbaudran for analysis. Lecoq, therefore, was more than likely the first person to measure germanium's spark spectrum and, although he, too, momentarily entertained the idea that germanium should fit between antimony and bismuth (a slot that actually does not exist!), he quickly saw that it should fit between silicon and tin. His reasoning was based on germanium's atomic weight, which was roughly midway between Si and Sn; Lecoq also remarked that John Alexander Reina Newlands (1837–1898) had also postulated the existence of an element there. His second reason came from the results of germanium's spark spectrum. Lecoq discerned two remarkably brilliant lines, one blue and one violet, at 468 and 422 nm, respectively. These lines lay midway between similarly determined lines for silicon and tin. So we may say that Lecoq "found" germanium spectroscopically, not by way of discovery but by way of confirmation [63].[18]

[18]It is interesting to note that in this publication Lecoq used the symbol "Gr" for germanium.

Clemens Alexander Winkler remained at the Bergakademie for the rest of his life although many prestigious universities, such as Göttingen, tried to entice him away. He continued his work with germanium production and isolation, but complained about lack of material since most of the argyrodite retrieved from the original mine had only a superficial coating of the precious mineral [64]. He also worked on the pressing problem of sulfuric acid production, perfecting the contact process that he had invented in 1875. He wrote many works on topics other than element discovery, including gas analysis and other practical handbooks. Realizing that his health was failing, he retired in 1902 and passed away in Dresden on 8 October 1904.

After germanium's discovery, more than fifty years elapsed before its first commercial use saw the light of day. It is chiefly useful because of its electrical properties midway between those of an insulator and those of a metal: it is a semiconductor, although well into the 1930s, it was believed to be a poorly conducting metal [65]. Its first broad application was as a point contact in devices called Schottky diodes, used for radar reception in World War II. It soon became essential in the manufacture of transistors and gamma-ray detectors [66]. Besides the fact that germanium is an intrinsic semiconductor, it also has four other very important properties: it is transparent to the infrared region of the electromagnetic spectrum; it is a glass-former, that is, it can form three-dimensional networks of somewhat ordered germanium-oxygen tetrahedra; it has a very high refractive index; it exhibits low chromatic dispersion. A combination of these properties makes it very useful in electronics, solar power arrays, night vision devices, and optical lens systems [67]. As for the properties predicted by Mendeleev, please see Table 10.3.

Germanium has four stable isotopes and a very long-lived one (with a half-life around 10^{21} years); at least 27 other radioisotopes have been identified, most of which are very short-lived. Its relative crustal abundance is 1.8 ppm, ranking it as number 53 among the elements, between tantalum and molybdenum and slightly more abundant than arsenic. A new ferro-germanium mineral species, eyselite, $Fe^{3+}Ge_3^{4+}O_7(OH)$, has been found in Namibia [68]. Germanium's presence has also been noted out there in the Solar System, produced, it is thought, by slow neutron capture [69]. Its dominant oxidation number is 4+. The world's total production in 2011 was about 118 metric tons. It costs about US$1 per gram, relatively cheap in comparison with some metals.

10.3 Discovery: By Speculation, by Detection, or by Isolation?

When it comes to the discovery of the elements, one must consider three ways or modes of discovery, all of which have played a role in the priority disputes that have now become so legendary.

Table 10.3 Selected properties of germanium (eka-silicon)

Property	Predicted	Measured
Atomic mass	72	72.63
Valence	4	4
Specific gravity	5.5	5.47
Specific heat	0.073	0.076
Specific gravity of dioxide	4.7	4.703
Boiling point of tetrachloride (°C)	Under 100	86

Can we give credit to Mendeleev for the discoveries of gallium, scandium and germanium? Did Mendeleev ever claim these elements as his own? Can we really assign their discoveries to, in turn, to Paul-Émile Lecoq de Boisbaudran, Lars Fredrik Nilson, and Clemens Alexander Winkler? Would these elements have been found when they were if Mendeleev had not pointed the way? These are not idle questions. Nowadays, people tend to give the guide just as much credit as the explorer. But Mendeleev made no such claim—all he wanted, and somewhat ferociously, was credit for the organizing principle that chiefly bears his name, although he, too, stands on the shoulders of many others—and even perhaps should share the platform with Julius Lothar Meyer.

We would like to call Mendeleev's role in the discovery of these elements "*discovery by speculation*," at least in the cases of eka-boron and eka-silicon. Here is another famous case of this type of discovery, but its outcome was not so easily resolved. In 1913, two great scientific advances, atomic number and the discovery of isotopes, revealed a clearer relationship of the chemical elements to one another. In the decade that followed, chemists used these two new tools so cleverly that by 1925, only one rare earth element remained unidentified: the recalcitrant element number 61. Its existence had been surmised by Bohuslav Brauner (1855–1935) in Prague as early as 1902 [70]:

> Apart from the 10 elements already listed...and more or less accurately studied by me, about seven to ten additional elements could be placed in this group...It is not impossible that one would be able to split neodymium, Nd = 143.8, into at least one element with a smaller atomic weight, and into another element with a higher atomic weight of about 145 and, similarly, some more gaps lying in the area between Ce and Ta could be filled.

Later work by Moseley showed definitively that an element should exist between neodymium and samarium [71]. Researchers took up the challenge, and many were in hot pursuit, assuming that this element lay hidden in very small amounts among the other rare earths.

When it was subsequently asserted (incorrectly, as it turned out), that element 61 did indeed exist between Nd and Sm, Brauner subtly claimed credit for the discovery in a letter to *Nature* [72].

> I arrived at the conviction that the gap between the neodymium and samarium was abnormally large. In my paper...read in St. Petersburg in 1902, I came to the conclusion, not reached by any chemist before—that the following seven elements, possessing now the

atomic numbers 43, 61, 72, 75, 85, 87, and 89, remained to be discovered. As regards element No. 61, the difference between atomic weights of Sm–Nd = 6.1, and it is greater than that between any other two neighboring elements.

Did Brauner ever receive credit for the discovery of element 61? It can be shown definitively that his claim was almost completely overlooked [73]. In part this was because his predictions had been superseded by Moseley's atomic numbers, but also because his 1902 paper actually proposed that there would be 20 rare earth elements (four more than actually exist), including two (not one) yet to be discovered between Nd and Sm.

Brauner was not new to being burned by dismissal of his speculations in at least one other case, that of the separation of didymium into praseodymium and neodymium by Carl Auer von Welsbach (1858–1929). The story of the separation in 1885 is told in detail with reference to the original literature in *Carl Auer von Welsbach: Chemist, Inventor, Entrepreneur* [74].

However, the principle of "*discovery by speculation*" was never resolved.

A second type of elemental discovery is "*discovery by detection*." The earliest and perhaps the most famous cases of this type of discovery was that of cesium (L. sky-blue) and rubidium (L. deep red), discovered in 1860 and 1861, respectively, by Robert Wilhelm Bunsen and Gustav Kirchhoff. Using an instrument of their own confection, Bunsen and Kirchhoff discerned spectral evidence of the presence of two elements never seen before. It was actually Kirchhoff who, in 1859, realized that the observed frequencies of the emission lines in bright-line spectra corresponded to the frequencies observed in Fraunhofer's dark line spectra [75]. He concluded that the dark lines were due to the absorption of the characteristic frequencies of the elements present in the cooler outer layers of the sun's atmosphere, and that these would be the same frequencies that these elements would emit when excited by an energy source such as a flame. A further conclusion was that each element should exhibit a line spectrum characteristic of that element, enabling not only chemical analysis, but analysis long distance—as much as eight light years away [76]!

They had done earlier studies of the characteristic colors of heated elements, and a burner that had been in use in Bunsen's laboratory since 1855 was ideal for this purpose since it gave a virtually colorless, soot-free flame of constant size [77]. In the summer of 1859, Kirchhoff suggested to Bunsen that they systematize their studies and try to develop a device that would form spectra of these colors by using a prism. By October of that year they had invented an appropriate instrument, an improved version of Fraunhofer's spectroscope.

Using this improved spectroscope, they were able to identify the characteristic spectra of sodium, lithium, and potassium. After numerous laborious purifications, Bunsen proved that highly pure samples gave unique spectra. In the course of this work, Bunsen detected previously unknown blue spectral emission lines in samples of brine water from Bad Dürkheim and other well-known German spas. He hypothesized that these lines indicated the existence of a hitherto-unknown element. After careful distillation of 600 quintals (about forty-four tons) of this water, in the spring of 1860 he was able to isolate and discern two blue spectral lines in the several liters of residue, and recognized them as the signature of a new element. He named the element "cesium," after the Latin word for "deep blue." The following year Bunsen and Kirchhoff discovered rubidium by a similar process, by identifying the spectroscopic signature of this new element in the mineral lepidolite. Spectroscopy had come a long way from looking at and contemplating the spectrum [22, 78].

Once detected, Bunsen and Kirchhoff continued to work on concentrating solutions of these alkali metals, once detected, in an attempt to isolate them but succeeded only in producing some of their intractable compounds [79]. It was only in 1881 that Carl Setterberg (1853–1941), a Swedish student working in Bunsen's laboratory, was able to isolate cesium and describe some of the properties of the metal: melting point 26–27 °C, density 1.88 g cm^{-3} (today's values: 28.4 °C and 1.93 g cm^{-3}, respectively) [80]. This is a prime example of "*discovery by isolation*."

So, who gets the credit for the discovery? Credit for detection, or credit for isolation? Apparently, the majority of the chemical community places the accolades at the feet of the detectors. Virtually any piece of literature describing the elements cesium and rubidium bestow credit for their discovery on the indomitable Heidelberg pair. But not all! In a 2010 article by Alan Dronsfield [81], Setterberg's painstaking work is described and the author clearly states that this near-forgotten chemist should indeed receive, or at least share in, the credit for the discovery of cesium. But, in this case, discovery by detection, not isolation, wins out.

From a different point of view we could say that the criteria in giving credit for the discovery of a new element have undergone profound changes over the years. In 1976, Rancke-Madsen suggested the concept of "effective discovery of an element" [82]. In laying out this idea, he defined a preliminary stage, an effective stage, and subsequent (or confirmatory) developments. These criteria seem to work for elements discovered before 1800 and for many of the rare earth elements, to which he limited his discussion. However, he left the question of collaborators, a most important consideration for developments in the twentieth century, unaddressed.

First of all, many years ago, the contribution of a scientist's collaborators went unrecognized (Gustave Bémont (1857–1937), for radium, Otto Berg (1874–1939), for rhenium, Carl Setterberg, for cesium, to name a few). It should also be said that many researchers were used to working individually, a scientific practice that has proven unsuccessful and impracticable in today's world. Teamwork is presently the most common and practical way to carry out good science, although there are some outstanding exceptions.

So to discern the identity of the "real discoverer" is an increasingly difficult task, as demonstrated by the exponential growth in the number of authors in the most recent scientific publications. Who gets the credit: the team leader, the researcher, or the one who correctly interprets the data? Or do they all share the laurels?

Secondly, over the last two centuries, discovery norms continually evolved due not only to constantly developing technology, but also to the scientific community's changing perceptions. In addition, the number of investigators involved in reaching the same goal within a limited period of time grew rapidly, making it more and more difficult to reach consensus regarding the validity of a discovery claim against the background of sometimes contentious priority disputes. Compromises among scientists from different cultural and social backgrounds had to be factored in as well; they often favored those from a culturally dominant nation.

Thirdly, only with the foundation of the national chemical societies, and then with the formation of a supranational institution of chemical societies (IUPAC), was it possible to establish comprehensive, universally agreed-upon criteria for the discovery of a chemical element. Even these criteria—far from being unchangeable—have undergone numerous adjustments over the years. For example, the 1976 paper that appeared in *Science*, "Criteria for the Discovery of Chemical Elements," [83] cites as a basic criterion proof that the atomic number of the new element be different from that of all other known elements. This requirement, of course, would not have been possible until after the discovery of atomic number. After the synthesis of the transuranides, additional criteria had to come into play [84] as part of an evolving consciousness that new knowledge requires new rules. These developments could be discussed with profit in another paper that might even take into account other types of discoveries such as new bodies in the cosmos or subatomic particles, but they are clearly beyond the scope of this chapter.

Other cases that could be examined with profit are those of helium, thallium, and protactinium. The case of the latter was complicated by the fact that element 91 possesses several radioisotopes of widely varying half-lives.

We must say that the difficulty in isolating the elements has been a perennial problem: for example, over much of the 225-year history of the rare earth elements, the isolation of the free element was simply not possible. However, the existence of these elements entered the chemical literature and, with them, the names of their discoverers. Holding everyone to the higher bar of element isolation would clearly have been impossible. Let us revert to the "famous three" (Table 10.4) to see who really discovered what using the criteria discussed above.

Clearly, one could construct a similar table encompassing all of the elements. It would be large, perhaps cumbersome, and perhaps controversial, but it would tell a more complete tale, giving credit where credit is due. Furthermore, when one considers that spectroscopes were not employed in the discovery of elements until

Table 10.4 Discovery profile of gallium, scandium, and germanium

Element discovery by	Speculation	Spectroscopic Detection	Isolation (compound)	Isolation (free element)
Gallium	Mendeleev	Lecoq de Boisbaudran	Lecoq de Boisbaudran	Lecoq de Boisbaudran
Scandium	Mendeleev	Thalén	Nilson	Fischer, Bürger, Grieneisen
Germanium	Mendeleev, Newlands	Lecoq de Boisbaudran[a]	Winkler	Winkler

[a]Confirmatory data

Bunsen and Kirchhoff did so in 1860, and even after that the chemical separation of new elements remained challenging, one can only admire the ingenuity and expertise of the early element discoverers. Their workhorses, emission, and visible spectroscopy, were much improved in the 1930s [85, 86].

10.4 Conclusion

The time of discovery of gallium is known: year, date, and time. Friday 27 August 1875, at 4 p.m. [87]. The discovery of the 31st element and its identification with Mendeleev's eka-aluminum indeed represents a milestone in the history of discovery of the elements. However, Per Teodor Cleve, who demonstrated the equivalence of Nilson's scandium and Mendeleev's hypothetical eka-boron, never alluded to the "periodic table" in his lectures. The scientific genius, Robert Bunsen, once stated that "to classify elements is the same thing as to search for regularities in the stock-exchange" [87]. In addition to the triad of elements being discussed in this chapter, Mendeleev predicted the existence of other unknown elements: eka-manganese, dvi-manganese, tri-manganese, eka-iodine, eka-cesium, eka-barium, eka-lanthanum, and eka-tantalum. In reaching out to the unknown, Mendeleev predicted the existence of newtonium and coronium, two elements lighter than hydrogen [88], and, in fact, ten other elements that turned out to be incorrect guesses [89]. He also reiterated his belief in the chemical nature of the ether [90, 91].[19] His success rate at prediction was a bit under 50%. In these cases, we are tempted to say that prediction of new elements by *speculation* could help further research, and it even might appear to be a prerequisite—albeit not the only one—for getting a clear route to finding a new element.

Theoretical speculations and predictions may put someone on the right path, but from a chemical point of view, the true discoverer should be that scientist who is able to isolate (or create) at least a small handful of new atoms and demonstrate that they are unquestionably new entities.

The latter decades of the twentieth century saw numerous cases of insisting on this new criterion for discovery, that is, proof that one actually held in one's hand, or test tube, or microscopic vial, at least a few atoms of an element—even if, after a few seconds, those atoms had already transformed themselves into a different element. These cases are not simple, they took decades to resolve, and in some cases, there are still some doubts. Part of the confusion in the early part of the twentieth century arose from the discovery of radioactive decay products of species with similar chemistry but with different atomic weights and half-lives. We refer the reader to our article on the history of the discovery of the actinides and transactinides [92], our book, *The Lost Elements: The Periodic Table's Shadow Side* [24], and the always-fresh memoir by Glenn Seaborg, *The Transuranium People* [93].

[19]This work is now available in a modern rendition in a collection edited by William Jensen [91].

On the other hand, the whole significance of "discovery" has changed. Up to the middle of the twentieth century, discovery of an element meant uncovering the presence of the pre-existent element in the universe, whether in the earth's crust or the sun's corona. Those that were missing were the objects of hunts, confounding the searchers for elements 43 and 61 who wrongly assumed that they were to be found among the naturally occurring elements. The accepted discovery criteria were detection and wet-chemical isolation. After 1913, the hunters were guided by Moseley's law of atomic number and Soddy's concept of the isotope. With the advent of the cyclotron in 1930, the game changed radically: the new criterion became demonstrable proof of synthesis accompanied by a unique atomic number, as noted above. At the same time, it became impossible for a single individual to seek scientific immortality by claiming priority for discovery of an element: synthesis was a big-budget, team effort, most of the time along nationalistic lines.

Thus, enshrinement in the periodic table is all but closed, at least for the time being. Furthermore, in its list of the top ten scientific experiments of all time, the *Discover* magazine's website [94] lists only one related to the discovery of an element (predictably, Marie Curie), but refers to her finding that radioactivity is an (intra)atomic process, not her discoveries of radium and polonium. So why look for new elements?

For some, ego-satisfaction, glory (and hopefully riches) seem to have been major motivating factors among the list of element discoverers. Witness, for example, the rather obstreperous priority disputes among some of the rare earth element researchers, to say nothing of the international and intranational naming rights disputes that characterized the syntheses of the transuranic elements. However, there were also enough discoveries driven by curiosity, love of science, the drive to understand, the joy of learning, and the satisfaction derived from comradeship and patriotism to balance out the scales. These latter are the qualities we should celebrate; those who practice them we should revere.

Acknowledgements Our great gratitude goes to Sally B. Mitchell, our wonderful colleague in the ACS Division of the History of Chemistry, for helping make the entire panorama of the periodic system accessible to us over the entire time it took us to develop this chapter. We also thank Dr. Soňa Štrbaňová and Dr. William B. Ashworth, Jr. for helping us access some important reference material. We are also grateful to the editors of this volume, Gregory S. Girolami, Carmen Giunta, and Vera V. Mainz for the opportunity to undertake this challenging task and for their painstaking guidance. Special thanks are due to Dr. Girolami and the anonymous reviewers whose very helpful and detailed criticism made this chapter so much better.

References

1. UNESCO. International Year of the Periodic Table of Chemical Elements 2019 (2009) https://en.unesco.org/commemorations/iypt2019. Accessed 27 Apr 2019
2. Meyer JL (1864) Die modernen Theorien der Chemie und ihre Bedeutung für die chemische Mechanik. Verlag von Maruschke & Berendt, Breslau
3. Boeck G (2019) Julius Lothar (von) Meyer (1830–1895) and the periodic system. Substantia 3 (2) suppl 4:13–25. https://doi.org/10.13128/substantia-503

4. Meyer JL (1870) Die Natur der chemischen Elemente als Function ihrer Atomgewichte. Annalen Chem Pharm, sup. 7:354–364
5. Mendeleev DI (1869) Sootnoshenie svoistv s atomnym vesom elementov [The correlation of properties with the atomic weights of the elements]. Zh Russ Fiz-Khim Ova 1:60–77
6. Moseley HGJ (1913) The high frequency spectra of the elements. Philos Mag 26:1025–1034
7. Moseley HGJ (1914) High frequency spectra of the elements, Part II. Philos Mag 27:703–713
8. van den Broek A (1911) The number of possible elements and Mendeléeff's "cubic" periodic system. Nature 87:78
9. Scerri E (2015) The discovery of the periodic table as a case of simultaneous discovery. Phil Trans R Soc A 373:20140172. http://dx.doi.org/10.1098/rsta.2014.0172. Accessed 19 Feb 2020
10. Rabinovich D (2019) International Year of the Periodic Table (IYPT): A Midyear Philatelic Report. Philatelia Chim Phys 40(2):56–65
11. Ball P (2019) "Whose periodic table is it anyway? Reproduced by permission from the royal society of chemistry. https://www.chemistryworld.com/opinion/whose-periodic-table-is-it-anyway/3009968.article. Accessed 1 Mar 2020
12. Gordin MD (2007) D. I. Mendeleev: reflecting on his death in 1907. Angew Chem Int Ed 48:2758–2765
13. Mendelejeff D (1869) Ueber die Beziehungen der Eigenschaften zu den Atomgewichten der Elemente [On the relationship of the properties of the elements to their atomic weights]. Z Chem 12:405–406
14. Giunta C. Elements and atoms: Chapter 12. Mendeleev's first periodic table. http://web.lemoyne.edu/~GIUNTA/EA/MENDELEEVann.HTML. Accessed 17 Feb 2019
15. Freund I (1904) The study of chemical composition: an account of its method and historical development. Cambridge University Press, Cambridge, UK, p 474. As quoted in Opitz DL (2019) Cupcakes and chemical composition: Ida Freund's legacy. In: Lykknes A, Van Tiggelen B (eds) Women in their element: selected women's contributions to the periodic system. World Scientific Publishing, Singapore, pp 457–467 at 463
16. Van Tiggelen B, Lykknes A, Martinez-Moreno L (2019) The periodic system, a history of shaping and sharing. Substantia 3(2), suppl 4:9–12
17. Mendeléeff D (1889) The periodic law of the chemical elements. J Chem Soc (London) 55:634–656
18. Stewart PJ (2019) Mendeleev's predictions: success and failure. Found Chem 21:3–9
19. Gordin MD (2018) A well-ordered thing: Dmitri Mendeleev and the shadow of the periodic table, rev edn. Princeton University Press, Princeton, NJ, p 202
20. Niaz M, Rodríguez MA, Brito A (2004) An appraisal of Mendeleev's contribution to the development of the periodic table. Stud Hist Philos Sci A 35(2):271–282
21. Weeks ME (1932) The discovery of the elements XV. Some elements predicted by Mendeleeff. J Chem Educ 9(9):1605–1619
22. Weeks ME (1968) Discovery of the elements, 7th edn., completely revised and new material added by Leicester HM. Journal of Chemical Education, Easton, PA, pp 643–665
23. Marché M (1958) Souvenir d'un grand chimiste cognaçais. Bulletin de l'Institut d'Histoire et d'Archéologie de Cognac et du cognaçais 1(3):1–19
24. Fontani M, Costa M, Orna MV (2014) The lost elements: the periodic table's shadow side. Oxford University Press, New York, pp 211–213
25. Wagner GH, Gitzen GH (1952) Gallium. J Chem Educ 29:162–167
26. Lecoq de Boisbaudran P-É (1869) Sur la constitution des spectres lumineux. C R Séances Acad Sci 69:445–451, 606–615, 657–664, 694–700
27. Kirchhoff G, Bunsen R (1860) Chemische Analyse durch Spectralbeobachtungen. Annalen Phys Chem 110:161–189
28. Lecoq de Boisbaudran P-É (1877) On the new metal—gallium. Chem News 35(907):148–150

29. Lecoq de Boisbaudran P-É (1874) Spectres lumineux: spectres prismatiques et en longueurs d'ondes destinés aux recherches de chimie. Gauthier-Villars, Paris. (Reprinted as a Classic Reprint by Forgotten Books, London, in 2018)
30. Lecoq de Boisbaudran P-É (1875) Caractères chimiques et spectroscopiques d'un nouveau métal, le Gallium, découvert dans une blende de la mine de Pierrefitte, vallée d'Argelès (Pyrénées). C R Séances Acad Sci 81:493–495
31. Lecoq de Boisbaudran P-É (1876) Sur le spectre du gallium. C R Séances Acad Sci 82:168
32. Lecoq de Boisbaudran P-É (1876) Nouvelles recherches sur le gallium. C R Séances Acad Sci 82:1036–1039
33. Lecoq de Boisbaudran P-É, Jungfleisch É-C (1878) Extraction du gallium. C R Séances Acad Sci 86:475–478
34. Mendeleev D (1875) Remarques à propos de la découverte du gallium. C R Séances Acad Sci 81:969–972
35. Lecoq de Boisbaudran PÉ (1875) Sur quelques propriétés du gallium. C R Séances Acad Sci 81:1100–1105
36. Gordin MD (2004) The short happy life of Mendeleev's periodic law. In: Rouvray DH, King RB (eds) The periodic table: into the 21st century. Research Studies Press Ltd, Baldock, Hertfordshire, England, pp 40–90 at 62
37. Strathern P (2001) Mendeleyev's dream. St. Martin's Press, New York, p 291
38. Lecoq de Boisbaudran P-É (1876) Sur les propriétés physiques du gallium. C R Séances Acad Sci 83:611–613
39. van Spronsen JW (1969) The periodic system of chemical elements: a history of the first hundred years. Elsevier, Amsterdam, p 221
40. Ramsay W (1913) Obituary notice: Paul Émile (dit François) Lecoq de Boisbaudran. J Chem Soc Trans 103:742–744
41. Bruker Metaljet D2 Plus. https://tinyurl.com/y8jmca9g. Accessed 27 May 2020
42. Excillum Metaljet X-ray sources. https://www.excillum.com/products/metaljet/. Accessed 27 May 2020
43. Pettersson O (1900) Nilson memorial lecture. J Chem Soc Trans 77:1277–1294
44. Scheerer Th (1840) Ueber den Euxenit, eine neue Mineralspecies. Ann Phys (Berlin) 50:149–153
45. Berlin NJ (1852) Ueber die Thorerde und die Donarerde. Ann Phys (Berlin) 85:554–557
46. Bahr JF, Bunsen RW (1866) Über Erbinerde und Yttererde. Ann Chem Pharm 137 (neue Reihe 61, Januar):1–33 at pp 26–28
47. Galissard de Marignac JC (1878) Sur l'ytterbine, nouvelle terre contenue dans la gadolinite. C R Séances Acad Sci 87:578–581
48. Nilson LF (1879) Sur l'ytterbine, terre nouvelle de M. Marignac. C R Séances Acad Sci 88:642-645
49. Nilson LF (1879) Om scandium, en ny jordmetall. Öfvers Kongl Vetensk-Akad Förh 36 (3):47–51
50. Nilson LF (1879) Sur le scandium, élément nouveau. C R Séances Acad Sci 88:645–648
51. Cleve PT (1879) Sur le scandium. C R Séances Acad Sci 89:419–422
52. Cleve PT (1879) Om Skandium. Öfvers Kongl Vetensk-Akad Förh 36(7):3–10
53. Goldschmidt VM (1922) Kristallographie und Metallkunde. Chemisches Zentrallblatt 93(Pt. 1, No. 1):251–252
54. Fischer W, Brünger K, Grieneisen H (1937) Über das metallische Scandium. Z Anorg Allg Chem 231(1–2):54–62
55. Brunck O (1906) Clemens Winkler. Ber Dtsch Chem Ges 39(4):4491–4548
56. Hinrichs CG (1904) Clemens Alexander Winkler. Am Jour Pharm 76:532–534
57. McCay L (1930) My student days in Germany. J Chem Educ 7:1081–1099
58. Pasteur L (1954) Inaugural lecture, University of Lille, 7 December 1854. In: Peterson H (ed) A treasury of the world's great speeches. Simon & Schuster, New York, p 473
59. Witzke T. Argyrodit. https://tw.strahlen.org/typloc/argyrodit.html. Accessed 29 Feb 2020

60. Weisbach A (1886) Argyrodit, ein neues Silbererz. Neues Jahrbuch für Mineralogie, Geologie und Paläontologie II:67–71
61. Winkler CA (1886) Germanium, ein neues nonmetallische Element. Ber Dtsch Chem Ges 19:210–211
62. Weinert CS (2020) Die Chemie ist schwierig: Winkler and the discovery of germanium. Bull Hist Chem 45(1):8–15
63. Lecoq de Boisbaudran P-É (1886) Sur le poids atomique et sur le spectre du germanium. C R Séances Acad Sci 102:1291–1295
64. Winkler CA (1887) Mittheilungen über das Germanium. J Prakt Chem 36:177–208
65. Haller EE (2006) Germanium: from its discovery to SiGe devices. Mater Sci Semicond Process 9:408–422
66. Haller EE (2006) Germanium: from its discovery to SiGe devices. US Department of Energy: Office of Scientific and Technical Information. https://www.osti.gov/biblio/922705. Accessed 29 Feb 2020
67. Shanks WCP III, Kimball BE, Tolcin AC, Guberman DE (2017) Germanium and indium. Chapter 1 in: Schulz KJ, DeYoung JH Jr, Seal RR II, Bradley DC (eds) Critical mineral resources of the United States—economic and environmental geology and prospects for future supply, USGS. https://pubs.er.usgs.gov/publication/pp1802I. Accessed 29 Feb 2020
68. Roberts AC, Seward TM, Reusser E, Carpenter GJC, Grice JD, Clark S, Marcus MA (2004) Eyselite, Fe^{3+} Ge^{4+} $_3O_7(OH)$, a new mineral species from Tsumeb, Namibia. Can Mineral 42:1771–1776
69. Cowan J (2003) Elements of surprise. Nature 423 (1 May):29
70. Brauner B (1902) Über die Stellung der Elemente der seltenen Erden im periodischen system von Mendelejeff. Z Anorg Chem 32:1–30
71. Moseley HGJ to James C, James Collection UNH Archives, UA 8/3/1, S.1, B.1, F.11, 27 May 1914
72. Brauner B (1926) The new element of atomic number 61: illinium. Nature 118(2959):84
73. Fontani M, Costa M, Orna MV (2014) The lost elements: the periodic table's shadow side. Oxford University Press, New York, p 308, footnote 258
74. Adunka R, Orna MV (2018) Carl Auer von Welsbach: Chemist, inventor, entrepreneur. Springer, Cham, pp 35–46
75. Kirchhoff G (1860) Über das Verhältniss zwischen dem Emissionsvermögen und dem Absorptionsvermögen der Körper für Wärme und Licht. Ann Phys 185(2):275–301
76. Kirchhoff G (1860) Über die Fraunhofer'schen Linien. Ann Phys 185(1):148–150
77. Jensen WB (2005) The origin of the Bunsen burner. J Chem Educ 82:518
78. Kirchhoff G, Bunsen R (1861) Chemische Analyse durch Spectralbeobachtungen. Ann Phys (Berlin) 189:337–381
79. Weeks ME (1932) The discovery of the elements. XIII. Some spectroscopic discoveries. J Chem Educ 9:1413–1434
80. Setterberg C (1881) Über die Darstellung von Rubidium- und Cäsium Verbindungen. Ann Phys (Berlin) 211:100–116
81. Dronsfield A (2010) Look who discovered caesium. Educ Chem (London) July 1. https://eic.rsc.org/feature/look-who-discovered-caesium-/2020183.article. Accessed 16 Jan 2019
82. Rancke-Madsen E (1976) The discovery of an element. Centaurus 19(4):299–313
83. Harvey BG et al (1976) Criteria for the discovery of chemical elements. Science 193:1271–1272
84. Wapstra AH (1991) Criteria that must be satisfied for the discovery of a new element to be recognized. Pure & Appl Chem 63(6):879–886
85. Taylor JK (1986) The impact of instrumentation on analytical chemistry. In: Stock JT, Orna MV (eds) The history and preservation of chemical instrumentation, D. Reidel, Dordrecht, pp 1–10
86. Altemose I (1986) Evolution of instrumentation for UV-visible spectrophotometry, part 1. J Chem Educ 63(9):A216–A223

87. Trifonov DN, Trifonov VD (1982) Chemical elements. How they were discovered. MIR Publisher, Moscow, pp 156–164
88. Fontani M, Costa M, Orna MV (2019) Ipotesi sui limiti della Tavola Periodica di Mendeleev: dalla nascita agli albori della meccanica quantistica. Rendiconti Accademia Nazionale delle Scienze detta dei XL Memorie di scienze fisiche e naturali, 137° Vol. XLIII, Parte II, Tomo II, pp 129–138
89. Lente G (2019) Where Mendeleev was wrong: predicted elements that have never been found. ChemTexts 5:17. https://doi.org/10.1007/s40828-019-0092-5. Accessed 30 May 2020
90. Mendeléeff D (1904) An attempt towards a chemical conception of the ether. Longmans, Green & Co., London
91. Jensen WB (ed) (2005) Mendeleev on the periodic law: selected writings 1869–1905. Dover Publications, Mineola, NY, pp 227–252
92. Orna MV, Fontani M (2019) Discovery of the actinide and transactinide elements. In: Evans WJ, Hanusa TP (eds) The heaviest metals: science and technology of the actinides and beyond. Wiley, New York, pp 3–30
93. Hoffman DC, Ghiorso A, Seaborg GT (2000) The transuranium people. Imperial College Press, London, pp 400–425
94. Hadhazy A (2019) The top 10 scientific experiments of all time. https://www.discovermagazine.com/the-sciences/the-top-10-science-experiments-of-all-time. Accessed 24 Mar 2020

11 The Rare Earths, a Challenge to Mendeleev, No Less Today

Simon A. Cotton

Abstract

Mendeleev's first periodic table (1869) included just five rare earth elements. This article traces the discovery and isolation of the rare earths—scandium, yttrium and the lanthanides—a process that extended until 1947, as well as the efforts to locate them in the periodic system. The striking similarity of the chemistry of the metals has made their isolation in a pure state a challenge. Developments in their chemistry extending up to the present day are considered, including extending the range of their coordination numbers, an increasing number of compounds in unusual oxidation states, and most recently lanthanide-containing enzymes, raising questions about their role in living systems.

11.1 Introduction

When Mendeleev published his first periodic table in 1869, only five rare earths were known, and even their atomic masses were incorrect, so that positioning them was a challenge. This article traces the discovery and isolation of the rare earths and their positioning in successive periodic tables [1], along with later key developments.

S. A. Cotton (✉)
School of Chemistry, University of Birmingham, Edgbaston, Birmingham B15 2TT, UK
e-mail: s.cotton@bham.ac.uk

C. J. Giunta et al. (eds.), *150 Years of the Periodic Table*, Perspectives on the History of Chemistry, https://doi.org/10.1007/978-3-030-67910-1_11

11.2 Discovery of the Lanthanides

Although its significance was not appreciated at the time, the year 1787 was momentous. That December, Captain William Bligh left Spithead in the Solent, accompanied by 46 men, *en route* for Tahiti, to collect a cargo of breadfruit. His ship was named HMS *Bounty*. Four months earlier, and eight hundred miles to the east in Vienna, Wolfgang Amadeus Mozart completed his composition "*Eine kleine Nachtmusik*." Meanwhile, nearly a thousand miles to the northeast, a Swedish army officer named Carl Axel Arrhenius (1757–1824) discovered a black stone at a mine near Ytterby in Sweden; perhaps displaying a lack of imagination, he called this mineral by the name "ytterbite." A few years later, in 1797, ytterbite was renamed gadolinite, after the Finnish chemist Johan Gadolin. Today we know that its composition is (Ce, La, Nd, Y)$FeBe_2Si_2O_{10}$; despite its name, it only contains a trace of gadolinium. Nevertheless, the mineral played a key role in the discovery of the lanthanide elements.

11.2.1 Discoveries by Gadolin and Berzelius

In 1794, Johan Gadolin (1760–1852) investigated a sample of ytterbite, given to him by Carl Axel Arrhenius. Arrhenius's profession led him to an interest in gunpowder, which broadened into chemistry as a whole, which for some 40 years he combined with the army, where he attained the rank of lieutenant-general, specializing in gunpowder—though he would have preferred to be a chemist. Gadolin, the pioneer Finnish chemist, separated the silica content from the other elements (Fe, Be, Y) by dissolution of the latter in mineral acid (either HCl or HNO_3). He found that treatment of the resulting solution with potassium carbonate precipitated the iron and beryllium (which he took to be aluminum), leaving the yttrium in solution, though that could be precipitated (as the hydroxide) with ammonia solution. He also found that alternatively the beryllium and iron could be precipitated out with potassium ferrocyanide, after which the dissolved yttrium again could be precipitated as the hydroxide with KOH, with any remaining beryllium being soluble as $[Be(OH)_4]^{2-}$, as we now know. He went on to describe how the "unknown earth" (which he termed *yttria*) would dissolve in all three common mineral acids, finding that oxalic acid precipitated the yttrium, and of course it was subsequently recognized that the oxalates of the rare earths are very insoluble. He obtained "sharp, rhombic" crystals of the sulfate, which he described as having a sweet taste; crystals of the acetate were also obtained [2, 3].

The historical importance of gadolinite (ytterbite) lies in the fact that it was the first mineral containing the new elements, the rare earths, to be identified. Some 40 years earlier in 1751, Axel Fredrik Cronstedt (1722–1765), another Swedish mineralogist, had discovered cerite, but he misidentified it as just an iron-containing mineral, not recognizing its content of lanthanum and cerium; its constitution is $(Ce, La, Ca)_9(Mg, Fe^{III})(SiO_4)_6(SiO_3OH)(OH)_3$.

In 1803–1804, Jöns Jacob Berzelius (1779–1848) and Wilhelm Hisinger (1766–1852) examined a number of minerals and claimed to have discovered a new element they called cerium, from a mineral that they named cerite. Berzelius and Hisinger published their discovery in the German journal *Neues allgemeines Journal der Chemie* [4] and this was rapidly reproduced in French (Klaproth 1804) and in two English magazines. Quite independently, Martin Heinrich Klaproth (1743–1817) isolated an "earth" from cerite, terming it *ceria*, also publishing a report in *Neues allgemeines Journal der Chemie* [5]; it was also translated into French [6]. Louis Nicholas Vauquelin (1763–1829) then published an account of his own experiments on the isolation of cerium too [7, 8].

11.2.2 Mosander and His Discoveries

For over 30 years, rare earth chemistry remained largely quiescent. It should be remembered that this was the time of rapid development in chemistry, so that from the middle of the nineteenth-century chemists came to understand organic structure and synthesis, as well as spectroscopy and separation techniques. The next advance was due to Carl Gustav Mosander (1797–1858), who during his medical studies worked in Berzelius's laboratory at the Karolinska Institute, and subsequently succeeded Berzelius as professor there [9]. He started work on lanthanides in 1825, when Berzelius asked him to investigate cerium sulfide; he synthesized it and also anhydrous $CeCl_3$, reducing it with potassium to isolate (impure) metallic cerium for the first time. Over a decade later, between 1838 and 1841, he found that what he had taken to be cerium oxide also contained another metal. He was able to obtain lanthanum oxide from what we now know to be CeO_2, taking advantage of the ability of cerium to be oxidized to a "higher" oxide. Mosander was able to make the sulfide and anhydrous chloride, reducing the latter to the metal (as with cerium) [10]. A year later he found that his lanthanum sulfate was still impure, separating it into two new earths, *lanthana* and *didymia*, making use of both precipitation and fractional crystallization. He was also able to obtain much purer ceria, though the pure element was not isolated until 1875. *Didymia*, with an amethyst-violet color which we now know is characteristic of neodymium, proved to contain five different lanthanides, praseodymium, neodymium, samarium, europium and gadolinium, but their separation was to lie well in the future.

Meanwhile, in 1843, Mosander showed that *yttria* was a mixture. He found that it could be separated by precipitation with dilute aqueous ammonia into three different earths, a purer, white, *yttria* as well as a yellow *erbia* and rose *terbia* (the labels of the latter were later to be exchanged) [11].

11.2.3 Spectroscopy Makes a Difference

Intense investigations took place over the next half century or so, particularly after Bunsen and Kirchhoff's development of a more precise spectroscope in 1859.

Because the chemistry of the lanthanide elements, particularly in aqueous solution, is dominated by the +3 oxidation state, separations relied on difficult fractional crystallizations, and it was the fact that each element had its own characteristic spectroscopic lines that showed how fractions were being enriched. From this time, eight lanthanides, plus scandium and yttrium, were identified within half a century [12].

In 1876, Marc Delafontaine (1837–1911) noticed that the absorption spectrum of didymium obtained from cerite differed from that of didymium from samarskite and deduced that didymium was a mixture [13]. Carl Auer von Welsbach (1858–1929) resolved didymium into praseodymium and neodymium in 1885 by fractional crystallization of the double ammonium nitrates [14]. In 1879 Paul Émile "François" Lecoq de Boisbaudran (1838–1912) isolated (impure) samarium from samarskite [15, 16], though europium was obtained from it by crystallization of magnesium samarium nitrate by Eugène-Anatole Demarçay (1852–1903) in 1901 [17].

It was a period of at times confusing endeavor; for example erbia and terbia swapped identities, becoming known as new erbia (old terbia) and new terbia (old erbia), respectively, owing to Delafontaine in 1873. New terbia was separated into gadolinium and terbium in 1880, thanks largely to Jean Charles de Marignac (1817–1894) [18, 19], while new erbia went through a succession of separations. It was first split into purer erbia and ytterbia by Marignac in 1878 [20].

In 1879 Per Cleve obtained holmium and thulium from erbia [21, 22] while dysprosium was got from holmia in 1886 [23]. Ytterbia yielded not only ytterbium to Marignac but also scandium to Lars F. Nilson (1840–1899) [24] and later—after a long delay—lutetium, due to the combined efforts of Georges Urbain, Auer von Welsbach and Charles James (1907). More of them later!

Predicted by Mendeleev just a decade earlier, scandium was isolated by Nilson in 1879 [24]. He extracted *erbia* from the mineral euxenite, separating ytterbium from it, which he was further able to separate to give some 0.3 g of a new "earth" which proved to be the oxide of Mendeleev's "ekaboron" with an atomic mass around 44. Later that year, Cleve separated 0.8 g of the same oxide from 4 kg of gadolinite, then managed to get 1.2 g of the same material from a different mineral, kielhauite, preparing a number of salts, including the nitrate, sulfate, oxalate [25, 26].

11.2.4 Ytterbium and Lutetium

After Marignac's separation and naming of ytterbium in 1878, its atomic mass was determined as 173 by Nilson in 1880 [27]. This was not the end of the story, as some researchers suspected that what had been isolated was not a pure element, though on the basis of thin evidence [28]. Auer von Welsbach, who had been a student of Bunsen, and who had already involved himself in the separation of didymium into praseodymium and neodymium by fractional crystallization of their ammonium double nitrates, separated samples of ytterbium into two elements, using the ammonium double oxalate salts, naming them cassiopeium and aldebaranium [29]. He made a preliminary announcement of this finding in 1905–1906 (lacking in some important details), with a formal announcement, complete with atomic masses, on December 19, 1907 [30]. But he had been overtaken by the Frenchman Georges Urbain (1872–1930), who carried out the separation of these last two lanthanides by fractional crystallization of double nitrates. As a good Frenchman, Urbain suggested that one of the elements should be called lutecium, based on Lutecia, a Latin name for Paris, and that the other should be neo-ytterbium. He read a paper summarizing his findings to the Paris Academy on November 4, 1907 [31]. Von Welsbach and Urbain were not alone. On the other side of the Atlantic, in the United States, Charles James (1880–1928) had separated a substantial amount of Urbain's lutecium by the summer of 1907, probably by fractional crystallization of the double magnesium nitrates, but his caution in announcing his discovery meant that he was pipped at the post, and his findings were never published. In 1949, controversy about the names of the two elements was settled in favor of retaining ytterbium and Urbain's suggestion of lutecium (although the spelling of the latter was changed to lutetium). The Commission on Nomenclature of Inorganic Chemistry decided that priority in discovery belonged to Urbain, though nowadays von Welsbach normally shares the credit.

11.2.5 Perspective

To anticipate the isolation of the radioactive promethium, the last lanthanide was not identified until 1947; as has been remarked [32], "The period of 153 years of fairly continuous research directed towards the attainment, at last achieved, of a limited and specific objective, must be one of the longest such in the history of science."

The discovery of the lanthanides [33] is summed up in a timeline (Fig. 11.1). If the preceding narrative sounds confusing, it represents the difficulties that the nineteenth-century researchers had to surmount. Unlike the separation of transition metals, or non-metals, where pronounced differences in chemical behavior between neighboring elements could be harnessed, similarities in properties of the lanthanides made the separations, particularly of adjacent elements, extremely difficult, depending as they did on techniques such as fractional crystallization. A famous example of this situation was reported by Charles James (a British born chemist carrying out his research in the USA) and a leading rare earth researcher, in reporting the isolation of a pure thulium compound, separating it from yttrium, erbium and ytterbium [34]:

> Thulium [bromate] separated fairly rapidly and collected in the fractions between erbium and ytterbium as the crystallization was continued. The less soluble fractions consisted of erbium with some yttrium. The more soluble portion was then taken and again submitted to fractional crystallization. This portion was nearly colorless. It showed the absorption bands of thulium together with a very weak spectrum of erbium. After about 15,000 operations the absorption spectrum underwent no change.

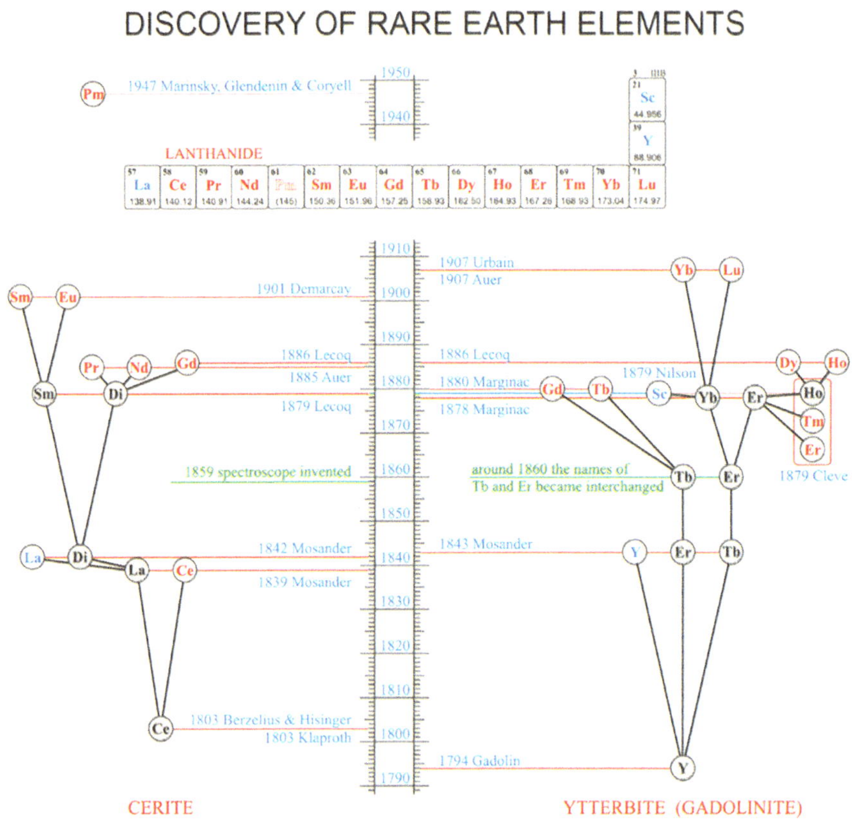

Fig. 11.1 Rare Earth discovery timeline. *Source* From E. Generalic, https://www.periodni.com/history_of_rare_earth_elements.html

This brings to mind the comment famously attributed to Sir Isaac Newton: "If I have seen further it is by standing on the shoulders of giants"—though a similar remark was made by Bernard of Chartres in the twelfth century.

Not the least of these problems was the fact that, before 1913, no one knew how many of these rare earth elements there were, and this was a problem that affected even Mendeleev.

11.2.6 Positioning the Lanthanides

Placing the lanthanides in the periodic table was a challenge that ultimately defeated Mendeleev. Within the main body of the table, elements in a particular group showed gradations in chemical properties and roughly incremental increases of atomic mass on descending a group, and could be arranged on that basis; a similar approach could not be carried out with the rare earths.

One anomalous feature of his pioneering 1869 publication [35], which arranges the elements in order of ascending atomic mass (Fig. 11.2), is that seven elements are positioned to interrupt this sequence—indium, thorium, and the five known rare earths, namely yttrium, lanthanum, cerium, erbium and "didymium" (at that time believed to be an individual element). There was, however, a big gap in the table in elements with atomic masses between barium (137) and tantalum (182).

This anomaly was rooted in the belief that the very basic rare earth oxides were analogous to the corresponding oxides of the alkaline earth metals, so that—like the rest of the chemical community at that time—Mendeleev followed Berzelius in assuming that metals like La, Ce and Y had oxidation states of +2 (as we would now say) in their compounds and formed oxides MO, leading to incorrect atomic weights and problems with positioning them in the periodic table.

Ueber die Beziehungen der Eigenschaften zu den Atomgewichten der Elemente. Von D. Mendelejeff. — Ordnet man Elemente nach zunehmenden Atomgewichten in verticale Reihen so, dass die Horizontalreihen analoge Elemente enthalten, wieder nach zunehmendem Atomgewicht geordnet, so erhält man folgende Zusammenstellung, aus der sich einige allgemeinere Folgerungen ableiten lassen.

			Ti = 50	Zr = 90	? = 180
			V = 51	Nb = 94	Ta = 182
			Cr = 52	Mo = 96	W = 186
			Mn = 55	Rh = 104,4	Pt = 197,4
			Fe = 56	Ru = 104,4	Ir = 198
			Ni = Co = 59	Pd = 106,6	Os = 199
H = 1			Cu = 63,4	Ag = 108	Hg = 200
	Be = 9,4	Mg = 24	Zn = 65,2	Cd = 112	
	B = 11	Al = 27,4	? = 68	Ur = 116	Au = 197?
	C = 12	Si = 28	? = 70	Sn = 118	
	N = 14	P = 31	As = 75	Sb = 122	Bi = 210?
	O = 16	S = 32	Se = 79,4	Te = 128?	
	F = 19	Cl = 35,5	Br = 80	J = 127	
Li = 7	Na = 23	K = 39	Rb = 85,4	Cs = 133	Tl = 204
		Ca = 40	Sr = 87,6	Ba = 137	Pb = 207
		? = 45	Ce = 92		
		?Er = 56	La = 94		
		?Yt = 60	Di = 95		
		?In = 75,6	Th = 118?		

1. Die nach der Grösse des Atomgewichts geordneten Elemente zeigen eine stufenweise Abänderung in den Eigenschaften.

2. Chemisch-analoge Elemente haben entweder übereinstimmende Atomgewichte (Pt, Ir, Os), oder letztere nehmen gleichviel zu (K, Rb, Cs).

3. Das Anordnen nach den Atomgewichten entspricht der *Werthigkeit* der Elemente und bis zu einem gewissen Grade der Verschiedenheit im chemischen Verhalten, z. B. Li, Be, B, C, N, O, F.

4. Die in der Natur verbreitetsten Elemente haben *kleine* Atomgewichte

Fig. 11.2 Mendeleev's periodic table of 1869, from [35]. *Source* Image from a private collection

Tabelle II.

Reihen	Gruppe I. — R^2O	Gruppe II. — RO	Gruppe III. — R^2O^3	Gruppe IV. RH^4 RO^2	Gruppe V. RH^3 R^2O^5	Gruppe VI. RH^2 RO^3	Gruppe VII. RH R^2O^7	Gruppe VIII. — RO^4
1	H = 1							
2	Li = 7	Be = 9,4	B = 11	C = 12	N = 14	O = 16	F = 19	
3	Na = 23	Mg = 24	Al = 27,3	Si = 28	P = 31	S = 32	Cl = 35,5	
4	K = 39	Ca = 40	— = 44	Ti = 48	V = 51	Cr = 52	Mn = 55	Fe = 56, Co = 59, Ni = 59, Cu = 63.
5	(Cu = 63)	Zn = 65	— = 68	— = 72	As = 75	Se = 78	Br = 80	
6	Rb = 85	Sr = 87	?Yt = 88	Zr = 90	Nb = 94	Mo = 96	— = 100	Ru = 104, Rh = 104, Pd = 106, Ag = 108.
7	(Ag = 108)	Cd = 112	In = 113	Sn = 118	Sb = 122	Te = 125	J = 127	
8	Cs = 133	Ba = 137	?Di = 138	?Ce = 140	—	—	—	— — — —
9	(—)	—	—	—	—	—	—	
10	—	—	?Er = 178	?La = 180	Ta = 182	W = 184	—	Os = 195, Ir = 197, Pt = 198, Au = 199.
11	(Au = 199)	Hg = 200	Tl = 204	Pb = 207	Bi = 208	—	—	
12	—	—	—	Th = 231	—	U = 240	—	— — — —

Fig. 11.3 Mendeleev's periodic table of 1872, from [37]. *Source* Image from a private collection

Over the next two years, Mendeleev realized that these masses were anomalous and revised his opinion, determining the specific heat of cerium and applying Dulong & Petit's Law to show that its atomic weight was around 140 [36]. In the next few years others were to obtain similar values for lanthanum and "didymium." Mendeleev now proposed that the rare earth metals formed oxides M_2O_3, and that cerium formed a "higher oxide" CeO_2, in addition to Ce_2O_3 (Fig. 11.3). He had thus corrected the anomaly of 1869 [37]. He did believe that lanthanum was tetravalent, with an atomic weight around 180.

The years around 1880 were to see the isolation of nine more lanthanides, plus scandium. These clearly went a long way to fill the gap between barium/cerium and tantalum, and posed questions about their arrangement. Henry Bassett pioneered an arrangement which came very near seeing the lanthanides in a block of their own [38] and the idea was continued by the Czech chemist Bohuslav Brauner (1855–1935), who spent much of his life accurately determining the atomic masses of elements. Brauner proposed in 1902 that the rare earths occupied an "interperiodic group" of their own in the periodic table following on from lanthanum (Fig. 11.4) [39].

Brauner and Cleve also closely examined "didymium," before Auer von Welsbach completed the process of separating it into its constituents [40–42]. It was Brauner—he looked like a fuller-faced version of Colonel Harland Sanders—who described a test for fluorine which involved inhaling the gas and exhaling "white clouds of hydrofluoric acid" (he lived to the age of 79, however).

Hot on the heels of Brauner was Alfred Werner, who in a paper of 1905 was to arrange the lanthanides into a horizontal block of their own [43], between barium and the space left for the as-yet-undiscovered lutetium, the arrangement of the table familiar to us today (Fig. 11.5). It should be noted that Werner also put thorium

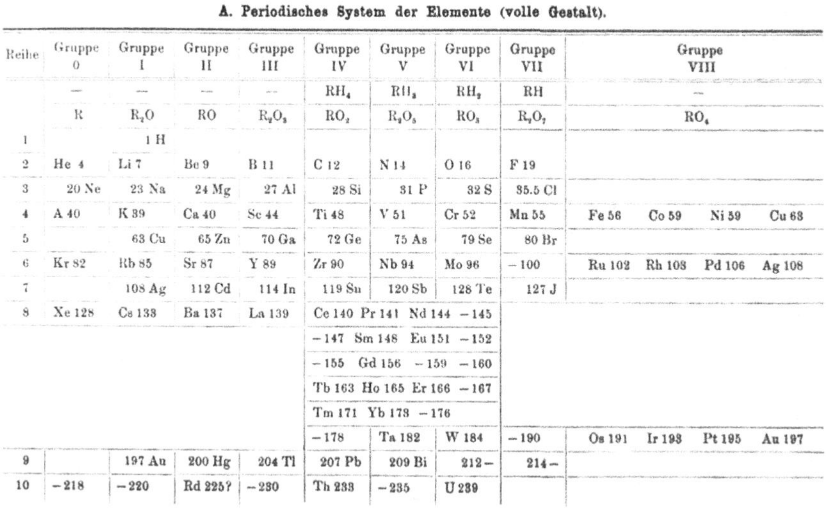

A. Periodisches System der Elemente (volle Gestalt).

Reihe	Gruppe 0	Gruppe I	Gruppe II	Gruppe III	Gruppe IV	Gruppe V	Gruppe VI	Gruppe VII	Gruppe VIII			
	—	—	—	—	RH_4	RH_3	RH_2	RH	—			
	R	R_2O	RO	R_2O_3	RO_2	R_2O_5	RO_3	R_2O_7	RO_4			
1		1 H										
2	He 4	Li 7	Be 9	B 11	C 12	N 14	O 16	F 19				
3	20 Ne	23 Na	24 Mg	27 Al	28 Si	31 P	32 S	35.5 Cl				
4	A 40	K 39	Ca 40	Sc 44	Ti 48	V 51	Cr 52	Mn 55	Fe 56	Co 59	Ni 59	Cu 63
5		63 Cu	65 Zn	70 Ga	72 Ge	75 As	79 Se	80 Br				
6	Kr 82	Rb 85	Sr 87	Y 89	Zr 90	Nb 94	Mo 96	–100	Ru 102	Rh 103	Pd 106	Ag 108
7		108 Ag	112 Cd	114 In	119 Sn	120 Sb	128 Te	127 J				
8	Xe 128	Cs 133	Ba 137	La 139	Ce 140	Pr 141	Nd 144	–145				
					–147	Sm 148	Eu 151	–152				
					–155	Gd 156	–159	–160				
					Tb 163	Ho 165	Er 166	–167				
					Tm 171	Yb 173	–176					
					–178	Ta 182	W 184	–190	Os 191	Ir 193	Pt 195	Au 197
9		197 Au	200 Hg	204 Tl	207 Pb	209 Bi	212–	214–				
10	–218	–220	Rd 225?	–230	Th 233	–235	U 239					

Fig. 11.4 Brauner's periodic table of 1902, from [39]. *Source* Image from a private collection

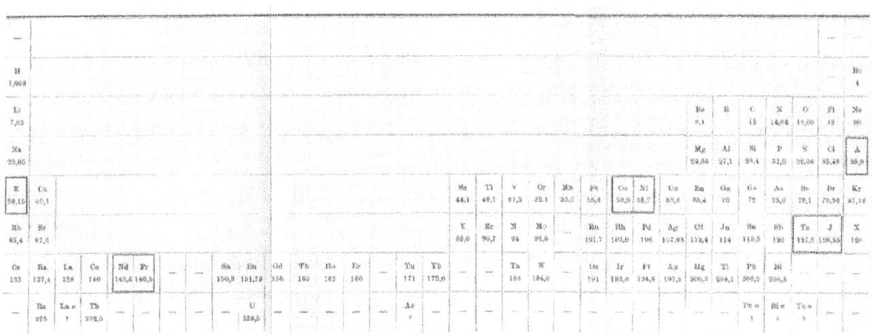

Fig. 11.5 Werner's periodic table of 1905, from [43]. *Source* Image from a private collection

under cerium and uranium (incorrectly) under europium, an approach toward an actinide series which was fully enunciated forty years later by Glenn Seaborg.

11.2.7 Element 61

Lutetium was the last stable lanthanide to be separated and identified (1907), but no one knew that at the time, just as no one knew how many of these elements there were. It was not until the researches of Henry Moseley on the x-ray spectra of metallic elements that this question was answered. He found that there was a linear relationship between the characteristic x-ray frequency (i.e., highest line in K group) and the square of the atomic number, $\nu = a\ (Z{-}b)^2$, i.e., $\sqrt{\nu} \propto Z$ (where ν = frequency; Z = atomic number; a, b are both constants). Thus, if $\sqrt{\nu}$ was plotted versus Z, then each element is a point on the resulting straight line. A gap represented an element yet to be discovered (Fig. 11.6). Gaps were found for elements 43, 61, 72, 75, 87, Moseley commenting: "Vacant lines have been left for … an element between Nd and Sa [samarium] …." [44, 45]. Thus the only rare earth remaining to be found was element 61.

Following this discovery, several groups of researchers studied rare earth ores in the hope of discovering element 61 [46, 47].

At the University of Munich, Prandtl and Grimm examined the x-ray spectra of Ce group rare earths, finding no clear evidence for missing elements [48]. Likewise some years later Ida Noddack also examined spectra of rare earth ores, with similar negative results [49]. One team, led by B. Smith Hopkins at the University of Illinois, fractionated neodymium and samarium salts, finding spectroscopic lines, largely in the infrared and red regions of the spectrum, which they claimed were evidence of element 61. They claimed to have discovered the element, even though they did not isolate it. They also said that their x-ray spectra showed lines in positions expected for element 61 [50, 51]. They named this new element illinium. (Il, after Illinois). Prandtl examined this evidence and said that the absorption bands were probably not due to a new element; the L_α and L_β lines in the x-ray spectrum were due to platinum, barium and bromine. He further said that he could reproduce their spectra by varying the amounts of samarium and neodymium in the mixture [52, 53]. Following the publication of the Illinois group's work, researchers under the direction of Luigi Rolla at the Royal University of Florence reported their examination of monazite ore from Brazil, and said that, based on x-ray and absorption spectra, they had discovered element 61, which they named florentium (Fl). Rolla said that he had made this discovery two years previously, in 1924, depositing his results in a sealed vault at the Accademia dei Lincei [54–57].

Neither of the results from Illinois nor Florence convinced the scientific community, and we now know that none of these discoveries could have been correct, because element 61 has no stable isotopes. Today, this conclusion can be predicted by applying Mattauch's rule [58], namely that two isobars (atoms with the same mass number but different atomic number) with consecutive atomic numbers Z are never stable. Neodymium ($Z = 60$) has stable isotopes of mass numbers 142, 143, 144, 145, 146, 148 and 150, while samarium ($Z = 62$) has stable isotopes with mass numbers 144, 147, 148, 149, 150, 152 and 154. Thus the mass numbers in the range expected for promethium are all "spoken for."

So the search for element 61 moved to the physicists Pool and Quill at Ohio State University, who claimed to have made the isotope ^{144}Il by neutron bombardment of neodymium [59]. The discovery of element 61 was made as a spin-off from the Manhattan project [46]. Identification of the fission products of ^{235}U required their separation, and a team of Coryell, Glendenin and Marinsky in the Clinton laboratories at Oak Ridge, Tennessee, used [60] the brand-new technique of

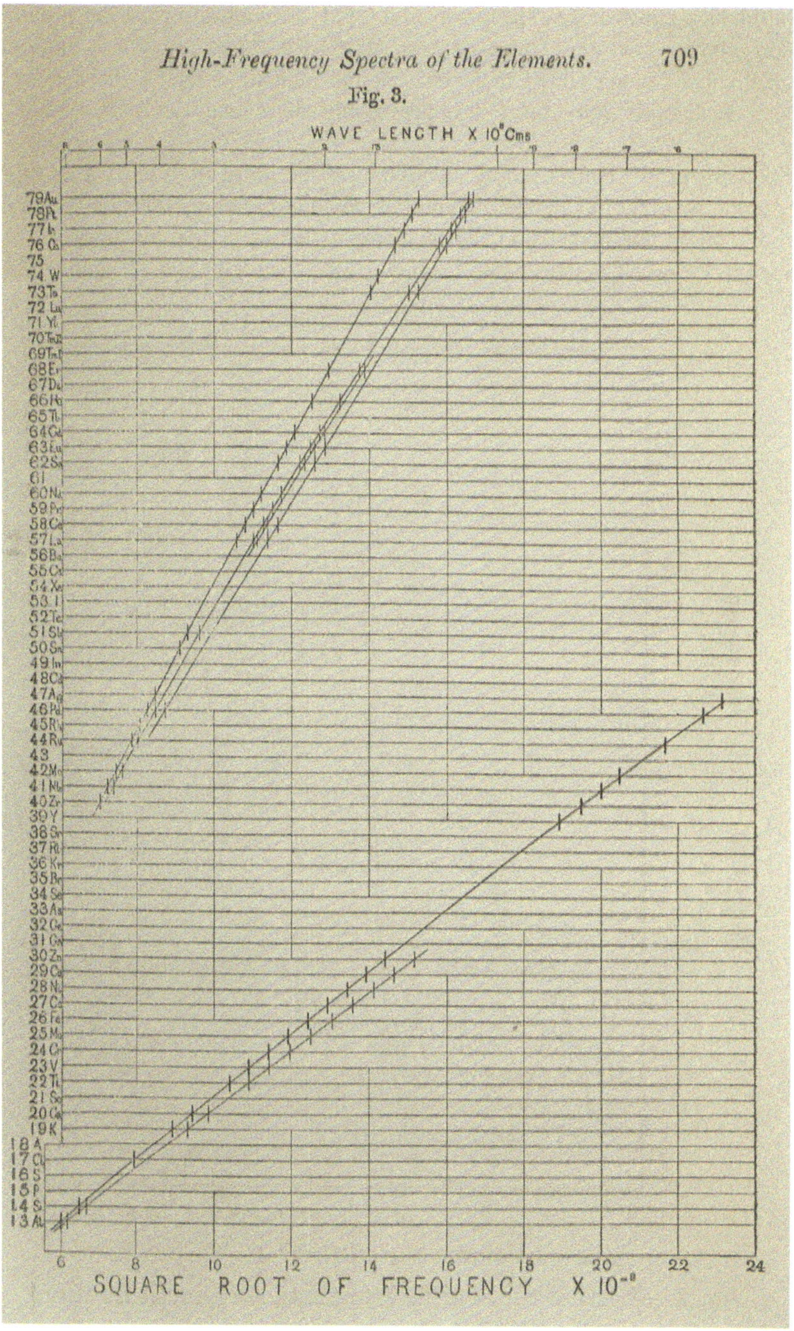

Fig. 11.6 Moseley plot of Atomic Number against $\sqrt{\nu}$, showing a gap for element 61, from [45]. *Source* Image from a private collection

ion-exchange chromatography to separate $^{147}61$. Having adsorbed the mixture of fission products onto the Amberlite resin, they used citrate solution to elute them, the lanthanides being eluted in reverse atomic number order, as the heavier and smaller lanthanide ions form stronger citrate complexes and are eluted first (Fig. 11.7).

So element 61 eluted before neodymium and praseodymium. The original suggestion was that the element would be named clintonium, after the laboratories (foreshadowing the names adopted for some of the actinides) but Coryell's wife, Grace Mary Coryell, suggested prometheum, after the mythological Titan who stole

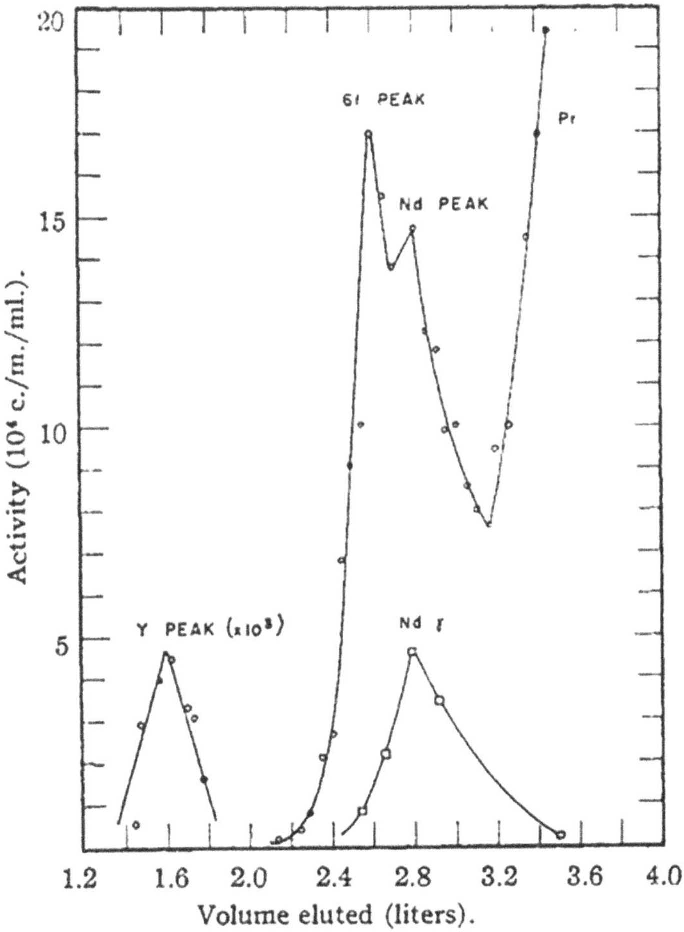

Fig. 2.—Elution curve of intermediate rare earth fraction.

Fig. 11.7 Elution of rare earth fission products from an Amberlite resin using citrate, showing element 61 eluted before neodymium, from [60]

fire from heaven for the use of mankind [61]. (The spelling was changed to promethium by the Commission on Nomenclature of Inorganic Chemistry in 1949, at the same time they changed the spelling of lutecium.) Promethium was discovered in nature—in pitchblende—during the 1960s, but at an extremely low level [62]. A sample of Congo pitchblende was found to contain the minuscule amount of $(4 \pm 1) \times 10^{-15}$ grams of ^{147}Pm per kg of ore, formed by spontaneous fission of ^{238}U.

11.2.8 Separating Lanthanides

But what were the techniques by which these elements were discovered? As Sidgwick remarked: "The history of the discovery and separation of the rare earth

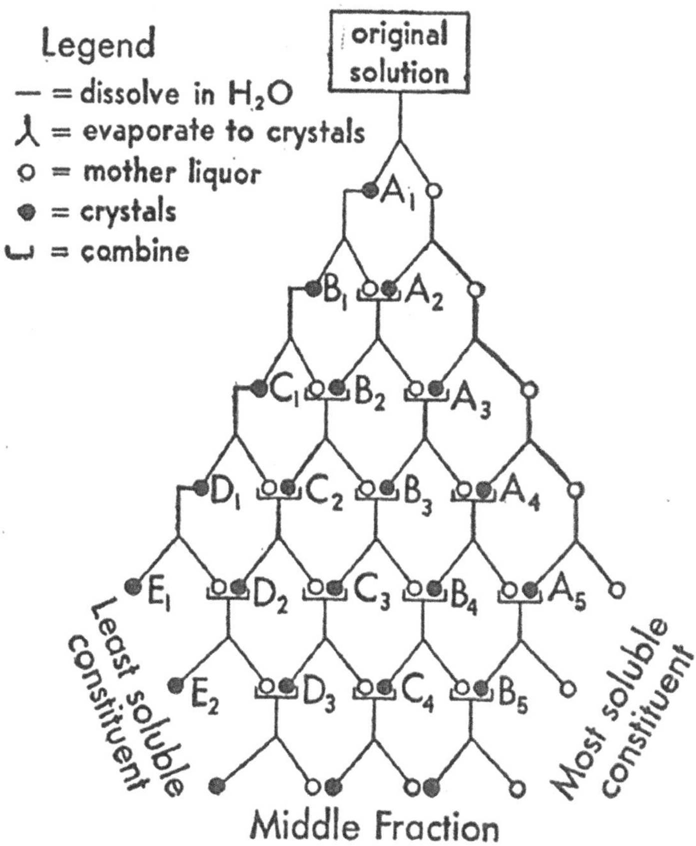

Fig. 11.8 Separation of the rare earths by fractional crystallization, from [67]

elements reflects the greatest credit on the chemists who took part in it. The work is extremely delicate and incredibly tedious" [63].

11.2.8.1 Fractional Crystallization

Classically, the most common separation method used was fractional crystallization of salts, taking advantage of small solubility differences between the compounds of neighboring metals. The solubilities of simple lanthanide(III) compounds can be categorized into three groups.

Soluble: chloride, bromide, iodide, bromate, perchlorate, nitrate and acetate

Slightly soluble: sulfate.

Insoluble: fluoride, hydroxide, carbonate, oxalate, phosphate. Oxalates are quantitatively precipitated, with their thermal decomposition to the oxide Ln_2O_3, used in analysis of lanthanides.

Very many soluble compounds were studied with the aim of developing an efficient method of separation based on crystallization. In other words, when a solution of a mixture of the same salt of a number of lanthanides is crystallized, the mother liquor is enriched in the more soluble salts and the solid is enriched in the less soluble salts.

As has already been noted, James used the bromate salt in the separation of thulium. Double sulfates $Na_2Ln_2(SO_4)_4{\cdot}2H_2O$ were widely used. Possibly the most common substances used for fractional crystallization around 1900 were the ammonium or magnesium double nitrates, $(NH_4)_2Ln(NO_3)_5{\cdot}4H_2O$ and $Mg_3Ln_2(NO_3)_{12}{\cdot}24H_2O$. These salts are easy to synthesize by dissolving the rare earth oxide in dilute nitric acid and adding the calculated amount of the other metal nitrate, then crystallizing [64, 65]. The ammonium salts have particularly been used for separation of lanthanum or praseodymium from neodymium. The double salts have the advantage of a high temperature coefficient of solubility, so that on cooling a boiling solution nearly all the solute crystallizes. The solubility of the double salt increases with increasing atomic number [66].

Fractional crystallization, which could take many stages, was done on the largest possible scale. A concentrated boiling aqueous solution of the salt is cooled rapidly, with crystallization occurring readily. Once the solution is cooled, it is filtered. The crystals are redissolved in fresh solvent and the process repeated; after the new crystals are separated, the mother liquor is used to redissolve crystals resulting from a second crystallization of the mother liquor from the first crystallization process. Then these processes are repeated [67]. This procedure is shown diagrammatically in Fig. 11.8.

11.2.8.2 Redox Chemistry

In exceptional cases, advantage can be taken of the fact that certain lanthanides can adopt unusual oxidation states in aqueous solution.

(a) Cerium. If an oxidizing agent such as alkaline $KMnO_4$ or bleaching powder is in an aqueous solution containing Ce^{3+}(aq) (with the pH above around 3), the cerium is precipitated as $Ce(OH)_4$, with the other lanthanides remaining in solution as Ln^{3+}(aq). The precipitate of $Ce(OH)_4$ can be redissolved in HCl(aq) and the process repeated giving a precipitate of high cerium purity. Manganese contamination can then be removed by a further dissolution in HCl(aq), followed by precipitation of cerium(III) carbonate [66].

Alternative purification processes include dissolving $Ce(OH)_4$ in nitric acid, followed by extraction into tributylphosphate as $Ce(NO_3)_4(TBP)_2$, under conditions where Ln^{3+} ions are not extracted. Another method dissolves $Ce(OH)_4$ in nitric acid, followed by addition of the stoichiometric amount of NH_4NO_3 to form $(NH_4)_2[Ce(NO_3)_6]$. On concentration of the solution by evaporation, $(NH_4)_2[Ce(NO_3)_6]$ crystallizes first, with hardly any contamination by the trivalent lanthanides.

(b) Europium Eu^{3+}(aq) is reduced to Eu^{2+}(aq) by zinc amalgam under a CO_2 atmosphere, while other lanthanides are unaffected. Addition of sulfate ions affords essentially quantitative precipitation of the insoluble $EuSO_4$ (isomorphous with $BaSO_4$), leaving the trivalent lanthanides, whose sulfates are soluble, in solution [66].

11.3 Separations of Lanthanides

High-purity lanthanides—and on a large scale—became available after World War II. The Manhattan project spurred improved separations of lanthanides. Nuclear reactors needed high-purity uranium, which meant removing rare earth impurities capable of absorbing the neutrons used to cause and maintain the fission reaction in ^{235}U. This need led to the development of cation-exchange resins. A very early application of these resins lay in the separation of element 61 from other lanthanides, and in 1947 Frank H. Spedding and his researchers pioneered the use of ion-exchange methods as a way of separating pure lanthanides on a large scale [68].

11.3.1 Separating the Lanthanides by Ion-Exchange

The mixture to be separated is loaded onto the top of the (long) column containing the cation-exchange resin, which absorbs the Ln^{3+} cations. The eluting agent is then gradually added to the column. This complexing agent binds the lanthanide ions, entering into an equilibrium with the column. The lanthanides forming the stronger complexes are preferentially removed, while the more weakly complexed Ln^{3+} tend to be concentrated on the column. The equilibration continues as the eluent goes down the column, with the eluent front becoming more concentrated in the later lanthanides, because they form the stronger complexes; the different metals get spread out, with those of highest atomic number at the front [69, 70].

Originally ammonium citrate was used as eluent, when a complex $[Ln(citrate)_2]^{3-}$ was involved over the pH range 5.0–8.0; subsequently the more strongly complexing EDTA was employed.

Since the advent of solvent extraction processes, ion-exchange methods are used to obtain only a few heavy lanthanides, and that on a small scale, for very high purity.

11.3.2 Separation by Solvent Extraction

Solvent extraction, another spin-off from the Manhattan project, is the principal method used to separate and produce the rare earths. It depends upon the differential partitioning of metal complexes between two phases, immiscible aqueous and organic phases. This method usually involves extracting metal nitrate complexes with a suitable solvent; among the extractants that have been studied are organic phosphates such as tri-*n*-butylphosphate (TBP), phosphoric acid diesters such as bis (2-ethylhexyl)phosphoric acid (HDEHP, $\{C_4H_9CH(C_2H_5)CH_2O\}_2P{=}O(OH)$), carbamoylphosphine oxides $R_2P(O)CH_2C(O)NR_2$) and carboxylic acids such as 2-(dibutylcarbamoyl)benzoic acid.

It was first found by J. C. Warf in 1949 that Ce^{4+} could easily be separated from Ln^{3+} ions by extraction from a solution in nitric acid into tributyl phosphate [71], then the process was extended to a more general separation of lanthanides from the early 1950s, using tributylphosphate in kerosene. It was subsequently found that bis (2-ethylhexyl)phosphoric acid in kerosene was more resistant to hydrolytic decomposition than tributylphosphate, and also a more efficient extractant, which also gave better separation factors for adjacent lanthanides [72]. Organic solvents are used as diluents of the extractants, reducing viscosity. This involves formation of a complex such as $[Ln(NO_3)_3(TBP)_3]$; for the heavy lanthanides, they are more soluble in the aqueous layer. In the case of TBP, the extraction equilibrium can be summarized as

$$Ln^{3+}(aq) + 3\ NO_3^-(aq) + 3\ TBP(org) \rightleftharpoons [Ln(NO_3)_3(TBP)_3]\ (org)$$

Using bis(2-ethylhexyl)phosphoric acid (HL) in kerosene, where the acid is dimeric in this solvent, the extraction reaction can be represented as

$$Ln^{3+} + 3H_2L_2 \rightleftharpoons Ln(HL_2)_3 + 3H^+$$

After the two immiscible solvents have been agitated together and separated, the organic layer is treated with acid and the lanthanide extracted. The solvent is recycled and the aqueous layer is put through further stages.

For a lanthanide Ln_A distributed between two phases, the distribution coefficent D_A is:

$$D_A = [Ln_A \text{ in organic phase}]/[Ln_A \text{ in aqueous phase}]$$

A measure of how good a system is at separating neighboring lanthanides A and B is a "separation factor," β_B^A; the larger is β_B^A, the better the separation.

$$\beta_B^A = D_A/D_B$$

An average separation factor of 1.5 for adjacent lanthanides was reported for the 15-M nitric acid-TBP system, while a factor of 2.5 has been reported using bis (2-ethylhexyl)phosphoric acid. Because the separation is relatively slight, in practice the separation is carried out over 1000 or more cycles, which enables the separation of individual lanthanides with 99.9% purity level [69, 70].

11.4 Stability of Lanthanide Complexes

Separating mixtures of the lanthanides relies upon complex formation. For the reaction between a metal ion and a ligand:

$$M^{n+} + L^{y-} \rightleftharpoons ML^{(n-y)+}$$

a stability constant may be defined approximately

$$\beta_1 = [ML^{(n-y)+}]/[M^{n+}]\,[L^{y-}]$$

In aqueous solution the lanthanides form weak complexes with unidentate ligands such as Cl^-, owing to their being unable to compete with water; Ln^{3+} ions have a large hydration energy, making the complexation process endothermic. They do however prefer to complex with the lighter halogens and O-donor ligands. This preference is characteristic of "hard" acids [73] and is exemplified by the stability constants of their complexes with fluoride and chloride [73]:

$$La^{3+}(aq) + F^-(aq) \rightleftharpoons LaF^{2+}(aq)\text{: } \log \beta = 2.67$$
$$La^{3+}(aq) + Cl^-(aq) \rightleftharpoons LaCl^{2+}(aq)\text{: } \log \beta = -0.11$$

La^{3+} ions will substitute for Ca^{2+} in biological settings, the two ions being of comparable size, but the higher charge and greater charge density of the La^{3+} ions means that their β values are larger.

Initially, separations of the lanthanides by ion-exchange used citrate as the eluting agent. This anion forms more stable complexes than do simpler ligands (thus log β for the yttrium citrate complex is 3.6) but it was soon realized that polydentate amine ligands such as the hexadentate EDTA (EDTA = ethylenediaminetetraacetate) form even more stable complexes (log β for the EDTA complex of yttrium is 18.08) so such ligands came into use for ion-exchange separations.

Values of the stability constant for complexes of the same unidentate ligand tend to increase with increasing atomic number and decreasing size of the lanthanide ion in question; this trend reflects the increased charge density of the metal ion in question and concomitant stronger attraction for the ligand. Thus, for fluoride complexes, log β values increase steadily from 2.67 (La) to 3.61 (Lu) [73]. Values for ligands of higher denticities are greater, partly because of entropy factors, as with the EDTA complexes below, and also on statistical grounds, as once a ligand is attached by one donor atom, there is a greater chance of other donor atoms in that molecule becoming attached too.

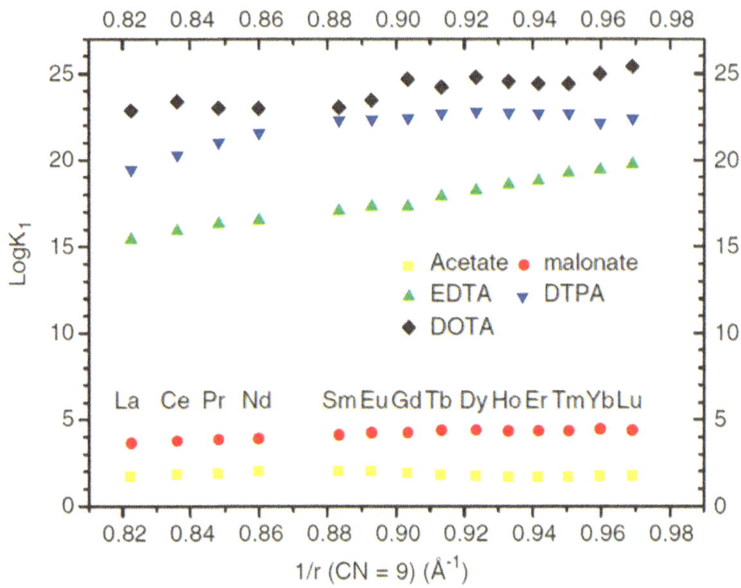

Fig. 11.9 Stability of rare earth complexes of acetate, malonate, EDTA, DTPA and DOTA. From [70]

For these complexation reactions, ΔH has small values, either exothermic or endothermic; the main driving force for complex formation, particularly where multidentate ligands are involved, is the large positive entropy change. Thus [74] for:

$$[La(OH_2)_9]^{3+}(aq) + [EDTA]^{4-}(aq) \rightarrow [La(EDTA)(OH_2)_3]^{-}(aq) + 6H_2O(l)$$
$$\Delta G = -87 \text{ kJ mol}^{-1}; \Delta H = -12 \text{ kJ mol}^{-1}; \Delta S = +125 \text{ J K}^{-1} \text{ mol}^{-1}.$$

The stability constants of the EDTA complexes again increase with increasing atomic number of the lanthanides, log β values rising from 15.46 (La) to 19.80 (Lu), an increase of 4.34, with the value for Y^{3+} of 18.08 being intermediate.

Complexes of the octadentate DTPA (DTPA = diethylenetriaminepentaacetate), as expected, have larger stability constants than do the EDTA complexes, log β values rising from 19.48 (La) to 22.44 (Lu). This synopsis masks the fact that the log β values peak at Dy^{3+} (22.82) thereafter manifesting a steady small decrease (Fig. 11.9). This behavior appears to reflect the inability of the ligand to wrap fully round the smaller metal ions, with resulting steric strain.

Similar effects can be seen in crown ether complexes, where the most stable complexes are formed by the lanthanide ions whose size best matches the cavity in the crown. Likewise, the stability constants of Ln^{3+}-cryptate complexes show little dependence upon the size of the lanthanide [75, 76].

11.5 Obtaining Lanthanides from Their Ores

Despite the difficulties that they faced, nineteenth-century chemists could extract small amounts of a lanthanide from a piece of rock, but it was quite another matter for the chemical industry to obtain these elements in the quantities needed today. Even as late as the 1960s, few significant applications for the rare earths existed—in zeolite catalysts used in petrochemical refining, as polishing agents for lenses. Then in the mid-1960s, color TV sets came on the scene, with Eu^{3+}-based red phosphors. Since then, increasing exploitation of the spectroscopic and magnetic properties of the lanthanides has led to an explosion in demand, whether in smartphones, LEDs or immunoassays, as well as catalytic converters.

Some applications only require small quantities (e.g., gadolinium in MRI); others, such as catalysts, either in car exhausts or in the petrochemical industry—use many thousands of tons. A wind turbine uses over 500 kg and an electric vehicle battery around 10 kg of rare earths. 60% of global consumption of rare earths comes under three headings—catalysts (car exhaust and industrial petrochemical cracking), alloys and magnets. In the year 2008, automobile catalysts accounted for 6840 t (metric tons) of CeO_2, 380 t of La_2O_3 and 228 t of Nd_2O_3, with catalysts for the petrochemical industry accounting for an additional 17800 t of La_2O_3 [77]. Together, lanthanum and cerium represent 63% of rare earth consumption. After these metals, neodymium (and to a lesser extent Pr, Dy and Gd) are important because of increasing use in strong and lightweight magnetic materials in applications such as wind turbines and motors, which account for some 18200 t of Nd_2O_3 consumption in 2008. Overall demand is increasing steadily, at 5% or more a year [78].

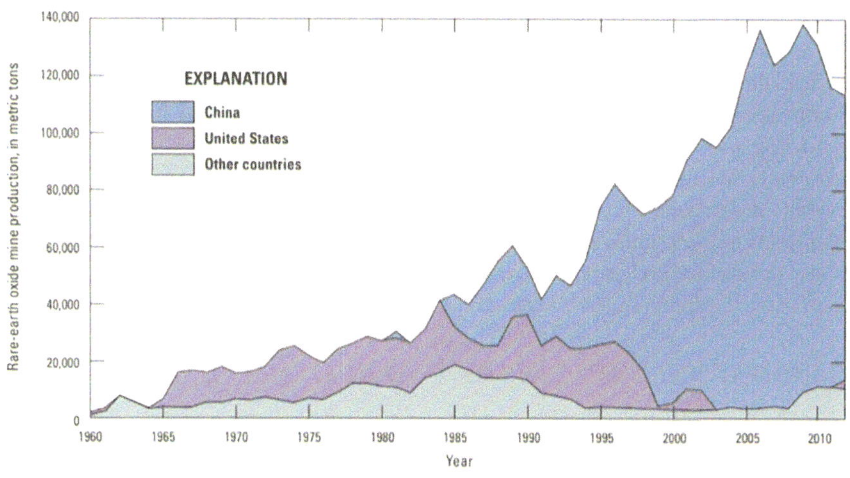

Fig. 11.10 Graph showing world mine production of rare earth oxides, by country and year, from 1960 to 2012

Lanthanide ores typically contain a number of elements, so that increasing production to meet increased demand of one metal also produces increased amounts of some less "popular" elements [79]. The "traditional" rare earth minerals have been the following three ores: bastnasite, $LnFCO_3$ (typically 30% Ce, 20% La, 7% Nd); monazite, $(Ln,Th)PO_4$ (richer in earlier lanthanides, typically 20% La, 40% Ce, 16% Nd and 9% Th) and xenotine $(Ln,Y)PO_4$ (richer in later lanthanides). Monazite processing and extraction presents an additional hazard, owing to the presence of radioactive thorium.

Bastnasite is first concentrated by crushing, grinding and froth flotation, then is treated with 10% HCl to remove calcite, by which time the mixture contains around 70% lanthanide oxides. On roasting, the cerium is oxidized to Ce^{IV}, as CeO_2; on further leaching with HCl, cerium stays as insoluble CeO_2, but the other lanthanides, which are in the (+3) state, dissolve. This solution can then be separated into the individual elements by ion-exchange or solvent extraction. After initial processing—gravity, electrostatic and magnetic separation—monazite is usually treated with NaOH at 150 °C, turning phosphate into soluble Na_3PO_4. The insoluble lanthanide oxides are dissolved in boiling HCl at pH 3.5, leaving behind insoluble ThO_2.

Since the 1980s, Chinese mines have come to dominate the market [80]. The largest of these, with the largest rare earth deposits in the world, is at Bayan Obo in Inner Mongolia. Open-pit mining began there in 1957. The deposits are complicated, with an estimated 1500 million metric tons of iron (around 35%) and 48 million tons of rare earth oxides (ca. 6%) as well as significant amounts of other metals, especially niobium. The rare earths are present as both bastnasite and monazite. At Bayan Obo, rare earth concentrates are heated at 300 °C with sulfuric acid, which results in leaching of the resulting sulfates. The dissolved sulfates are precipitated as double sulfates. These salts are then converted into the hydroxides before dissolution with HCl(aq), the resulting chlorides being purified by solvent extraction.

The unique ion-absorption ores are particularly found in Jianxi province in Southeastern China; the low rare earth content is compensated by their ease of extraction and low radioactivity. These ores are weathered granites with lanthanides adsorbed onto the surface of aluminum silicates. Much of the extraction has been carried out by open-pit mining in cheap unregulated undertakings, with serious resulting environmental damage. The minerals are easy to obtain, a meter or so below the surface and are put into pits then treated with ammonium sulfate solution, displacing the lanthanides by ion-exchange:

$$\left[(\text{Kaolin})^{3-}\text{Ln}^{3+}\right](\text{s}) + 3\text{NH}_4^+(\text{aq}) \rightarrow [(\text{Kaolin})^{3-}(\text{NH}_4^+)_3](\text{s}) + \text{Ln}^{3+}(\text{aq})$$

More recently, the ammonium sulfate solution has been applied directly through holes in the rock, the extract being tapped out through holes drilled at a lower level. The solution of lanthanide ions can be treated with oxalic acid or ammonium carbonate, the precipitated oxalates or carbonates yielding the lanthanide oxides on heating [81].

These processes give a product concentrated in the rare earths, which is ready to be separated into the individual elements.

Figure 11.10 indicates production trends in the latter part of the twentieth century.

Alternative deposits are being sought, not only on land, with attention focused on the western North Pacific Ocean; notably deep-sea mud, with over 5000 ppm rare earth content, was discovered off Minamitorishima Island, Japan, in 2013 [82, 83].

As China became the largest producer of rare earths from the 1980s, resultant swings in price made it difficult for other suppliers to be competitive, resulting in closure of some mines, as it has been difficult to supply increasing quantities without driving prices down. They have been under more environmental scrutiny than their rivals. At the moment little use is made of the significant reserves in the USA, Australia, Brazil and Canada. There has also been a greater tendency for China to move from being purely a miner to being a country which produces lanthanide-containing consumer products, increasing the value-added factor.

11.6 Oxidation States Other Than +3

11.6.1 Introduction

Mendeleev had known from the 1860s that cerium exhibited two oxidation states, and came to see them as (+3) and (+4), so that it formed oxides Ce_2O_3 and CeO_2. Chemistry in the (+4) state was slow to develop for cerium, and not until early 1900s were compounds established in the +2 state for any lanthanides. Chemistry in the (+3) oxidation state was seen to be the norm for the rare earths, witness a celebrated textbook quote (1971): "Lanthanum has only one important oxidation state in aqueous solution, the +3 state. With few exceptions, this tells the whole boring story about the other 14 lanthanides" [84].

Given that most lanthanide/rare earth chemistry is associated with the (+3) oxidation state, what determines the adoption of the (+2) and (+4) states for a particular lanthanide? The (+4) state is particularly associated with $Ce^{4+}(f^0)$; $Tb^{4+}(f^7)$ and $Dy^{4+}(f^8)$, the (+2) state with $Sm^{2+}(f^6)$; $Eu^{2+}(f^7)$ and $Yb^{2+}(f^{14})$, but there is more to it than thinking just of terms of empty, half and fully filled f subshells.

The stability of the (+3) state in a lanthanide halide (for example) is due to the interrelationship of a number of factors, including the ionization enthalpies of the lanthanide and its enthalpy of atomization, the energy of atomization of the halogen and the lattice enthalpy of the halide [85–88].

It is possible to calculate the enthalpy of formation of LnX_2, LnX_3 and LnX_4 for each combination of lanthanide and halogen. Thus, while $\Delta H_f(LaF_3) = -1726$ kJ mol^{-1}, the corresponding calculated values for LaF_2 and LaF_4 are −880 and −600 kJ mol^{-1}, respectively, showing that the di- and tetra-fluorides are

Table 11.1 Enthalpies of disproportionation of LnX_2 (X = F, Cl, Br, I) in kJ mol^{-1}, from [89]

	La	Ce	Pr	Nd	Pm	Sm	Eu	Gd	Tb	Dy	Ho	Er	Tm	Yb
F						47	132							125
Cl				13	9	119	200			33	8		49	161
Br				21	43	153	226			16			10	97
I				11	27	137	229			15			35	146

thermodynamically stable with respect to the *elements*, but unstable with respect to the *trifluoride*. Since they cannot be isolated, they are evidently kinetically unstable too.

The calculated values of ΔH_f for CeF_3, CeF_2 and CeF_4 are−1733,−950 and −1946 kJ mol^{-1}, respectively. Here CeF_2 is considerably less stable than the other two compounds, both of which can be isolated. The reason for CeF_4 being more stable than CeF_3 is mainly found in the smaller value of the fourth ionization potential (I_4) for cerium, 3547 kJ mol^{-1}, compared to 4819 kJ mol^{-1} for lanthanum. In contrast, $CeCl_4$ cannot be isolated. There are two main reasons that $CeCl_4$ is less stable than CeF_4: principally the lattice energy of $CeCl_4$ is considerably less exothermic, owing to the greater size of the chloride ion, and, to a lesser extent, the bond energy in Cl_2 is greater.

A similar examination can be made of the stability of LnX_2 (X = halogen)

$$3\ LnX_2(s) \rightleftharpoons Ln(s) + 2\ LnX_3(s)$$

Values are shown in Table 11.1 for the enthalpies of disproportionation of LnX_2, giving the values only where these are endothermic. Unsurprisingly, these values indicate that the most stable LnX_2 compounds are found for samarium, europium and ytterbium [89].

11.6.2 The (+4) State

The area has been well reviewed [89–91]. The cerium(IV) aqua ion is kinetically stable (though thermodynamically unstable). The reduction potential depends upon the anion present; E = 1.61 V in 1 M HNO_3, 1.44 V in 1 M H_2SO_4 and 1.70 V in 1 M $HClO_4$. The cerium(IV) aqua ion has been suggested to be $[Ce(OH_2)_8]^{4+}$, though ceric ammonium nitrate in strongly acidic nitrate solution appears to be a dinuclear Ce^{IV} complex with a bridging oxo ligand, formulated as $[(H_2O)_xCe^{IV}\text{-O-}Ce^{IV}(OH_2)_x]^{6+}$ (x = 6 or 7) [92, 93].

Cerium(IV) compounds have been known for over a hundred years, with [Ce(acac)$_4$] having been reported in 1897 and salts of the hexa(nitrato)cerate(IV) ion, $M_2[Ce(NO_3)_6]$ (M_2 = $(NH_4)_2$, K_2, Rb_2, Cs_2, Mg, Zn, Ni, Co, Mn) having been synthesized in 1901 [94, 95]. Cerium(IV) ammonium nitrate (CAN) along with cerium(IV) ammonium sulfate (CAS) are particularly well-established reagents in organic synthesis [96]. CeF_4, the only binary cerium(IV) halide, was first reported

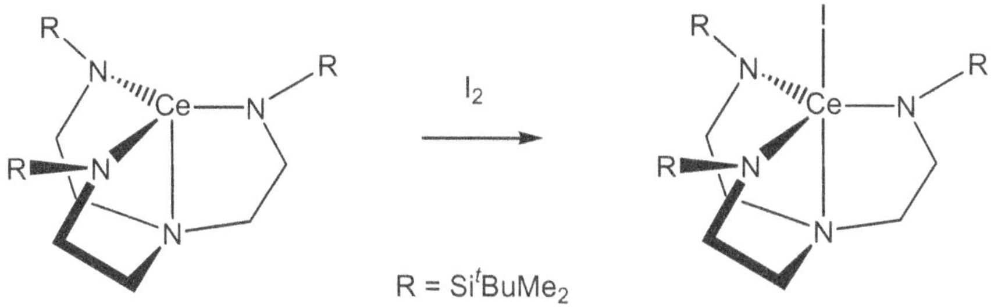

Fig. 11.11 Synthesis of a compound with a Ce^{IV}-I bond

in 1934 [97]. Although $CeCl_4$ does not exist, salts of $[CeCl_6]^{2-}$ can be isolated with large cations (Me_4N, Rb, Cs, but not Li, K).

With the use of non-aqueous solvents, an extensive coordination chemistry of cerium(IV) opened up. Reaction of $[CeX_6]^{2-}$ with O-donor ligands in non-polar solvents leads to complexes such as $[Ce(NO_3)_4(Ph_3PO)_2]$ and $[CeCl_4(R_3PO)_2]$ (R = Ph, NMe_2 etc.) [98]. Even an exceptional fluoride complex, $[CeF_4(Me_2SO)_2]$, has been described [99].

More than that, Don Bradley used non-aqueous solvents to lead the way to an extensive molecular chemistry of Ce^{IV} alkoxides [100], more recently extended into siloxides. These are usually oligomeric [101], with examples such as $[Ce(OCH^t_2Bu)_4]_3$ and $[Ce(O^tBu)_4]_2$. There is an increasing chemistry of Ce^{IV} amides, whether homoleptic, such as $[Ce(N(SiHMe_2)_2)_4]$ [102] or the family $[CeX\{N(SiMe_3)_2\}_3]$ (X = F, Cl, Br, I) [103]. The stability of the Ce^{IV}-I linkage is a surprise, also evinced in the tripodal compound (Fig. 11.11) prepared by I_2-oxidation of a Ce^{III} precursor [104].

The reason for Ce^{IV} being bound to "soft" ligands, also found in the dithiocarbamate $[Ce(S_2CNEt_2)_4]$ [105], is presumably to be found in the sensitivity of the Ce^{4+}/Ce^{3+} redox potential upon the donor atoms in the coordinated ligands, already noted in the case of the aqua species.

Molecular chemistry is being found for other lanthanides in the (+4) state. With tetrafluorides PrF_4 and TbF_4 being well established for praseodymium and terbium, molecular chemistry in this state is starting to be established, given recent reports [106, 107] of siloxides $[Pr(OSiPh_3)_4(MeCN)_2]$ and $[Tb(OSi^tBu_3)_4]$ (like its Ce analogue), as well as $[Tb(NP(1,2\text{-bis-}^tBu\text{-diamidoethane})(NEt_2))_4]$ [108]. There is a limited organometallic chemistry in the (+4) state for cerium [92].

The importance of cerium(IV) is linked with several present-day technological applications, notably the role of CeO_2 as an oxidation catalyst for self-cleaning ovens and automotive catalytic converters, as well as in solid oxide fuel cells.

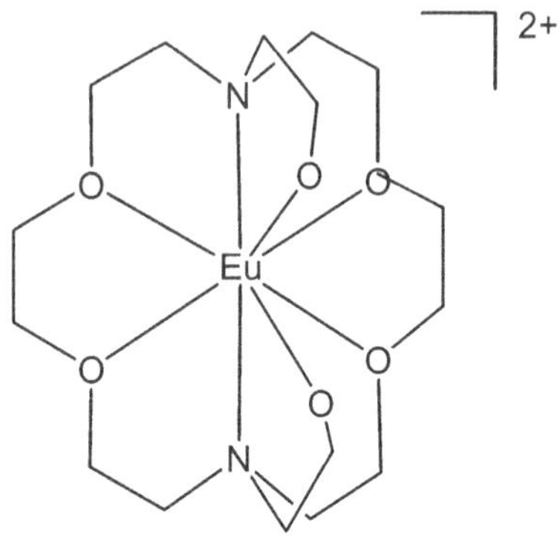

Fig. 11.12 A stable Eu(II)-cryptate complex

11.6.3 The (+2) State

Due to its f^7 electron configuration, Eu^{2+} is the most stable divalent lanthanide with the most extensive chemistry [109]. Beyond the solid state, it tends to oxidize, with $E^0(Eu^{3+}/Eu^{2+}) = -0.35$ V. It is the only lanthanide that exhibits significant aqueous chemistry in the (+2) state. Eu^{2+}(aq) is stable for some hours in the absence of air. EXAFS results indicate that it is very similar in dimensions to Sr^{2+}(aq), with $[Eu(H_2O)_7]^{2+}$ as the principal species, with a small amount of $[Eu(H_2O)_8]^{2+}$ in equilibrium [110]. Cryptates have been used to stabilize europium(II) in aqueous solution (Fig. 11.12), taking advantage of the good fit of the Eu^{2+} ion in the cavity [111].

The first binary halides were made a century ago. $SmCl_2$ was the first dihalide to be made [112], followed by $EuCl_2$ (1911), $YbCl_2$ (1929) and the other LnX_2 compounds (X = F, Cl, Br, I; Ln = Sm, Eu, Yb). At that time, reduction of the heated trihalide with H_2 was the synthetic route:

$$2\ SmCl_3 + H_2 \rightarrow 2\ SmCl_2 + 2\ HCl$$

So for many years, the +2 state was associated with Sm^{2+} (f^6), Eu^{2+} (f^7), Yb^{2+} (f^{14}), ions that have (or nearly have) half or completely filled f-shells.

It was with the adoption of a new synthetic route, metallothermic reduction, that more dihalides became accessible through comproportionation reactions.

$$Nd + 2\ NdCl_3 \rightarrow 3\ NdCl_2\ (950\ °C)$$
$$Tm + 2\ TmI_3 \rightarrow 3\ TmI_2\ (550\ °C)$$

Subsequently this route was extended to other metals including Nd^{2+} (f^4), Dy^{2+} (f^{10}) and Tm^{2+} (f^{13}), enabling the isolation of further dihalides, notably NdI_2 and DyI_2. Certain diiodides (e.g., LaI_2) have properties such as metallic conductivity that are best explained in terms of the presence of the (+3) state, together with free electrons [113].

Owing to the use of non-aqueous solvents, molecular chemistry in the (+2) state expanded greatly in the 1980s, when amido complexes such as $[Eu\{N(SiMe_3)_2\}_2$

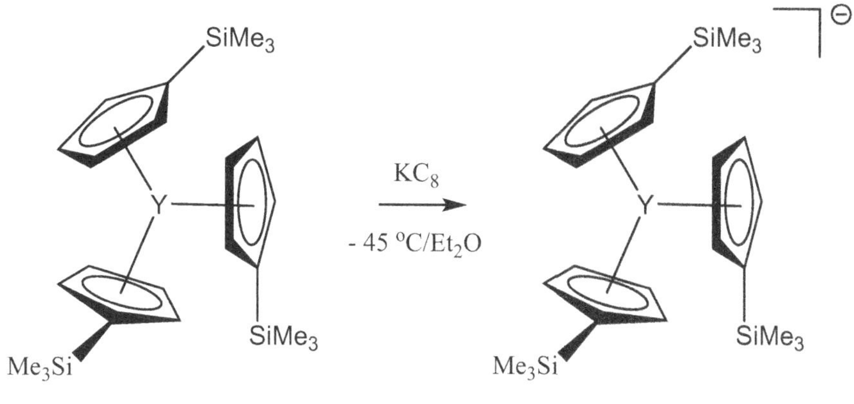

Fig. 11.13 Synthesis of a yttrium(II) tris{(trimethylsilyl)cyclopentadienyl} complex

$(DME)_2]$ and $[Eu\{N(SiMe_3)_2\}_2(bipy)]$ (DME = $MeOCH_2CH_2OMe$; bipy = 2,2′-bipyridyl); $[Sm\{N(SiMe_3)_2\}_2(THF)_2]$ and $[Yb\{N(SiMe_3)_2\}_2\}_2]$ were discovered [114]. These amide ligands are insufficiently bulky to afford linear two-coordinate monomers, but this goal has more lately achieved by the use of bulkier groups in compounds such as $[Ln\{N(Si^iPr_3)_2\}_2]$ (Ln = Eu, Sm, Tm, Yb) [115].

LnI_2 salts (Ln = Tm, Dy, Nd) have been converted into molecular compounds of Tm^{2+} (f^{13}), Dy^{2+} (f^{10}) and Nd^{2+} (f^4), namely seven coordinate $NdI_2(THF)_5$, eight coordinate $DyI_2(DME)_3$ and seven coordinate $TmI_2(DME)_3$ (the coordination numbers of the dimethoxyethane complexes reflecting the smaller radius of the Tm^{II} ion) [116, 117].

The most remarkable development in the (+2) state has taken place in organometallic chemistry. The first discovery, by Lappert's group in 2008, was that metallothermic reduction of Cp''_3La (Cp = cyclopentadienyl, C_5H_5; Cp″ = $C_5Me_3(SiMe_3)_2$−1,3) afforded La^{II} compounds such as [K(2.2.2-cryptand)][Cp''_3La]. A significant point is that, at 2.632 Å, the La-C distance is only slightly longer than that in the Ln^{III} analogue [Cp''_3La], at 2.600 Å. This behavior has been explained by La^{2+} being a $5d^1$ ion and not a $4f^1$ ion [118].

Subsequently the Evans group carried out an extensive investigation in this area, synthesizing a complete family [K(2.2.2-cryptand)][Cp'_3Ln] ($C_5H_4(SiMe_3)$ = Cp′; Ln = Y, La-Lu except Pm) (Fig. 11.13). Some similar compounds have been made for certain metals with other cyclopentadienyl ligands—the choice of the cyclopentadienyl ligand is important, as the corresponding $[Cp''_3Ln]^-$ ion ($C_5H_3(SiMe_3)_2$ = Cp″) has only been obtained for Ln = La, Ce, Pr and Nd [119, 120].

Probably the key factor in isolating these compounds containing several lanthanide ions in the (+2) state is that the ligand set creates a ligand field with a splitting pattern with the $5d_{z2}$ orbital lowest (Fig. 11.14), and of comparable energy to the 4f orbitals, so that the $5d_{z2}$ accommodates one electron in these compounds. For the compounds containing $[LnCp'_3]^-$ ions, the Ln^{2+} ions in these compounds

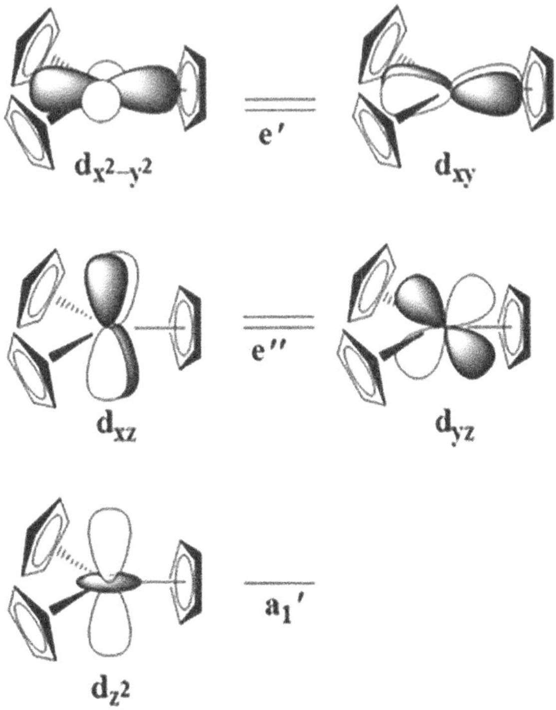

Fig. 11.14 Splitting diagram caused by a trigonal ligand field of three cyclopentadienyl ligands, from [121]

have [Xe]$4f^n$ configurations for Sm^{2+}, Eu^{2+}, Tm^{2+} and Yb^{2+}, and [Xe]$4f^n$ $5d^1$ electron configurations for all the others.

The crystal structures of the Ln^{II}-containing $[LnCp'_3]^-$ ions and the Ln^{III} species $[LnCp'_3]$ show that for the metals Sm, Eu, Tm and Yb, the difference in metal-ring (center) distance between compounds in the two oxidation states is in the range 0.123–0.156 Å, typically what is seen in a comparison of other compounds of these metals, with $4f^n$ electronic configurations. For all the other metals, the difference in distance on changing oxidation state is much smaller, in the range 0.027–0.032 Å, consistent with the addition of an electron to a d orbital (rather than to a f orbital).

Nd^{2+} and Dy^{2+} have $4f^9 5d^1$ and $4f^3 5d^1$ configurations, respectively, in the $[LnCp'_3]^-$ ions, in contrast to the $4f^{10}$ and $4f^4$ configurations they have in previously known compounds of Nd^{2+} and Dy^2 such as the dihalides; remarkably the ligand set determines the electronic ground state of a lanthanide ion in these compounds [120].

11.7 Coordination Numbers in Lanthanide Complexes

Alfred Werner's coordination theory (1893) relied on the adoption of octahedral six coordination in the complexes of many transition metal ions, notably cobalt(III). In the absence of the structural information which today we take for granted, he was able to use the stability and inertness of these complexes to make deductions on the basis of observations such as isomerism in complexes, as well as conductivity behavior, to make structural predictions which later generations were able to verify. A pioneering study in 1921 was able to verify the regular octahedral coordination of platinum in $(NH_4)_2[PtCl_6]$ by x-ray diffraction [121]. At the time that this paper was published, few lanthanide complexes were established, principally the double nitrates, sulfates and oxalates used in separating rare earths, and even by 1953 it was possible to review the field with just 60 citations [122].

11.7.1 High Coordination Numbers

In the early 1960s it was still generally assumed that complexes of yttrium and the lanthanides were generally six coordinate. This assumption ignored two pioneering x-ray diffraction studies reported just before World War II, when tricapped trigonal prismatic 9-coordination was established in the hydrated ethylsulfates $[Ln(H_2O)_9](EtSO_4)_3$ (Ln = Y, La, Ce, Pr, Nd, Sm, Gd, Dy) [123] and in hydrated neodymium bromate, $[Nd(H_2O)_9](BrO_3)_3$, [124].

The late 1950s and 1960s saw two important developments. Inorganic chemistry underwent a renaissance—particularly in coordination chemistry, and this reached the lanthanides. With greatly increased use of non-aqueous solvents, this meant that a bigger range of complexes was being synthesized. Thus Alan Hart's group at Queen Mary College London made two series of complexes $[LnL_2(NO_3)_3]$ (Ln = La-Lu; L = 2,2′-bipyridyl, 1,10-phenanthroline); subsequently it was confirmed that they were all ten coordinate, with bidentate nitrate groups, which showed that this high coordination number could be obtained even with the smallest metals [125, 126].

The second development was the increased use of x-ray diffraction as a structural tool, which was to benefit from the increasing availability and speed of computing, as well as automatic diffractometers.

One important early discovery was that of 12-coordinate $[Ce(NO_3)_6]^{3-}$ units in $Ce_2Mg_3(NO_3)_{12}{\cdot}24H_2O$ in 1963 [127]. The ability of the coordinated nitrate group with its small bite angle to afford high coordination numbers is now well recognized.

However, the first compounds to really challenge the prevailing orthodoxy were the EDTA complexes. Complexes of the potentially hexadentate EDTA (EDTA = ethylenediaminetetraacetate) had become important in the separation of the lanthanides, but it was not until 1965 with the report [128] of the structures of 9-coordinate $K[La(EDTA)(OH_2)_3]{\cdot}5H_2O$ and 10-coordinate $[La(HEDTA)(OH_2)_4]$ that the denticity of the ligand was established, together with the coordination of water molecules [129].

Another significant discovery concerned diketonate complexes. Transition-metal acetylacetonates $[M(acac)_3]$ (M = Ti-Co) were well-established examples of octahedral coordination. While hydrated acetylacetonates $M(acac)_3{\cdot}xH_2O$ (M = La, Gd) had first been reported by Georges Urbain in the nineteenth century [94], it was not until 1967–1968 that $[Y(acac)_3(H_2O)_2]$ was shown to be eight coordinate;

the same coordination number was established in $La(acac)_3{\cdot}3H_2O$, which is $[La(acac)_3(H_2O)_2]{\cdot}H_2O$. This result confirmed the ability of the lanthanides to pick up small solvent molecules to expand their coordination spheres [130].

So by 1970, coordination numbers greater than 6 were seen to be the norm for the lanthanides, receiving further confirmation [131] in structures such as 9-coordinate $[Eu(terpy)_3](ClO_4)_3$ (terpy = 2,2′:6′,2″-terpyridyl). As might be expected, high coordination numbers are associated with borohydride complexes, due to the ability of large numbers of small hydrogens to pack around lanthanide atoms, with 12 coordination found in examples such as $Y(BH_4)_3$ [132]. Even higher numbers are found in certain *N,N*-dimethylaminodiboranates $Ln(H_3BNMe_2BH_3)_3$ (Ln = Pr–Lu) and their thf solvates $Ln(H_3BNMe_2BH_3)_3(thf)$, due to B − H···Ln bridging. The structures of $Ln(H_3BNMe_2BH_3)_3$ show coordination numbers decreasing from 14 (Pr) through 13 (Sm) to 12 (Dy, Y, Er), showing the influence of the decreasing size of the lanthanide ions [133].

11.7.2 Low Coordination Numbers

By 1970, it had been shown that planar three coordination was enforced in an transition metal complex by the use of a bulky amide ligand in $[M(N(SiMe_3)_2)_3]$ (M = Fe), subsequently extended to the whole series for Ti-Co. It was soon found that the whole family of $[Ln(N(SiMe_3)_2)_3]$ (Ln = La–Lu except Pm; Sc, Y), could be synthesized, these all being three coordinate, though unlike the transition metal analogues these have pyramidal coordination geometries [134, 135]. The crystal structure of $[Sm\{N(SiMe_3)_2\}_3]$ shows that one methyl group in each ligand is close to the samarium, calculations indicating β-Si-C agostic interactions (intramolecular interactions between electron-deficient metals and "inert" bonds such as C–H) between the samarium atom and all three alkyl substituents [136, 137]. The corresponding three-coordinate alkyls $[Ln\{CH(SiMe_3)_2\}_3]$ (Ln = Y, La, Pr, Nd, Sm, Er and Lu) likewise have pyramidal structures [138]. Complexes of even bulkier amides, such as $[Ln\{N(Si^tBuMe_2)(SiMe_3)\}_3]$ (Ln = La, Ce) and $[La\{N(Si^tBuMe_2)_2\}_3]$ have been found to have planar three coordination [139]. Very bulky alkyls have been used to make the two coordinate lanthanide(II) compound $[Yb\{C(SiMe_3)_3\}_2]$ and likewise the three-coordinate $[Li(thf)_4][YbR_3]$ (R = $CH(SiMe_3)_2$) [140, 141]. Deviations from linearity are found in the two-coordinate compounds, due to agostic interactions, though near-linear compounds are known, such as the amide $[Sm\{N(Si^iPr_3)_2\}_2]$ [142].

Four coordination, a "low" coordination number in the case of the lanthanides, also requires bulky ligands, such as the 2,6-dimethylphenyl groups in $[Li(THF)_4]^+[Lu(2,6\text{-}Me_2C_6H_3)_4]^-$ and the Yb analogue [143], and the Ph_3PO adducts of lanthanide tris(bis(trimethylsilyl)amide) complexes, $[M(N(SiMe_3)_2)_3(Ph_3PO)]$ (Ln = La, Eu, Lu) [144].

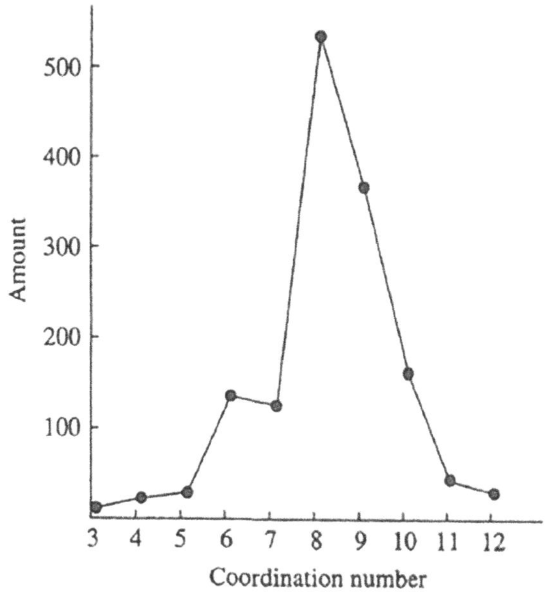

Fig. 11.15 Frequency of coordination numbers in lanthanide complexes (data from 1391 structures, 1935–1995), from [152]

11.7.3 The Role of the Counter-Ion

J. L. Hoard showed that KLa(EDTA)•8H_2O contains nine-coordinate $[La(EDTA)(OH_2)_3]^-$ ions, noting that MLn(EDTA)•8H_2O (M = K, Ln = La, Nd, or Gd; M = Na, Ln = Nd, Tb or Er; M = NH_4, Ln = Nd or Gd) are isomorphous, arguing the persistence of nine-coordinate $[La(EDTA)(OH_2)_3]^-$ ions over a range of ionic radii [129]. A subsequent study showed that later lanthanides could form eight-coordinate $[La(EDTA)(OH_2)_2]^-$ ions and that in some cases, the counter-ion could influence which complex crystallized. Thus $Na[Ho(EDTA)(H_2O)_3]\cdot 5H_2O$ and $K[Ho(EDTA)(H_2O)_3]\cdot 2H_2O$ contain nine coordinate $[Ho(EDTA)(H_2O)_3]^-$ ions but $K[Ho(EDTA)(H_2O)_2]\cdot 3H_2O$ has eight coordinate $[Ho(EDTA)(H_2O)_2]^-$ ions. Solid $[C(NH_2)_3]_2$ $[Er(EDTA)(H_2O)_2]$ $ClO_4{\cdot}6H_2O$ contains eight coordinate Er^{3+}, while $Na[Er(EDTA)(H_2O)_3]\cdot 5H_2O$ contains nine coordinate Er^{3+}.

Several series of salts $[Ln(H_2O)_9]X_3$ (X = CF_3SO_3, BrO_3, $C_2H_5SO_4$), as well as the iodides of the earlier lanthanides $LnX_3{\cdot}9\ H_2O$ (Ln = La–Ho), contain tricapped trigonal prismatic lanthanide ions with coordination numbers of nine [145–148]. However other aqua ions can be isolated. It has long been known that the hydrated lanthanide perchlorates $Ln(ClO_4)_3\cdot 6\ H_2O$ contain octahedral $[Ln(H_2O)_6]^{3+}$ ions in the solid state (Ln, e.g., La, Er, Tb) [149] and other species have been identified. Thus terbium is eight coordinate in $[terpyH_2]_2[Tb(H_2O)_8]_7Cl_7\cdot 8/3\ H_2O$ [150]. Small factors can tip the balance in favor of one coordination number at the expense of another.

11.7.4 Overall Factors

Analysis of 1391 crystal structures determined over the period 1935–1995 (Fig. 11.15) shows that the most common coordination number for lanthanides is 8, followed by 9 [151].

Consideration of the structures of lanthanide complexes indicates that the coordination number is determined by saturation in the coordination sphere, which can occur in two ways, as follows [152, 153]:

1. *"First order effects"*

If ligands of small bulk such as water or a halide bind to a lanthanide, the coordination number is determined by how many donor atoms can pack round the central metal ion, depending upon repulsion between the donor atoms directly in contact with the metal, a so-called "first-order" effect.

2. *"Second order effects"*

If ligands that feature a small donor atom attached to bulky substituents such as bis(trimethylsilyl)amido [-$N(SiMe_3)_2$] and the isolobal alkyl [-$CH(SiMe_3)_2$] bind to a lanthanide, their size restricts how many ligands can bind to the metal through second-order steric effects that generate crowding round the lanthanide.

Species such as $LnCl_6^{3-}$ and $[Ln(H_2O)_9]^{3+}$ are examples in which first-order effects determine the coordination number, while $[Ln\{CH(SiMe_3)_2\}_3]$ and $[Ln\{N(SiMe_3)_2\}_3{\cdot}(Ph_3PO)]$ are cases in which second-order steric effects are generated by the bulky substituents on the amide and alkyl ligands.

The ionic radii of La^{3+}, Y^{3+} and Lu^{3+} ions in octahedral coordination are 1.032 Å, 0.900 Å and 0.861 Å, respectively, in contrast to the value of 0.670 Å for a six coordinate Ti^{3+} ion, the largest 3d metal ion. Steric effects require that the lanthanides can accommodate more than six ligands in their coordination sphere. Because the f-orbitals of the lanthanides are inner orbitals that cannot participate in directional bonding, the absence of strong ligand-field effects familiar with transition metals means that the energetic reasons favoring particular coordination numbers are absent.

Simply, the coordination number adopted by a lanthanide ion in a complex is generally determined by how many ligands can be packed round the central metal ion, so that—in the absence of steric crowding—coordination numbers of 8 or more are common.

11.8 Trends in Lanthanide Radii and the "Lanthanide Contraction"

In the course of his pioneering investigations in crystal chemistry, the Swiss-born Norwegian geochemist Victor Moritz Goldschmidt (1888–1947) made numerous remarkable discoveries; possibly the most significant was his study of the rare earth sesquioxides, Ln_2O_3. He found that the unit cell dimensions of the lanthanide oxides decreased regularly with increasing atomic number of the element, in contradistinction to the trend observed on descending a group—thus the cell dimensions of Sc_2O_3 and Y_2O_3 are 9.79 and 10.60 Å, respectively. This showed that the ionic radii of the lanthanide ions, Ln^{3+}, decrease with increasing atomic number. This research was published in 1925 as the fifth volume of the series *Geochemische Verteilungsgesetze der Elemente* (The laws of the geochemical distribution of the elements) [154].

The cause of the decrease in radii with increasing atomic number is that the 5 s and 5p orbitals penetrate the 4f subshell and are not shielded from increasing nuclear charge; thus as the atomic number increases and the series La-Lu is traversed, the 5 s and 5p orbitals contract because of the increasing nuclear charge so there is a decrease in both the atomic radii and in the radii of the Ln^{3+} ions, this is more marked at the start of the series. As Goldschmidt found for comparable compounds with different lanthanides, the length of the bonds involving the lanthanide becomes shorter as atomic number increases. As the size of the metal ion

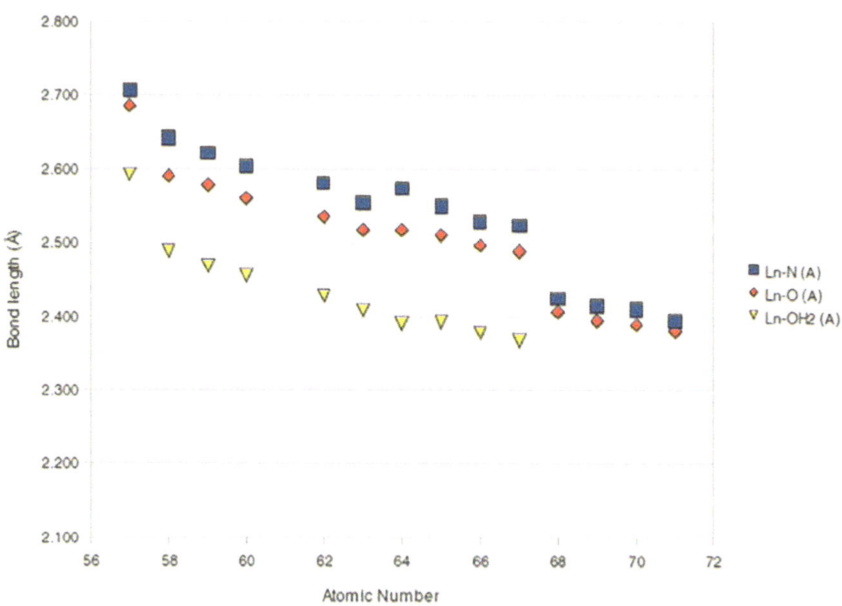

Fig. 11.16 Bond lengths in the family $[Ln(terpy)(NO_3)_3(H_2O)_n]$. From [157]

decreases, it would be expected, on a simple packing model, that fewer anions could be packed round the central metal ion, so that the coordination number will decrease with increasing *Z*, and that similarly the coordination number would decrease with increasing size of the anion.

Examining simple binary compounds first, this behavior is demonstrated in the halides and oxides of the lanthanides. Thus the trifluorides of the early lanthanides (La-Pm) adopt structures in which the metal ion has nine near-neighbor fluorides but with two more a little further away, in what might be termed 9 + 2 coordination. In trifluorides of later lanthanides (Sm–Lu), there are eight near-neighbor fluorides and a ninth a little further away, giving an 8 + 1 coordination pattern. Lanthanum is nine coordinate in $LaCl_3$ but lutetium is six coordinate in $LuCl_3$; in the bromides, $LnBr_3$, nine coordination is found for La-Pr, eight coordination for Nd–Eu and six coordination for Gd–Lu. Looking at the effect of increasing anion size, lanthanum is 11-coordinate in LaF_3, nine coordinate in the chloride and bromide, and eight coordinate in the iodide. In Ln_2O_3 three structures are found, depending upon the lanthanide and the temperature. La_2O_3 to Sm_2O_3 inclusive adopt the A-type structure, with capped octahedral seven coordination of the lanthanide, at room temperature. At higher temperatures they have the B-type structure, with both six and seven coordinate sites. The oxides of heavier lanthanides have the C-type structure, with distorted octahedral six coordination.

In solution, the aqua ions of the early lanthanides are $[Ln(H_2O)_9]^{3+}$, with regular tricapped trigonal prismatic structures, for La-Nd. From samarium onwards, a distortion appears in the three capping waters, with one of the three more strongly bound than the other two, as congestion appears in the coordination sphere, and by holmium it is not possible to maintain full occupancy of the capping positions, and a "water deficit" results with the holmium aqua ion regarded as $[Ho(H_2O)_{8.91}]^{3+}$; the trend continues until at lutetium the aqua ion can be represented as $[Lu(H_2O)_{8.2}]^{3+}$ [155].

As noted in Sect. 11.6, in the solid state the counter-ion can affect the ion that crystallizes. Thus many $[Ln(H_2O)_9]X_3$ (Ln = La-Lu, Y; X e.g., BrO_3, CF_3SO_3, $C_2H_5SO_4$) are known, all of which contain the $[Ln(H_2O)_9]^{3+}$ ion; however, the perchlorates are $Ln(ClO_4)_3.6H_2O$, with octahedral $[Ln(H_2O)_6]^{3+}$ ions. In some compounds with very bulky ligands, particularly some alkoxides, aryloxides and alkylamides, where a small donor atom such as O or N is attached to very bulky substituents, such as bis(trimethylsilyl)amido $[-N(SiMe_3)_2]$, and the isolobal alkyl $[-CH(SiMe_3)_2]$, steric interaction between the substituents determines the coordination number, so that all $[Ln\{N(SiMe_3)_2\}_3]$ (Ln = Y, La-Lu) are all three coordinate—as indeed they are for Sc-Co, U and Pu. When ligands of higher denticity are involved, more factors intrude. Thus $Ln(bipy)_2(NO_3)_3$ (bipy = 2,2′-bipyridyl) and $Ln(phen)_2(NO_3)_3$ (phen = 1,10-phenanthroline) are all 10 coordinate in the solid state, with bidentate nitrates, across the series from La to Lu. When the terdentate ligand 2,2′:6′,2″-terpyridyl (terpy) is added to lanthanide nitrates in CH_3CN solvent, three types of complexes are isolated, depending upon the lanthanide: 11-coordinate $[La(terpy)(NO_3)_3(H_2O)_2]$; 10-coordinate $[Ln(terpy)(NO_3)_3(H_2O)]$ (Ln = Ce–Dy) and 9-coordinate $[Ln(terpy)(NO_3)_3]$ (Ln = Ho–Lu). This trend is predictable, with bond lengths involving the lanthanides in line with expectations

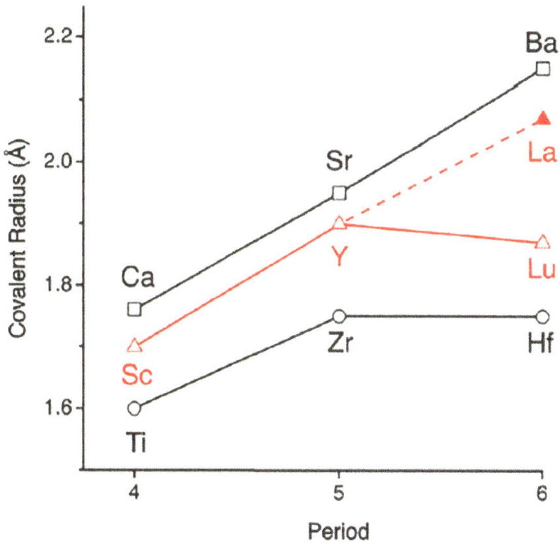

Fig. 11.17 Variation of the covalent radii down groups 2 (squares) and 4 (circles) compared to the group 3 variations when either Lu (empty triangles) or La (dashed line) is included in that group. From [168]

on the basis of the decrease in the lanthanide radii as the atomic number increases (Fig. 11.16) [156, 157].

However, one factor that does emerge is that within each series the coordination of the nitrate groups becomes more asymmetric as the ionic radius of the lanthanide decreases, evidently as a way of responding to increasing congestion in the coordination sphere. The similar terdentate ligand 2,4,6-tri-α-pyridyl-1,3,5-triazine (tptz) forms an isostructural 10-coordinate family $[Ln(tptz)(NO_3)_3(H_2O)]\cdot 2C_2H_5OH$ (Ln = Y, La-Yb). This family also exhibits increased asymmetry in nitrate coordination as Z increases [158].

It should also be remembered, as Goldschmidt pointed out [155], that similar contractions are seen elsewhere, as in families of transition elements, such as the 3d series. They are more clearly manifested in the 4f elements, as the preponderance of the +3 oxidation state means that it is easier to study related families across the series. Goldschmidt applied the term "lanthanide contraction" to the unexpectedly small radii of the 5d elements, which leads to pronounced similarities between zirconium and hafnium, and even later on, with notable resemblances between palladium and platinum, for example. However in the chemical literature today, the term is generally applied to the decrease in size across the 4f series.

11.9 Which Are the Lanthanides?

For over half a century, there was no issue with the arrangement of the elements in Group 3 (IIIA) of the periodic table: scandium above yttrium, yttrium above lanthanum. In the 1960s, physicists began to point out that replacing lanthanum with lutetium gave better periodic trends: the sequence Sc, Y, Lu was favored by a comparison of properties such as the crystal structures, superconductivity and the melting points, as well as the structures of their conductivity bands [159, 160]. In an influential paper, Jensen assembled this and other evidence, also pointing out that scandium and yttrium closely resemble lutetium, rather than lanthanum, in that the structures of their oxides and chlorides (as well as the bromides and iodides) are all six coordinate [161]. It was noted that trends in atomic radii, electronegativity, melting point, and the sum of the first and second ionization energies in the triad Sc, Y and Lu, rather than Sc, Y and La, resemble the succeeding triad Ti, Zr and Hf. The question has subsequently been intensively debated [162–166].

However, comparisons can be odious. As Alvarez has recently pointed out, if we place Lu in Group 3, below Y, and look at the evolution of atomic radii down the group, it can indeed be seen that the trend is very similar to that for the Group 4 transition metals (Ti, Zr and Hf). However, if we include La in Group 3 instead, the variation in atomic radii down the group is identical to that presented by the alkaline earth metals [167] (Fig. 11.17).

The preceding debate has been largely based upon physical properties, but similarities between lutetium and scandium and yttrium should not be overemphasized. The decrease in radius from La^{3+} (radius 1.172 Å for six coordination) to Lu^{3+} (radius 1.001 Å) causes lutetium to be smaller than Y^{3+} (1.040 Å), the latter having virtually the same ionic radius as Ho^{3+} (1.041 Å) and thus very similar chemistry to holmium [168].

So where does Sc^{3+} fit in? Although it has a significantly smaller ionic radius (0.885 Å) than yttrium or even lutetium, it is sometimes compared to Lu^{3+}. While MX_3 (M = Sc, Y, Lu; X = Cl, Br, I) are all six-coordinate, the fluorides differ—

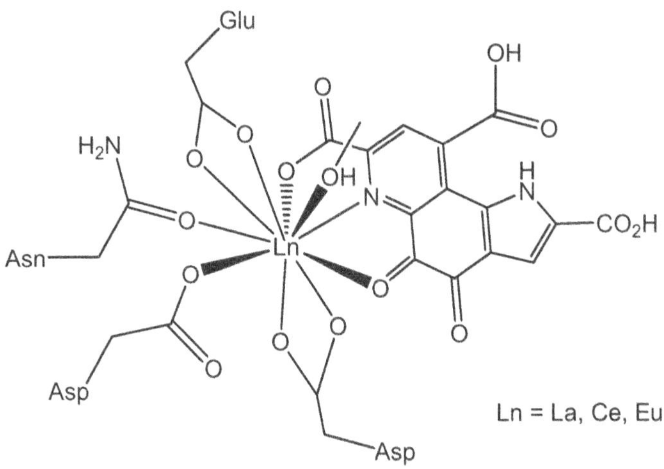

Fig. 11.18 The coordination sphere in the lanthanide-dependent MDH enzymes, with methanol substrate bound to the lanthanide (MDH = methanol dehydrogenase)

ScF_3 has the six coordinate WO_3 structure while LuF_3 adopts the nine coordinate YF_3 structure and LaF_3 is 11-coordinate. An excellent example of the effect of metal ion size upon coordination number is provided by the complexes of 2,2′:6′,2″-terpyridyl (terpy) with the metal chlorides. Scandium exhibits *mer*-octahedral coordination in *mer*-[$ScCl_3$(terpy)], while the greater size of lutetium additionally permits a water molecule to coordinate in seven-coordinate [$LuCl_3$(terpy)(OH_2)]. The even larger lanthanum does not coordinate a second water molecule; instead it has a dimeric structure with two bridging chlorides, attaining eight coordination in [{La(terpy)(OH_2)Cl_2}$_2$(μ-Cl)$_2$]. Clearly, coordination number increases in line with increasing ionic radius of the metal [169].

Sometimes the differences are more subtle. Tribenzyls are formed by scandium and all the lanthanides, all isolated as THF adducts with the formula [M(benzyl)$_3$(THF)$_3$]. While [Sc(benzyl)$_3$(THF)$_3$] and [Lu(benzyl)$_3$(THF)$_3$] are both six-coordinate with sigma-bonded benzyls (η^1), the greater size of lanthanum means that this structure for [La(benzyl)$_3$(THF)$_3$] would be coordinatively unsaturated. Instead, three additional La-C interactions (with La-C distances some 0.3 Å longer than the other La-C bonds) are present that result in the benzyl groups being bound in an η^2 fashion [170, 171]. Scandium and lutetium also form [M(benzyl)$_3$(THF)$_2$] (M = Sc, Lu) complexes with different structures. All the benzyl groups remain η^1 in the five coordinate Sc compound, but one benzyl group has an *ipso*-interaction in the lutetium compound, evidently on account of the slightly greater size of lutetium [172].

A comparison can be made in the change in coordination behavior within some 30 types of complexes formed by Sc, La and Lu with the same ligands in each type. Of the thirty "families," almost half the scandium and lutetium members exhibit different coordination numbers, showing that the significant difference in ionic radius between Sc^{3+} and Lu^{3+} does often result in a significant difference in structure [173].

11.10 Lanthanides in Biological Systems

It was believed until very recently that—unlike many transition metals, alkali metals and alkaline earth metals—lanthanides play no biological role. Their ionic radii are similar to certain metals such as Ca^{2+} and, as a result, lanthanides have often been used as bioprobes; they could substitute for "silent" metal ions such as calcium in proteins, and their spectra, like the luminescence of Eu^{3+}, could give information about the binding site. In 2011, however, a report appeared that a methanol dehydrogenase (MDH) protein XoxF that converts methanol into methanal as an energy source was lanthanide-dependent, and could be expressed when lanthanides were added to a culture medium [174]. It was then discovered that a thermoacidophilic bacterium *Methylacidiphilum fumariolicum SolV*, isolated from a volcanic mudpot in Italy, needed lanthanides to exist. The crystal structure of XoxF from *Methylacidiphilum fumariolicum SolV* revealed that its active site contained Ce^{3+} ions, and subsequently the structures of the La^{3+} and Eu^{3+} analogues were determined [175]. As expected for "hard" lanthanide ions, they are bound to "hard" donor atoms, one nitrogen and eight oxygens. The coordination sphere of the lanthanides (Fig. 11.18) involves four amino acids, as well as the redox cofactor pyrroloquinoline quinone (PQQ), plus an active site that binds the methanol.

The lanthanide environment is very similar to that in Ca^{2+}-dependent MDH enzymes, but because of the greater size of the lanthanide ions there is one more carboxylate ligand, so that the coordination number increases from 7 in Ca-MDH to 9 in the lanthanide-containing XoxF. The role of the lanthanides in XoxF seems to be as a Lewis acid; although the later and smaller lanthanides are stronger Lewis acids than the early metals, these organisms prefer the early lanthanides. This preference may be linked with the higher natural abundance of the early lanthanides La and Ce.

In 2018, a protein was identified in the lanthanide-dependent *Methylobacterium extorquens*, and was given the name Lanmodulin. It has considerable selectivity for lanthanides even in the presence of high concentrations of other ions, outperforming many synthetic chelators [176]. Lanmodulin binds lanthanides at picomolar levels, with 10^8 fold selectivity for La^{3+} over Ca^{2+}. As lanmodulin binds lanthanides, it undergoes a considerable conformational change from a very disordered state to a very compact one [177]. One suggestion is that there may be an uptake system for lanthanides which employs a chelating ligand, in the same way that bacteria use siderophores to take up Fe^{3+}.

No one knows how this area will develop [178, 179]—and who can tell in which direction rare earth chemistry will go next?

Acknowledgements The author is grateful to the editors, especially Professor Gregory Girolami, for a great deal of assistance during the production of this chapter.

References

1. Thyssen P, Binnemans K (2011) Accommodation of the rare earths in the periodic table: A historical analysis. Handb Phys Chem Earths 41:1–93
2. Gadolin J (1794) Undersökning af en svart tung Stenart ifrån Ytterby Stenbrott i Roslagen. Kongl Vetenskaps Acad Nya Handlingar 15:137–155
3. Pyykkö P, Orama O (1996) What did Johan Gadolin actually do? In: Evans CH (ed) Episodes from the history of the rare earths. Kluwer, Dordrecht, pp 1–12
4. Hisinger W, Berzelius J (1804) Cerium, ein neues Metall aus einer schwedischen Steinart. Bastnas Tungstein genannt, Neues allg J Chem 2:397–418
5. Klaproth MH (1804) Chemische Untersuchung des Ochroïts. Neues allg J Chem 2:303–316
6. Klaproth MH (1804, printed in 1807) Analyse chimique de l'ochroïte. Mémoires de l'Académie royale des Sciences et Belles-Lettres: 155–164
7. Vauquelin LN (1804) Essais sur l'ocroïte de M. Klaproth, lus à l'Institut le 12 geminal an 12. Ann Chim 50:140–143
8. Trofast J (1996) The discovery of cerium—A fascinating story. In: Evans CH (ed) Episodes from the history of the rare earths. Kluwer, Dordrecht, pp 13–36

9. Tansjö L (1996) Carl Gustav Mosander and his research on rare earths. In: Evans CH (ed) Episodes from the history of the rare earths. Kluwer, Dordrecht, pp 37–54
10. Mosander CG (1842) Något om Cer och Lanthan. Skand Naturf Forhandl 3:387–398
11. Mosander CG (1844) On the new metals, lanthanium and didymium, which are associated with cerium; and on erbium and terbium, new metals associated with yttria. Report of the thirteenth meeting of the British Association for the Advancement of Science held at Cork in August 1843, pp 25–32
12. Szabadvary F, Evans CH (1996) The fifty years following Mosander. In: Evans CH (ed) Episodes from the history of the rare earths. Kluwer, Dordrecht, pp 55–66
13. Delafontaine MA (1876) Le didyme de la cérite est probablement un mélange de plusieurs corps. CR Hebd Séances Acad 87:634–635
14. Auer von Welsbach C (1885) Die Zerlegung des Didyms in seine Elemente. Monatsh Chem 6:477–491
15. Lecoq de Boisbaudran PE (1879) Nouvelles raies spectrales observées dans les substances extradites de la samarskite. CR Hebd Séances Acad Sci 88:322–324
16. Lecoq de Boisbaudran PE (1879) Recherches sur le samarium, radical d'une terre nouvelle extradite de la samarskite. CR Hebd Séances Acad Sci 89:212–214
17. Demarcay E (1901) Sur un nouvel élément, l'europium. CR Hebd Séances Acad Sci 132:1484–1486
18. Marignac J-C (1880) Sur les terres de la samarskite. Arch Sci Phys Nat 64:97–107
19. Lecoq de Boisbaudran PE (1886) Le Yα der M. de Marignac est définitivement nomme gadolinium. CR Hebd Séances Acad Sci 102:902
20. Marignac MC (1878) Sur l'ytterbine, nouvelle terre contenue dans la gadolinite. CR Hebd Séances Acad Sci 87:578–581
21. Cleve PT (1879) Om tillvaron af tvänne nya grundämnen i erbinjorden. Öfvers Kongl Vetensk-Akad Förh 36(7):11–14
22. Cleve PT (1879) Sur deux nouveaux éléments dans l'erbine. CR Hebd Séances Acad 89:478–481
23. Lecoq de Boisbaudran PE (1886) Sur le dysprosium. CR Hebd Séances Acad 102:1005–1006
24. Nilson LF (1879) Om scandium, en ny jordmetall. Öfvers Kongl Vetensk-Akad Förh 36(3), 47–51; published in translation (1879) Über Scandium, ein Neues Erdmetall. Ber Dtsch Chem Ges 12:554–557
25. Cleve PT (1879) Om Skandium. Öfvers Kongl Vetensk-Akad Förh 36(7):3–10
26. Vickery RC (1960) The chemistry of yttrium and scandium. Pergamon, Oxford, p 4
27. Nilson LF (1880) Om ytterbiums atomvigt. Öfvers Kongl Vetensk-Akad Förh 37(6):45–52
28. Kragh H (1996) Elements 70, 71 and 72: Discoveries and controversies. In: Evans CH (ed) Episodes from the history of the rare earths. Kluwer, Dordrecht, pp 67–89
29. Baumgartner E, E (1996) Carl Auer von Welsbach a pioneer in the industrial application of rare earths. In: Evans CH (ed) Episodes from the history of the rare earths. Kluwer, Dordrecht, pp 113–130
30. Auer von Welsbach C (1907) Die Zerlegung des Ytterbiums in seine Elemente. Sitzungsberichte der mathematisch-naturwissenschaftliche Klasse der kaiserlichen Akademie der Wissenschaften Abteilung IIb 116:1425–1469
31. Urbain G (1907) Un nouvel element: le lutécium, résultant du dédoublement de l'ytterbium de Marignac. CR Hebd Séances Acad 145:759–762
32. Cotton SA, Hart FA (1975) The heavy transition elements. Macmillan, Basingstoke, p 188
33. Emsley J (2001) Nature's building blocks—An A-Z guide to the elements. OUP, Oxford
34. James C (1911) Thulium I. J Am Chem Soc 33:1332–1344
35. Mendeleev D (1869) Ueber die Beziehungen der Eigenschaften zu den Atomgewichten der Elemente. Zeitschrift für Chemie 12:405–406
36. Mendeleev D (1870) Über die Stellung des Ceriums im System der Elemente. Bull Cl Phys-Math Acad Imp Sci St Petersb 16:45–51

37. Mendelejeff D (1872) Die periodische Gesetzmässigkeit der chemischen Elemente. Ann Chem Pharm Suppl 8:133–229
38. Bassett H (1892) A tabular expression of the periodic relations of the elements. Chem News 65(3–4):19
39. Brauner B (1902) Über die Stellung der Elemente der seltenen Erden im periodischen System von Mendelejeff. Z Anorg Allg Chem 32:1–30
40. Cleve P-T (1882) Note préliminaire sur le didyme. CR Hebd Séances Acad 94:1528–1530
41. Brauner B (1882) Sur le didyme. CR Hebd Séances Acad 94:1718–1719
42. Adunka R, Orna MV (2018) Carl Auer von Welsbach: Chemist, inventor, entrepreneur. Springer, Cham
43. Werner A (1905) Beitrag zum Aufbau des periodischen systems. Chem Ber 38:914–921
44. Moseley HGJ (1913) The high-frequency spectra of the elements. Phil Mag 26:1024–1034
45. Moseley HGJ (1914) The high-frequency spectra of the elements. Part II. Phil Mag 27:703–713
46. Fontani M, Costa M, Orna MV (2014) The lost elements: The periodic table's shadow side. OUP, New York
47. Scerri E (2020) The periodic table: Its story and its significance, 2nd edn. OUP, New York
48. Prandtl W, Grimm A (1924) Über die Aufsuchung des Elementes Nr. 61. Z Anorg Allgem Chem 136:283–288
49. Noddack I (1934) Das Periodische System der Elemente und seine Lücken. Angew Chem 47:301–305
50. Harris JA, Hopkins BS (1926) Observations on the rare earths XXIII. Element no. 61. Part one. Concentration and isolation in impure state. J Am Chem Soc 48:1585–1594
51. Harris JA, Yntema LF, Hopkins BS (1926) Observations on the rare earths XXIII. Element No. 61. Part two. X-ray analysis. J Am Chem Soc 48:1594–1598
52. Prandtl W (1926) Auf der Suche nach dem Element Nr. 61. Angew Chem 39:897–898
53. Prandtl W, Grimm A (1926) Auf der Suche nach dem Element Nr. 61 (II). Angew Chem 39:1333
54. Rolla L, Fernandes L (1926) Über das Element der Atomnummer 61. Z Anorg Allg Chem 157:371–381
55. Rolla L, Fernandes L (1927) Über Das Element der Atomnummer 61 (Florentium). Z Anorg Allg Chem 160:190–192
56. Rolla L, Fernandes L (1927) Florentium. Z Anorg Allg Chem 163:40–42
57. Rolla L, Fernandes L (1928) Florentium II. Z Anorg Allg Chem 169:319–320
58. Mattauch J (1934) Zur Systematik der Isotopen. Z Physik 91:361–371
59. Pool ML, Quill LL (1938) Radioactivity induced in the rare earth elements by fast neutrons. Phys Rev 53:437
60. Marinsky JA, Glendenin LE, Coryell CD (1947) The chemical identification of radioisotopes of neodymium and of element 61. J Am Chem Soc 69:2781–2785
61. Marinsky JA (1996) The search for element 61. In: Evans CH (ed) Episodes from the history of the rare earths. Kluwer, Dordrecht, p 101
62. Atreep M, Kuroda PK (1968) Promethium in pitchblende. J Inorg Nucl Chem 30:699–703
63. Sidgwick NV (1950) The chemical elements and their compounds, vol I. OUP, Oxford, p 456
64. Jantsch G, Wigdorow S (1911) Zur Kenntnis der Doppelnitrate der seltenen Erden. 1. Mitteilung. Über die Doppelnitrate der seltenen Erden mit den Alkalielementen. Z Anorg Allgem Chem 69:221–231
65. Jantsch G (1912) Zur Kenntnis der Doppelnitrate der seltenen Erden. II. Mitteilung. Z Anorg Allgem Chem 76:303–323
66. Callow RJ (1966) The rare earth industry. Pergamon, Oxford, pp 41–58
67. Yost DM, Russell H, Garner CS (1947) The rare-earth elements and their compounds. Wiley, New York, p 43

68. Spedding FH, Voigt AF, Gladrow EM, Sleight NR (1947) The separation of rare earths by ion exchange. I. Cerium and yttrium. J Am Chem Soc 69:2777–2781
69. Moeller T (1963) The chemistry of the lanthanides. Reinhold, New York, pp 80–91
70. Bünzli J-CG, McGill I (2018) Rare earths. Ullmann's encyclopedia of industrial chemistry. Wiley-VCH, Weinheim
71. Warf JC (1949) Extraction of cerium(IV) nitrate by butyl phosphate. J Am Chem Soc 71:3257–3258
72. Peppard DF, Mason GW, Maier JL, Driscoll WL (1957) Fractional extraction of the lanthanides as their di-alkyl orthophosphates. J Inorg Nucl Chem 4:334–343
73. Cotton SA (2006) Lanthanide and actinide chemistry. Wiley, Chichester p 36
74. Cotton SA (1991) Lanthanides and actinides. Macmillan, Basingstoke, p 51
75. Martell AE, Smith RM (1974) Critical stability constants, vols. 1–6. Plenum, New York
76. Bünzli J-CG (2014) Lanthanides. Kirk-Othmer encyclopedia of chemical technology. Chichester, Wiley
77. Goonan TG (2011) Rare earth elements-end use and recyclability. U.S. Geological Survey, Reston, VA, p 5
78. Atwood DA (2012) Sustainability of rare earth resources. In: Atwood DA (ed) The rare earth elements: Fundamentals and applications. Wiley, Chichester, pp 21–25
79. Binnemans K, Jones PT, Van Acker K, Blanpain B, Mishra B, Apelian D (2013) Rare-earth economics: The balance problem. JOM 65:846–848
80. Xie Y, Hou Z, Goldfarb RJ, Guo X, Wang L (2016) Rare earth element deposits in China. Rev Econ Geol 18:115–136
81. Yang XJ, Lin A, Li X-L, Wu Y, Zhou W, Chen Z (2013) China's ion-adsorption rare earth resources, mining consequences and preservation. Environ Dev 8:131–136
82. Kato Y, Fujinaga K, Nakamura K, Takaya Y, Kitamura K, Ohta J, Toda R, Nakashima T, Iwamori H (2011) Deep-sea mud in the Pacific Ocean as a potential resource for rare-earth elements. Nat Geosci 4:535–539
83. Takaya Y et al (2018) The tremendous potential of deep-sea mud as a source of rare-earth elements. Sci Rep 8:5763
84. Pimentel GC, Spratley RD (1971) Understanding chemistry. Holden-Day, San Francisco, p 862
85. Kim Y-C, Oishi J (1979) On the valence changes of lanthanide elements in compounds and the enthalpies of formation and stabilities of their dihalides. J Less Common Met 65:199–210
86. Smith DW (1986) A simple empirical analysis of the enthalpies of formation of lanthanide halides and oxides. J Chem Educ 63:228–231
87. Smith DW (1990) Inorganic substances. CUP, Cambridge
88. Platt AWG (2012) Variable valency. In: Atwood DA (ed) The rare earth elements: Fundamentals and applications. Wiley, Chichester, pp 35–42
89. Sroor FMA, Edelmann FT (2012) Tetravalent chemistry: Inorganic. In: Atwood DA (ed) The rare earth elements: Fundamentals and applications. Wiley, Chichester, pp 313–320
90. So Y-M, Leung W-H (2017) Recent advances in the coordination chemistry of cerium(IV) complexes. Coord Chem Rev 340:172–197
91. Anwander R, Dolg M, Edelmann FT (2017) The difficult search for organocerium(IV) compounds. Chem Soc Rev 46:6697–6709
92. Persson I (2010) Hydrated metal ions in aqueous solution: How regular are their structures? Pure Appl Chem 82:1901–1917
93. Demars TJ, Bera MK, Seifert S, Antonio MR, Ellis RJ (2015) Revisiting the solution structure of ceric ammonium nitrate. Angew Chem Int Ed 54:7534–7538
94. Urbain G, Budischovsky E (1897) Recherches sur les sables monazités. CR Hebd Séances Acad 124:618–621
95. Meyer RJ, Jacoby R (1901) Die Doppelnitrate des vierwertigen Ceriums und des Thoriums. Z Anorg Allg Chem 27:359–389

96. Imamoto T (1996) Lanthanides in organic synthesis. Academic Press, London, pp 119–123
97. Klemm W, Henkel P (1934) Messungen an zwei- und vierwertigen Verbindungen der seltenen Erden. V. Certetrafluorid. Z Anorg Allg Chem 220:180–182
98. Caughlan CN, Mazhar-ul-Haque FA, Hart FA, VanNice R (1971) Crystal and molecular structure of tetranitratobis(triphenylphosphine oxide)cerium(IV). Inorg Chem 10:115–122
99. Champion MJD, Levason W, Reid G (2014) Synthesis and structure of $[CeF_4(Me_2SO)_2]$—A rare neutral ligand complex of a lanthanide tetrafluoride. J Fluorine Chem 157:19–21
100. Bradley DC, Chatterjee AK, Wardlaw W (1956) Structural chemistry of the alkoxides. Part VI. Primary alkoxides of quadrivalent cerium and thorium. J Chem Soc: 2260–2264
101. Friedrich J, Schneider D, Bock L, Maichle-Mössmer C, Anwander R (2017) Cerium(IV) neopentoxide complexes. Inorg Chem 56:8114–8127
102. Crozier AR, Bienfait AM, Maichle-Mössmer C, Törnroos KW, Anwander R (2013) A homoleptic tetravalent cerium silylamide. Chem Comm 49:87–89
103. Williams UJ, Carroll PJ, Schelter EJ (2014) Synthesis, bonding, and reactivity of a cerium (IV) fluoride complex. Inorg Chem 53:6338–6345
104. Morton C, Alcock NW, Lees MR, Munslow IJ, Sanders CJ, Scott P (1999) Stabilization of cerium(IV) in the presence of an iodide ligand: Remarkable effects of Lewis acidity on valence state. J Am Chem Soc 121:11255–11256
105. Hitchcock PB, Hulkes AG, Lappert MF, Li Z (2004) Cerium(III) dialkyl dithiocarbamates from $[Ce\{N(SiMe_3)_2\}_3]$ and tetraalkylthiuram disulfides, and $[Ce(\kappa^2\text{-}S_2CNEt_2)_4]$ from the Ce^{III} precursor; Tb^{III} and Nd^{III} analogues. J Chem Soc Dalton Trans 129–136
106. Willauer AR, Palumbo CT, Fadaei-Tirani F, Zivkovic I, Douair I, Maron L, Mazzanti M (2020) Accessing the +IV oxidation state in molecular complexes of praseodymium. J Am Chem Soc 142:5538–5542
107. Palumbo CT, Zivkovic I, Scopelliti R, Mazzanti M (2019) Molecular complex of Tb in the +4 oxidation state. J Am Chem Soc 141:9827–9831
108. Rice NT, Popov IA, Russo DR, Bacsa J, Batista ER, Yang P, Telser J, La Pierre HS (2019) Design, isolation, and spectroscopic analysis of a tetravalent terbium complex. J Am Chem Soc 141:13222–13233
109. Garcia J, Allen MJ (2012) Developments in the coordination chemistry of europium(II). Eur J Inorg Chem: 4550–4563
110. Moreau G, Helm L, Purans J, Merbach AE (2002) Structural investigation of the aqueous Eu^{2+} ion: comparison with Sr^{2+} using the XAFS technique. J Phys Chem A 106:3034–3043
111. Garcia J, Kuda-Wedagedara ANW, Allen MJ (2012) Physical properties of Eu^{2+}-containing cryptates as contrast agents for ultrahigh-field magnetic resonance imaging. Eur J Inorg Chem: 2135–2140
112. Matignon C, Cazes EC (1906) Le chlorure samareux. Ann Chim Phys 8:417–426
113. Meyer G (2012) The divalent state in solid rare earth metal halides. In: Atwood DA (ed) The rare earth elements: Fundamentals and applications. Wiley, Chichester, pp 161–173
114. Tilley TD, Zalkin A, Andersen RA, Templeton DH (1981) Divalent lanthanide chemistry. Preparation of some four- and six-coordinate bis[(trimethylsilyl)amido] complexes of europium(II). Crystal structure of bis[bis(trimethylsilyl)amido]bis(1,2-dimethoxyethane) europium(II). Inorg Chem 20:551–554
115. Ortu F, Mills DP (2019) Low coordinate rare earth and actinide complexes. In: Bünzli J-C G, Pecharsky VK (eds) Handbook on the physics and chemistry of rare earths 55:1–87
116. Evans WJ, Allen NT, Ziller JW (2000) The availability of dysprosium diiodide as a powerful reducing agent in organic synthesis: Reactivity studies and structural analysis of $DyI_2((DME)_3$ and its naphthalene reduction product. J Am Chem Soc 122:11749–11750
117. Bochkarev MN (2004) Molecular compounds of “new” divalent lanthanides. Coord Chem Rev 248:835–851
118. Hitchcock PB, Lappert MF, Maron L, Protchenko AV (2008) Lanthanum does form stable molecular compounds in the +2 oxidation state. Angew Chem Int Ed 47:1488–1491

119. MacDonald MR, Ziller JW, Evans WJ (2011) Synthesis of a crystalline molecular complex of Y^{2+}, $[(18\text{-crown-6})K][(C_5H_4SiMe_3)_3Y]$. J Am Chem Soc 133:15914–15917
120. Evans WJ (2016) Tutorial on the role of cyclopentadienyl ligands in the discovery of molecular complexes of the rare-earth and actinide metals in new oxidation states. Organometallics 35:3088–3100
121. Wyckoff RWG, Posnjak E (1921) The crystal structure of ammonium chloroplatinate. J Am Chem Soc 43:2292–2309
122. Moeller T (1967) Coordination chemistry of the lanthanide elements—One hundred years of development and understanding. In: Kaufmann GB (ed) Werner centennial. Adv Chem Ser 62:306–317
123. Ketelaar JAA (1937) The crystal structure of the ethyl sulphates of the rare earths and yttrium. Physica 4:619–630
124. Helmholz L (1939) The Crystal structure of neodymium bromate enneahydrate, Nd(Br O3) 3·9H2O. J Am Chem Soc 61:1544–1550
125. Hart FA, Laming FP (1965) Lanthanide complexes—II: Complexes of 1:10-phenanthroline with lanthanide acetates and nitrates. J Inorg Nucl Chem 27:1605–1610
126. Kepert DL, Semenova LI, Sobolev AN, White AH (1996) Structural systematics of rare earth complexes. IX. Tris(nitrato-O, O′)(bidentate-N, N′)lutetium(III), N, N′-bidentate equals 2,2′-bipyridine or 1,10-phenanthroline. Austral J Chem 49:1005–1008
127. Zalkin A, Forrester JD, Templeton DH (1963) Crystal structure of cerium magnesium nitrate hydrate. J Chem Phys 39:2881–2891
128. Hoard JL, Lee B, Lind MD (1965) On the structure-dependent behavior of ethylenediaminetetraacetato complexes of the rare earth Ln^{3+} ions. J Am Chem Soc 87:1612–1613
129. Janicki R, Mondry A (2014) A new approach to determination of hydration equilibria constants for the case of $[Er(EDTA)(H_2O)_n]^-$ complexes. Phys Chem Chem Phys 16:26823–26831
130. Binnemans K (2005) Rare-earth beta-diketonates. In: Gschneidner KA, Bünzli, J-C G, Pecharsky VK (eds), Handbook on the physics and chemistry of rare earths 35:107–272
131. Frost GH, Hart FA, Heath C, Hursthouse MB (1969) The crystal structure of tris-(2,2′,6′,2″-terpyridyl)europium(III) perchlorate. J Chem Soc Chem Comm 1421–1422
132. Ravnsbæk DB, Filinchuk Y, Černý R, Ley MB, Haase D, Jakobsen HJ, Skibsted J, Jensen TR (2010) Thermal polymorphism and decomposition of $Y(BH_4)_3$. Inorg Chem 49:3801–3809
133. Daly SR, Kim DY, Girolami GS (2012) Lanthanide N, N-dimethylaminodiboranates: Highly volatile precursors for the deposition of lanthanide-containing thin films. Inorg Chem 51:7050–7065
134. Bradley DC, Ghotra JS, Hart FA (1973) Low co-ordination numbers in lanthanide and actinide compounds. Part I. The preparation and characterization of tris{bis(trimethylsilyl)-amido}lanthanides. J Chem Soc Dalton Trans: 1021–1023
135. Ghotra, JS, Hursthouse MB, Welch AJ (1973) Three-co-ordinate scandium(III) and europium(III); crystal and molecular structures of their trishexamethyldisilylamides. J Chem Soc Chem Comm: 669–670
136. Brady ED, Clark DL, Gordon JC, Hay PJ, Keogh DW, Poli R, Scott BL, Watkin JG (2003) Tris(bis(trimethylsilyl)amido)samarium: X-ray structure and DFT study. Inorg Chem 42:6682–6690
137. Perrin L, Maron L, Eisenstein O, Lappert MF (2003) γ Agostic C-H or β agostic Si–C bonds in $La\{CH(SiMe_3)_2\}_3$? A DFT study of the role of the ligand. New J Chem 27:121–127
138. Avent AG, Caro CF, Hitchcock PB, Lappert MF, Li Z, Wei XH (2004) Synthetic and structural experiments on yttrium, cerium and magnesium trimethylsilylmethyls and their reaction products with nitriles; with a note on two cerium β-diketiminates. J Chem Soc Dalton Trans 1567–1577

139. Goodwin CAP, Joslin KC, Lockyer SJ, Formanuik A, Morris GA, Ortu F, Vitorica-Yrezabal IJ, Mills DP (2015) Homoleptic trigonal planar lanthanide complexes stabilized by superbulky silylamide ligands. Organometallics 34:2314–2325
140. Eaborn C, Hitchcock PB, Izod K, Smith JD (1994) A monomeric solvent-free bent lanthanide dialkyl and a lanthanide analog of a grignard reagent. Crystal structures of Yb{C$(SiMe_3)_3\}_2$ and [Yb{C$(SiMe_3)_3$}I.OEt_2]. J Am Chem Soc 116:12071–12072
141. Hitchcock PB, Khvostov AV, Lappert MF (2002) Synthesis and structures of crystalline bis (trimethylsilyl)methanidocomplexes of potassium, calcium and ytterbium. J Organomet Chem 663:263–268
142. Chilton NF, Goodwin CAP, Mills DP, Winpenny REP (2015) The first near-linear bis (amide) f-block complex: A blueprint for a high temperature single molecule magnet. J Chem Soc Chem Comm 51:101–103
143. Cotton SA, Hart FA, Hursthouse MB, Welch AJ (1972) Preparation and molecular structure of a σ-bonded lanthanide phenyl. J Chem Soc Chem Comm 1225–1226
144. Bradley DC, Ghotra JS, Hart FA, Hursthouse MB, Raithby PR (1977) Low co-ordination numbers in lanthanoid and actinoid compounds. Part 2. Syntheses, properties, and crystal and molecular structures of triphenylphosphine oxide and peroxo-derivatives of [bis (trimethylsilyl)-amido]lanthanoids. J Chem Soc Dalton Trans 1166–1172
145. Hubbard CR, Quicksall CO, Jacobsen RA (1974) A neutron-diffraction study of holmium ethylsulfate enneahydrate by the white-radiation method. Acta Crystallogr B 30:2613–2619
146. Chatterjee A, Maslen EN, Watson KJ (1988) The effect of the lanthanoid contraction on the nonaaqualanthanoid(III) tris(trifluoromethanesulfonates). Acta Crystallogr B 44:381–386
147. Kurisaki T, Yamaguchi T, Wakita H (1993) Effect of temperature on the structure of hydrated lanthanide(III) ions in crystals and in solution. J Alloys Compd 192:293–295
148. Lim KC, Skelton BW, White AH (2000) Structural systematics of rare earth complexes. XXII. ('Maximally') hydrated rare earth iodides. Austral J Chem 53:867–873
149. Glaser J, Johanson G (1981) Crystal structures of the isomorphous perchlorate hexahydrates of some trivalent metal ions (M = La, Tb, Er, Tl). Acta Chem Scand 35A:639–644
150. Kepert CJ, Skelton BW, White AH (1994) Structural systematics of rare earth complexes. VII. Crystal structure of bis(2,2′/6′,2-terpyridinium) octaaquaterbium(III) heptachloride hydrate. Austral J Chem 47:391–396
151. Huang C, Bian Z (2010) Introduction. In: Huang C (ed) Rare earth coordination chemistry: Fundamentals and applications. Wiley, Singapore, p 18
152. Feng X-Z, Guo A-L, Xu Y-T, Li X-F, Sun P-N (1987) The packing saturation rule and the packing centre rule: Structural characteristics in lanthanide coordination compounds. Polyhedron 6:1041–1048
153. Marçalo J, Pires de Matos A (1989) A new definition of coordination number and its use in lanthanide and actinide coordination and organometallic chemistry. Polyhedron 8:2431–2437
154. Goldschmidt VM, Barth T, Lunde G (1925) Geochemische Verteilungsgesetze der Elemente. Die Lanthaniden-Kontraktion und ihre Konsequenzen, Skrifter Norske Videnskaps-Akademi i. Oslo, Norway, Part V Isomorphie und Polymorphie der Sesquioxyde
155. D'Angelo P, Spezia R (2012) Hydration of lanthanoids(III) and actinoids(III): An experimental/theoretical saga. Chem Eur J 18:11162–11178
156. Cotton SA (2005) Establishing co-ordination numbers for the lanthanides. Compt Rend Chimie 8:129–145
157. Cotton SA, Raithby PR (2017) Systematics and surprises in lanthanide coordination chemistry. Coord Chem Revs 340:220–231
158. Cotton SA, Franckevicius V, Mahon MF, Ooi LL, Raithby PR, Teat SJ (2006) Structures of 2,4,6-tri-α-pyridyl-1,3,5-triazine complexes of the lanthanoid nitrates: A study in the lanthanoid contraction. Polyhedron 25:1057–1068
159. Hamilton DC (1965) Position of lanthanum in the periodic table. Amer J Phys 33:637–640

160. Merz H, Ulmer K (1967) Position of lanthanum and lutetium in the periodic table. Phys Lett 26A:6–7
161. Jensen WB (1982) The positions of lanthanum (actinium) and lutetium (lawrencium) in the periodic table. J Chem Educ 59:634–636
162. Lavelle L (2008) Lanthanum (La) and actinium (Ac) should remain in the d-block. J Chem Educ 85:1482–1483
163. Jensen WB (2009) Misapplying the periodic law. J Chem Educ 86:1186
164. Lavelle L (2009) Response to "misapplying the periodic law". J Chem Educ 86:1187
165. Scerri ER (2009) Which elements belong in group 3? J Chem Educ 86:1188
166. Jensen WB (2015) The positions of lanthanum (actinium) and lutetium (lawrencium) in the periodic table: An update. Found Chem 17:23–31
167. Alvarez S (2020) The transition from 4f to 5d elements from the structural point of view. Cryst Eng Comm. https://doi.org/10.1039/d0ce00029a
168. Shannon RD (1976) Revised effective ionic radii and systematic studies of interatomic distances in halides and chalcogenides. Acta Crystallogr Sect A 32:751–767
169. Curnock E, Levason W, Light ME, Luthra SK, McRobbie G, Monzittu FM, Reid G, Williams RN (2018) Group 3 metal trihalide complexes with neutral N-donor ligands—exploring their affinity towards fluoride. J Chem Soc, Dalton Trans 47:6059–6068
170. Bambirra S, Meetsma A, Hessen B (2006) Lanthanum tribenzyl complexes as convenient starting materials for organolanthanum chemistry. Organometallics 25:3454–3462
171. Meyer N, Roesky PW, Bambirra S, Meetsma A, Hessen B, Saliu K, Takats J (2008) Synthesis and structures of scandium and lutetium benzyl complexes. Organometallics 27:1501–1505
172. Saliu KO, Takats J, McDonald R (2018) Crystal structure of tribenzylbis (tetrahydrofuran-κ*O*)lutetium(III). Acta Crystallogr E Crystallogr Commun 74:88–90
173. Cotton SA, Harrowfield JM, Raithby PR (2020), manuscript in preparation
174. Hibi Y, Asai K, Arafuka H, Hamajima M, Iwama T, Kawai K (2011) Molecular structure of La^{3+}-induced methanol dehydrogenase-like protein in methylobacterium radiotolerans. J Biosci Bioeng 111:547–549
175. Pol A, Barends TRM, Dietl A, Khadem AF, Eygensteyn J, Jetten MSM, Op den Camp HJM (2014) Rare earth metals are essential for methanotrophic life in volcanic mudpots. Environ Microbiol 16:255–264
176. Cotruvo JA, Featherston EA, Mattocks JA, Ho JV, Larmore TN (2018) Lanmodulin: A highly selective lanthanide-binding protein from a lanthanide-utilizing bacterium. J Am Chem Soc 140:15056–15061
177. Deblonde GJ-P, Mattocks JA, Park DM, Reed DW, Cotruvo JA, Jiao Y (2020) Selective and efficient biomacromolecular extraction of rare-earth elements using lanmodulin. Inorg Chem 59:11855–11867
178. Cotruvo JA (2019) The chemistry of lanthanides in biology: Recent discoveries, emerging principles, and technological applications. ACS Cent Sci 5:1496–1506
179. Daumann LJ (2019) Essential and ubiquitous: The emergence of lanthanide metallobiochemistry. Angew Chem Int Ed 58:12795–12802

The History (and Pre-history) of the Discovery and Chemistry of the Noble Gases

12

Jay A. Labinger

Abstract

The existence of the then-so-called inert gases was discovered over a short period of time in the 1890s; the demonstration that they were in fact not inert took place during an even shorter period of time, in 1962. This paper surveys the key events, as well as some of the earlier work that led up to them, of those two crucial episodes in the history of chemistry.

12.1 Introduction

Mendeleev's periodic table, initially proposed in 1869, was improved and generally accepted over the ensuing quarter-century—a period of "prediction and accommodation" [1] that did *not* include any thought of the noble gases. And why should it have? Predictions made by Mendeleev and others were aimed mainly to fill in a missing single member of a group, such as Ga between Al and In; some less-than-perfectly-regular atomic weight sequences also caused some concern. But there was nothing to suggest that an *entire* group might be missing! Well, not quite *nothing*: a couple of earlier observations—one almost a century earlier—might have started chemists down the right track sooner. As it happened, though, it was a project with a quite different goal that led to the discovery of the first of the "inert" or "noble" gases in the early 1890s. Characterization of the entire group was completed in less than a decade, a remarkably short period for such a far-reaching development.

J. A. Labinger (✉)
Beckman Institute, California Institute of Technology, Pasadena, CA 91125, USA
e-mail: jal@its.caltech.edu

C. J. Giunta et al. (eds.), *150 Years of the Periodic Table*, Perspectives on the History of Chemistry, https://doi.org/10.1007/978-3-030-67910-1_12

The chronology of the chemistry of the noble (*not* inert, therefore) gases is somewhat analogous. After their discovery there was an extended period of unsuccessful—or at best ambiguous—experimentation to establish reactivity, accompanied by argumentation for and against. But when the dam finally broke in 1962, a vast flood of positive findings followed quickly, such that most of the principles of noble gas chemistry were established in fairly short order.

I will focus here on the work preceding and comprising the discovery of the elements, as well as some of the more important studies leading up to the demonstration, in 1962, that they *do* have chemistry. I will not say much about subsequent developments of noble gas chemistry; exhaustive surveys may be found in the *Comprehensive Inorganic Chemistry* series [2, 3].

12.2 The Pre-history of the Discovery

12.2.1 Cavendish's Residual Air

In a paper read to the Royal Society on June 2, 1785, and subsequently published [4], Henry Cavendish described some experiments carried out to test his proposal—based not on his own work but that of others—that the volume decrease seen upon "phlogistication" of ordinary air by sparking was not due to the generation of "fixed air"—CO_2—as had previously been suggested, but rather to the "burning of some inflammable matter in the apparatus." The first such experiment he described was simply to pass a spark through "common air" contained in a small tube in contact with an aqueous solution of litmus; he found, as Priestley had previously observed, that the amount of air diminished, and the water took on a red color, indicating the formation of an acid.

Several additional experiments found that if the gas was "confined" by an alkaline solution—lime-water or (better) "soap-lees" (primarily KOH?)—the "diminution" of air was more pronounced; and the extent of that in turn depended strongly on the nature of the air; "perfectly dephlogisticated air" (O_2) showed no such decrease at all. On trying various proportions, he found that "when five parts of pure dephlogisticated air were mixed with three parts of common air, almost the whole of the air was made to disappear." In an earlier paper he had found that "when nitre is detonated with charcoal, the acid is converted to phlogisticated air" (in modern language, NO_3^- + C gives N_2), whence he concluded that "phlogisticated air is nothing else than nitrous acid united to phlogiston." Accordingly he deduced here that "phlogisticated air ought to be reduced to nitrous acid by being deprived of its phlogiston," and that is what is happening in these experiments: $N_2 + O_2$ gives "nitrous acid" (actually a mixture of oxides of nitrogen). Of course, what he meant by "reduced" was completely different from our present usage!

Being appropriately cautious, however, Cavendish acknowledged that our knowledge of "the nature of the phlogisticated part of our atmosphere" is limited, and hence "though it was reasonable to suppose, that part at least of the phlogisticated air of the atmosphere consists of this acid united to phlogiston, yet it might fairly be doubted whether the whole is of this kind, or whether there are not in reality many different substances confounded together by us under the name of phlogisticated air." To investigate that possibility, he repeated the exhaustive sparking of the 5:3 mixture, then added a little more dephlogisticated air (O_2), continued sparking until no further change could be observed, and added some "liver of sulphur" (a mixture of sulfides, polysulfides, etc. obtained by fusing K_2CO_3 with sulfur) to absorb any remaining O_2, "after which only a small bubble of air remained unabsorbed, which certainly was not more than 1/120 of the bulk of the phlogisticated air let up in the tube; so that if there is any part of the phlogisticated air of our atmosphere which differs from the rest...we may safely conclude, that it is not more than 1/120 part of the whole." He refrained from speculating about what that residue might be, or even whether it was really there, or just represented the limited precision of his measurements. His observation attracted little attention over the next century.

12.2.2 An Extraterrestrial Element?

The solar eclipse of August 1868 provided an opportunity for spectroscopic examination of solar prominences. One observer, Jules Janssen, has been given credit by many for the discovery of helium during the course of this eclipse, but that is untrue [5]. English astronomer Norman Lockyer had ordered a new, powerful instrument, which was not ready in time to use on the eclipse; but in October of the same year he found that a yellow line in the solar spectrum close to, but not identical with, the well-known sodium "D" line was bright enough to be seen even without an eclipse [6, 7]. Initially he thought it was a new feature in the spectrum of hydrogen, too weak to be observed by any concentration of the gas that could be achieved under laboratory conditions; but on further consideration (in collaboration with chemist Edward Frankland) he changed his mind, noting that it did not move in parallel with known hydrogen lines [8]:

> I found that the orange line behaved quite differently...so then we knew that we were not dealing with hydrogen; hence we had to do with an element which we could not get in our laboratories, and therefore I took upon myself the responsibility of coining the word helium....I did not know whether the substance...was a metal like calcium or a gas like hydrogen, but I did know that it behaved like hydrogen and that hydrogen, as Dumas had stated, behaved as a metal.

This proposal was apparently not published anywhere at the time of the observation, but must have circulated informally, since Thomson mentioned it in a footnote to the published version of his presentation to the 1871 meeting of the British Association for the Advancement of Science [9]. Kragh has provided a detailed account of the early history of helium [10].

As with Cavendish's finding, there was no follow-up with respect to possible implications for the periodic table until the 1890s—although someone came close.

Lockyer relates his correspondence with American geochemist William Hillebrand, who in 1888 dissolved a uranium ore called uraninite in sulfuric acid and observed evolution of a gas, which he characterized by spectroscopy and concluded it was nitrogen, although he and a collaborator *did* find some extra unaccountable lines [11]. But Hillebrand comments "The well-known variability in the spectra of some substances...led me to ascribe similar causes for these anomalous appearances, and to reject the suggestion made by one of us in a doubtfully serious spirit, that a new element might be in question" [12]. There is an object lesson here: pay attention to those not-so-serious suggestions!

12.3 The History of the Discovery

12.3.1 Rayleigh's Anomaly

In the late 1880s and early 1890s Lord Rayleigh (né John William Strutt) embarked on a program of measuring gas densities, aimed at a more reliable and precise set of atomic weights. This may *seem* like a confirmation of the common "wisdom" around the turn of the century: that there was not much left to discover in science beyond the next figures after the decimal point. But one commentator suggests that would be unfair: Rayleigh was inspired by "the numerical coincidences that had led much earlier to the formulation of Prout's Law" and hence had a much more fundamental aim [13]. In any case, the eventual outcome of his work amply contributed to the soon-to-be-obvious correction of that drastic misconception. All went smoothly until he came to nitrogen, as he wrote in a letter to *Nature* [14]:

> I am much puzzled by some recent results as to the density of *nitrogen*, and shall be obliged if any of your chemical readers can offer suggestions as to the cause. According to two methods of preparation I obtain quite distinct values. The relative difference, amounting to about 1/1000 part, is small in itself; but it lies entirely outside the errors of experiment, and can only be attributed to a variation in the character of the gas.

The first method involved removing O_2 from air by passing it through a hot copper tube, the "ordinary way." For the other (a suggestion from William Ramsay, about whom we will hear much more shortly), air was bubbled through liquid ammonia, followed by exposure to hot copper to both oxidize the ammonia to N_2 and scavenge the remaining O_2. The latter procedure gave "N_2" that was lighter by one part in a thousand. Rayleigh considered explanations based on contamination: either "the first nitrogen would be too heavy, if it contained residual oxygen;" or "can the ammonia-made nitrogen be too light from the presence of impurity?"—most probably H_2. But additional experiments ruled those out (mostly) to his satisfaction, and he ended by wondering "Is it possible that the difference is independent of impurity, the nitrogen itself being to some extent in a different (dissociated) state?"

In a full paper the following year [15], Rayleigh described his experimental methodology in considerable detail, reporting densities for O_2 and N_2; the latter value was obtained using the method of removing O_2 from air. But he inserted a comment, beginning with "Although the subject is not yet ripe for discussion…" about the discrepancy reported in the *Nature* letter. He further noted that if pure O_2 instead of air were used to oxidize the ammonia, the difference could be as large as 0.5%, and repeated his belief that "everything suggests that the explanation is to be sought in a dissociated state of the nitrogen itself."

By the next year the subject *had* ripened, sufficiently to merit a paper in its own right [16]. There Rayleigh elaborated upon the experiments he had carried out to exclude the possibility of contamination (in either direction). He also carried out the chemical preparation of N_2 by additional methods: reduction of NO or NO_2, and decomposition of NH_4NO_3. All gave the same density as the NH_3-derived preparation, about 1/200 smaller than that from air. He reported two further studies to test his proposal of an alternate state of N_2: exposing both air- and NH_3-derived samples to electric discharge; and storing (what he took to be) the anomalous NH_3-derived version for 8 months. Neither had any detectable effect.

12.3.2 The Discovery of Argon

As mentioned in Rayleigh's 1892 letter in *Nature*, Rayleigh and Ramsay had been corresponding on the problem. Ramsay replied to the call for ideas in the *Nature* letter, at some point calling Rayleigh's attention to the century-old Cavendish result, which he had read about in a text [17]. Having previously found that N_2 reacts with hot magnesium turnings, Ramsay thus repeatedly treated the air-derived N_2, and found a steady increase in the density of the residual gas. At the end he obtained a gas sample of about 1/80 of the original volume, which no longer attacked hot Mg, exhibited a density of 19.086 (relative to $H_2 = 1$), showed no reactivity upon sparking with O_2 or Cl_2, and exhibited novel spectral lines [18]. Ramsay seems to have been open to the possibility of a new element more quickly, as he wrote to Rayleigh in May 1894 (Fig. 12.1) [19]:

> Has it occurred to you that there is room for gaseous elements at the end of the first column of the periodic table? Thus

Li	Be	B	C	N	O	F	XXX
–	–	–	–	–	–	Cl	–
–	–	–	–	–	–	Mn	Fe Co Ni
–	–	–	–	–	–	Br	–
–	–	–	–	–	–	?	Pd Ru Rh

> etc. Such elements should have the density 20 or thereabouts, and 0.8 pc (1/120 about) of the nitrogen in the air could so raise the density of nitrogen that it would stand to pure [chemical] nitrogen in the ratio 230:231.

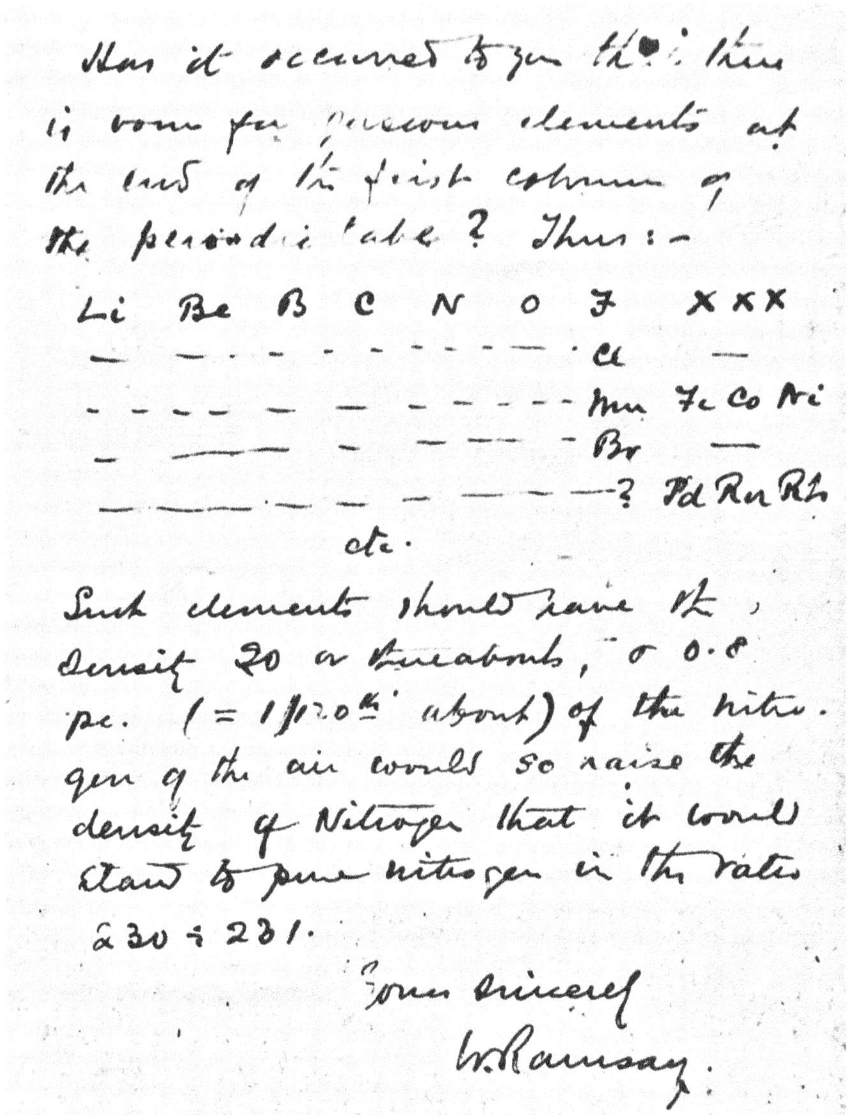
Has it occurred to you that there is room for gaseous elements at the end of the first column of the periodic table? Thus:—

Li Be B C N O F XXX

— — — — — — — Cl —

— — — — — — — Mn Fe Co Ni

— — — — — — Br —

— — — — — — ? Pd Ru Rh

etc.

Such elements should have the density 20 or thereabouts, & 0.8 p.c. (= 1/120th about) of the nitrogen of the air would so raise the density of Nitrogen that it would stand to pure nitrogen in the ratio of 230 : 231.

Yours sincerely

W. Ramsay.

Fig. 12.1 Letter from Ramsay to Rayleigh of 24 May 1894 *Source* Image courtesy of University College London Library Services, Special Collections, Ramsay papers

He was referring to a form of the periodic table in which the halogens lined up with Mn, so that the postulated new elements would lie above the group VIII metals. Of course, the proposal of density 20 implies that they would all (like N, O, and F) be diatomic gases, not metals; but once one accepts the halogens and manganese being placed in the same column, this doesn't seem such an insurmountable obstacle.

By August of 1894 Rayleigh and Ramsay were convinced that they had in fact discovered "a new Gaseous Component of the Atmosphere," as Rayleigh reported (but making it clear these were joint findings) at the annual meeting of the British Association for the Advancement of Science in Oxford. This new species comprised around 1% of air, was more inert than N_2, and exhibited a density between 18.9 and 20 as well as new spectral lines (the latter determined by William Crookes). They refrained from proposing a name for the substance, or even definitively identifying it as an element. Disagreement ensued almost immediately. James Dewar opined (in a letter to the London *Times*: not the sort of forum in which we would expect such a scientific dispute today!) that it could not be an entirely new species, since he was confident that he would have noticed an unsuspected one-per-cent constituent during in his work with liquid air; he felt it must be an allotrope of nitrogen generated by the separation processes used by the two researchers, and suggested they test this possibility by exposing "pure" samples to their conditions [13, 20]. (As we have seen, Rayleigh had already done so.) Note that if this new allotrope were N_3, its density would be 21—not quite within the experimental range reported, but not too far off.

At a special meeting of the Royal Society in January 1895, the full story to date was presented (by Ramsay) [21], and subsequently published at considerable length [22]. It is worth going through the contents in some detail. They summarized the evidence for the discrepant densities, including a new study on N_2 generated by oxidizing urea with NaOBr (which after additional treatment to remove some contaminant "smelling like a dead rat" eventually gave a consistent result). Perhaps in allusion to Rayleigh's original motivation, they commented that the ratio of densities between "chemical nitrogen" and oxygen was almost precisely integral (14.003:16), which was not the case for "atmospheric nitrogen."

They described experiments (mentioned in Rayleigh's 1894 paper) that addressed the possibility (Dewar's theory) that one or more of their procedures could have chemically changed N_2, by exposing "pure" chemical N_2 to the same conditions, and again reported no effect. As a further check, they constructed a device for diffusing air through a series of clay pipes ("atmolysis"), a process that was known to afford a gas mixture enriched in a heavier component but should not cause any chemical change, and found that the gas thus obtained (after removal of O_2) was indeed denser than ordinary atmospheric nitrogen.

From all this they concluded with confidence that there is in fact another chemically inert component of the atmosphere, no artifact of experimental procedures, which they called argon. The name appears not only in the title but *many* times in the body of the paper before they explain the choice, near the very end (it was taken from the Greek α-εργον, meaning no-work or idle).

Then they turned to the isolation of a large-scale sample of pure argon, both for further study and to establish as precisely as possible its concentration in the atmosphere. Using Cavendish's "oxygen method" (Fig. 12.2) proved problematic

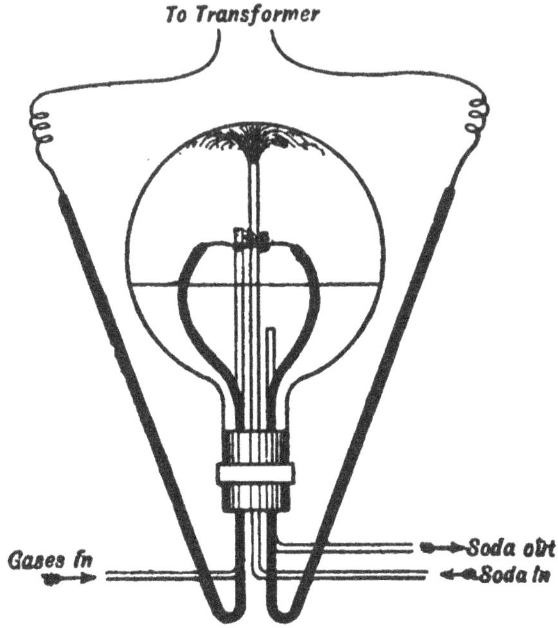

Fig. 12.2 Apparatus for isolating Ar by the sparking method. A mixture of air and O_2 in the proper proportions for complete consumption of N_2 is fed into a bulb equipped with Pt electrodes; the NO_x produced is removed by the continuously circulating solution of caustic soda. Excess O_2 is subsequently removed from the residual gas by passage over hot Cu *Source* Reproduced from [75]

from a quantitative point of view (although they were *highly* complimentary about Cavendish's work). They observed: "In all the large-scale experiments, an attempt was made to keep a reckoning of the air and oxygen employed, in the hope of obtaining data as to the proportional volume of argon in air, but various accidents too often interfered." Their best estimate was that Ar comprises between 0.986 and 1.11% of what had previously been thought to be atmospheric nitrogen; they determined that Ar is fairly soluble in water, and discussed how that would affect the conclusion. But they were able to obtain samples of mostly pure Ar (some N_2 was still present, by spectroscopy) in amounts of 75 cc or more. They did not have enough to completely fill the gas bulb used to determine density, so they used Ar-O_2 mixtures for that purpose, and calculated (after correcting for the residual N_2) a density of 19.7. Using Ramsay's method of removal of N_2 by hot Mg (Fig. 12.3), which was perhaps more reliable and faster but presented its own problems, they *were* able to produce samples large enough to fill the bulb, allowing a direct measurement of density; they took the best value to be 19.88.

These preparations were used to determine spectroscopic properties (a full account was given by Crookes in a paper immediately following [23]); the boiling, freezing, and critical points (also detailed in a companion paper, by Karol Olszewski [24]); and—perhaps most importantly—the specific heat ratio (determined from the velocity of sound), which was virtually exactly that expected for a *monatomic* gas. They attempted to demonstrate reactivity with a wide variety of chemicals, with no success—although they did mention that working with F_2 was too challenging for them, and recommended that experiment be carried out. (Which Moissan did, later the same year, and again found no reaction [25].)

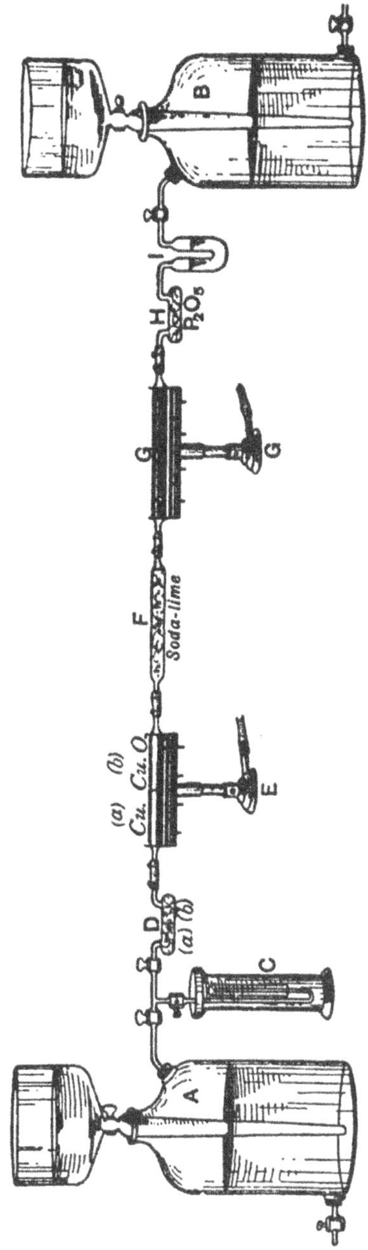

Fig. 12.3 Apparatus for isolating Ar by the Mg method. N_2 (obtained from air by passage over hot Cu) is passed back and forth between reservoirs **A** and **B** over hot Mg in tube **G**. Additional tubes contain hot CuO to oxidize any carbonaceous impurities, a soda-lime mixture (**F** and **I**) to adsorb the resulting CO_2, and P_2O_5 (**D** and **H**) to remove water *Source* Reproduced from [76]

Summing up, they claimed strong evidence that a hitherto unrecognized species is a natural component of the atmosphere, at a level around 1% that of N_2; that it is chemically inert; that it has a density around 20 relative to hydrogen; and that it is monatomic, implying an atomic weight around 40. They did allow for the possibility that what they called argon might still be a mixture of two or more species, but felt that the melting/boiling/critical behavior all argued strongly for a single substance. If so, they argued, "there is reason to doubt whether the periodic classification of elements is complete; whether, in fact, elements may not exist which cannot be fitted among those of which it is composed." If argon was a mixture, the lighter component could come after Cl (and the heavier after Br) as part of the eighth group (as in Fig. 12.1), suggesting that by extension, the series Si-P-S-Cl-? "might be expected to end with an element of monatomic molecules, of no valency, *i.e.*, incapable of forming a compound."

Not surprisingly, the report elicited considerable controversy (a more detailed exposition of which may be found in ref. [26]), both at the meeting itself and shortly thereafter. Much of it focused on the fact that a new element of atomic weight 40 *wouldn't* fit right after Cl, given the then-accepted values of Cl = 35.5, K = 39.1, Ca = 40.1. It was suggested by some that the correlation of atomicity and specific heat ratio might not be all that universal (the only known monatomic gas at the time was vaporized Hg). Perhaps argon could be diatomic, with the absence of chemistry signaling an abnormally strong bond, such that it could effectively behave almost like a spherical molecule? Mendeleev—who was *most* reluctant to accept irregularities of atomic weight progression in his table—speculated on "an inverse correlation between the magnitude of the specific heat ratio and the chemical reactivity of a given gas; the extreme inertness of argon might be responsible for a specific heat ratio higher than would otherwise be expected for a molecule containing two or three atoms." His first choice was that "argon" was actually N_3, as Dewar and others had previously proposed, but he was also open to the idea that it was a new species, diatomic with atomic weight 20, or even hexatomic (!) with atomic weight 6.5 [13, 27].

Later the same year Rayleigh gave a lecture to the Royal Institution, replete with actual demonstrations of some of the experimental procedures. The conclusions were essentially the same as in the full paper; he specifically addressed the N_3 proposal, arguing that it was inconsistent with their observations, as well as being highly unlikely to be a stable molecule; but (generously) ended with "The balance of evidence still seems to be against the supposition that argon is N_3, but for my part I do not wish to dogmatize." [28]. In his full-length book *The Gases of the Atmosphere* which appeared the following year, Ramsay considered the atomic weight issue at length [29]. On the one hand, he argued that this was not necessarily a real problem: "If the numbers in the table actually showed regular intervals… argon might be regarded as of wholly exceptional behaviour. But this is not so…. similar divergences, though not of equal magnitude, are common." He did speculate —we might now say rather wildly—on the possibility that atomic weight or mass might somehow depend on properties such as reactivity. Nonetheless, the book includes a periodic table showing argon to the right of chlorine with an atomic weight of 39.9 [30].

12.3.3 The Discovery of the Others

Still in 1895, Ramsay learned of Hillebrand's observation of gas from uraninite (described above), as well as the fact that a related mineral, cleveite, exhibited similar behavior. He thought it exceedingly unlikely that the gas could be N_2, as Hillebrand had proposed, and repeated the experiment on cleveite. Crookes carried out spectroscopy on the resultant gas, finding "a brilliant yellow line...identical with the line D_3, to which Mr. Lockyer many years ago gave the name 'helium,' from its occurrence in the spectrum of the sun's chromosphere." Ramsay reported the density of the gas to be no higher than 3.9; the specific heat ratio to be consistent with monatomicity (although he was not entirely happy with the quality of the data); and chemical inertness similar to that of argon, which "makes the inference probable that they belong to the same natural group." But he also noted that, assuming the atomic weight of argon to be 40, that of He would be 8 on the same basis; whereas an atomic weight of 4, more consistent with placing it before Li, would suggest that Ar = 20, "a supposition which may be supported by some lines of argument," ending with "Which of these views is correct time must decide." [31]

In a subsequent 1895 paper [32] Ramsay measured the density of gases similarly obtained from a large number of minerals, and got a rather large range of results, from around 2.04 to as high as 3.75. While the last was most consistent with the previous report, he was able to show that the higher numbers were due to contamination. He settled on a lower limit of 2.13, implying an atomic weight of 4.26 (assuming monatomicity), thus obviating the dilemma expressed in the preceding paragraph. The periodic table in his 1896 book includes He above Ar—with a space marked "?" between them, and several more question marks below argon. The atomic weight of He is given as 4.2 [32].

That first question mark was highlighted in a talk Ramsay gave to the British Association for the Advancement of Science meeting in 1897, held in Toronto, with the title "An Undiscovered Gas" [33]. Ramsay must have had something of a sense of humor: he began by asking to be "excused if I take this opportunity of indulging in the dangerous luxury of prophecy....The subject of my remarks to-day is a new gas. I shall describe to you later its curious properties; but it would be unfair not to put you at once in possession of the knowledge of its most remarkable property—it has not yet been discovered." He briefly recounted the history of the periodic table, beginning with Döbereiner's triads and ending with Mendeleev and Lothar Meyer (to whom he gave pretty much equal credit), followed by summarizing his and Rayleigh's work on argon—with the interesting revelation that they had at first thought it "was probably a mixture of three gases, all of which possessed nearly the same atomic weights, like iron, cobalt and nickel. Indeed, their names were suggested, on this supposition, with patriotic bias, as Anglium, Scotium and Hibernium." But the strong evidence for monatomicity, along with the subsequent findings for helium, convinced them to postulate a new periodic group. By analogy to the pervasive appearance of triads in the table, "There should, therefore, be an undiscovered element between helium and argon, with an atomic weight 16 units higher than that of helium, and 20 units lower than that of argon, namely 20....And pushing the analogy still farther, it is to be expected that this element should be as indifferent to union with other elements as the two allied elements." Ramsay went on to describe some attempts to find the new gas—mostly as a component of helium samples—but admitted failure; and he ended with a discussion of the problem of atomic weight irregularities, which he also felt unable to resolve.

All but one of the remaining (naturally occurring) noble gases were discovered by Ramsay and Travers in fairly short order, during the first part of 1898, by cryogenic experiments on liquid air. In June Ramsay read a paper to the Royal Society on their preliminary results: after evaporating all but 10 cc of a 750 cc sample of liquid air, followed by removal of O_2 and N_2 in the usual manner, they obtained a gas that exhibited new spectral lines (in addition to those of argon), was monatomic according to the speed of sound, and had a measured density of 22.5. That they felt must be a minimum value, since there was certainly some Ar still present; they argued for a most likely density around 40 and hence an atomic weight of 80. They proposed to call it "krypton," for "hidden" [34]. Shortly thereafter they reported more careful fractionation studies, which led to a *more* volatile species, again with new spectral lines, and an upper limit on the density estimated to be 14.7. They named it "neon" or "new," and predicted the density would turn out to be 10; thus it would correspond to the "undiscovered gas" predicted by Ramsay in 1897. (He refrained from congratulating himself on that score!). They also noticed that a solid condensed out at low temperatures; on warming it turned into a gas with the same density but *not* the same spectrum as Ar, so they believed it must be a new element which they called "metargon" [35]. The latter was eventually recognized to be due to contamination, although the realization took some time [13].

Xenon ("stranger") was announced later that year [36], and in November 1900 a paper on the full group was presented to the Royal Society [37]. After acknowledging their blunder with metargon, they described the experimental apparatus and procedures used to isolate and characterize each of the gases in considerable detail; the latter studies include extensive pressure-volume-temperature data. At this point their best density values were given as: He, 1.98; Ne, 9.96; Ar, 19.96; Kr, 40.78; Xe, 64.0. From those they showed how the elements are well placed as a new group, following the halogens, with atomic weights, respectively 4, 20, 40, 82, and 128. All this impressive success notwithstanding, they remained rather modest about their accomplishment in their general conclusions, excerpts of which are worth quoting. (It seems a bit ironic that they now give credit for the periodic table to *three* inventors, since their work could be viewed as the final nail in the coffin of Newlands's "Law of Octaves." But surely nobody was taking seriously the idea of an analogy between chemistry and music long before the discovery of an 8th group.)

> The great value of Newland's [*sic*], Mendeleef's, and Lothar Meyer's generalisation, known as the periodic arrangement of the elements, is universally acknowledged. But a study of this arrangement, it must be allowed, is a somewhat tantalising pleasure; for, although the properties of elements do undoubtedly vary qualitatively, and, indeed, show approximate quantitative relations to their position in the periodic table, yet there are

inexplicable deviations from regularity, which hold forth hopes of the discovery of a still more far-reaching generalisation....When we began the search for the elements of which the physical properties are described in the foregoing pages, we were not without a strong hope that their discovery would solve the problem....But our hope has been fruitless. While the same rough quantitative correspondence between the order in the periodic table and the physical properties is manifest, as with other similar series of elements, we have failed to trace any simple mathematical expressions which would make it possible to predict with accuracy the physical properties of any one of these elements, from a knowledge of those of its congeners. It is possible that such expressions exist; we venture to hope that others, more mathematically gifted than we are, may succeed where we have failed.

The sixth member of the noble gases was discovered in a different manner. In 1899, the American physicist Robert Bowie Owens had discovered that thorium salts steadily generated a radioactive material that could be removed from the containing vessel by passing air through it [38]. Three years later, Rutherford and Soddy, reinvestigating the radioactivity of thorium compounds, observed that they "continuously emit into the surrounding atmosphere, under ordinary conditions, something which, whatever its real nature may be, behaves in all respects like a radioactive gas" [39]. They tried passing it through various solutions and over hot solids, and found no evidence for reactivity, commenting "It will be noticed that the only known gases capable of passing in unchanged amount through all the reagents employed are the recently discovered gases of the argon family." It took some time —not surprisingly—to obtain it in sufficient quantity and purity for full characterization. A lengthy paper by Ramsay [40] summarized some earlier efforts and then described their own, which required particular ingenuity to deal with the complication that the radioactive decay continually generates He. They concluded that the "molecular weight" (later in the paper they refer to it as atomic weight) of the new gas, which they called "niton," was 218.

By 1915 Ramsay's *Gases of the Atmosphere* was up to a fourth edition [41]. There the periodic table (p 221) showed all six members of the last column; the atomic weights were essentially unchanged from the 1901 paper discussed above, with the addition of (still called) niton at 222.4. (A question mark was inserted between xenon and niton, presumably the consequence of not yet understanding how the lanthanides fit into the table, but there was no discussion thereof.) The group was simply called "the inactive elements" (p 259); the name "noble gases" has been credited to German chemist H. Erdmann [13]. "Radon" was finally chosen for its heaviest member in the 1920s. In 1904 both Ramsay and Rayleigh were Nobel laureates—but not by sharing a single prize: Ramsay won for chemistry "in recognition of his services in the discovery of the inert gaseous elements in air," while Rayleigh's physics prize was "for his investigations of the densities of the most important gases and for his discovery of argon in connection with these studies." And that pretty much completes the story.

12.3.4 But Wait, There's (One) More!

The extension of the periodic table into the trans-uranium region of entirely synthetic elements proceeded at a fairly constant rate over the last half of the twentieth century [42], so it must have seemed just a matter of time until atomic number 118, which would fall in the noble gas group, was reached at the end of the next period. The first such claim was reported in 1999, when a group at the Lawrence Berkeley lab claimed [43] to have produced three atoms of element 118 by bombarding a lead target with krypton (atomic numbers 82 + 36); but it was retracted the following year, and later found to be not just erroneous but fraudulent [44]. The successful synthesis required fusion of two less common isotopes—$^{249}Cf + ^{48}Ca$—to get an isotope of element 118 whose atomic weight was closer to the "Island of Stability." A report was issued in 2002, co-authored by a large team at the Joint Institute for Nuclear Research in Dubna (Russia), led by Yuri Oganessian, and several scientists from the Lawrence Livermore Laboratory, but the first "real" publication [45] appeared a few years later, in 2006. Final acceptance of the discovery took another nine years and some additional results; on 12/30/2015 IUPAC's website [46] announced "the verification of the discoveries of four new chemical elements: The 7th period of the periodic table is now complete," crediting the Dubna/Livermore group. In 2016 the group held a conference call to decide on a name, and—after asking Oganessian to hang up—proposed to name it after him: oganesson [47].

The half-life of ^{294}Og is less than a millisecond, and only a handful of atoms have ever been produced, so clearly there is no experimental evidence available as to whether it in fact *does* have the chemical and physical properties expected for a noble gas. Relativistic effects become increasingly important with atomic weight, so that expectation is by no means a given. Calculations (carried out before it was definitively synthesized!) suggest that Og should be considerably more reactive than Rn, due in part to the magnitude of spin-orbit coupling [48]; others, that it might well be a liquid or even a solid at room temperature [49]. Those of us who like to see experimental confirmation before buying into computational predictions are not likely to be satisfied anytime soon.

12.4 The Pre-history of the Chemistry

The unreactivity of argon toward all common reagents had of course been recognized during the course of the experiments leading to its isolation, and (as already noted) Moissan quickly extended that to the most reactive known species, F_2. Generally similar behavior (or, rather, the absence of any) was established for the other members of the group as they became available. There *was* one positive report: dean of French chemists Marcellin Berthelot sparked Ar with benzene vapor and observed formation of a yellow solid, which he believed to be a compound of Ar [50], but that was soon discredited. One explanation is that no "Ar" was present at all: there is evidence suggesting that the vessel which Ramsay had sent to Berthelot had been opened at French customs, so it would have contained air instead [13]. Nonetheless, H. G. Wells was quick to postulate a compound of argon —and maybe another of the new gases?—as a weapon used by the invading Martians in his 1898 *War of the Worlds*: "Spectrum analysis of the black powder points unmistakably to the presence of an unknown element with a brilliant group of three lines in the green, and it is possible that it combines with argon to form a compound which acts at once with deadly effect upon some constituent in the blood." (That wasn't the only reference to the recent discovery: elsewhere it appears Wells believed the atmosphere of Mars to be rich in argon.)

Laszlo and Schrobilgen [51] discuss some further attempts, referring to a correspondence between Italian chemist Giuseppe Oddo and Ramsay. Oddo suggested that the heavier elements should be more reactive than Ar [52]; Ramsay agreed but was dubious about being able to obtain enough for meaningful experimentation. Walther Kossel proposed, on electronegativity grounds, that fluorides of Kr and Xe should exist [53]. German chemist Andreas von Antropoff offered a similar argument in 1924, and then tried to confirm it by sparking Kr with Cl_2 and Br_2; initially they claimed a new red solid [54], but subsequently found it was an artifact, and retracted their claim.

In the same year Fritz Paneth wrote a short essay on the periodic table [55] in which he mentioned Antropoff's work, but concluded "Die Sonderstellung der nullwertigen Edelgase im periodischen System gehört zu den sichersten Ergebnissen des chemischen Experimentes und der physikalischen Interpretation, und es scheint mir ein Anachronismus, sie der äusserlichen Gleichförmigkeit der acht Gruppen zuliebe preiszugeben." Or: "The special position of zero-valued noble gases in the periodic system is one of the safest results of chemical experimentation and physical interpretation, and it seems to me anachronistic to give them up for the sake of the uniformity of the eight groups." (That was obtained from Google Translate; the only important change needed is "zero-valued" to "zerovalent." Not bad for a machine translation: they must have gotten much better since I last tried one!)

A 1933 paper by Linus Pauling, ostensibly about antimonates, proposed that XeO_6^{4-} should form isolable silver salts, based on ionic radius considerations and analogy to known species. He also "predicted formulas KrF_6 and XeF_6, with XeF_8 as an unstable compound which might be capable of existence" [56]. But Pauling was already pushing beyond predictions: the previous year he had written to his former teacher Fred Allen, now at Purdue, asking for a sample of xenon [57]:

> I should like to do some work (with Professor Yost) in an attempt to prepare certain compounds of Xenon suggested by theoretical arguments. No doubt your xenon is precious; if, however, you could lend us 10 cc. or so (of not necessarily pure stuff), we would try to return it to you either as such or in some compound (I hope), and we would be properly grateful. If this is asking too much, or if you can't lend it, could you give us advice as to where we might possibly obtain some?

The sample was duly received, and given to Pauling's colleague Don Yost. Working with his graduate student Albert Kaye, they carried out a series of studies of sparking mixtures of Xe with either F_2 or Cl_2, using an old Ford coil (photoirradiation was also tried with Cl_2, but not F_2). Unfortunately, they saw nothing beyond some attack on the quartz vessel, and submitted an account of their efforts to *JACS*—where it was published, perhaps surprisingly, considering they had only negative findings [58].

The version of this attempt in the Laszlo and Schrobilgen essay—which was based in large part on an interview with Kaye, whom they managed to track down many years after the event—is quite different. According to Kaye (presumably) it was *Yost* who took the initiative, with Pauling nothing more than "a most interested on-looker." Kaye was not even aware that Pauling had obtained the Xe sample: when he gave a departmental seminar on his work, he was surprised at how much Pauling seemed to know about the subject [53]! This is pretty much diametrically opposed to all other accounts of the episode—both those of Pauling himself and other commentators—and I have to believe that the latter are much closer to the truth. True, Pauling was often ready to reconstruct stories to put himself in a better light. For that matter, so was Yost: his own retrospective, in a collection of articles published shortly after Bartlett's breakthrough (see below), omitted any mention of Pauling whatsoever. It is well known that Pauling and Yost came to detest one another, but exactly when that began is not so clear [59]. But Kaye's version, as recounted by Laszlo and Schrobilgen, just doesn't ring true—whether due to his own faulty and/or selective memory, or because Yost misled him about Pauling's role, or both. It is notable that Kaye described—and "drew from memory"—the "copper reaction vessel" used for the study, whereas the original *JACS* article explicitly refers to "an all-quartz apparatus provided with copper electrodes." Laszlo and Schrobilgen seem to cast some doubt on the latter, noting that "Pauling and Kaye both recall the use of metal" (but according to Pauling, many years later, it was a *nickel* reactor), but it seems far more probable to me that the contemporaneous report was correct.

In any case, success was not achieved. After Bartlett's 1962 paper, a number of people (including Pauling) offered possible reasons for the failure. I have discussed these in detail elsewhere [60], including my own interpretation—that they would have succeeded had they used a higher pressure of Xe and/or longer reaction times (Laszlo and Schrobilgen have a different opinion here too)—and will not repeat them here. Over the next three decades there were a few more claims and predictions, but none of those stood up either [53]. Pauling in particular ceased to be an enthusiast, whether because of Yost and Kaye's failure or for other reasons. The 1947 edition of his general chemistry textbook has the following language [61]:

> Helium, the second element, is a gas with the striking chemical property that it forms no chemical compounds….The congeners of helium—neon, argon, krypton, xenon, and radon—are also chemically inert. The failure of these inert elements to form chemical compounds is similarly due to the great stability of their electronic structures.

And so things (mostly) remained, until 1962.

12.5 The History of the Chemistry

The story of Neil Bartlett's discovery is well known, and needs little elaboration here. As he reports in a very short (three paragraphs!) communication, having previously found that O_2 is oxidized by PtF_6 to give a salt, and recognized that O_2 and Xe have almost the same ionization potential, he predicted the analogous reaction, Xe + PtF_6 = $Xe^+PtF_6^-$, and indeed found the two substances reacted to give a solid that evolved gaseous Xe (along with O_2 and HF) upon hydrolysis [62]. Subsequently it has been recognized that the formula was not correct; a later study indicated the *presence* of $XeF^+PtF_6^-$ in a complex mixture whose composition depends on initial stoichiometry [63].

Bartlett's paper was received on May 4, and was published in June. A huge amount of work quickly ensued; indeed, some of it was already underway. Rudolf Hoppe had begun trying to react Xe with F_2 the previous year—using almost exactly the same methodology as Yost and Kaye had tried nearly 30 years earlier—and "in the last week of July 1962...were already certain that they had pure, crystalline XeF_2 in hand," as told in a letter [64] (Figs. 12.4, 12.5, and 12.6) from Hoppe's crystallographer colleague W. Klemm to Herbert Hyman (who led a group of noble gas chemistry researchers, inspired by Bartlett's paper, at Argonne National Laboratory). Hoppe's paper [65] —like Bartlett's, only three paragraphs long—was received on October 8 and appeared in the November 21 issue of *Angewandte Chemie*. That *followed* a report of XeF_4 (from the Argonne group) which had a received date of August 20 [66]; another preparation of XeF_2, by a different procedure, was published in a paper received on October 29 [67]. Klemm notes "The American work in this field was...first begun on August 2nd, when Professor Hoppe already had prepared xenon difluoride. These results were not immediately published...because they had to wait for the mass spectrometric studies. In Münster there were at that time no mass spectrometers, and moreover because August is the holiday month, this research was considerably delayed." So Hoppe lost priority for the discovery of binary noble gas compounds due, in large part, to European vacation practices! In any case, though, there appears to be no question, from the chronology of receipt and publication dates, that Bartlett's demonstration of reactivity was indeed the first, in terms of both when it was carried out and when it was published.

By April of the following year so much had already been accomplished that the Argonne group felt a conference was in order; the proceedings, published in book form [68], contained over 50 papers on xenon fluorides, oxyfluorides, and oxides, including several theoretical treatments (one of them oxymoronically titled "Theory of Binding in Inert-Gas Molecules"!), and prefaced by "Historical Remarks on the Discovery of Argon" (by Erwin Hiebert) and Don Yost's afore-mentioned reminiscences. One paper reported a radon fluoride [69], and another claimed the isolation of KrF_4 [70]; the latter work proved irreproducible, and it is believed that KrF_2 was actually obtained [71].

Herrn Prof.Dr. Bailar zur Kenntnisnahme!

Anorganisch-Chemisches Institut der Universität
[illegible]
Hindenburgplatz [illegible] — Telefon 40739

10. Juni 1964

Prof.Dr.W. Klemm

Herrn
Professor Dr. H.H. H y m a n
Argonne National Laboratories
9700 South Cass Avenue
A r g o n n e / Illinois 60440

Sehr geehrter Herr Hyman!

Ich möchte heute kurz in einer Angelegenheit an Sie schreiben, die mich schon seit längerer Zeit beschäftigt. Es handelt sich um den Anteil, den Herr Professor Hoppe an der Entdeckung von binären Edelgasverbindungen, insbesondere Xenondifluorid gehabt hat. Sie schreiben auf Seite 34 Ihres Buches 'Noble Gas-Compounds' darüber einiges, wobei Sie am Schluß schreiben:

> "On learning of the Argonne discovery of xenon tetrafluoride, the München group proceeded with their own approach and published some inadequately established preliminary observations. Their contribution to this volume is a revised and somewhat enlarged version of this earlier communication."
>
> Bei München ist wohl gemeint Münster.

Diese Darstellung gibt ohne Zweifel den Sachverhalt nicht richtig wieder und mit Rücksicht darauf, daß die Gefahr besteht, daß bei einer späteren historischen Darstellung die Verhältnisse unrichtig wiedergegeben werden, möchte ich Ihnen kurz schildern, wie sie wirklich gewesen sind.

1.) Wie Sie richtig schreiben, bestand der Plan, Fluoride des Xenons darzustellen, bei Herrn Professor Hoppe schon sehr lange[1]. Die Schwierigkeit bestand in der Beschaffung des flüssigen Fluors, das in Deutschland industriell nicht hergestellt wird. Ein Versprechen der Allied Chemical Coorporation (im August 1961!), Herrn

[1] Die Arbeiten von Herrn Professor Hoppe sind also nicht durch die schönen Untersuchungen von BARTLETT über $XePtF_6$ angeregt worden, denn die Pläne zur Darstellung von Xenonfluoriden sind viel älter und gehen bis 1949 zurück, wie ich bezeugen kann.

Fig. 12.4 Letter from W. Klemm to H. Hyman of June 10, 1964, page 1. (Kindly provided by G. Girolami [66]). See the Appendix for the translation

- 2 -

Professor Hoppe einige Bomben flüssigen Fluors zu schicken, konnte nicht realisiert werden, da die amerikanischen Ventile in Deutschland nicht zugelassen waren und die deutschen Ventile nicht in Amerika. Herr Professor Hoppe mußte daher mit Laboratoriumsmethoden zunächst einmal flüssiges Fluor herstellen und hat dann Anfang Juli 1962 - wie durch viele Zeugen belegt werden kann - das erste Xenondifluorid hergestellt und analysiert; in der letzten Juliwoche war man bereits sicher, reines, kristallines XeF_2 in Händen zu haben. Dieses Ergebnis wurde nicht sofort veröffentlicht, da noch die massenspektrometrischen Untersuchungen abgewartet werden sollten. In Münster befand sich damals kein Massenspektrometer, und da~~ß~~ der August zudem Ferienmonat ist, verzögerten sich diese Untersuchungen erheblich. Außerdem war ja auch nicht bekannt, daß von anderer Seite Arbeiten auf diesem Gebiet in Angriff genommen werden würden.

2.) Die amerikanischen Arbeiten auf diesem Gebiet sind, wie sich aus Ihrem Buch, Seite 31, ergibt, erst am 2. August begonnen worden, als Herr Professor Hoppe bereits Xenondifluorid dargestellt hatte. Die Veröffentlichung der Darstellung von Xenontetrafluorid ist aber schneller erfolgt (H.H. Claassen, H. Selig, and J.G. Malm, J.Am.Chem.Soc. 84, 3593, 1962). Diese Mitteilung wurde hier etwa am 1. Oktober bekannt.

3.) Am 8. Oktober hat dann Professor Hoppe in der Zeitschrift ANGEWANDTE CHEMIE 74, 903 (1962) das Xenondifluorid beschrieben. Bald darauf hat dann auch D.F. Smith in der Zeitschrift J.Chem.Physics 38, 270, 1963, das Xenondifluorid beschrieben - Eingangsdatum 29. Oktober 1962 - Bei der Darstellung des durch Herrn Professor Hoppe beschriebenen Xenonfluorids handelt es sich um ein anderes Verfahren.

Fig. 12.5 Letter from W. Klemm to H. Hyman of June 10, 1964, page 2. (Kindly provided by G. Girolami [66]). See the Appendix for the translation

\- 3 -

Aus dem Vorstehenden ergibt sich, daß die erste Publikation über Xenontetrafluorid und damit über binäre Edelgasverbindungen überhaupt durch die amerikanischen Forscher erfolgt ist, daß aber Xenondifluorid von Herrn Professor Hoppe dargestellt worden ist, ehe diese ihre Arbeiten begonnen haben, und daß Herr Professor Hoppe über die Darstellung von XeF_2 publiziert hat, ehe die erste amerikanische Mitteilung über XeF_2 erfolgte.

Eine geschichtlich korrekte Darstellung müßte also hervorheben, daß praktische gleichzeitig und unabhängig die Darstwllung von binären Edelgasverbindungen erfolgt ist, und daß das XeF_4 zuerst angeregt durch BARTLETT's Arbeit von den Amerikanern, XeF_2 dagegen zuerst von deutscher Seite (unabhängig von BARTLETT's Untersuchung) dargestellt und publiziert worden ist.

Wie Sie sehen, entspricht dies nicht der Darstellung, die Sie in Ihrem Buch[x)] geben, und ich darf der Erwartung Ausdruck geben, daß Sie bei einer Neuauflage die Verhältnisse richtig wiedergeben.

Ich weiß, daß unsere amerikanischen Kollegen größten Wert darauf legen, Fragen persönlicher Verdienste in fairer Weise zu behandeln. Ich möchte Ihnen die unkorekte Darstellung des zeitlichen Ablaufs, den Sie infolge der etwas verwickelten Verhätlnisse nicht übersehen konnten, natürlich nicht zum Vorwurf machen, umso weniger, als auch in einem deutschen Buch von WIBERG (1964) der Name HOPPE nicht einmal erwähnt ist.

Ich bin aber überzeugt, daß es Ihnen angenehm ist, daß ich den historischen Ablauf korrekt dargestellt habe.

Mit verbindlichen Grüssen

Ihr sehr ergebener

W. Klemm

---

x) In einer späteren Arbeit (J.Chem. Education 41, 174, 1964) wird bei der Besprechung von XeF_2 der Name HOPPE nicht einmal erwähnt.

Fig. 12.6 Letter from W. Klemm to H. Hyman of June 10, 1964, page 3. (Kindly provided by G. Girolami [66]). See the Appendix for the translation

In the half-century since, the chemistry of the noble gas elements has been greatly expanded. Compounds of the lighter noble gases still exist only in the gas phase or in matrices, but those of the heavier ones are no longer limited to combinations with strongly electronegative elements such as F and O, the rationale that governed most of the early work. These so-called "atypical" compounds [72] include organoxenon species such as $[Xe(C_6F_5)]^+$ (a quite electronegative carbon-centered bonding partner, to be sure). Quite a number of compounds—some quite stable—involve noble gas centers acting as ligands to transition metal centers; for example, square-planar $[AuXe_4]^{2+}$ and linear $[(F_3As)AuXe]^+$ have been isolated and crystallographically characterized [73]. A review [74] calls these coordination compounds "perhaps the most surprising, interesting and thought provoking observations" in the history of the field; but given that noble gas atoms are iso-electronic to halide ions—among the most common ligands—in retrospect perhaps we should not have been all that surprised. It will be interesting to see whether greater surprises still await us.

12.6 Appendix

The following is a translation of the letter from W. Klemm to H. Hyman of June 10, 1964, provided by G. Girolami [66].

June 10, 1964

Herr Professor Dr. H. H. Hyman

Argonne National Laboratories

9700 South Cass Avenue

Argonne, Illinois 60440

Dear Mr. Hyman:

I wanted to write briefly to you about a matter that has occupied me for some time. It concerns the share which Herr Professor Hoppe had in the discovery of binary noble gas compounds, especially xenon difluoride. You write on page 34 of your book "Noble Gas Compounds" concerning this, and at the end you write:

> "On learning of the Argonne discovery of xenon tetrafluoride, the München group proceeded with their own approach and published some inadequately established preliminary observations. Their contribution to this volume is a revised and somewhat enlarged version of this earlier communication."

By München is actually meant Münster.

This account without doubt does not give the correct facts, and in this regard there is a risk that later historical accounts of the circumstances will again be incorrectly given. I would like to describe to you briefly how it really was.

1.) As you correctly write, Professor Hoppe had long had a plan to prepare fluorides of xenon.[1] The difficulty existed in the procurement of liquid fluorine, which was not prepared industrially in Germany. A promise by Allied Chemical Corporation (in August 1961!) to send Professor Hoppe several bombs of liquid fluorine could not be realized, because the American valves were not permitted in Germany and the German valves were not permitted in America. Therefore, Professor Hoppe had to prepare some liquid fluorine via laboratory methods, and by the beginning of July 1962—as many witnesses can attest—had prepared and analyzed the first xenon difluoride; in the last week of July 1962 they were already certain that they had pure, crystalline XeF_2 in hand. These results were not immediately published, because the mass spectrometric studies were still being awaited. In Münster there were at that time no mass spectrometers, and moreover because August is the holiday month, this research was considerably delayed. Also it was not known that on the other side [of the Atlantic] work attacking this field was being undertaken.

2.) The American work in this field was, as you prove in your book on page 31, first begun on August 2nd, when Professor Hoppe already had prepared xenon difluoride. However, the publication of the preparation of xenon <u>tetra</u>fluoride took place more quickly (H. H. Claassen, H. Selig, and J. G. Maim, J. Am. Chem. Soc. 84, 3593, 1962). This communication became known here about October 1st.

3.) On October 8th, Professor Hoppe described xenon <u>di</u>fluoride in the journal Angewandte Chemie 14, 903 (1962). Soon thereafter, D. F. Smith also described xenon difluoride in the journal J. Chem. Physics, 38, 270, 1963—receipt date October 29, 1962. It concerns a different procedure for the preparation of xenon difluoride than that described by Professor Hoppe.

From the preceding it follows that the first publication concerning xenon <u>tetra</u>fluoride, and therewith concerning binary noble gas compounds generally, resulted owing to the American researchers, but that xenon difluoride was prepared by Professor Hoppe before their work had begun, and that Professor Hoppe had published the preparation of XeF_2 before the first American communication on XeF_2 took place.

A historically correct account must also emphasize that the preparation of binary noble gas compounds occurred practically <u>at the same time and independently</u>, and that, stimulated by Bartlett's work, XeF_4 was first prepared and publicized by the Americans, while XeF_2 was first prepared and publicized by the German side (independently of Bartlett's research).

As you see, the account given in your book[2] does not correspond to this, and I should give the expected expression that you will correctly recount the circumstances in a new edition.

[1]The work of Professor Hoppe was not stimulated by the beautiful investigations of Barlett on $XePtF_6$, because the plans for the preparation of xenon fluorides were much older and go back to 1949, as I can testify.

[2]In a later work (J. Chem Education 41, 174, 1964) during the discussion of XeF_2 the name Hoppe is not mentioned once.

I know our American colleagues place great value in treating questions of personal credit in a fair way. I naturally do not wish to blame you for the incorrect account of the past course of events, which you could not survey due to the somewhat complicated circumstances; much less, as also in a German book by Wiberg (1964) the name Hoppe is not mentioned once. I am however convinced that you will accept that I have depicted the historical course of events correctly.

With obliging greetings
Yours very devotedly

W. Klemm
Inorganic-Chemical Institute of the University
44 Münster (W)
Hindenburgplatz 65—Telephone 40739

[Translated by G. S. Girolami, 26 Dec 1992.]

Acknowledgements I thank Greg Girolami for calling my attention to the Klemm letter and providing his translation thereof; I also am indebted to a reviewer and the editors for valuable suggestions.

References

1. Scerri E (2007) The periodic table: Its story and its significance. Oxford University Press, Oxford, chap. 5
2. Bartlett N, Sladky FO (1973) The chemistry of krypton, xenon and radon. In: Bailar JC Jr, Emeleus HJ, Nyholm R, Trotman-Dickinson AF (eds) Comprehensive inorganic chemistry, vol. 1. Pergamon Press, Oxford, Chapter 6, pp 213–330
3. Brock DS, Schrobilgen GJ, Žemva, B (2013) Noble-gas chemistry. In: Reedijk J, Poeppelmeier K (eds) Comprehensive inorganic chemistry II, vol. 1. Elsevier, Oxford, Chapter 25, pp 755–822
4. Cavendish H (1785) Experiments on air. Phil Trans Roy Soc (London) 75:372–384
5. Launay F (2008) The astronomer Jules Janssen—a globetrotter of celestial physics. Springer, New York, p 45
6. Lockyer JN (1868) Spectroscopic observations of the sun—No. II. Phil Trans Roy Soc (London) 159:425–444
7. Lockyer WJS (1920) Helium: Its discovery and applications. Nature (London) 105:360–363
8. Lockyer JN (1897) The sun's place in nature. Macmillan, London, pp 32–33
9. Thomson W (1871) Inaugural address of Sir William Thomson, LL.D., F.R.S., President. Nature (London) 4:262–270 (on p 268)
10. Kragh H (2009) The solar element: a reconsideration of helium's early history. Ann Sci 66:157–182
11. Hillebrand WF (1891) On the occurrence of nitrogen in uraninite and on the composition of uraninite in general. Bull US Geol Survey 78:43–79
12. Lockyer JN (1897) The sun's place in nature. Macmillan, London, pp 36–38
13. Wolfenden JH (1969) The noble gases and the periodic table: Telling it like it was. J Chem Ed 46:569–576
14. Lord Rayleigh (1892) Density of nitrogen. Nature (London) 46:512–513
15. Lord Rayleigh (1893) On the densities of the principal gases. Proc Roy Soc (London) 53:134–149

16. Lord Rayleigh (1894) On an anomaly encountered in determinations of the density of nitrogen gas. Proc Roy Soc (London) 55:340–344
17. Brock WH (1992) The Norton history of chemistry. Norton, New York, pp 331–340
18. Ihde AJ (1964) The development of modern chemistry. Harper & Row, New York, pp 370–371
19. Travers MW (1956) A life of Sir William Ramsay, K.C.B., F.R.S. E. Arnold, London, p 110
20. Dewar, J (1894 August 16) The new element. The Times (London), p 3
21. Lord Rayleigh, Ramsay W (1895) Argon, a new constituent of the atmosphere. Proc Roy Soc London 57:265–287
22. Lord Rayleigh, Ramsay W (1895) Argon, a new constituent of the atmosphere. Phil Trans Roy Soc (London) 186:187–241
23. Crookes W (1896) On the spectra of argon. Phil Trans Roy Soc (London) 186:243–251
24. Olszewski K (1896) The liquefaction and solidification of argon. Phil Trans Roy Soc (London) 186:253–257
25. Moissan H (1895) Action du fluor sur l'argon. C R Séances Acad Sci 120:290–296
26. Giunta CJ (2001) Argon and the periodic system: the piece that would not fit. Found Chem 3:105–128
27. Mendeleeff D (1895) Professor Mendeleeff on argon. Nature 51:543
28. Lord Rayleigh (1895) Argon. Roy Inst Proc 14:524–538
29. Ramsay W (1896) The gases of the atmosphere, 1st edn. Macmillan, London, pp 232–240
30. Ramsay W (1896) The gases of the atmosphere, 1st edn. Macmillan, London, p 219
31. Ramsay W (1895) Helium, a gaseous constituent of certain minerals. Part I. Proc Roy Soc (London) 58:80–89
32. Ramsay W, Collie JN, Travers M (1895) Helium, a constituent of certain minerals. J Chem Soc Trans 67:684–701
33. Ramsay W (1897) An undiscovered gas. Science (New Series) 6:493–502
34. Ramsay W, Travers MW (1898) On a new constituent of atmospheric air. Proc Roy Soc (London) 63:405–408
35. Ramsay W, Travers MW (1898) On the companions of argon. Proc Roy Soc (London) 63:437–440
36. Ramsay W, Travers MW (1898) On the extraction from air of the companions of argon and neon. Report of the sixty-eighth meeting for the British Association for the Advancement of Science, Bristol, September 1898. John Murray, London, pp 828–830
37. Ramsay W, Travers MW (1901) Argon and its companions. Phil Trans Roy Soc (London) 197:47–89
38. Owens RB (1899) Thorium radiation. Phil Mag 48:360–387
39. Rutherford E, Soddy F (1902) The radioactivity of thorium compounds. I. An investigation of the radioactive emanation. J Chem Soc, Trans 81:321–350
40. Gray WW, Ramsay W (1911) The density of niton ("radium emanation") and the disintegration theory. Proc Roy Soc A 84:536–550
41. Ramsay W (1915) The gases of the atmosphere, 4th edn. Macmillan, London
42. Ball P (2019) On the edge of the periodic table. Nature 565:552–555
43. Service, RF (1999) Berkeley crew bags element 118. Science 284:1751
44. Dalton R (2002) The stars who fell to earth. Nature 420:728–729
45. YuTs Oganessian et al (2006) Synthesis of the isotopes of elements 118 and 116 in the ^{249}Cf and ^{245}Cm + ^{48}Ca fusion reactions. Phys Rev C 74:044602
46. IUPAC (2015) Discovery and assignment of elements with atomic numbers 113, 115, 117 and 118. https://iupac.org/discovery-and-assignment-of-elements-with-atomic-numbers-113-115-117-and-118/. Accessed 31 Jan 2020
47. Chapman K (2016) What it takes to make a new element. Chemistry World: https://www.chemistryworld.com/what-it-takes-to-make-a-new-element/1017677.article. Accessed 31 Jan 2020
48. Nash CS (2005) Atomic and molecular properties of elements 112, 114, and 118. J Phys Chem A 109:3493–3500
49. Giuliani SA et al (2019) Colloquium: superheavy elements: oganesson and beyond. Rev Mod Phys 91:011001

50. Berthelot M (1895) Essais pour faire entrer l'argon en combinaison chimique. C R Séances Acad Sci 120:581–585
51. Laszlo P, Schrobilgen GJ (1988) One or several pioneers? The discovery of noble-gas compounds. Angew Chem Int Ed Engl 27:479–489
52. Oddo G (1933) Sul potere di combinarsi dei cripton e dello xenon: due lettere inedite del prof. W. Ramsay. Gazz Chim Ital 63:380–395
53. Kossel W (1916) Über Molekülbildung als Frage des Atombaus. Ann Phys (Leipzig) 49: 229–362
54. Von Antropoff A, Weil K, Frauenhof H (1932) Die Gewinnung von Halogenverbindungen der Edelgase. Naturwissenschaften 20:688–689
55. Paneth F (1924) Über die heutige Schreibweise des periodischen Systems der Elemente. Angew Chem 37:421–422
56. Pauling L (1933) The formulas of antimonic acid and the antimonates. J Am Chem Soc 55:1895–1900
57. Letter from L. Pauling to F. Allen, September 13, 1932. Courtesy of the Ava Helen and Linus Pauling Papers, Oregon State University Libraries
58. Yost DM, Kaye AL (1933) An attempt to prepare a chloride or fluoride of xenon. J Am Chem Soc 55:3890–3892
59. Labinger JA (2013) Up from generality: How inorganic chemistry finally became a respectable field. Springer, Dordrecht, pp 33–48
60. Labinger JA (2015) Why isn't noble gas chemistry 30 years older? The failed (?) 1933 experiment of Yost and Kaye. Bull Hist Chem 40:29–36
61. Pauling L (1947) General chemistry: An introduction to descriptive chemistry and modern chemical theory. W.H. Freeman, San Francisco, p 67
62. Bartlett N (1962) Xenon hexafluoroplatinate(V) $Xe^+PtF_6^-$. Proc Chem Soc London: 218
63. Graham L, Graudjys O, Jha NK, Bartlett N (2000) Concerning the nature of $XePtF_6$. Coord Chem Rev 197:321–334
64. Letter from W. Klemm to H. H. Hyman, June 10, 1964, translated by G. S. Girolami
65. Hoppe R, Dähne W, Mattauch H, Rödder K (1962) Fluorination of xenon. Angew Chem Int Ed Engl 1:599
66. Claasen HH, Selig H, Malm JG (1962) Xenon tetrafluoride. J Am Chem Soc 84:3593
67. Smith DF (1963) Xenon difluoride. J Chem Phys 38:270–271
68. Hyman HH (ed) (1963) Noble-gas compounds. University of Chicago Press, Chicago
69. Fields PR, Stein L, Zirin MH (1963) Radon fluoride: further tracer experiments with radon. In: Hyman HH (ed) Noble-gas compounds. University of Chicago Press, Chicago, pp 113–119
70. Streng AG, Kirshenbaum AD, Streng LV, Grosse AV (1963) Preparation of rare-gas fluorides and oxyfluorides by the electric-discharge method and their properties. In: Hyman HH (ed) Noble-gas compounds. University of Chicago Press, Chicago, pp 73–80
71. Lehman JF, Mercier HPA, Schrobilgen GJ (2002) The chemistry of krypton. Coord Chem Rev 233–234:1–39
72. Grochala W (2007) Atypical compounds of gases, which have been called "noble". Chem Soc Rev 36:1632–1655
73. Seppelt K (2003) Metal-xenon complexes. Z Anorg Allg Chem 629:2427–2430
74. Hope EG (2013) Coordination chemistry of the noble gases and noble gas fluorides. Coord Chem Rev 257:902–909
75. Ramsay W (1896) The gases of the atmosphere, 1st edn. Macmillan, London, p 169
76. Ramsay W (1896) The gases of the atmosphere, 1st edn. Macmillan, London, p 158

Element Discovery and the Birth of the Atomic Age

13

Kit Chapman

Abstract

The story of the first synthetic elements is one interlinked with the development of nuclear science and atomic weapons. As such, its impact is often overlooked or underestimated, particularly in how the hunt for elements affected research priorities and contributed to the prestige of the discoverers, many of whom would go on to hold influential positions throughout the cold war. This review focuses on the first attempts to synthesize transuranic elements in the 1930s until the discovery of einsteinium and fermium in 1952. It charts the discovery of the elements, gives context to the character of the discoverers, describes some of their more colorful adventures and provides wider context in terms of the political and scientific changes occurring at the start of the atomic age, which set the scene for the so-called "transfermium wars" and the hunt for the superheavy elements.

13.1 The Via Panisperna Boys

Few events have so dramatically altered the course of human history as the advent of the nuclear age. In the space of a decade, the world went from a cycle of perpetual conflict to the possession of weapons so terrible they were only used in anger twice, discovery of a new source of energy and invention of a pathway to medicines and technologies that have saved millions of lives. It also found itself moving into a stage of geopolitical conflict between superpowers not seen since the Seven Years' War between the British and French empires in the eighteenth century. Perhaps the most remarkable thing about this scientific revolution is that, save

K. Chapman (✉)
University of Sunderland, Sunderland, UK

C. J. Giunta et al. (eds.), *150 Years of the Periodic Table*, Perspectives on the History of Chemistry, https://doi.org/10.1007/978-3-030-67910-1_13

for visionaries such as H. G. Wells in his 1913 tale *The World Set Free* [1], few saw it coming. And, of those that did, no one imagined it could lead to elements heavier than those already known.

The story of synthetic elements—those created by humans rather than isolated in nature[1]—begins in the most unlikely of places: a villa of the Via Panisperna, in the heart of Rome. Here, in 1934, Italian physicist Enrico Fermi had decided to experiment with the new phenomenon of induced radioactivity, discovered by the wife and husband duo Irène and Frédéric Joliot-Curie [2]. The Joliot-Curies' experiments involved bombarding stable targets with alpha particles; as these were positively charged, a particle accelerator was required to push beyond the electrostatic repulsion of the nucleus, known as the Coulomb barrier.

Fermi did not have the space nor funding for a particle accelerator. Indeed, his team were so impoverished they could not afford proper radiation shielding and would instead hide down the corridor while performing their experiments, running to their equipment after they were concluded. Considering alternative approaches, Fermi decided not to use alpha particles, but instead to try and induce radioactivity using another new discovery, the neutron, a particle first identified by James Chadwick in 1932 [3]. Neutrons do not have a charge, and therefore did not experience resistance from the Coulomb barrier—meaning a particle accelerator was not required.

The Via Panisperna Boys, as Fermi's team were known, set to work. A neutron beam was created, first by piping radon gas, an alpha particle emitter, given off by a sample of radium contained in a colleague's safe. This was then mixed with beryllium powder, causing the alpha particles to strike beryllium atoms and give off neutrons. Team member Emilio Segrè was dispatched across Rome to purchase samples of every element available, and the beam was fired at each element in sequence. When the Boys fired their beam at uranium (atomic number, $Z = 92$), the heaviest element known to exist, they discovered they had transmuted their sample into two elements they could not identify. Making only a cursory chemical check to determine it was not a lighter element, they concluded that they had produced elements with atomic numbers 93 and 94. It was the first time most had considered the possibility an element heavier than uranium could exist, let alone be made [4].

The purported mechanism for Fermi's discovery was neutron capture. The neutron had been added to the nucleus of the target, and, in undergoing beta decay, had turned into a proton, thus increasing the atom's atomic number. This discovery, alongside Fermi's brilliant observation that neutrons could be slowed by passing them through water (proved, in part, by leaping into the villa's fountain with his neutron beam), became an overnight sensation. With the notable exception of Ida Noddack [5], who considered it more likely Fermi had caused his atom to break apart, the Via Panisperna elements were accepted by the scientific community. For Mussolini, Fermi's sponsor, the discovery was a coup that showed "fascist victories

[1]Throughout this chapter, I will refer to the discovery of synthetic elements; while some philosophers of science have questioned whether synthesis counts as "discovery" in the traditional sense, that debate is outside the purview of this discussion.

in the field of culture"—and the dictator soon began to suggest naming the elements after the lictors of ancient Rome, whose badge of office gave fascism its name [6].

In 1937, the first discovery of a synthesized element was made [7]. Now working at the University of Palermo in Sicily, Segrè requested any radioactive spare parts from the 37-inch cyclotron, a new type of particle accelerator designed by Ernest Lawrence at the University of California, Berkeley. Lawrence, having no use for the parts, obliged, sending Segrè a molybdenum ($Z = 42$) plate. Analyzing the plate by boiling with sodium hydroxide and potassium peroxide, Segrè and colleague Carlo Perrier discovered an unknown substance. Some of the plate's atoms had transmuted into the then-undiscovered $Z = 43$ [8]. This was a critical missing piece of the periodic table jigsaw (by then the only other gaps were $Z = 61$, $Z = 85$ and $Z = 87$). Segrè's discovery was later named technetium, from the Greek word for artificial.

It would soon emerge—in dramatic fashion—that Segrè had in fact succeeded where his mentor had, in fact, failed. In Rome, Enrico Fermi was now desperately trying to escape Europe, worried for the safety of his wife Laura (who was Jewish) and their children. On November 10, 1938, the couple awoke to news of Kristallnacht in Nazi Germany, and new anti-Semitic laws in Italy [6]. Shortly after, Fermi received a phone call informing him that he had been awarded the 1938 Nobel Prize in physics for [9]:

> ... his demonstrations of the existence of new radioactive elements produced by neutron irradiation, and for his related discovery of nuclear reactions brought about by slow neutrons.

Fermi used the prize as an excuse to flee Italy with his family, first to Sweden and then to the United States. It is ironic that, for a scientist widely regarded as one of the most brilliant minds of the twentieth century, his only Nobel win was for the one feat he *did not* do. A month later, in December 1938, German chemist Otto Hahn corresponded to his colleague, Austrian physicist Lise Meitner, that he had been attempting to replicate Fermi's experiment, but could only produce barium. Meitner, who was also Jewish and had fled Germany for the safety of Sweden, worked on the problem with nephew Otto Frisch. The duo concluded that neutron bombardment had, far from creating a new element, resulted in Fermi's uranium atoms breaking apart. The discovery, dubbed nuclear fission, would prove to be the cornerstone of nuclear power and the birth of the atomic age [10].

And yet neutron capture was possible—albeit far less probable than fission. The elements beyond uranium were waiting to be found.

13.2 The Element in the Attic

The news of nuclear fission arrived at Berkeley in notable style. On January 31, 1939, physicist Luis Alvarez went for a haircut. Mid-way through his trim, he noticed the discovery in the *San Francisco Chronicle* and immediately rushed out of the barbershop, making his way to Lawrence's Radiation Laboratory [11]. Here, he insisted colleague Philip Abelson lie down on the workbench before he explained the concept to him. Abelson was distraught: he had noticed similar findings to Hahn and was probably moments from discovering fission himself.

The news quickly spread across campus, with many of the researchers taking an active interest in the new phenomenon. Chief among them was Edwin McMillan, who had been lured by Lawrence back to his native California from Princeton in 1933. McMillan began to experiment by bombarding uranium with neutrons, and soon noted, along with the fission products reported by Hahn, an unusual isotope with a half-life of around 2.3 days. Keen to investigate further, he approached element synthesis's only expert: Segrè, now resident at Berkeley.[2] Segrè was dismissive of McMillan's results, going so far as to publish a paper: *an unsuccessful search for transuranic elements* [12].

Fortunately, McMillan was not so easily dissuaded. In 1940 he consulted Abelson, and the duo successfully identified the unusual isotope that had been produced. It was $Z = 93$—the first transuranic element [13].[3] Fanfare of the discovery was muted due to the ongoing second world war; although the United States had not yet joined the conflict, its scientists had already begun preparation for that eventuality and it was considered revealing the existence of neptunium could aid the Axis powers [8].

Throughout late 1940, McMillan and collaborators worked on $Z = 93$, believing it could be possible for beta decay to result in the unknown $Z = 94$. One scientist who quickly took an interest was the 28-year-old chemist Glenn Seaborg, who would pursue McMillan around campus (including in the shower) asking about his work [11]. Seaborg had grown up in Ishpeming, Michigan, before moving to Los Angeles. There, he had added the second "n" to his first name (because he thought it looked cool), and befriended fellow chemist and future collaborator Stanley Thompson. The duo had both attended the University of California, Los Angeles—Thompson helping the impoverished Seaborg with tuition fees—before Seaborg moved to Berkeley to work with physical chemist Gilbert N. Lewis, most famous for explaining and representing covalent bonds with shared pairs of electrons. During his time with Lewis, he had been walking across the campus when a member of Lawrence's laboratory had asked for help with a chemical separation. This act of serendipity led Seaborg to take a deep interest in radioactivity and would go on to define his career. In 1938 he and Segrè isolated the metastable ^{99m}Tc. The latter is the most used medical radioisotope in the world, accounting for 80% of diagnostic nuclear medicine procedures [14].

[2]Segrè was Jewish, and had been en route to California in 1938 when anti-Semitic laws were passed banning Jewish academics from holding professorships in Italy. Rather than return home, he decided to continue his career in the United States.

[3]While McMillan and Abelson are credited with the discovery of neptunium, it is worth noting that in Japan Yoshio Nishina, using a cyclotron based on Lawrence's design, almost certainly produced neptunium too. He was, however, unable to prove his discovery.

In November 1940, McMillan was asked to stop his research immediately and move to the Massachusetts Institute of Technology to work on developing microwave radar. Seaborg wrote to McMillan and asked permission to continue his attempt to synthesize $Z = 94$, and became head of the project. Given the potential use of transuranic elements in nuclear fission, Seaborg's team was assembled in secret, performing their chemical analysis at night in a small attic laboratory in Berkeley's Gilman Hall. In February 1941, Seaborg's graduate student Arthur Wahl, working late and—according to legend—amid a thunderstorm, confirmed the team had made $Z = 94$. As the elements followed uranium, named in 1789 in honor of the discovery of the planet Uranus, McMillan and Seaborg later decided their new elements would be neptunium and plutonium. In one of the few jokes on the periodic table, Seaborg recalled how terrible the attic space, filled with reagents, smelled and insisted plutonium's symbol would be Pu (*pee-eew*) [15].

The discovery of plutonium marked the start of an intense period of research for the Berkeley team, and Segrè was drafted into help—a move that created more problems than it solved. As an Italian national, Segrè was technically on the opposing side of the second world war, even though the United States was still technically neutral. Seaborg therefore had to ask him to perform separations without being allowed to tell him what he was doing or why [11]! Despite such hindrances, Seaborg's group soon realized plutonium's potential, particularly the fissile ^{239}Pu as a candidate for nuclear weapons. On December 6, 1941, the S-1 Section of the US Office of Scientific Research and Development, responsible for the US nuclear weapons program, met and agreed to accelerate the project. Here, Arthur Compton and Lawrence proposed the investigation of plutonium as a potential weapon core. Although the chair of the committee, James Conant, was skeptical (infamously stating "Glenn Seaborg is a very competent young chemist, but he isn't that good" [11]), permission was given to investigate plutonium. Less than 24 h later, the Japanese attacked Pearl Harbor and the United States was at war.

13.3 The Metallurgical Laboratory

When Glenn Seaborg moved, at Compton's request, to the Metallurgical Laboratory at the University of Chicago in 1942 to mastermind plutonium production, he had little concept of what would come to pass. In less than three years, working to exhaustion and beyond, the men and women of his team would, with others, lay the foundations for the defining characteristics of the second half of the twentieth century. Nuclear weapons, geopolitical divisions, the possibilities of atomic energy and the concept of "Big Science"—carried over from Lawrence and now exemplified by the Manhattan Project—would all have consequences that continue to affect our lives.[4]

Seaborg's task was monumental. If plutonium was to be used for a weapon, its production had to be scaled up one billion times. The chemistry of the new element was largely unknown, save that it was radioactive and therefore required extensive precautions in the laboratory. There were virtually no radiochemists available. And, with the majority of known talent already at work on different aspects of the project, Seaborg had to recruit a team of the greatest undiscovered chemists in the country, all while being unable to tell them what they were working on. A typical letter from Seaborg was as follows [15]:

> The work here is extremely important, perhaps the number one war research project in the country, and it is of such a character that it will almost certainly have post-war significance and develop into a large industry … unfortunately I cannot divulge to you the nature of the work but, knowing the nature of my activities in the past, you are in a fair position to guess. It is research work of the most interesting type; it is the most interesting problem upon which I have ever worked.

Of those recruited, two require special mention given their later contributions to element discovery. The first is Seaborg's childhood friend Stanley Thompson. Recruited from Standard Oil, Thompson was expected to be "solid" but unremarkable. Instead, he used the knowledge he had gained in the oil industry to revolutionize plutonium separation, inventing the so-called bismuth phosphate process. After his death, Seaborg praised him as "the best experimental chemist I have ever known" [11].

The second individual was an unlikely draftee: perhaps the ultimate renegade and archetypal anti-authoritarian inventor. Albert Ghiorso was an electrical engineer who occasionally made Geiger counters for Berkeley. A tinkerer, he had met his wife Wilma, a secretary at the Radiation Laboratory, while wiring the intercom, and she had encouraged him to contact Seaborg—a man he barely knew—for a reference so he could join the US Navy. Seaborg, at the prompting of his wife Helen (a friend of Wilma and Lawrence's former personal secretary), instead offered Ghiorso a position at Chicago. It was the start of what blossomed into the most successful element discovery team in history.

It is worth remembering the age of these pioneers. Seaborg arrived in Chicago on April 19, 1942—his 30th birthday. Isadore Perlman, who would go on to be Seaborg's invaluable right hand, had turned 27 a week earlier; the still unrecruited Thompson was less than a month older than Seaborg, while Ghiorso was a youthful 26. Indeed, Fermi—destined to play a further role in element discovery since his experiments eight years earlier—was at that time still only 41.

[4]Seaborg's personal influence, which would only grow throughout his lifetime, cannot be understated. In 1998, members of the American Chemical Society voted Seaborg the third greatest chemist of the past 75 years, behind organic chemistry doyen Robert Burns Woodward and two-time Nobel laureate Linus Pauling [8].

The scale of the challenge was staggering. In particular, the team had to design a method of plutonium processing that could be scaled up, all while less than a microgram of plutonium was available. The team eventually created a new field, "ultra-microchemistry," a term coined by Paul Kirk [15], with the team using a Salvioni balance—effectively weighing sub-microgram samples on a quartz fiber. It was a method Seaborg characterized as "invisible material being weighed with an invisible balance" [15]. The amount of plutonium hydroxide was so limited that, when General Leslie Groves, in command of the Manhattan Project, visited the lab, he was unable to see the sample when it was shown to him under a microscope.

Ultimately, it was Fermi whose brilliance was able to come up with a way to scale up production, with the creation of the world's first nuclear reactors, which could result in neutron capture on a larger scale and convert uranium rods into plutonium. This was first the Chicago Pile, built on a racquets court under the bleachers of the University of Chicago's football stadium, and then the X-10 Graphite Reactor at Oak Ridge in Tennessee [8]. Thus nuclear reactors—which today power millions of homes around the world—were originally used solely for production of the transuranic elements. This work continued throughout the war, with Thompson eventually moving to the Hanford Site in Washington to oversee plutonium production, and would eventually conclude with plutonium used in the Trinity test, the world's first nuclear detonation, on July 16, 1945, and the atomic bombing of Nagasaki on August 9, 1945.

The work was grueling, with team members working 12-hour days and six-day weeks. Accidents happened. At one point, a brick fell onto a beaker, leaving a quarter of the world's supply of plutonium soaked into a copy of the *Chicago Tribune* [11]. Seaborg's anxiety turned into a fever that hospitalized him, cured only by long walks and golf—a habit he would continue for the rest of his life. But eventually the pressure eased and, as it did so, the team decided it would be worthwhile to look for yet more elements. At Oak Ridge, scientists discovered $Z = 61$, later called promethium and the last missing piece of the periodic jigsaw puzzle, although it was considered so insignificant to the war effort its discovery was not announced until 1947. Seaborg's team, however, set their sights on the transuranics. If neptunium and plutonium existed, why not heavier elements, too?

In December 1943, work began on the transplutonium elements, with Seaborg assigning chemist Ralph James to tackle the problem, later joined by Leon Morgan and Ghiorso. Using samples of ^{239}Pu from Oak Ridge, the team bombarded them with deuterons in January 1944, at Washington University in St. Louis, before expanding the work to the 60-inch cyclotron at Berkeley. At the same time, samples of ^{239}Pu were bombarded with neutrons at Oak Ridge. This work was not done in isolation: at the same time, a rival team under John Gofman were conducting similar experiments in Los Alamos. These experiments all ended in failure [15].

It was here Seaborg made arguably his most significant contribution to the periodic table. At the time, the elements actinium through plutonium ($Z = 89$–94) were considered part of the seventh row of the periodic table (i.e., with uranium a homologue of tungsten), and not a series equivalent to the lanthanides. Seaborg believed that, instead, the electron orbitals followed the lanthanide series, and that the reason for the failure to isolate heavier elements was bad chemistry. Proposing an "actinide" series instead, he proposed his team work on the assumption that $Z = 95$ and $Z = 96$ would have +3 and +4 oxidation states with insoluble fluorides. By changing the chemical extraction process, in July 1944 the team discovered their first signs of the new elements, later confirmed as $Z = 96$ [16].[5] The discovery of $Z = 95$ followed soon after.

The elements have since become some of the most important in modern life; ^{241}Am is used in modern ionization smoke detectors and has thus saved countless lives around the world, while curium is used in radiotherapy. It was also used in the Alpha Particle X-Ray Spectrometer (APXS) on the Mars Curiosity Rover [17].

The announcement of these new elements is also of note. Seaborg had intended to reveal the new elements at a symposium of the American Chemical Society at Northwestern University on November 16, 1945. However, fate intervened. On 11 November, Seaborg was invited to be the guest on *Quiz Kids*, an NBC radio program. During the show, Seaborg was asked by one of the children, Richard Williams,[6] whether any new elements had been discovered. Seaborg's reply was as follows [15]:

> Oh yes, Dick. Recently there have been two new elements discovered—elements with the atomic numbers 95 and 96 out at the Metallurgical Laboratory here in Chicago. So now you'll have to tell your teachers to change the 92 elements in your schoolbook to 96 elements.

It remains the only time a new element has been announced on children's radio, and this, along with Seaborg's appearance on *Adventures in Science* on 15 December 1945, led to numerous suggestions as to what the new elements should be called, including Latin root words, astronomical bodies, famous scientists and US presidents (including "roosium" after Franklin Roosevelt and "washingtonium" after George Washington). Morgan, reflecting on how hard their separation had been, wanted to name the elements "pandemonium" and "delirium" [8]. However, to cement his controversial actinide concept, Seaborg chose names that reflected their elements' lanthanide homologues: thus $Z = 95$, a homologue of europium, became americium, while $Z = 96$, whose homologue was named (albeit indirectly) after chemist Johan Gadolin, became curium after the chemists Marie and Pierre Curie.

Here, it is worth noting an unexpected impact of the silence around the team's work. In April 1942, a lieutenant in the USSR's Volunteer Air Force, Georgy Flerov, noted the lack of any discussion of fission-related subjects in physics journals [18]. A physicist specializing in fission before the second world war, Flerov immediately realized this must mean the other Allied powers were working on an atomic weapon, and that the USSR was falling behind in a scientific arms

[5]The discovery, analyzed in Chicago, was based on a bombardment of ^{239}Pu by helium ions at Berkeley.

[6]Richard Williams would become a career diplomat and served as the first US ambassador to Mongolia and later Consul General in Hong Kong.

race. Unable to convince his superiors to take the matter seriously, he decided to gamble and wrote directly to Stalin [18]:

> Ten months have already elapsed since the beginning of the war, and all the time I have felt like a man trying to break through a stone wall with his head. Where did I go wrong? Am I overestimating the significance of the "uranium problem"? No, I am not. What makes the uranium projects fantastic are the enormous prospects that will open up if a successful solution to the problem is found.

The letter concluded with a request to present his plan in person. As fortune would have it, Flerov's letter arrived virtually at the same time as intelligence on nuclear weapons from Lavrentiy Beria, Stalin's head of the NKVD, the state secret police. The letter effectively kick-started the USSR's own atomic program, and led to Flerov's rise in the Soviet system. He would go on to become Seaborg's greatest rival throughout the cold war [8].

13.4 Returning to Berkeley

The team that returned to Berkeley in 1945 and 1946 had, in the space of years, gained experience most scientists take decades to acquire. Lawrence gave his former charges free rein to explore the world as they saw fit, while Seaborg was able to retain many of the key personnel that had helped him in Chicago—particularly Ghiorso, Thompson and Perlman [15].

Even so, the team's work (at Building 5 of what is today Lawrence Berkeley National Laboratory) took several years to bear fruit. The team was ill-supplied, their workspace was cramped, and available material was limited. It took three years for the team to get ready to make their first experiments aimed at the synthesis of elements beyond curium, largely as the team were limited to using helium ion bombardment and the availability of americium and curium for targets was limited. In addition, they had to develop new methods for chemical separation (based on ion exchange) and develop lab safety protocols.

The work ethic of the team was staggering. Thompson, who had returned to Berkeley to complete his Ph.D., worked extensively on developing ion exchange, often with Burris Cunningham. During one such experiment, in 1947, the duo worked 36 h without sleep or a break. When they finally left the lab, on the second night, they were so exhausted Cunningham accidentally put on both his own and Thompson's jacket—a fact that neither man initially realized [15].

It was therefore not until December 19, 1949 when the team would strike again and discover the next of the transuranic elements. A 7 mg target of ^{241}Am was bombarded with helium ions, and, after the bulk of the americium had been removed, the sample was carried on lanthanum fluoride, dissolved and separated. As predicted by Seaborg's actinide concept, atoms of the unknown $Z = 97$ eluted before the curium and americium. Unlike its predecessors, the new element was immediately named berkelium; once again, the team wanted to insert a joke onto the periodic table and give it the symbol "Bm" (bum), "because it had been such a stinker in resisting identification for so long" [15].

The discovery of $Z = 98$ followed soon after, on February 9, 1950, with an 8 mg target of ^{242}Cm bombarded and extracted using the same process [19]. In fact, choosing a name for the element was considered more difficult than its synthesis—its lanthanide homologue, dysprosium, was named from the Greek dysprositos, or "hard to obtain." Names such as cyclotronium, lewisium (after Seaborg's mentor), radlabium and colonium were all considered before the team decided on californium [15].

As with all history, it is easy to view this work in hindsight, rather than as the discoverers themselves viewed it at the time. It is important to remember that a mere 15 years earlier, the idea of elements existing beyond uranium had not been seriously considered. The discoverers themselves regularly viewed their creations as the new edge of the periodic table; before settling on neptunium and plutonium, Seaborg and McMillan had considered extremium and ultimum as possible names. On the basis of current nuclear theory, the liquid drop model, the periodic table was predicted to end around element 100. This idea would soon be updated with the nuclear shell model of Hans Jensen and Maria Goeppert-Mayer [20], which caused a refinement of nuclear physics that would result in the duo winning the Nobel Prize with Eugene Wigner in 1963. This led, in 1955, to John Wheeler concluding that nuclei twice as heavy as those then known could exist, which he termed "superheavy nuclei" [21].

In 1951, the Berkeley element team also celebrated a Nobel, when McMillan and Seaborg were awarded the 1951 chemistry prize. As their citation could not match Fermi's, they were instead given the prize "for their discoveries in the chemistry of the transuranium elements" [22].

It was a fitting point for the award to be made. Despite the discovery, the team's work was only beginning—macroscopic quantities of berkelium and californium were not synthesized until 1958. To date, californium has proved to be the heaviest element that can be created in large quantities using neutron capture, and its production is limited to only two facilities in the world: the High Flux Isotope Reactor in Oak Ridge, and the Research Institute of Atomic Reactors in Dimitrovgrad, Russia. While berkelium has never found a use beyond the laboratory (to date its greatest contribution has been as a target for the synthesis of $Z = 117$, tennessine [8]) californium has become a valuable resource as a neutron source for the petrochemical industry, and is also used in metal detectors and identifying gold deposits.

It was also the last of the elements discovered at Berkeley whose priority would be uncontested by rival groups.[7]

[7]Strictly speaking, priority was contested over californium by the Soviet physicists Znoyko and Semishin on the basis they had predicted the elements' properties two years prior to its synthesis. However, this claim was considered spurious and ignored.

13.5 The Children of the Bomb

The final elements discussed in this chapter, $Z = 99$ and $Z = 100$, are unique among the transuranic elements, in that they were not first created in the laboratory. Rather, they were discovered as an unexpected scientific benefit from an atomic weapons test. On November 1, 1952, the US military detonated "Mike" as part of Operation Ivy, on the island of Elugelab in the Marshall Islands. This was the first full test of a thermonuclear weapon (the so-called hydrogen bomb), which resulted in a far more complete fission reaction and thus a far more powerful detonation than any before it—equivalent to some 10 megatons of TNT. The Ivy Mike test demolished the island of Elugelab, so much so that today the bomb's crater can be seen on satellite images as simply a deep blue hole in the ocean!

As part of the scientific analysis of the mission, the US Air Force ordered several flights of modified F-84 Thunderjets to pass through the mushroom cloud created by the blast. These planes all contained a suite of monitoring equipment, as well as filters on their wingtips to scoop up debris for analysis. This was perilous work, with the official report based on the testimony of lead pilot Virgil Meroney describing the cloud's interior thus [23]:

> Inside the cloud Colonel Meroney was impressed with the color. It cast a dull red glow over the cockpit. His radiac instruments all "hit the peg." The hand on the integron, which showed the rate at which radioactivity was being accumulated, "... went around like the sweep second hand on a watch ... and I had thought it would barely move" the Colonel reported. With "everything on the peg" and the red glow like the inside of a red hot furnace, Colonel Meroney made a 90-degree tum and left the cloud. He had spent about five minutes in radiation over one roentgen intensity.

The work was so hazardous that, during the flight, one of the planes, flown by Captain Jimmy Robinson, stalled inside the cloud. Although Robinson recovered, he had used his limited fuel reserves and was forced to ditch in the sea and was killed. The returning planes had their filters removed and sent back to the United States for testing. They first arrived at Los Alamos, where the coral debris reacted poorly to dissolution with nitric acid and often caught fire, requiring the team, led by Roderick Spence, to work outdoors in tents. Here, the team discovered previously unknown, neutron-rich isotopes of plutonium, including ^{244}Pu, with a half-life of some 12 million years (the most neutron-rich isotope detected was ^{246}Pu).

At Berkeley, Seaborg received a secret teletype informing him of the highly classified discovery. This was thanks to his involvement with a program known as "napkin ring," aiming to make heavier plutonium isotopes at the Materials Testing Reactor in Idaho. Seaborg informed Ghiorso and Thompson, who had experience analyzing both US and Soviet nuclear tests. Los Alamos's discovery had been made with a mass spectrograph, which due to the detection sensitivity meant around 0.1% of the plutonium fraction was ^{244}Pu. Based on this and his expertise of bomb yields, he calculated that around 10^{-9} of the uranium yield would have captured 16 neutrons (i.e., would be ^{254}U) which would beta decay into elements as heavy as $Z = 100$. The neutron flux was later calculated at around 10^{25} neutrons cm^{-2} s^{-1} [15].

Although Thompson and Ghiorso were convinced, Seaborg and Perlman remained skeptical. Even so, the team obtained half of one of the filters from the newly established laboratory at Livermore (today Lawrence Livermore National Laboratory) from colleague Kenneth Street, who had also worked on the discovery of californium. Following careful analysis, along with additional lanthanum fluoride precipitate fractions received from Tracerlab (a local company under contract to provide weapons test analysis), the team soon confirmed both $Z = 99$ and $Z = 100$.

This discovery was contested, although surprisingly not from a group based outside the US. On 18 December, the Berkeley team contacted Los Alamos for additional samples so they could verify their discovery. Spence informed them that around 90% of all material had been transferred to Argonne Laboratory in Chicago; Berkeley contacted the second lab, only to be told their request would be taken "under consideration." On 7 January, the team at Argonne informed Berkeley that they believed the Berkeley data to be in error, and in fact an Argonne team had discovered both $Z = 99$ and $Z = 100$.

This debate between the two labs continued, entirely in secrecy until 1955, when it was finally agreed (as Seaborg noted with "an abundance of cocktails") that priority would be given to Berkeley, with Spence and some of the Los Alamos team also credited as co-discoverers. At this point a Swedish team based at the University of Stockholm had also made a claim for the priority of $Z = 100$; although the Berkeley paper was published 10 days afterward, it referenced "unpublished information" that was revealed when the Ivy Mike findings were declassified a year later [24].

The names chosen for the elements were einsteinium and fermium, celebrating two of the preeminent physicists of the twentieth century, both of whom died after the discovery but before the element names were announced; although Segrè was asked to inform the dying Fermi he would have an element named after him, he declined to do so [8].

13.6 Reflections on the Atomic Age

This phase of element discovery is one of the richest in history, equivalent to the advent of electrolysis or the exploration of the miraculous ore at Ytterby, Sweden, that led to multiple new rare earth elements. It is also a point where the periodic table found itself, perhaps more than any other, at the heart of world consciousness and geopolitics. The discovery of nuclear fission, tied inextricably to Fermi's erroneous claim for $Z = 93$, gave us nuclear power and all that came with it—most notably the atomic bomb. Rightly, the focus of history has always remained fixed on the monumental consequences of this weapon and the host of characters involved in the upper echelons of the Manhattan Project [25].

For the chemist, however, there is a deeper story to be told. The discovery of the synthetic elements—first technetium and then promethium and the transuranics—speaks to the wider changes of how science is performed. While Segrè and Perrier are recognized as the discoverers of technetium, this is simply because Lawrence and his Berkeley cyclotrons failed to notice what they had created. The story of the transuranics is therefore a story of the transition of science from small-scale teams and the dedicated individual to Big Science, epitomized by Lawrence and realized by Seaborg and others during the Manhattan Project. This came to an early apotheosis with the discovery of einsteinium and fermium, only made possible by a combination of a military research program, incredible personal bravery on the part of pilots and a research collaboration—albeit with some rivalry—that spanned four national laboratories. Although the Berkeley research machines would continue to get bigger, with the HILAC linear accelerator emerging in 1957, partnership on this scale would not reappear until after the cold war [8].

The discoveries of einsteinium and fermium are a key turning point in the history of element discovery, as they mark the end of bulk production of new elements: the elements beyond would all be produced one atom at a time, starting with mendelevium at Berkeley in 1955 and completing the actinide series before beginning the hunt for the "superheavy" elements of $Z = 104$ and beyond. They also mark the point where Berkeley was no longer alone in the search; as mentioned above, a Swedish–British collaboration would soon lay claim to the discovery of $Z = 102$ (although they are not recognized as the discoverers, their name for the element, nobelium, is used today). And, with the creation of the Joint Institute for Nuclear Research at Dubna in 1956, Berkeley would soon be in direct competition with Flerov's team for the discovery of new elements.

This new phase of element discovery, which persisted for 40 years and is known as the "transfermium wars," would lead to an acrimonious split along the Iron Curtain and, for a time, two different periodic tables as each group claimed priority over the other. This dispute was settled in the 1990s through a joint working group by the International Unions of Pure and Applied Chemistry (IUPAC) and Physics (IUPAP), first with the determination of priority for the elements [26] and then—after multiple protests from both sides—the elements' names [8].

One of the most controversial of these names was that chosen for element 106, largely because, for the first time in history, it celebrated a scientist who was still alive at the time of naming.[8]

It was called seaborgium.

References

1. Wells HG (1913) The world set free. Macmillan, London
2. Ellis CD, Henderson WJ (1934) Artificial radioactivity. Proc Roy Soc Lond A 146:206–216. https://doi.org/10.1098/rspa.1934.0149
3. Chadwick J (1932) Possible existence of a. Nature 129:312. https://doi.org/10.1038/129312a0

[8]Seaborg died in 1999. He remained the only person to have an element named after them while still alive until 2016, when Flerov's successor Yuri Oganessian was recognized with element 118, oganesson. Lawrence (103), Meitner (109) and Flerov (114) would also have elements named after them.

4. Fermi E (1934) Possible production of elements of higher than 92. Nature 113:898–899. https://doi.org/10.1038/133898a0
5. Noddack I (1934) Das periodische System der Elemente und seine Lücken. Angew Chem 47:301–305. https://doi.org/10.1002/ange.19340472002
6. Fermi L (1954) Atoms in the family: My life with Enrico Fermi. University of Chicago Press, Chicago
7. Chapman K (2019) The first synthetic element. Nature 565:570–571. https://doi.org/10.1038/d41586-019-00236-4
8. Chapman K (2019) Superheavy. Bloomsbury Sigma, London
9. The Nobel Prize in Physics 1938. https://www.nobelprize.org/prizes/physics/1938/summary/. Accessed 30 May 2020
10. Meitner L, Frisch O (1939) Disintegration of uranium by: a new type of nuclear reaction. Nature 143:239–240. https://doi.org/10.1038/143239a0
11. Seaborg GT, Seaborg E (2001) Adventures in the atomic age. New York, Farrar, Straus and Giroux
12. Segrè E (1939) An unsuccessful search for. Phys Rev 55:1104–105. https://doi.org/10.1103/physrev.55.1104
13. McMillan E, Abelson P (1940) Radioactive element 93. Phys Rev 57:1185–1186. https://doi.org/10.1103/physrev.57.1185.2
14. Segrè E, Seaborg GT (1938) Nuclear isomerism in element 43. Phys Rev 54:772. https://doi.org/10.1103/physrev.54.772.2
15. Hoffman DC, Ghiorso A, Seaborg GT (2000) The transuranium people: The inside story. Imperial College Press, London
16. Seaborg GT, James RA, Morgan LO (1949) Vol 14 B The transuranium elements: research papers, National nuclear energy series vol 14B, Paper 22.1. McGraw-Hill Book Co, New York
17. Mars Curiosity Rover: APXS. https://mars.nasa.gov/msl/spacecraft/instruments/apxs/. Accessed 30 May 2020
18. Cochran T, Norris R, Bukharin O (1995) Making the Russian bomb: From Stalin to Yeltsin. Westview Press, Boulder CO
19. Thompson SG, Seaborg GT, Ghiorso A et al (1950) The new element (98). Phys Rev 80:790–796. https://doi.org/10.1103/physrev.80.790
20. Goeppert-Mayer M (1949) On closed shells in nuclei. II. Phys Rev 75:1969–1970. https://doi.org/10.1103/physrev.75.1969
21. Wheeler JA (1955) Nuclear fission and nuclear stability. In: W (ed). Niels Bohr and the development of physics, Pergamon Press, London
22. The Nobel Prize in Chemistry 1951. https://www.nobelprize.org/prizes/chemistry/1951/summary/. Accessed 30 May 2020
23. US Air Force (1963) History of Air Force atomic cloud sampling. US Air Force, Washington, DC
24. Ghiorso A, Thompson SG, Higgins GT et al (1955) The new elements and 99 and 100. US Atomic Energy Commission, Berkeley
25. Rhodes R (1987) The making of the atomic bomb. Simon & Schuster, London
26. Barber RC, Greenwood NN, Hrynkiewicz AZ et al (1993) Discovery of the elements. Part II: Introduction to discovery profiles. Part III: Discovery profiles of the transfermium elements. Pure Appl Chem 63:1757–1814. https://doi.org/10.1351/pac199365081757

Mary Elvira Weeks and *Discovery of the Elements* 14

Vera V. Mainz

Abstract

Mary Elvira Weeks (1892–1975) was the author of the highly successful book, *The Discovery of the Elements*. She was born in Lyons, Wisconsin, and attended Ripon College, Ripon, Wisconsin (B.A. 1913), the University of Wisconsin (M.A. 1914), and the University of Kansas (Ph.D. 1927). In 1932, Weeks began publishing a series of more than twenty-one articles in the *Journal of Chemical Education*, and in 1933 she combined these articles in book form and published the first edition of *The Discovery of the Elements*. The book went through six further editions, ending with the 7th edition in 1968 which was co-authored with Henry M. Leicester. In 1944 Weeks moved to Wayne State University in Detroit to accept a post as a scientific researcher at the Kresge-Hooker Science Library. She retired in 1954 and stayed in Detroit until her death. In addition to these and other biographical details, the present paper contains an analysis of the changes made in the various editions of *Discovery of the Elements*, along with a study of how the book was received, as illustrated by excerpts from book reviews over the years. Much of the material in *Discovery of the Elements* is as enjoyable and informative today as it was when first published beginning in 1932.

14.1 Introduction

It seems appropriate during the celebration of the 150th anniversary of Mendeleev's publication of the periodic law that we remember Mary Elvira Weeks (1892–1975) (Fig. 14.1). Her publication, *The Discovery of the Elements*, introduced many

V. V. Mainz (✉)
School of Chemical Sciences, University of Illinois at Urbana-Champaign, Urbana, IL, USA

C. J. Giunta et al. (eds.), *150 Years of the Periodic Table*, Perspectives on the History of Chemistry, https://doi.org/10.1007/978-3-030-67910-1_14

Fig. 14.1 Mary Elvira Weeks in the 1930s [2] (Courtesy of the Spencer Research Library, University of Kansas)

scientists (and non-scientists) to the stories of how the elements of the periodic table were discovered. It remains widely read even today, and continues to serve as an important source text in the history of chemistry. She began the first article in the series with the following [1]:

> The story of the disclosure, one by one, of the chemical elements has never been told as a connected narrative. The reports of these discoveries and the life stories of the discoverers are recorded for the most part in old chemical journals, biographical dictionaries, old letters, and obsolete textbooks that are seldom read by the busy modern chemist. It is hoped, therefore, that these chapters may not only render tribute to the honored men and women who helped to reveal the hidden chemical elements, but that they may also serve to acquaint chemists and others with these great achievements [2].

Mary Elvira Weeks created a lasting tribute to all those who discovered the elements and she told those stories in an accessible way.

14.2 Mary Elvira Weeks's Family and Early Years

The Weeks family traces its roots back to Robert de Wrey, an Englishman who lived in the early twelfth century [3]. The name is of Norman origin, so the family likely immigrated to England with William the Conqueror after 1066. A later descendant, George Weeks (1600–1659), "came to Dorchester, MA, [now a Boston neighborhood] in the same ship with the Rev. Richard Mather in 1635, five years after the original settlement [of Dorchester] in 1630" [4]. A descendant of George Weeks, Spencer Weeks (1797–1859), was born in Norwich, MA, roughly 110 miles west of Dorchester. He was a farmer in Norwich and moved his family,

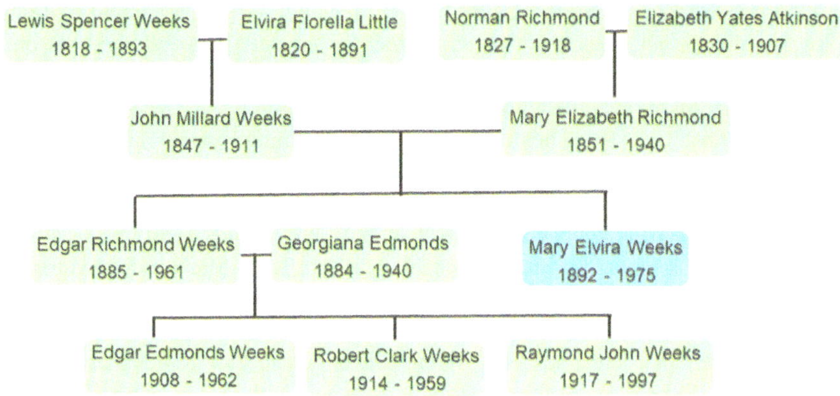

Fig. 14.2 Recent family members of Mary Elvira Weeks. Elvira's great nephew, Wayne D. Weeks (Edgar Edmonds Weeks's son) was a member of the Mayflower Society and was very interested in the family history

including Weeks's grandfather, Lewis Spencer Weeks, to Lyons, WI, in 1843 [5]. On 1 July 1846, Lewis Spencer Weeks bought 80 acres of farmland near Lyons from the US government. A brief summary of the more recent descendants in Weeks's family is given in Fig. 14.2.

Weeks's maternal grandmother, Elizabeth Yates Richmond, born in Liverpool, England, in 1830, was brought to Montreal when young. As a teenager she moved to Appleton, WI with her father and in 1850 she married Norman Richmond, who shortly thereafter started the first paper mill in the Fox River Valley of Wisconsin. Elizabeth became a noted writer and published one book, *Poems of the Western Land*, in 1878 [6]. This collection of poems about the American West was based on Indian folklore she had collected and translated. She corresponded with such eminent poets of the period as Henry Wadsworth Longfellow and John Greenleaf Whittier, who expressed great appreciation of her work [7]. Two of her poems were later successfully set to music by Elizabeth's daughter, Mary Elizabeth ("Minnie"), the mother of Elvira.

Elvira's father, John Millard Weeks, attended Lawrence College in Appleton, WI from 1869–1873. At Lawrence College he met his future wife, Mary Elizabeth Richmond, who was also a student there. He graduated in 1873 with a B.A. degree. Mary Elizabeth Richmond graduated with a B.S. and an M.S. degree [8] and attended Wellesley College for one year just prior to her marriage.

The Weeks family obviously valued education. In 1869, only 1.3% of the 18–24-year-old population in the US were enrolled in institutions of higher education [9]. Both of Elvira's parents obtained college educations at a time when that was rare, especially in rural America. In addition, Elvira's later scholarly interests may have been inspired by the literary accomplishments of her grandmother, Elizabeth Yates Richmond, who died when Elvira was 13.

John Millard Weeks married Mary Elizabeth Richmond on November 6, 1884. John "taught district schools a few terms and for several years followed the profession of civil engineer, his health breaking down from over-study and close application to his work; after which he engaged in farming at Lyons, moving in 1895 to Oshkosh …" [10]. Thus, John retired from farming at age 47, and the move to Oshkosh presumably occurred because of poor health. By the time of the 1900 US Census, the family lived at 100 Fulton Ave, Oshkosh, WI.

Mary Elvira Weeks, who was known to family and friends as Elvira, was born on her parent's farm near Lyons, Wisconsin, April 10, 1892. She was named for her mother Mary Elizabeth (Minnie) Richmond and her paternal grandmother, Elvira Little. Weeks attended elementary and secondary public schools in Oshkosh, WI [11]. She noted that she had been "interested in science for as long as she can remember [2]." She took all the scientific classes available and received a copy of Charles Darwin's *Origin of Species* as a Christmas present from a high-school "chum" [2].

14.3 Mary Elvira Weeks's Education and Early Employment. 1910–1921

At age 17, in January 1910, Weeks (Fig. 14.3) enrolled at Ripon College in Ripon, WI. In the years since her parents had attended college, matriculation rates had increased only slightly: in 1910, 2.8% of the 18–24-year-old population in the US were enrolled in institutions of higher education [9]. Weeks had inherited $5000 from her grandfather Lewis Spencer Weeks upon his death in 1893. The money was held in trust for her education and/or support or until she reached the age of 21 [12].

Weeks majored in chemistry. She was mentored by the chemist Albert Franklin Gilman[1] (1871–1951) [13] (Fig. 14.4), and in her senior year assisted him in the laboratory [11]. She completed her B.A. in June 1913 [14], although her graduation picture appears only in the 1914 Ripon Yearbook, the *Crimson* [15] (Fig. 14.5). During her time at Ripon College Weeks was a member of the Y. W. C. A. She did not join the Science Club at Ripon, which seemed to be mostly a male bastion, according to the evidence in the *Crimson*. We do not know what classes Weeks took at Ripon, but it is likely that she expanded her knowledge of foreign languages, a talent that was later to prove of great value to her. German, French, Latin, and Greek were all taught at Ripon at the time.

[1]Albert Franklin Gilman received his B.A. from Amherst College 1897, A.M. Amherst 1901, and his. Ph.D. from the University of Denver 1913, with the thesis *Some Reactions of Oxycellulose*. From 1897–1906, he taught science in Maine, New Hampshire, and Tennessee. He was Professor of Chemistry at Ripon College from 1906–1917. Gilman later taught at Huron College (SD), Illinois Wesleyan, Carroll College (WI), and the Central YMCA College of Arts and Sciences, Chicago. In 1908 he published *A Laboratory Outline for Determinations in Quantitative Chemical Analysis*, and in 1913 *The Origin of the Republican Party*.

Fig. 14.3 Mary Elvira Weeks, likely a high-school picture from around 1909 (Courtesy of Sandra Weeks Brangan)

Fig. 14.4 Albert Franklin Gilman (1871–1951) [16] (Courtesy of Ripon College Archives)

In 1914, Weeks was the only woman to earn a B.A. in chemistry (out of 12 women in her class); for comparison, 5 of the 25 men in her class majored in chemistry. In the January 1914 *Ripon College Bulletin*, which listed all of those who graduated in June 1913 and their current occupations, the eleven other women in Weeks's class listed themselves as teachers, whereas Weeks was the only female graduate student [14]. From 1910–1915, the average number of women receiving B.A. degrees in science at Ripon College (chemistry, physics, mathematics, and

Fig. 14.5 Mary Elvira Weeks, Ripon College Crimson, 1914 [15] (Courtesy of Ripon College Archives)

biology) was 14% (9 out of 63) while the average number of men receiving B.A. degrees in science was 56% (67 out of 119).

Weeks attended the University of Wisconsin at Madison where she worked with J. Howard Mathews[2] (1881–1970) [17] (Fig. 14.6). She earned an M.A. in 1914, for her thesis *The Effect of Various Negative Catalyzers on the Atmospheric Oxidation of Sodium Sulphite in Ultraviolet Light*. This work also resulted in her first publication [18]. In 1920, at the University of Wisconsin, women constituted about 22% of chemistry course enrollments [19].

After obtaining her M.A. degree in 1914, Weeks taught high school in Wisconsin, but during and after WWI, she left teaching and worked as an analyst in the mining laboratories of the New Jersey Zinc Corp., Franklin, NJ, and the Carborundum Company in Niagara Falls, NY [11, 20]. The exact dates of her industrial

[2]Joseph Howard Mathews was one of the early researchers in colloid chemistry in the US. Mathews, who obtained his B.S. at the University of Wisconsin, joined the Wisconsin faculty after completing his Ph.D. under the physical chemist T. W. Richards at Harvard in 1908, with the thesis *A Study of Compressibility and its Relation to Various Other Physical Properties of Certain Organic Compounds*. Mathews was one of the original nine undergraduates at the University of Wisconsin who organized and founded Alpha Chi Sigma, the professional chemistry fraternity, in 1902. His gravestone contains visual references to the organization, showing his continued interest. Mathews collaborated with Wisconsin colleagues Farrington Daniels (1889–1972) and John Warren Williams (1898–1988) to publish *Experimental Physical Chemistry* in 1929. That book, through seven editions and with added authors from the Wisconsin chemistry faculty, was the market leader among physical chemistry laboratory textbooks until the 1970 s and was used by the author for her undergraduate physical chemistry laboratory work. Mathews was also an expert on scientific methods of crime detection. During his long career, he conducted research in this field and lectured on it at the University of Wisconsin and nationwide. He published his authoritative three-volume work *Firearms Identification* in 1962.

Fig. 14.6 Joseph Howard Mathews (1881–1970) [23]. Used with permission of the University of Wisconsin at Madison (Courtesy of the University of Wisconsin-Madison Archives [ID 2020s00027])

employment are not known, but the *Ripon College Bulletin* dated March 1916, reports that she was then employed by the New Jersey Zinc Corp [21]. A publication from 1937 [22] gives her dates of employment at the Carborundum Company as 1918–1920.

This career change, from teaching to working as a chemical analyst, came about because employment opportunities opened up for women chemists, especially as the US started deploying men for the battlefields of WWI. As Marelene and Geoffrey Rayner-Canham observe [19]:

> In our view and that of others, the inter-war years did, in fact, see an increase in discrimination against women. Before the First World War, women were tolerated in male professions in so far as they were not perceived as a threat but more as a curiosity. During the war, however, women filled a significant portion of skilled occupations. As a result, with the end of the war, as [economic historian Elizabeth] Roberts comments: "They [women] were often regarded with open hostility by men who had realized for the first time that women were fully capable of carrying out jobs previously perceived as men's, thus presenting a real challenge" [24].

It would not be unusual if Weeks encountered this kind of hostility and if her employment at the Carborundum Company ended under circumstances similar to those given below [25]:

> For example, the Illinois Steel Company announced in 1919 that: "The women chemists of the Illinois Steel Company not only made good as chemists but showed their fine spirit by resigning in order to make places for the men returning from war work." The company did not elaborate on whether the resignations were given with enthusiasm or under coercion.

14.4 University of Kansas. 1921–1944

In 1921, after leaving the Carborundum Company, Weeks (Fig. 14.7) went to the University of Kansas (KU) in Lawrence, KS, as an instructor in chemistry at a salary of $1900/year [21]. As a comparison, C. S. (Speed) Marvel, who later became a well-known polymer chemist, was paid $1700/year as an instructor in chemistry at the University of Illinois in 1920 [26], so Weeks's salary was not atypical for the time. Weeks worked in Bailey Hall, known to the students as Bailey's Barn [27], and taught quantitative analysis there for the next 22 years. She also started working on her Ph.D. degree with Prof. Hamilton P. Cady[3] (1874–1943) (Fig. 14.8). Cady had become chair of the Chemistry Department in 1920, just before Weeks's arrival.

Weeks was teaching full time and working in the Cady laboratory when her heavy teaching duties permitted. After six years of work, she was able to complete her Ph.D. thesis in 1927 on *The Role of Hydrogen Ion Concentration in the Precipitation of Calcium and Magnesium*, which also resulted in her second publication [30]. She earned the twelfth Ph.D. degree in chemistry granted by KU and the first in chemistry granted to a woman. One result of this distinction was her appointment as KU's first female assistant professor of chemistry in 1927, at a salary of $2000 per year [31].

Weeks's achievement was part of a short-lived national trend in which larger numbers of women were earning Ph.D. degrees in chemistry [32]:

> A maximum was reached in 1929 when 10% of chemistry PhD degrees were awarded to women, dropping to 5% by 1933, reaching a minimum of 2% in the 1940s. The 10% figure was reached again in the abnormal year of 1946 (which came after five years in the 2-4% range) and not regained until 1972, when women started consistently to exceed the 10% figure.

Weeks was a shy woman by all accounts, and there is one good contemporary description of her character by Doris Brewster Swift [33–35]. Ms. Swift was the daughter of Ray Q. Brewster[4] (1892–1983), who started as an assistant professor in chemistry at KU in 1919, shortly before Weeks's arrival on campus. The reminiscences that involved Weeks are compiled in Appendix 2. Ms. Swift summarized her recollections in 1999 [31]:

---

[3]Hamilton Perkins Cady received his B.A. in 1897 from KU. He returned to Kansas in 1899 as an assistant professor and worked on his Ph.D. under Edward Curtis Franklin (1862-1937), a pioneer in the chemistry of ammonia and related compounds. Cady completed his Ph.D. in 1903 with the thesis *Concentration Cells in Liquid Ammonia*. In 1905, Cady discovered with fellow KU chemistry professor David F. McFarland that there was a significant amount of helium in a sample of Kansas natural gas. When they published their complete findings in November 1907 [28, 29], Cady commented that their work "assures the fact that helium is no longer a rare element, but a common element, existing in goodly quantity for uses that are yet to be found for it." He served as chair of the KU Chemistry Department from 1920-1940.

[4]Ray Q. Brewster was promoted to associate professor in 1921, and full professor in 1927, becoming chair of the Department of Chemistry in 1940, after Cady stepped down.

Fig. 14.7 Mary Elvira Weeks, in the early 1920s (Courtesy of Sandra Weeks Brangan)

Fig. 14.8 Hamilton P. Cady (1874–1943) (Courtesy of the Spencer Research Library, University of Kansas)

Elvira was most often described as extremely timid and shy. She lived with her mother in a small yellow bungalow with a well kept lawn and hedge and brightly colored flowers. The bungalow was located on the northeast corner of 18th and Mississippi Streets, just south of the campus. Her house was the most colorful one on the block – in contrast to her lack of color in clothing, which was typical of the period. Her house was several doors down from the Brewster home and catty-corner from Miss Daum, a librarian at Lawrence Public Library who shared common interests. Miss Weeks never had a car and either walked or rode the bus. When Elvira's mother was alive, the Brewsters sometimes took them to church. Miss Weeks was the first person on the block to have a radio and frequently invited the neighbors in to listen to important broadcasts.

After being promoted to assistant professor in 1927, Weeks tried to continue experimental research while also carrying a heavy teaching load. Her real interests were, however, shifting to languages and humanistic studies and she started to devote her time to studies of the history of chemistry [36]. It is at this point that her story crosses paths with two men who were to play significant roles in her future: Franklin Dains and Neil Gordon. We will take a moment to summarize their careers before proceeding with Weeks's story.

Franklin Burnett Dains (1869–1948) (Fig. 14.9) [37] received his undergraduate education at Wesleyan University in Middletown, CT. He completed his B.A. in 1890 and then spent two years at Wesleyan as an assistant in chemistry. He had an early connection with KU in 1893–1894 when he was hired as a temporary lecturer. He then moved to the University of Chicago, where he finished his Ph.D. degree in 1898, working under two well-known organic chemists, Julius Stieglitz (1867–1937) and John Ulric Nef (1862–1915). In 1901–1902, Dains was in Freiburg, Germany, doing post-graduate work with Ludwig Gattermann (1860–1920) at the University of Freiburg. It was during this trip that Dains likely started collecting portrait medallions, old pewter, old books, and engravings, especially those related to chemistry. He returned to the US in 1902, where he worked as a professor of chemistry at Washburn College in Topeka, KS. In the fall of 1911, Dains was hired as an associate professor of chemistry at KU and promoted to full professor in 1914. He had an active research group and was the first professor to initiate research in organic chemistry at KU. But Dains also had a strong interest in the history of chemistry. In 1909, he served as the President of the Kansas Academy of Science and gave his first published paper on the history of chemistry as his presidential address [38]. As Weeks wrote about Dains, her colleague at KU [39]:

One of the first courses in history of chemistry was taught by Dr. Dains. His lectures are always superbly illustrated, for he has about two thousand lantern slides. Visitors to the Chemistry Building at the University of Kansas are always impressed by the many fine portraits of chemists which adorn the halls, offices, and lecture rooms and which were collected largely through the efforts of Dr. Dains. [Dains left the collection to KU. There are approximately 2,500 lantern slides and 1,700 printed images and related materials in the collection.] They are astonished, too, to find that this mid-western university contains in its library a wealth of rare treasures on the history of chemistry which were acquired largely through the vision and foresight of Dr. Dains.

Dains was a contributing editor for the *Journal of Chemical Education* [39]:

Dr. Dains believes that one of the most far-reaching recent advances in the teaching of chemistry was the founding of the Division of Chemical Education and its journal. Few realize how great have been his services to the *Journal of Chemical Education* or know how many hours he has spent in estimating the value of manuscripts submitted for publication, in searching his files for illustrative material for frontispieces and for the articles of other contributors, in writing careful abstracts of historical articles selected with great discrimination, and in preparing helpful book reviews.

Fig. 14.9 Franklin Burnett Dains (1869–1948). Wesleyan University graduation photograph (Courtesy of the Spencer Research Library, University of Kansas)

It was through the *Journal of Chemical Education* that Weeks established a lasting professional connection with Neil E. Gordon (1886–1949) (Fig. 14.10). Neil Elbridge Gordon received his B.S. (1911) and M.S. (1912) degrees in mathematics from Syracuse University. He then went to Johns Hopkins University and switched to the study of physical chemistry under Ebenezer Emmet Reid (1872–1973), finishing his Ph.D. degree in 1917. He joined the University of Maryland Chemistry Department faculty in 1920, going through the professorial ranks and ending as Department Chair before his departure in 1939. In 1932, he founded the Gibson Island Conferences, which later became known eponymously as the Gordon Research Conferences. Gordon was the founding Editor of the *Journal of Chemical Education* in 1924 [40] and continued in this role until 1933. He specifically intended to include history of chemistry articles in the new *Journal*, as he made clear in his editorial of December 1924 [41]:

> The importance of history in chemical education warrants us organizing a new department in THIS JOURNAL for 1925 known as Historical Chemistry. This department will be headed by Dr. Lyman C. Newell who has made this field a life study and who has a wealth of material from which to draw. Dr. Edgar F. Smith, Leader of Historical Chemistry in America, has consented to write the first series of articles, and it is hoped that this series may begin with the January issue.

The founding of the *Journal of Chemical Education* proved to be providential for Weeks, because the years 1929–1935 were her most productive period. She published:

Fig. 14.10 Neil E. Gordon (1886–1949) (Courtesy of the Chemistry Department, Wayne State University)

- 4 scientific research articles
- 30 articles in the history of chemistry, **24** of which appeared in 1932–1933 in the *Journal of Chemical Education* and later were published in book form as *The Discovery of the Elements*
- 2 books, one of which, *The Discovery of the Elements*, went through 3 editions in this period (1933, 1934, 1935); she also published *A Laboratory Manual of Quantitative Analysis*.

One cannot fail to be impressed by the number of papers that Weeks published in the *Journal of Chemical Education* in 1932–1933. The first four papers appeared in the January issue of volume 9 in 1932, and the last paper, the twenty-first, appeared in the May issue of volume 10 in 1933. An enormous amount of work was involved in this project: collecting the widely scattered information, writing all the papers, finding the illustrations (many of which came from the Dains collection), and preparing the articles for publication. Unfortunately, Weeks left no records relating to *Discovery of the Elements* from her time at KU, and none remain at the *Journal of Chemical Education*. The latter was likely lost during a change of editors.

Figures 14.11 and 14.12 show Weeks with her brother Edgar Richmond Weeks (1885–1961) at about the same time she was writing the first of these articles. The photographs were taken at the family farmhouse in Lyons in the 1930s, most likely when the family came together to move Weeks's mother, Minnie, to Lawrence, Kansas, to live with Weeks.

I came across one other image of Weeks from this time, taken on May 11, 1932 (Fig. 14.13) [42], when all the *Discovery* papers were rolling out. The description of the photo from the KU Archives reads as follows:

Fig. 14.11 Mary Elvira Weeks, her brother Edgar Richmond Weeks, and her mother Mary (Minnie) Weeks, around 1930 (Courtesy of Sandra Weeks Brangan)

The occasion was a meeting of the Kansas (Iota) Chapter of Sigma Xi, the national honorary scientific society. The meeting included the initiation of new members, a banquet, and a guest speaker. Some women members are performing the entertainment portion of the meeting.

The skit is a take-off on the 1928 novel *Ladies in Hades: A Story of Hell's Smart Set* by Frederic Arnold Kummer, in which Eve forms a woman's club of *femme fatales*—Cleopatra, Delilah, Scheherazade, and others—to talk about their love affairs. According to the Lawrence newspaper, the Sigma Xi women have adapted the story to "the wives of famous scientists telling their troubles to the king of Hades."

The women are all faculty at the University of Kansas, Lawrence. From left to right, the players are:

Cornelia Mitchell Downs, associate professor of bacteriology, as Hades

Beulah M. Morrison, associate professor of psychology

"Fritzie" Downs as Cerberus, the three-headed Hound of Hades

Kathleen Doering, assistant professor of entomology

Florence L. Blackman, assistant professor of mathematics

Fig. 14.12 From left to right: Mary Elvira, Robert C. (nephew), Raymond J. (nephew), Mary (Minnie) (mother), Edgar R. (brother), around 1930 (Courtesy of Marsha Weeks)

Fig. 14.13 Sigma Xi skit, May 11, 1932. Mary Elvira Weeks is second from the right (Courtesy of Walter P. Reuther Library, Archives of Labor and Urban Affairs, Wayne State University)

Selma Gottlieb, assistant professor of chemistry. At her feet is a monkey sitting on a tree branch and reading a book

Wealthy Babcock, assistant professor of mathematics

Ethel Ann Jones, assistant professor of chemistry

Lalia Viola Walling, assistant professor of physiology

Mary Elvira Weeks, assistant professor of chemistry

Mary Elizabeth Larson, assistant professor of zoology, as Hermes

Weeks in 1932 was just completing the first edition of *The Discovery of the Elements* which featured famous scientists, many of them ancients, and was extensively illustrated with images from Frank B. Dains's history of chemistry collection. Selma Gottlieb read portions of Weeks's manuscript and Mary Larson provided translations of Swedish manuscript sources.

I find it comforting that Weeks had women friends with whom she felt comfortable enough to participate in a skit in costume and have some fun. Several of these women spent their professional lives on the KU staff. Wealthy Babcock had a career path similar to Weeks's: she became an assistant professor of mathematics in 1926 and an associate professor in 1940. The photo caption does not say, but it is reasonable to assume (as judged from the costume and hairstyle) that Weeks was playing the role of Marie Lavoisier, who participated actively in the chemical researches of her husband Antoine.

The years 1936–1944 were also productive ones for Weeks:

- Published the 4th edition of *Discovery*, 1939
- Published 9 articles in the history of chemistry
- Promoted to associate professor in 1937, at a salary of $2025 (her salary had decreased 10% in 1933 as a result of the Great Depression) [21]

In addition to this activity, on July 14, 1939, Weeks wrote to Albert Einstein to ask for his help in arranging for the emigration of Max Speter to the US (1883–1942) [43]. Speter, a noted historian of chemistry, was particularly known for his work on the history of sugar and superphosphate. Born of Jewish parents in what was Austria-Hungary and is now Romania, he was a member of the founding team of the Deutsches Museum and had held several positions in chemical industry. Weeks wrote to Einstein [44]:

> A recent letter from Dr. Günther Bugge, author of "Das Buch der grossen Chemiker," has brought to me the sad news that Dr. Max Speter's situation in Germany is becoming more desperate each day and that, if his life is to be saved, he must be given some reasonable basis for hope.

Weeks noted that she had been trying for several years to arrange for Speter's emigration but things were now desperate as he had been forbidden to publish or give lectures. Einstein wrote back with his advice [45, 46], asking for names from chemical companies that he could approach. Weeks sent a series of six letters to Einstein, ending on 17 September 1939, noting that Speter's passport had expired. She wrote [44]:

> I, too, have reluctantly concluded that it must now be impossible for Dr. Speter to leave Germany. He has an expert knowledge of respirators, gas masks, and other kinds of life-saving equipment.

Max Speter was neither the first nor the last Jewish scientist for whom Weeks actively sought to arrange safe passage from Germany during the years immediately before World War II. Some of these efforts were successful: in 1941 a woman who had fled Hitler's Germany, Dr. Steinberger, was living in Weeks's home in Lawrence [47].

Several things happened in Weeks's life in 1940. Her mother died, and her Ph.D. advisor, H. P. Cady, stepped down as Chair of the Chemistry Department at KU. I believe that she realized then, if she had not before, that she would never attain the rank of full professor, likely because her work in the history of chemistry was not valued. She knew at least one other woman on the KU science faculty—Cornelia Mitchell Downs—who was appointed full professor of bacteriology in 1935. That action shows that a woman could be appointed to a full professorship at KU. But Weeks had stopped (or was prevented from) doing experimental chemistry, and (as was and often still is true in many US universities) scholarship in the history of chemistry was not sufficiently respected to be recognized by promotion to a full professorship in a chemistry department. No other woman would be appointed as professor of chemistry at KU until 1975 [31]. With her mother gone, Weeks (Fig. 14.14) was free to move somewhere else where she could do the type of work she was interested in and could be valued for. Figure 14.15 is a picture of the KU Chemistry Department staff in 1944, Weeks's last year on the faculty there.

14.5 The Kresge-Hooker Scientific Library and Later Life. 1944–1975

In 1944, Neil Gordon re-entered Weeks's life. In 1936, Gordon had moved to Central College in Fayette, MO, to become the Chair of the Chemistry Department. While at Central College, he became aware of a unique opportunity to acquire the library of Samuel Hooker[5] (1864–1935) [48, 49]. At the time of Hooker's death, the library contained 2470 monographs and 18,850 periodical volumes. Gordon negotiated to purchase the Hooker Library, *in toto*, for $100,000 and move it to Central College. As part of the purchase arrangements, he founded the Friends of the Hooker Library in 1939 to keep the library up-to-date and make the collection available for translation services, abstracts work, searches, and photocopies to those outside of Central College. In 1942, Gordon was offered the job of chair of the Department of Chemistry at Wayne State University in Detroit, MI. He accepted on

[5]Hooker had emigrated to the U.S. in 1885 after getting his Ph.D. in Munich. He began his career as chief chemist with the Sugar Refining Company of Philadelphia. When he retired in 1915, Hooker spent a lot of his time enlarging his library. His objective was "the building of a world-famous library… [it] was known in 1935 as the most complete chemical library in the world" [49].

Fig. 14.14 Mary Elvira Weeks, in the early 1940s (Courtesy of Sandra Weeks Brangan)

the condition that the Hooker Library would move to Wayne State as well. The cost of this move was $200,000: $100,000 to buy the collection plus another $100,000 to bring it up-to-date. The Kresge Foundation[6] contributed $100,000 of the funds needed. With this donation, Gordon was able to raise the other $100,000 by subscription from a number of companies, organizations, and private individuals. Once at Wayne State, the collection became known as the Kresge-Hooker Library. An announcement at the time stated: "So today [1946] Dr. Hooker's dream is finally realized. Persons engaged in scientific research anywhere in the country may secure information and material on any subject in chemistry through the Library which he began" [50]. Now Gordon needed staff for his library.

In 1944, Weeks left KU at Gordon's invitation and joined the Kresge-Hooker Scientific Library in Detroit as a research associate of scientific literature and translator. The Kresge-Hooker Scientific Library was unique at the time. The services offered included carrying out searches of the scientific literature and preparing translations of scientific articles. According to the description of Weeks's papers at the Reuthers Library at Wayne State University: "Weeks focused on the translation of scientific, technical, and medical articles into German, French, Spanish, Italian, Swedish, Danish, Norwegian, Russian, and Portuguese with the supervision and assistance from linguists, as well as students and professors from various language

[6]Sebastian Spering Kresge (1867-1966) established the Kresge Foundation in Detroit in 1924 with an initial gift of $1.6 million. He did this to commemorate the 25th anniversary of the S.S. Kresge Company, which began in 1899 as a single 5-and-10-cent store – a revolutionary merchandising idea at the time – and grew into a nationwide chain of stores. Many years later, the enterprise became known as Kmart Corp. Owing to the Foundation's mission to benefit the city of Detroit, it was natural to approach them as a possible donor.

Fig. 14.15 KU Chemistry Staff, 1944. Front row only (left to right): Ray Q. Brewster (1892–1983), Mary Elvira Weeks (1892–1975), Franklin B. Dains (1869–1948), Robert Taft (1894–1955), Arthur W. Davidson, George W. Stratton. Brewster, an organic chemist, was chair of the department at the time (Courtesy of the Spencer Research Library, University of Kansas)

departments within the university" [51]. In most of these languages, Weeks was self-taught. A story told by her nephew Raymond [44] was that a colleague mentioned that they could not find a Russian translator for scientific documents, so she decided to learn Russian. Figure 14.16 shows the members of the staff of the Kresge-Hooker Library in the late 1940s, including Gordon and Weeks.

After her arrival in Detroit, Weeks also became an associate editor for the *Record of Chemical Progress*, which in 1942 had become the official journal of the American Association for the Advancement of Science section C (chemistry). She served in this capacity until 1971, when publication of the *Record* ceased.

Weeks issued a new edition of *Discovery* (the 5th) in 1945. In 1946, Charles A. Browne (1870–1947) invited Weeks to collaborate in writing *A History of the American Chemical Society: 75 Eventful Years (1870–1947)* [52]. Browne had started planning the volume in 1944, but his health started to fail before he could finish the work. He died in 1947, just as the collaboration was beginning. Weeks (Fig. 14.17) took on the major task of finishing the work, which was published in 1952.

Fig. 14.16 Left to right: Dr. W. H. Powers, V. Wallace, E. Erlandson, H. W. Pearsall, **N. E. Gordon, M. E. Weeks**, M. K. Shaw (Courtesy of Walter P. Reuther Library, Archives of Labor and Urban Affairs, Wayne State University)

Fig. 14.17 Mary Elvira Weeks, in the early 1950s (Courtesy of Sandra Weeks Brangan)

In 1954, Weeks retired from her position at the Kresge-Hooker Library. She was still busy as an associate editor for the *Record of Chemical Progress*, and also served as a consulting editor for *Chymia* from 1956–1967, when it too ceased publication. *Chymia* was an annual journal which published studies in the history of chemistry.

Weeks wrote the 6th edition of *Discovery* in 1956 [53]. Weeks decided she needed help writing a new chapter on elements discovered by atomic bombardment: the invention of the cyclotron and then the atomic pile had enabled elements to be synthesized artificially, with the latest, element 101–mendelevium—having been announced in 1955. She turned to Henry Leicester.

Henry Marshall Leicester (1906–1991) [54] (Fig. 14.18) was born in San Francisco, CA in 1906. He went to Stanford University at age 16, and got three degrees there: A.B., 1927; M.A., 1928, thesis *The Synthesis of Derivatives of Triphenyl Selenonium Hydroxide*; and Ph.D., 1930, *Aromatic Selenonium Salts and the Relative Electronegativities of Organic Radicals,* the latter two degrees under Francis W. Bergstrom (1897–1946). From 1930–1938, Leicester traveled and worked as a chemist in various places. At one of his stops, the Ohio State University, he discovered a complete run of the *Journal of the Russian Physico-Chemical Society*. This journal aroused his interest in Russian chemistry and he eventually became the undisputed American authority on the lives and works of Russian chemists. He corresponded with colleagues in the Soviet Union and amassed a large collection of Russian books on the history of science, which he later donated to Stanford. In 1941, he was appointed to a faculty position in the Biochemistry Department at College of Physicians and Surgeons, San Francisco (now the Dental School of the University of the Pacific). He served as Chair of the Department of Biochemistry and as Head of the Research Program and retired in 1977. He was an expert on water fluoridation and its utility in preventing dental caries.

In 1967 Weeks was chosen to receive the Dexter Award of the History of Chemistry (HIST) Division of the American Chemical Society [55], in recognition of her outstanding contributions to the field. As stated in a letter from Martin Levey

Fig. 14.18 H M. Leicester [54] (Reproduced with permission of the Division of the History of Chemistry of the American Chemical Society)

Fig. 14.19 Dr. Charles Harmison, chair of the Detroit Local ACS section, presenting the Dexter Award to Mary Elvira Weeks in 1967 (Courtesy of Walter P. Reuther Library, Archives of Labor and Urban Affairs, Wayne State University)

(then chair of the HIST Division) to Mary Elvira Weeks, 23 May 1967 [56], "By unanimous choice of the judges, you have been awarded the Dexter Prize . . . and a plaque signifying your unusual contributions to the history of chemistry. I am very happy that you were the one chosen for you certainly deserve this recognition."

One major problem, however, was the requirement that the Dexter prize winner travel to the fall national meeting of the American Chemical Society and give a presentation. Weeks's medical condition (she suffered from diabetes insipidus) did not allow her to travel, so she believed she should refuse the award. When notified that she had been selected, Weeks wrote ". . . in comparison to others who had received the Dexter Award I feel humble and unworthy . . . I would be happy to see this prize awarded instead, as an encouragement, to some younger person who is still actively engaging in research in the history of chemistry [57]." Levey responded, " ... we have decided that you most deserve the award, and in the best Emily Post tradition, you must accept it" [58]. Arrangements were made for the award plaque to be presented to Weeks in her home by Dr. Charles Harmison, chair of the Detroit local section of the American Chemical Society (Fig. 14.19).

In 1968, the 7th and last edition of *Discovery of the Elements* was published. When the time came to start work on the new edition, Weeks's health was such that she could not carry out the task and she asked Henry Leicester to serve as her co-author. He did the bulk of the work, but she was involved, as the letters between the two make very clear [59]. The page count, 910 pages, was the same as for the 6th edition, but much of the material was rearranged into a more logical order.

Previous editions had simply added new chapters as Weeks wrote additional articles. This additional material was now integrated into the relevant chapters. Weeks wrote to Leicester on March 26, 1968 [60]:

> Heartiest congratulations on your revision of our D of E, and many thanks for your kind words in the Foreword to the 7th ed. You have done remarkably well in condensing the 6th so that, in spite of the new material the book has not been lengthened.

Mary Elvira Weeks died on June 20, 1975. She was buried in the Weeks family plot, Hudson Cemetery in Lyons, WI. Her obituary noted she was a "writer, linguist, translator, scholar, historian and humanitarian. . . . In the span of one year, she contributed to 70 various humanitarian causes [61]." Her great-niece Sandra Weeks Brangan told the following story that illustrates Weeks's commitment to helping others [44]:

> My Dad met two men of color at her funeral that she had helped get through college. One was a doctor and one was a lawyer. Both of these individuals said she had changed their lives and had always supported them and others in their poor neighborhood. She had lived there for years and had few possessions [having given most of her money to worthwhile causes].

Members of her church, Greenfield United Methodist, Detroit, noted [44]:

> The number of various projects she supported was huge. She gave for relief to Africa, India, and Viet Nam; to the Meharry Medical College for Negroes in Nashville, Tennessee; to the American Bible Society and her own church, besides many others. She had a keen interest in civil rights for all peoples. . . . That Dr. Weeks was a humble, warm-hearted and generous person is readily seen. Much of her giving and interests were unknown to friends and relatives prior to her death.

14.6 Discovery of the Elements

One of the truly charming characteristics of *Discovery of the Elements* (or *Discovery*, as I will hereafter refer to it) is that the material is presented in such an interesting and accessible way, including using poems and extensive excerpts from letters written by the discoverers. I believe it is this feature, even more so than its wealth of information, that made it so popular, especially for chemistry teachers who used it as a handy source of information to enliven their lectures. I will give three quotes to illustrate this aspect of the book:

From Chap. II. Elements Known to the Alchemists [62]:

> No one knows what led this zealous alchemist [Hennig Brand] to hope that in human urine he might find a liquid capable of converting silver into gold, but it is well-known that his queer experiments made in the seventeenth century produced results that were both startling and strangely beautiful. Small wonder that he was so delighted with the white, waxy substance that glowed so charmingly in his dark laboratory. The method of obtaining this light-giving element, which is now called phosphorus, Brand kept secret, but the news of the amazing discovery soon spread throughout Germany.

From Chap. IX. Three Alkali Metals: Potassium, Sodium, and Lithium [63]:

> In a similar manner [to the discovery of phosphorus] Davy's isolation of sodium and potassium immediately fired the imagination of the nineteenth-century public and aroused intense interest. These elements, like phosphorus, made their entrance upon the chemical stage in a manner nothing short of dramatic, and the accompanying phenomenon of light helped to focus all eyes upon them. Lithium, however, entered the chemical world in a more quiet manner and was introduced by a scientist of lesser prominence, J. A. Arfwedson, a student of Berzelius.

From Chap. X. The Alkaline Earth Metals and Magnesium and Cadmium [64]:

> Sir Humphry's ardent nature could not rest content with his recent triumphs over sodium and potassium. With a conqueror's enthusiasm he pushed ahead toward the still more difficult task of decomposing the alkaline earths.

Another delightful aspect of *Discovery* was the copious illustrations. Images supplied by Dains provided the bulk of the illustrations in Weeks's articles on the discovery of the elements in the *Journal of Chemical Education*, as well as the subsequent book and its multiple editions. Each edition of *Discovery* acknowledged Dains's contribution on the title page. Weeks continued to hunt for images to add to each new edition of *Discovery*. In a letter to Leicester written while making the revisions leading to the 7th edition, Weeks stated "My chief treasure for the new edition is a 3 × 4" color print, sent me by Mr. Bernardo J. Caycedo of Bogotá, of a painting of Juan José d'Elhuyar. Mr. Caycedo is a descendant of Juan José" [59].

Discovery was an extraordinarily popular book for a history of science text, as shown by the seven editions it went through from 1933 through 1968, as well as the many printings made between editions. It was also translated into Spanish, French, Chinese, and Japanese.

Few popular science books have passed through multiple editions so quickly (four editions in six years from 1933–1939, during the Great Depression, no less) or continued to require multiple printings as *Discovery*. I could not find any information on the actual print runs or copies sold. There are no records at the *Journal of Chemical Education* offices or the Division of Chemical Education; the printer, Mack Printing Co. of Easton, PA, had been sold before the start of my research for this chapter, and apparently their records were not kept.

Here is a summary of the editions, their publication dates, and their page counts:

- 1st Edition—1933; 363 pages
- 2nd Edition—1934; 363 pages
- 3rd Edition—1935; 371 pages
- 4th Edition—1939; 470 pages
- 5th Edition—1945; 587 pages
- 6th Edition—1956; 910 pages
- 7th Edition—1968; 910 pages

The first four editions (Fig. 14.20) are all bound in dull green-brown boards with the title and author's name incised in gilt. The covers for the next three editions (Figs. 14.20 and 14.21) are all different from the first four and from one another. The 5th edition cover is dark blue and bears a cartouche of Hermes Trismegistus.[7] The title and author's name are relegated to the spine of the book. The 6th edition cover was different again, with a brick red cover, title (no author), the edition number, and a figure showing electron orbits surrounding a silhouette of a person holding a crucible emitting "rays." This figure might have been chosen to acknowledge the chapter written by Leicester and the many elements relatively recently discovered by atomic bombardment. The 7th and last edition had a green cover, title (no author), the edition number, and an embossed image of the periodic table (Fig. 14.22).

Figure 14.23 shows the Table of Contents from the 1st edition. The 1st edition consisted of the *Discovery* papers (I–XXI), as well as the Foreword and Chronology that were published in the *Journal of Chemical Education* in 1932 and 1933. These papers were reset because the page size and format differed between the journal and the book. The organization of the book is essentially chronological, with the elements presented (mostly) in the order of their discovery. However, many of the elements are also grouped according to their periodic relationships rather than strictly chronologically.

A comparison of the number of pages shows that the first two editions were in essence the same book. There was a slight page increase between the 2nd and 3rd editions, due to the addition of new material on recent discoveries in three chapters: XIII Some Spectroscopic Discoveries (1 page), XIX Radioactive Elements (3 pages), and XX Recently Discovered Elements (6 pages).

There was a significant (almost 100 pages) increase between the 3rd edition and 4th edition. The Table of Contents for the 4th edition is shown in Fig. 14.24. The black arrows indicate chapters that were newly added to that edition. These chapters came from articles that Weeks was continuing to publish in the *Journal of Chemical Education*. As reflected in the chapter titles, all of these new chapters focused on specific people involved in the discovery of elements: Daniel Rutherford, the de Elhuyar brothers, Martin Klaproth, Paul Kitaibel, Charles Hatchett, Don Andrés Manuel del Río, and Johan A. Arfwedson. The 4th edition also included an Appendix containing corrections and other addenda that didn't fit in elsewhere.

The page count for the 5th edition increased by 22%. This edition contains two new sections: a List of Illustrations and A List of Chemical Elements. The Index increased in length by 50% and the Chronology by 33%. Roughly 2/3 of the chapters/sections increased in length, so the expansion was fairly uniform.

[7]Hermes Trismegistus was associated with the Greek god Hermes and the Egyptian god Thoth. He was the purported author of the *Corpus Hermeticum*, a collection of philosophical works probably written in the second century CE. During the Middle Ages and the Renaissance, it inspired the development of alchemy, which became known as the "Hermetic Science."

Fig. 14.20 *The Discovery of the Elements*. Left 1st Edition—1933. The covers for the 2nd—1934, 3rd—1935, and 4th—1939 editions were the same as for the 1st edition; Right 5th Edition—1945

Fig. 14.21 *The Discovery of the Elements*. Left 6th Edition—1956; Right 7th Edition—1968

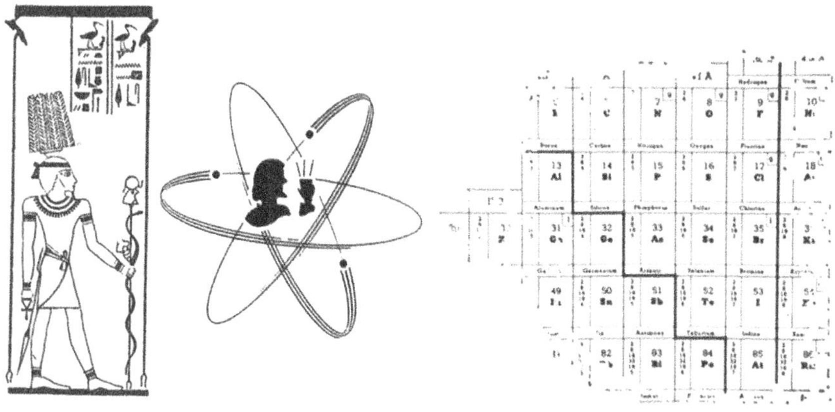

Fig. 14.22 Images from the half-title pages of the *Discovery of the Elements*. These images also appear on the covers of the books. Right 5th Edition; Center 6th Edition; Left 7th Edition

A major change occurred with the publication of the 6th edition. This edition was partly reorganized and a new chapter was written by Leicester. The page count increased by 55% to 910 pages and the increase in length was mentioned unfavorably in some of this edition's reviews. The number of chapters was increased from twenty-seven to thirty-one, and many of the chapters' titles were augmented by adding sub-headings showing the elements covered (such as Elements known to the ancient world—gold, silver, copper, iron, lead, tin, mercury, sulfur, carbon). Three new chapters were added: Chap. 2—Carbon and some of its compounds; Chap. 6—Old compounds of hydrogen and nitrogen; and Chap. 17—Some old potassium and sodium compounds. Chapter 31—Elements discovered by atomic bombardment, replaced Recently Discovered Elements.

The letters between Weeks and her publishers (the *Journal of Chemical Education*) in 1960 contain discussions about the necessity of reprinting the sixth edition as the stock would run out soon [59]. Another exchange of letters on the same topic took place in 1966, which resulted in the conclusion that it would be better to do a revision than a reprinting, a decision that led to Leicester's involvement in the 7th edition. As co-author and major writer for the 7th edition, Leicester completed the reorganization of all the material into a more coherent whole, which he accomplished without increasing the page count. The table of contents is shown in Fig. 14.25.

The reorganization mostly involved rearranging the existing text, rather than adding new material. The purpose of the reorganization was to group all material relevant to a specific element into one chapter. The chapters focusing on specific people were incorporated into the chapters discussing the elements they discussed, as were chapters focusing on compounds of hydrogen, nitrogen, and phosphorus. The two chapters on elements isolated with the aid of potassium and sodium were

TABLE OF CONTENTS

		Page
Foreword		1
I.	Elements Known to the Ancient World	2
II.	Elements Known to the Alchemists	9
III.	Some Eighteenth-Century Metals	20
IV.	Three Important Gases	29
V.	Chromium, Molybdenum, Tungsten, and Uranium	50
VI.	Tellurium and Selenium	65
VII.	Columbium, Tantalum, and Vanadium	77
VIII.	The Platinum Metals	99
IX.	Three Alkali Metals: Potassium, Sodium, and Lithium	117
X.	The Alkaline Earth Metals and Magnesium and Cadmium	128
XI.	Some Elements Isolated with the Aid of Potassium and Sodium: Zirconium, Titanium, Cerium, and Thorium	140
XII.	Other Elements Isolated with the Aid of Potassium and Sodium: Beryllium, Boron, Silicon, and Aluminum	153
XIII.	Some Spectroscopic Discoveries	180
	Supplementary Note on the Discovery of Thallium	202
XIV.	The Periodic System of the Elements	203
XV.	Some Elements Predicted by Mendeléeff	215
XVI.	The Rare Earth Elements	230
XVII.	The Halogen Family	253
XVIII.	The Inert Gases	278
XIX.	The Radioactive Elements	292
XX.	Recently Discovered Elements	316
XXI.	Supplementary Note on the Discovery of Phosphorus	336
Chronology		346

Fig. 14.23 *Discovery of the Elements*, 1st Edition, 1933, Table of Contents

combined. As an example of what was involved, the letter Leicester wrote to Weeks on Feb. 10, 1967, describes the process for the first four chapters [65]:

> I am making fairly good progress with the revision of the book, and I have made a copy of the first 132 pages of the revision for you to look over and criticize. This includes the first four chapters of the sixth edition. However, in view of our previous idea of combining the material relating to each element in one place, you will note that I have reduced this to three chapters. I did this by combining the material on carbon in chapter one with the chapter on carbon, making this a separate chapter two. The material in chapter four was then made part of chapter three, so that all the material on phosphorus was in one place. This has involved some rearranging of matter and remembering of the references in many case [sic], but I have tried to retain all the information you have put in.

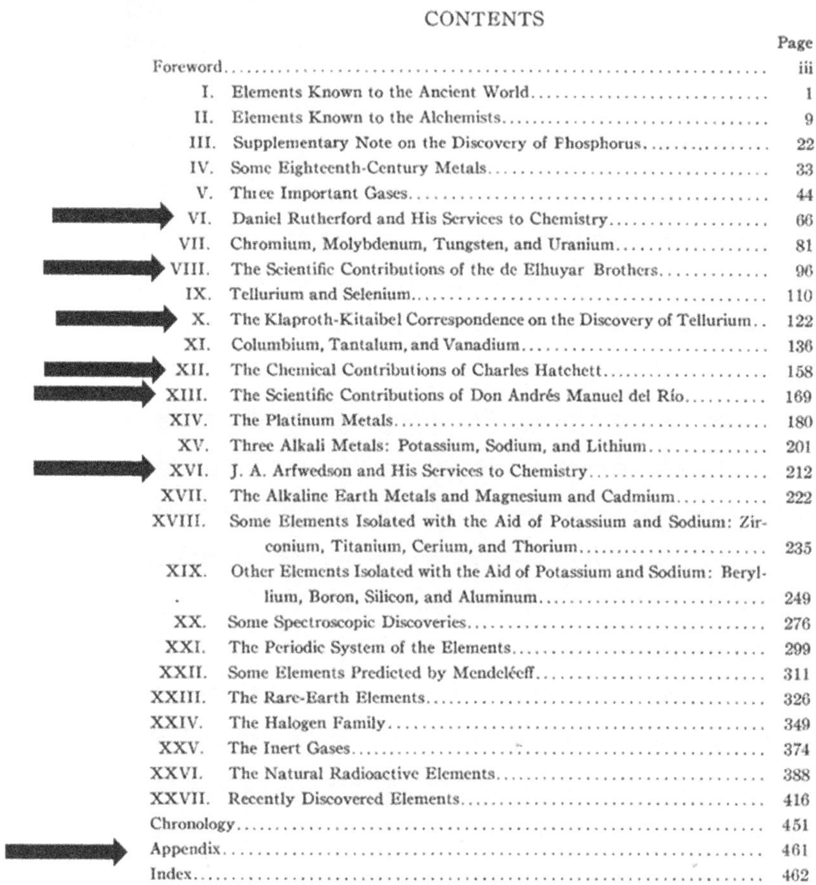

CONTENTS

		Page
Foreword		iii
I.	Elements Known to the Ancient World	1
II.	Elements Known to the Alchemists	9
III.	Supplementary Note on the Discovery of Phosphorus	22
IV.	Some Eighteenth-Century Metals	33
V.	Three Important Gases	44
VI.	Daniel Rutherford and His Services to Chemistry	66
VII.	Chromium, Molybdenum, Tungsten, and Uranium	81
VIII.	The Scientific Contributions of the de Elhuyar Brothers	96
IX.	Tellurium and Selenium	110
X.	The Klaproth-Kitaibel Correspondence on the Discovery of Tellurium	122
XI.	Columbium, Tantalum, and Vanadium	136
XII.	The Chemical Contributions of Charles Hatchett	158
XIII.	The Scientific Contributions of Don Andrés Manuel del Río	169
XIV.	The Platinum Metals	180
XV.	Three Alkali Metals: Potassium, Sodium, and Lithium	201
XVI.	J. A. Arfwedson and His Services to Chemistry	212
XVII.	The Alkaline Earth Metals and Magnesium and Cadmium	222
XVIII.	Some Elements Isolated with the Aid of Potassium and Sodium: Zirconium, Titanium, Cerium, and Thorium	235
XIX.	Other Elements Isolated with the Aid of Potassium and Sodium: Beryllium, Boron, Silicon, and Aluminum	249
XX.	Some Spectroscopic Discoveries	276
XXI.	The Periodic System of the Elements	299
XXII.	Some Elements Predicted by Mendeléeff	311
XXIII.	The Rare-Earth Elements	326
XXIV.	The Halogen Family	349
XXV.	The Inert Gases	374
XXVI.	The Natural Radioactive Elements	388
XXVII.	Recently Discovered Elements	416
Chronology		451
Appendix		461
Index		462

Fig. 14.24 *Discovery of the Elements*, 4th Edition, 1939, Table of Contents

Another example of the reorganization was that three chapters of the previous edition were combined into one: Niobium (columbium), tantalum, vanadium; Contributions of Charles Hatchett—niobium; and Contributions of Andrés Manuel del Rio—vanadium. One other strength that Leicester brought to the 7th edition was the addition of Russian material which he was uniquely suited to add.

14.6.1 Reviews

The reviews of *Discovery* that I have found are listed in Appendix 3. I will quote short extracts in this section.

Foreword v

1 Elements known to the ancient world 3
gold, silver, copper, iron, lead, tin, mercury, sulfur

2 Carbon and some of its compounds 69

3 Elements of the alchemists 91
arsenic, antimony, bismuth, phosphorus

4 Some eighteenth-century metals 137
zinc, cobalt, nickel, manganese

5 Three important gases 177
hydrogen, nitrogen, oxygen

6 Tungsten, molybdenum, uranium, chromium 241

7 Tellurium and selenium 289

8 Niobium (columbium), tantalum, vanadium 323

9 The platinum metals 385
platinum, rhodium, osmium, iridium, palladium, ruthenium

10 Three alkali metals 433
potassium, sodium, lithium

11 Alkaline earth metals, magnesium, cadmium 479
calcium, barium, strontium, magnesium, cadmium

12 Elements isolated with the aid of potassium and sodium . . . 517
zirconium, titanium, cerium, thorium, beryllium, boron, silicon, aluminum

13 Some spectroscopic discoveries 591
cesium, rubidium, thallium, indium

14 Periodic system of the elements 625

15 Some elements predicted by Mendeleev 643
gallium, scandium, germanium

16 The rare earth elements 667
ytterbium, cerium, lanthanum, neodymium, praseodymium, erbium, terbium, yttrium, scandium, holmium, thulium, samarium, gadolinium, dysprosium, europium, lutetium

17 The halogen family 701
fluorine, chlorine, bromine, iodine

18 The noble gases 750
helium, neon, argon, krypton, xenon

19 The natural radioactive elements 774
radium, polonium, uranium, radon, protactinium, actinium, thorium

20 Discoveries by X-ray spectrum analysis 816
hafnium, rhenium

21 Modern Alchemy 830
francium, technetium, promethium, astatine, neptunium, plutonium, americium, curium, berkelium, cialiforium, mendelevium, einsteinium, fermium, 102, lawrencium, 104

List of the chemical elements 858

Chronology of element discovery 860

Index 873

Fig. 14.25 *Discovery of the Elements*, 7th Edition, 1968, Table of Contents

The first review of the first edition of any book is an important one. The only review of the first edition of *Discovery* I could find was one by Tenney L. Davis.[8] Note that Davis calls the author "Miss Weeks" despite her possession of a Ph.D. degree and her listing on the title page as an assistant professor of chemistry at the University of Kansas. The choice of honorific reflects the standards of the time: in many ways, women in chemistry were simply treated differently.

Discovery of the Elements 1st Edition, Review by Tenney L. Davis [66]:

> . . . This book ought to be in the hands of every teacher of general chemistry and in every high-school library or other library which is designed to promote an interest in science. It is up to date, containing account of the discoveries of the newest elements and of the methods by which they have been made. It is profusely illustrated with portraits, many of them new, of contemporary chemists, with pictures of places and of apparatus, with facsimiles of title pages, letters, etc., and with ornamental reproductions of old cuts at the ends of the chapter. . . . This book will be of interest and of service to students of the history of culture. We are grateful to the publishers for bringing Miss Weeks's essays together in a manner to increase their accessibility and usefulness.

---

[8]Tenney Lombard Davis (1890-1949) was an outstanding organic chemist who spent his career at MIT. He was an authority on the chemistry of explosives (he acted as the director of research for the National Fireworks Company of Boston), including the early history of pyrotechnics [67]. Although he wrote on many historical topics, the area for which he is chiefly known is Chinese alchemy, carried out with a series of Chinese collaborators.

For the later editions, the following review is representative: *Discovery of the Elements* 4th Edition, Review by F. O. Koenig[9] [68]:

> The work is exactly documented; a separate bibliography is given for each chapter. The illustrations, consisting of portraits and of pictures of places, apparatus, title-pages, etc., are a valuable and enlivening feature of the work; they are very numerous and have evidently been collected – and selected – with zeal and taste. With regard to the literary style, the reviewer finds himself compelled to state that he has found reading the book an entertaining undertaking as well as merely a profitable one. . . . Nevertheless the book fails to realize to the full the possibilities for synthesis inherent in its subject matter, and this constitutes, for the reviewer, the chief criticism that can be made of the work. The failure consists in a lack of explicitness regarding the relation between the history of the discovery of the elements and the history of chemistry in general. These two histories are of course not coextensive, but the former is an indispensable part – one of the main currents, as it were – of the latter.

The next review is unusual as it appeared in a major newspaper (the Kansas City Times), as a full two page spread and accompanied by original artwork. *Discovery of the Elements* 4th Edition, Review by Hillier Krieghbaum[10] [70]:

> Dr. Mary Elvira Weeks, …, has compiled the human-interest background of the search for the 90-odd building blocks that comprise our present foundation for the chemical universe. A revised, fourth editions of her "Discovery of the Elements," . . . is said to be the first connected story of how scientists unraveled the mysteries of matter. . . . The fact that Dr. Weeks had to revise her book so soon is an indication that research is moving forward more rapidly than ever before.

Discovery of the Elements 5th Edition, Review by Tenney L. Davis [72]:

> The new fifth edition of Miss Weeks's book is considerably larger than the fourth edition which appeared in 1939. . . . The book remains unique in its class, scientific and broadly humanistic, scholarly and interesting to read. The fifth edition is the best yet because there is the most of it.

Discovery of the Elements, 6th Edition Review by Linus Pauling[11] [73]:

> I am sure that every chemist would find it interesting to read at least parts of the sixth edition, revised and enlarged, of the book "Discovery of the Elements." . . . I have thought

[9]Frederick Otto Koenig (1902-1974) spent 37 years doing research in thermodynamics and teaching at Stanford University. After he retired, he published articles on the history of science, especially the history of the second law of thermodynamics [69].

[10]Hillier Krieghbaum (1902-1993) "was a man who stood at the forefront of journalism education during its formative years, regarding it not just as a training ground for future reporters and editors but as a means of bettering the profession and the individuals who inhabit it" [71]. He worked as a science writer for United Press at their Washington DC bureau in the mid-1930s and remained interested in science for the rest of his career. He was an assistant professor of journalism in 1938 at Kansas State College (later University) in Manhattan. He would eventually join the journalism department at New York University (1948-1973).

[11]Linus Carl Pauling (1901-1994) received his Ph.D. degree in 1925 from the California Institute of Technology. He was on the faculty of CalTech from 1925-1964. He is the only person ever to win two unshared Nobel Prizes—for Chemistry (1954) and Peace (1962). Pauling was a pioneer in structural chemistry, quantum chemistry, and molecular biology. He introduced several indispensable scientific concepts including valence bond theory, the α-helix and β-sheet structures of proteins, and a new quantitative scale for electronegativity [74–76].

of this book as being of especial value to the young student of chemistry, in providing for him an account of those aspects of the history of chemistry that he can easily appreciate and understand, and that are likely to develop his interest in science. The sixth edition may be less valuable in this way than the earlier editions. With over 900 pages, it is nearly twice as large as the fourth edition (1939), and the various chapters, which may be looked upon as nearly independent essays, are in general about twice as long as in the earlier editions. . . . The tale of chemical discovery told in the new edition is not so straightforward and so easily followed as in the earlier editions. Nevertheless, the book is an excellent one, satisfying the real need for an account of the history of the discovery of the elements and of related aspects of the history of chemistry.

Note that thirty years after Davis's review of the 1st edition of *Discovery*, we are back to "Miss" Weeks in the following review of the 7th edition.

Discovery of the Elements, 7th Edition Review by William H. Brock[12] [77]:

Prompted by her series of isolated character sketches which were performed by the *Journal of Chemical Education* during the season 1932-1933, Mary Elvira Weeks' famous full-length history of the actors in this theater was first produced in 1933. It now appears for the seventh time with the capable assistance of Henry M. Leicester, who was the author of the final act on the transuranium elements in the sixth production of 1956. In this new performance Leicester has greatly improved the stagecraft by cutting the number of acts from 31 to 21, and (with little actual rewriting) he skillfully rearranged the script so that each player, or group of players, appears once only in a single, but longer, scene. . . . No doubt the book will be thought curiously old-fashioned for present tastes in history of chemistry. But we should remember with gratitude that during the past thirty years Miss Weeks' brilliant popularization of historical studies has stimulated many people to attend the theater of history of chemistry regularly, and our present freedom and ability to dramatize chemistry in a different way owes much to her pioneering efforts in the 1930's.

14.7 Conclusions

As another admirer of *Discovery* stated [23]: "This delightfully informative and copiously illustrated account is the only work in existence that tells the story of the discovery of each of the elements, one by one." It is a well-researched work that has stood the test of time. Although it is out of print, I continue to recommend it as an informative and entertaining guide to the stories behind the discovery of the various elements and as a lasting monument to its author.

At the beginning of Chapter I of the 1st edition of *Discovery*, Weeks quoted Clemens Winkler[13] in the original German [80] and then in the 6th and 7th editions in English translation [81]:

[12]William Hodson Brock (1936-) is a historian of chemistry and has concentrated on four main areas: the history of chemistry, the social history of Victorian science and mathematics, the development of scientific education, and the development of scientific periodicals [78].

[13]Clemens Alexander Winkler (1838-1904) [79] studied at the Freiburg School of Mines and early on developed new techniques for analyzing glass. In 1885 he discovered the element germanium, which proved to be the third of Mendeleev's predicted elements, eka-silicon.

> Die Welt der chemischen Vorgänge gleicht einer Bühne, auf welcher sich in unablässiger Aufeinanderfolge Scene um Scene abspielt. Die handelnden Personen auf ihr sind die Elemente.
>
> The world of chemical reactions is like a stage on which scene after scene is ceaselessly played. The actors on it are the elements.

I think she chose this quotation because the actors—the elements—fascinated her and she wanted to share her knowledge and enthusiasm about them with others. In terms of the analogy used by Brock in his review of the 7th edition of *Discovery*, Weeks could be considered not only the playwright but also the director of the piece. She spent much of her career putting together a detailed but compelling narrative of the history of the discovery of the elements and the stories of the discoverers. Her book not only educated its readers, it inspired them.

Acknowledgements I thank Greg Girolami and Carmen Giunta, the co-organizers of the symposium that led to this book, for all their support and encouragement. I thank an anonymous reviewer for many helpful comments. I thank Carol Bray, now retired from the University of Kansas, for her wonderful article on Mary Elvira Weeks and her helpful discussions. I'd also like to thank Susan Teague, Business Manager, Dept. of Chemistry, University of Kansas, Alison Stankrauff, University Archivist, Wayne State University, and Mrs. Marsha Weeks for all their help in providing access to material and images. Images from all editions of *Discovery of the Elements* were reprinted with permission from (Weeks, ME (1933–1968) *Discovery of the Elements*. Journal of Chemical Education, Easton, PA). Copyright (1933–1968) Division of Chemical Education, Inc. of the American Chemical Society.

Appendix 1: Publications of Mary Elvira Weeks

Books

Browne CA, Weeks ME (1952) A History of the American Chemical Society: Seventy-Five Eventful Years. American Chemical Society, Washington, DC

Weeks ME (1933) Discovery of the Elements; Collected Reprints of a Series of Articles Published in the Journal of Chemical Education. Journal of Chemical Education, Easton, PA

Weeks ME (1934) A Laboratory Manual of Qualitative Analysis. Edwards Brothers, Inc., Ann Arbor, MI, 1934. https://lccn.loc.gov/34010341. Accessed 13 Apr 2020

Weeks ME (1934) Discovery of the Elements. 2nd Edn. Journal of Chemical Education, Easton, PA

Weeks ME (1935) Discovery of the Elements. 3rd Edn. Journal of Chemical Education, Easton, PA

Weeks ME (1939) Discovery of the Elements. 4th Edn. Journal of Chemical Education, Easton, PA

Weeks ME (1945) Discovery of the Elements. 5th Edn. Journal of Chemical Education, Easton, PA

Weeks ME (1956) Discovery of the Elements: With a Chapter on Elements Discovered by Atomic Bombardment. 6th Edn. Journal of Chemical Education, Easton, PA

Weeks ME, Leicester HM (1968) Discovery of the Elements: Completely Rev. and New Material Added by Henry M. Leicester. Illus. Collected by F. B. Dains. 7th Edn. Journal of Chemical Education, Easton, PA.

Theses

Weeks ME (1914) The Effect of Various Negative Catalyzers on the Atmospheric Oxidation of Sodium Sulphite in Ultraviolet Light. Master's Dissertation, University of Wisconsin, Madison.

Weeks ME (1927) The Role of Hydrogen Ion Concentration in the Precipitation of Calcium and Magnesium Carbonates. PhD Dissertation, University of Kansas

Papers—Chemistry

Cady HP, Kemmerer G, Weeks ME (1929) Role of Hydrogen-ion Concentration in the Precipitation of Calcium and Magnesium Carbonates. J Phys Chem 33:1769–1780

Caldwell JM, Weeks ME (1934) Victoria Blue BX as an Internal Indicator in Ceriometry. Trans Kansas Acad Sci 37:117–118

Mathews JH, Weeks ME (1917) Effect of Various Substances on the Photochemical Oxidation of Solutions of Sodium Sulfite. J Am Chem Soc 39:635–646

Weeks ME (1931) *p*-Phenetidine and *p*-Anisidine as Oxidation Indicators. Trans Kansas Acad Sci 34:158–163

Weeks ME (1932) Bio-o-Anisidine as Internal Indicator in the Dichromate Method for Iron. Indus Eng Chem Anal Edn 4:127–128

Papers—History of Chemistry

Weeks ME (1932) The Discovery of the Elements. Foreword. J Chem Educ 9(1):3

Weeks ME (1932) The Discovery of the Elements. I. Elements Known to the Ancient World. J Chem Educ 9(1):4–10

Weeks ME (1932) The Discovery of the Elements. II. Elements Known to the Alchemists. J Chem Educ 9(1):11–21

Weeks ME (1932) The Discovery of the Elements. III. Some Eighteenth-Century Metals. J Chem Educ 9(1):22–30

Weeks ME (1932) The Discovery of the Elements. IV. Three Important Gases. J Chem Educ 9(2):215–235

Weeks ME (1932) The Discovery of the Elements. V. Chromium, Molybdenum, Tungsten and Uranium. J Chem Educ 9(3):459–473

Weeks ME (1932) The Discovery of the Elements. VI. Tellurium and Selenium. J Chem Educ 9(3):474–485

Weeks ME (1932) The Discovery of the Elements. VII. Columbium, Tantalum, and Vanadium. J Chem Educ 9(5):863–884

Weeks ME (1932) The Discovery of the Elements. VIII. The Platinum Metals. J Chem Educ 9(6):1017–1034

Weeks ME (1932) The Discovery of the Elements. IX. Three Alkali Metals: Potassium, Sodium, and Lithium. J Chem Educ 9(6):1035–1045

Weeks ME (1932) The Discovery of the Elements. X. The Alkaline Earth Metals and Magnesium and Cadmium. J Chem Educ 9(6):1046–1057

Weeks ME (1932) The Discovery of the Elements. XI. Some Elements Isolated with the Aid of Potassium and Sodium: Zirconium, Titanium, Cerium, and Thorium. J Chem Educ 9(7):1231–1243

Weeks ME (1932) Paracelsus and the Discovery of Hydrogen. J Chem Educ 9(7):1296–1296

Weeks ME (1932) The Discovery of the Elements. XII. Other Elements Isolated with the Aid of Potassium and Sodium: Beryllium, Boron, Silicon, and Aluminum. J Chem Educ 9(8):1386–1412

Weeks ME (1932) The Discovery of the Elements. XIII. Some Spectroscopic Discoveries. J Chem Educ 9(8):1413–1434

Weeks ME (1932) The Discovery of the Elements. XIV. The Periodic System of the Elements. J Chem Educ 9(9):1593–1604

Weeks ME (1932) The Discovery of the Elements. XV. Some Elements Predicted by Mendeleeff. J Chem Educ 9(9):1605–1619

Weeks ME (1932) The Discovery of the Elements. Vanadium (The Author Replies). J Chem Educ 9(9):1658–1659

Weeks ME (1932) The Discovery of the Elements. XVI. The Rare Earth Elements. J Chem Educ 9(10):1751–1773

Weeks ME (1932) The Discovery of the Elements. XVII. The Halogen Family. J Chem Educ 9(11):1915–1939

Weeks ME (1932) The Discovery of the Elements. XVIII. The Inert Gases. J Chem Educ 9(12):2065–2078

Weeks ME (1932) The Discovery of the Elements. XIII. Supplementary Note on the Discovery of Thallium. J Chem Educ 9(12):2078–2078

Weeks ME (1932) The Discovery of the Elements. III. A Correction. J Chem Educ 9(12):2127

Weeks ME (1933) The Discovery of the Elements. XIX. The Radioactive Elements. J Chem Educ 10(2):79–90

Weeks ME (1933) The Discovery of the Elements. XX. Recently Discovered Elements. J Chem Educ 10(3):161–170

Weeks ME (1933) The Discovery of the Elements. Chronology. J Chem Educ 10(4):223–227

Weeks ME (1933) The Discovery of the Elements. XXI. Supplementary Note on the Discovery of Phosphorus. J Chem Educ 10(5):302–306

Weeks ME (1933) Credit Where Due. J Chem Educ 10(8):510–510

Weeks ME (1934) Daniel Rutherford and the Discovery of Nitrogen. J Chem Educ 11(2):101–107.

Weeks ME (1934) Daniel Rutherford and the Discovery of Nitrogen. Correction. J Chem Educ 11(5):314

Weeks ME (1934) The Scientific Contributions of the de Elhuyar Brothers. J Chem Educ 11(7):413–419

Weeks ME (1934) The Author of "The Chemical Analysis of Bismuth. J Chem Educ 11(7):428

Weeks ME (1935) Frank Burnett Dains. Indus Eng Chem News Edn 13:118

Weeks ME (1935) The Discovery of Tellurium. J Chem Educ 12(9):403–409

Weeks ME (1935) The Scientific Contributions of Don Andres Manuel del Rio. J Chem Educ 12(4):161–166

Weeks ME (1936) Some Scientific Friends of Sir Walter Scott. J Chem Educ 13(11):503–507

Weeks ME (1936) M. and Mme. Joliot-Curie, Nobel Laureates, Chemistry, 1935. Homage from the Women Chemists of the United States Assembled at the Kansas City Meeting of the American Chemical Society, April 14, 1936. Publisher not identified, Place of publication not identified.

Weeks ME, Dains FB (1937) Mrs. A. H. Lincoln Phelps and her Services to Chemical Education. J Chem Educ 14(2):53–57

Weeks ME, Larson ME (1937) J. A. Arfwedson and his Services to Chemistry. J Chem Educ 14(9):403–407

Weeks ME (1938) The Chemical Contributions of Charles Hatchett. J Chem Educ 15(4):153–158.

Weeks ME (1938) Nils Gabriel Sefström—The Sesquicentennial of His Birth. Isis 29:49–57

Weeks ME, Amberg LO (1940) M.-E. Chevreul. The Fiftieth Anniversary of His Death. J Am Pharm Assoc (1912–1977) 29:89–96

Weeks ME (1943) An Exhibit of Chemical Substances Mentioned in the Bible. J Chem Educ 20(2):63–76

Weeks ME (1943) Max Speter (1883–1942). Isis 24:240–244

Weeks ME (1944) Don José Celestino Mutis, 1732–1808. J Chem Educ 21(2):55–56.

Weeks ME (1958) The Chemical Contributions of William Allen. J Chem Educ 35(2):70–73

Book Reviews

Weeks ME (1936) Antoine Lavoisier. The Father of Modern Chemistry by Douglas McKie. Isis 26:180–183

Weeks ME (1938) Historia de la física by Paul F. Schurmann. Isis 29:172–176

Weeks ME (1938) A Hundred Years of Chemistry by Alexander Findlay. Isis 29:176–179

Weeks ME (1957) Tobern Bergman as Pioneer in the Domain of Mineral Waters by Uno Boklund. J Chem Educ 34(10):A496–A497

Weeks ME (1957) Tobern Bergman: A Bibliography of His Works by Birgitta Mostrom. J Chem Educ 34(10):A500

Appendix 2: Quotes from Doris Brewster Swift—Life on Mississippi Street

In *Life on Mississippi Street*, Ms. Swift gives a diary entry made at the time, shown in italics. She then adds, with text not in italics, further comments, either to put the entry in context or adding other material. The volume is referenced in brackets at the end of the quote, followed by the page number. These are all the entries that mention Weeks in her three-volume work, as far as the author can determine.

October 31, Saturday [1925]. Harry and Charles dressed up to go frighten people. Charles said he was a goat, not a real goat, but a boy goat (ghost). Charles wanted us to be sure to lock both doors so no one would come into steal his pop corn which Miss Weeks gave him. At this time they would have knocked on doors, said "boo!", then waited for the people to guess who they were. It was rare for any one to give the children a treat. Miss Weeks was a Chemistry Professor who lived in the little house on the corner next to 18th Street—across from our former house at 1737 [33], p 24.

March 4, Sunday [1928]. After Sunday School and church we go to Week's for dinner. Listen to a good radio program in eve. While Doris goes to Durbin's. Doris wants to live at Durbin's. Elvira Weeks was a Chemistry professor, who lived with her Mother in the small cottage on the southeast corner of our block. They attended the Baptist Church and we frequently took them with us after we moved back to town. [33], p 124.

June 14, Thursday [1928]. Ray and I go down in the evening to hear the Republican convention over Miss Weeks' radio. Count nominating votes for Hoover. . . . In 1928 it was a new thing to have the party conventions broadcast nation-wide [33], p 134.

October 2, Sunday [1938]. Ray has a bad cold so stays in bed. All the neighborhood goes to a cabin on Uncle Jimmy's old place for a picnic. Lane's—4, McBeth's—5, Allen's—5, Elbel's—4, Lindstrom's—5, Maddox'—3, Storer's—6, Brewster's—4, Six'—4, and Miss Weeks'—1. A total of forty-one people made the trip to the wooded location for the picnic Mother had organized [34], p 265 [*Note* from the author: a picture showing "most" of the neighborhood women who were at the picnic does not include Weeks. She appears in very few group shots of organizations where she might be expected to appear, given her status in the organization and the fact that she was in attendance.]

March 2, Sunday [1941]. Doris stays in bed again today. In the evening Ray and I go down to Miss Weeks' house to meet Dr. Steinberger, a German refugee who is living with Miss Weeks. Dr. Steinberger was a Jew who had escaped Hitler's persecution. She was an object of idle curiosity, but I was busy with my own life and ignored her. The folks learned her background when they spent the evening at the home of Miss Weeks. Eventually, Mother thought Dr. Steinberger was tiresomely repetitious, for every time they would be together she insisted that something had to be done to help her people. The neighbors reaction was, Do what? And how? She could provide no answers [34], p 395.

July 3, Saturday [1943]. Get Dr.'s permit for Nita to go to camp. Girls carry out Helleberg's magazines. Mrs. Storer, Miss Weeks and I go to Helleberg's to help them get straightened around. Another rental house had been located for them in Lawrence. Mother and the other women helped them get settled along with some members of the sociology department [35], p 186.

February 19, Saturday [1944]. Nita stays out all night at Elbel's so she could go to Kansas City with them on an early bus. She is worn out when she gets back. Buys a red and also a white sweater. Ray and I attended a dinner as a farewell for Miss Weeks. It is held at the Hearth. Miss Elvira Weeks had received her doctorate in chemistry from K.U. in 1927 and she had been an instructor since 1921 and lived in a neat bungalow on the south corner of our block since before I was born. My cousin, Jay Junior Stewart, was able to provide the following information.

Miss Weeks was one of my teachers. She taught quantitative analysis, which was a hard laboratory course. I'm not sure that she taught anything else unless it might have been a class in the history of chemistry, which was really her strength. She was sweet but "mousy" and was dull when lecturing, talking in a monotone and looking either at the floor or at the ceiling (occasionally) but never in between at the level of the class. She left eventually for a position with a library in Detroit or Toledo. She was best known (and had an international reputation) for her "History of the Discovery of the Elements." This was a most thorough book which was translated into a number of other languages. I understood that she did much of the translating, even learning Japanese to do it [35], p 236.

March 12, Sunday [1944]. Mother had another heart spell about 6 in the morning. I fix chicken loaf to take to Miss Daum's for farewell party for Miss Weeks. Norman (Storer) is sick with a high temperature. We had a good time at Miss Daum's. Miss Daum, the city librarian and Miss Weeks lived on diagonal corners of 18th and Mississippi Streets and had been friends for years [35], p 242.

Appendix 3: Reviews of *Discovery of the Elements*

Davis TL (1933) The Discovery of the Elements by Mary Elvira Weeks. J Chem Educ 10:710

Davis TL (1939) The Discovery of the Elements by Mary Elvira Weeks. J Chem Educ 16:550

Krieghbaum H Book Review—Discovery of the Elements, 4th Edn; Important Chapters are Added to Romance of Chemical Research. Kansas City Times, Kansas City, MO, Jan. 11, 1940.

Koenig FO (1940) Discovery of the Elements by Mary Elvira Weeks. Isis 32:386–389

Davis TL (1945) The Discovery of the Elements by Mary Elvira Weeks. J Chem Educ 22:416

Browne CA (1945) Discovery of the Elements by Mary Elvira Weeks. Scientific Monthly 61:326–327

Pauling L (1957) Book Review—Discovery of the Elements 6th Edn. J Chem Educ 34:51

Ihde AJ (1958) Discovery of the Elements by Mary Elvira Weeks. Isis 49:86–87

Kauffman GB (1968) Discovery of the Elements by Mary Elvira Weeks, Henry M. Leicester and F. B. Dains. Science 162(3849):110–111

Brock WH (1969) Discovery of the Elements by Mary Elvira Weeks. Isis 60:113–114

References

1. Weeks ME (1932) The discovery of the elements. Forward. J Chem Educ 34:3
2. Hodgson A (1934) Her book on elements is already famous [Univ. Kansas] Graduate Magazine 32:10
3. Weeks RD (1885) Genealogy of the family of George Weekes of Dorchester, MA, 1635–1650. Press of L. J. Hardham, Newark, NJ, pp 18–19
4. Weeks RD (1885) Genealogy of the Family of George Weekes of Dorchester, MA, 1635–1650. Press of L. J. Hardham, Newark, NJ, p 27
5. Weeks, RD (1885) Genealogy of the family of George Weekes of Dorchester, Mass., 1635–1650. Press of L. J. Hardham, Newark, NJ, pp 46–48
6. Galpin, ER (1924) Mrs. Elizabeth Yates Atkinson Richmond. Sketches of Wisconsin in pioneer women. Hoard & Sons, Fort Atkinson, WI, pp 107–108
7. Mary Elvira Weeks Box 1, Folder 3: Richmond, Correspondence and Newspaper Clippings. University Archives, Wayne State University
8. 1857–1922 Lawrence College Alumni Record (1922) Castle-Pierce Printing Co., Oshkosh, WI, pp 268–269
9. Snyder TD (1993) 120 Years of American Education: A Statistical Portrait. U.S. Dept. of Education, Office of Educational Research and Improvement, National Center for Education Statistics: Washington, DC, 1993, p 76
10. (13 April 1911) Death of John Weeks. Lake Geneva Regional News, Lake Geneva, WI, p 1
11. Wayne State University Vertical Files—Biography File for Mary Elvira Weeks. On Weeks attending the Second Century Program of Ripon College in Ripon, WI. University Archives, Wayne State University
12. Last Will and Testament of Lewis S. Weeks. Wisconsin, Wills and Probate Records, 1800–1987. Ancestry.com, Provo, UT, No. 2523, Court Record, vol. 3, p 133; Record Book, vol. 42, p 489, and vol. 28, p 627
13. Albert Franklin Gilman. In: who's who in America, vol. 8. Leonard, JW, Marquis, AN, Eds. A. N. Marquis & Co., Chicago, IL, p 909
14. (January 1914) Ripon College Bulletin (Catalogue 1913–1914). Ripon, WI, p 88
15. Mary Elvira Weeks (1914) The Crimson, Ripon College, Ripon, WI, p 13
16. Albert Franklin Gilman A.M. (1913) The Crimson, Ripon College, Ripon, WI, p 17
17. Ihde AJ (1990). Chemistry, as viewed from Bascom's hill: a history of the chemistry department at the University of Wisconsin in Madison. Department of Chemistry, University of Wisconsin–Madison, Madison, WI, pp 373–379 and pp 478–483
18. Mathews JH, Weeks ME (1917) Effect of various substances on the photochemical oxidation of solutions of sodium sulfite. J Am Chem Soc 39:635–646

19. Rayner-Canham MF, Rayner-Canham GW (1996) Women in chemistry: participation during the early 20th Century. J Chem Educ 73:203–205
20. Mary Elvira Weeks BioFile, University of Kansas Archives
21. (March, 1916) Ripon College Bulletin, Ripon, WI, p 172
22. Goldblatt LA (ed) (1937) Collateral readings in inorganic chemistry. D. Appleton-Century Company Inc, New York, NY, p 225
23. Ihde AJ (1990) Chemistry, as viewed from Bascom's Hill: a history of the chemistry department at the University of Wisconsin in Madison. Department of Chemistry, University of Wisconsin-Madison, Madison, WI, p 430
24. Roberts E (1988) Women's work 1840–1940. Macmillan Education, London, p 67
25. Rayner-Canham MF, Rayner-Canham G (1998) Women in chemistry: their changing roles from alchemical times to the mid-twentieth century. American Chemical Society and Chemical Heritage Foundation, Washington, DC, p 171
26. Cunningham HE (1921) Transactions of the board of trustees, June 15, 1920 to June 13, 1921. University of Illinois, Urbana, IL, p 80
27. McCool, JH Bailey's Barn. https://kuhistory.ku.edu/articles/baileys-barn. Accessed 13 Apr 2020
28. Cady HP, McFarland DF (1907) Helium in Kansas natural gas. Trans Kansas Acad Sci 20(Part II):80–81
29. McCool, JH High on helium. https://kuhistory.ku.edu/articles/high-helium Accessed 13 Apr 2020
30. Cady HP, Kemmerer G, Weeks ME (1929) Role of hydrogen-ion concentration in the precipitation of calcium and magnesium carbonates. J Phys Chem 33:1769–1780
31. Bray C (1999) KU's first woman of chemistry. Mary Elvira Weeks. A history of our historian. Chemistry at the Univ Kansas, Department of Chemistry, KU, Lawrence, KS. https://kuscholarworks.ku.edu/handle/1808/26294 Accessed 20 Jan 2019
32. Rayner-Canham MF, Rayner-Canham G (1998) Women in chemistry: their changing roles from alchemical times to the mid-twentieth century. American Chemical Society and Chemical Heritage Foundation, Washington, DC, p 199
33. Swift DB (1992) Life on Mississippi street. The post war years 1919–1931. LOMS Publications, Tulsa, OK
34. Swift DB (1993) Life on Mississippi street. The depression years 1932–1941. LOMS Publications, Tulsa, OK
35. Swift DB (1996) Life on Mississippi street. The war years 1941–1945. LOMS Publications, Tulsa, OK
36. Bohning, JJ Mary Elvira Weeks (1892–1975). http://acshist.scs.illinois.edu/awards/Dexter%20Papers/WeeksDexterBioJJB1.pdf Accessed 13 Apr 2020
37. Neeley KL (2018) Frank Burnett Dains and his history of chemistry collections at the University of Kansas. https://kuscholarworks.ku.edu/handle/1808/26405?show=full. Accessed 14 Apr 2020
38. Dains FB (1911) Silliman, Hare and Cooke, and their relation to American science. Trans Kansas Acad Sci 23:22–33
39. Weeks ME (1935) Frank Burnett Dains. Indus Eng Chem New Edn 13:118
40. Gordon NE (1943) The section, division, and Journal of Chemical Education—a brief historical retrospect. J Chem Educ 20:369–372; 405
41. Gordon NE (1924) Editor's outlook. J Chem Educ 1:217–218
42. Mary Elvira Weeks Papers, Box 1, Folder 8, University Archives, Wayne State University
43. Weeks ME (1943) Max Speter (1883–1942). Isis 34:340–344
44. Personal communication, Sandra Weeks Brangan, 28 June 2020
45. Einstein, A Mary Elvira (Receiver) Weeks, and Max (Relevant for) Speter. Ihre Mitteilung über Hernn Speter Hat Grossen Eindruck. Nassau Point, NY

46. Einstein, A Mary Elvira (Receiver) Weeks, and Max (Mentioned In) Speter. Ich Habe Für Herrn Dr. Speter Nichts Mehr Unternehmen Können. Nassau Point, NY
47. Swift DB (1993) Life on Mississippi street. The depression years 1932–1941. LOMS Publications, Tulsa, OK, p 395
48. (1946) Dedication of Kresge-Hooker scientific library. Chem Eng News 24:1381
49. Gordon NE (1944) Kresge-Hooker scientific library. Rec Chem Progress 5:2–5
50. Gordon NE (1944) Kresge-Hooker scientific library. Rec Chem Progress 5:33–36
51. Mary Elvira Weeks Papers. http://reuther.wayne.edu/node/12456. Accessed 12 June 2020
52. Browne CA, Weeks ME (1952) A history of the American chemical society: seventy-five eventful years. American Chemical Society, Washington, DC
53. Weeks ME (1956) Discovery of the elements: with a chapter on elements discovered by atomic bombardment. 6th Edn. J Chem Educ, Easton, PA. Available at https://archive.org/details/discoveryoftheel002045mbp/mode/2up. Accessed 15 Apr 2020
54. Kauffman GB (1991) Henry Marshall leicester (1906–1991) a memorial tribute. Bull Hist Chem 10:15–21
55. (1967) Dexter Award Announcement. Chem Eng News 45:73
56. Mary Elvira Weeks Papers, Box 1, Folder: Levey Letters. University Archives, Wayne State University. Letter, Levey to Weeks, 23 May 1967
57. Mary Elvira Weeks Papers, Box 1, Folder: Levey Letters. University Archives, Wayne State University. Letter, Weeks to Levey, 27 May 1967
58. Mary Elvira Weeks Papers, Box 1, Folder: Levey Letters. University Archives, Wayne State University. Letter, Levey to Weeks, 31 May 1967
59. Mary Elvira Weeks Papers, Box 1, Folder 2, University Archives, Wayne State University
60. Mary Elvira Weeks Papers, Box 1, Folder 2, University Archives, Wayne State University. Letter, Weeks to Leicester, 26 March 1968
61. Weeks M (2019) Personal communication
62. Weeks ME (1933) Discovery of the elements; collected reprints of a series of articles published in the Journal of Chemical Education. J Chem Educ, Easton, PA, p 14
63. Weeks ME (1933) Discovery of the elements; collected reprints of a series of articles published in the Journal of Chemical Education. J Chem Educ, Easton, PA, p 117
64. Weeks ME (1933) Discovery of the elements; collected reprints of a series of articles Published in the Journal of Chemical Education. J Chem Educ, Easton, PA, p 129
65. Mary Elvira Weeks Papers, Box 1, Folder 2, Correspondence 1942–1974. Letter, Leicester to Weeks, 10 Feb 1967
66. Davis TL (1933) The discovery of the elements by Mary Elvira weeks. J Chem Educ 10:710
67. Leicester HM, Klickstein HS (1950) Tenney Lombard Davis and the history of chemistry. Chymia 3:1–16
68. Koenig FO (1940) Book review—discovery of the elements, 4th Edn. Isis 32:386–389
69. Frederick Otto Koenig. https://chemistry.stanford.edu/people/koenig. Accessed 16 Apr 2020
70. Krieghbaum H (1940) Book review—discovery of the elements, 4th Edn; Important Chapters are Added to Romance of Chemical Research. Kansas City Times, Kansas City, MO, Jan. 11, 1940
71. Stovall JG (1990) The practioners. In: Sloan WD (ed) Makers of the media mind: journalism educators and their ideas. L Erlbaum Associates, Hillsdale, NJ, pp 41–46
72. Davis TL (1945) Book review—discovery of the elements, 5th Edn. J Chem Educ 22:416
73. Pauling P (1957) Book review—discovery of the elements, 6th Edn. J Chem Educ 34:51
74. The Nobel Prize in Chemistry 1954. NobelPrize.org. Nobel Media AB 2020. https://www.nobelprize.org/prizes/chemistry/1954/summary/. Accessed 17 Apr 2020
75. Linus Pauling Biography. https://lpi.oregonstate.edu/about/linus-pauling-biography. Accessed 16 Apr 2020
76. Goertzel T, Goertzel B (1995) Linus Pauling—a life in science and politics. Basic Books, New York, NY
77. Brock WH (1969) Book review—discovery of the elements, 7th Edn. Isis 60:113–114

78. Bohning JJ (2006) William Brock (1936 -). http://acshist.scs.illinois.edu/awards/Dexter%20Papers/BrockDexterBioJJB.pdf. Accessed 16 Apr 2020
79. Asimov I (1982) Asimov's biographical encyclopedia of science and technology, 2nd edn. Doubleday, New York, NY, p 740
80. Weeks ME (1933) Discovery of the elements; collected reprints of a series of articles published in the Journal of Chemical Education. J Chem Educ, Easton, PA, p 2
81. Weeks ME (1956) Discovery of the elements: with a chapter on elements discovered by atomic bombardment. 6th Edn. J Chem Educ, Easton, PA, p 3

Part III
The Periodic Table from Other Perspectives

Astronomy Meets the Periodic Table, Or, How Much Is There of What, and Why?

15

Virginia Trimble

Abstract

A confluence of circumstances led to astronomers proposing three new elements in the 1860s (nebulium 1864, helium 1868, and coronium 1869) though the names came a bit later. Two of these proved to be familiar elements under unfamiliar conditions. Despite this rather unpromising start, astronomical observations were in due course required and accepted to quantify the amounts of the elements in the universe and both observations and calculations to sort out where it had all come from. These latter endeavors are still in progress today.

15.1 Introduction

"Interdisciplinarity," though 50 years old, is a current buzz word, useful for extracting dollars, euros and other support for scientific and other scholarly pursuits. But even casual examination of history of scientific questions and answers reveals that the important ones have all had input either from folks we would describe as unified natural philosophers or as practitioners of different sciences, or both. The filling-out of Mendeleev's 1869 table, sorting out which elements are

V. Trimble (✉)
Department of Physics & Astronomy, University of California, Irvine, CA 92697, USA
e-mail: vtrimble@uci.edu

V. Trimble
Queen Jadwiga Astronomical Observatory, Rzepiennik Biskupi 80, 33-163, Rzepiennik Strzyżewski, Poland

C. J. Giunta et al. (eds.), *150 Years of the Periodic Table*, Perspectives on the History of Chemistry, https://doi.org/10.1007/978-3-030-67910-1_15

common and which rare, and determining how they came to be that way required not just chemistry but help from the geo-sciences, meteoritics, observational astronomy, theoretical astrophysics, atomic, nuclear and particle physics, and even a bit of general relativity.

15.2 The Three New Elements

Several things contributed to these nearly-simultaneous discoveries. First was the large gap in atomic weights between hydrogen (=1) and lithium (=7) [1], and where else should you look for light things (since they are likely to float) besides the outer layers of the sun and in the space between the stars. Second was the work of Bunsen and Kirchhoff, which made spectroscopy a respectable way of discovering new elements like their cesium (1860, sky blue) and rubidium (1861, red) as well as William Crookes's 1861 thallium (green). They were also the ones who recognized the wavelengths of familiar sodium and iron among the absorption features in spectrograms of the sun. Sodium was a challenge to identify, because the pair of yellow lines (called Fraunhofer D in the solar spectrum), which seemed to come from everything, revealed a contaminant in the labs of salty-fingered spectroscopists. Kirchhoff had actually defined a scale of wavelengths for spectral features, which measured their refrangibility, giving larger numbers to bluer colors. I shall use modern wavelengths in Angstroms here. MKS devotees just need to move the decimal points one place to the left to get nanometers.

Next came a good deal of serendipity: William Huggins [2] and his London chemist friend William Allen Miller were the first to put the slit of their spectroscope (meaning you look through it rather than photographing something, as with a spectrograph) on a diffuse nebula. They called it 37 H IV Draconis; we write NGC 6543 or sometimes The Cat's Eye nebula. Instead of the rainbow with dark absorption features they were used to from the sun and stars, before their eyes appeared only a few sharp, bright features. One was a well-known hydrogen line, now called H-β (formerly Fraunhofer's feature C), and they attributed the other two to magnesium and nitrogen, that third line being the brightest. More careful measurements ruled out the nitrogen identification, leaving the close pair at 5007 and 4958 Å to be called "nebulium" [3]. These lines often kept the names "nebular" and N1 and N2 even after they were understood, and remain the brightest features in spectra of many kinds of nebulae, including famous ones like Orion.

The total solar eclipses of 1868 (in India) and 1869 (across the US) put many astronomers in the paths of totality, looking for various things. On a site in India, Jules Janssen realized he could catch light from the solar chromosphere without waiting for an eclipse by placing the slit of his spectrograph where he knew there was a prominence [4]. Norman Lockyer did the same back home in England soon after and decided that the emission feature near 5389 Å was not just the blended sodium-D lines. Norman R. Pogson (inventor of the modern magnitude scale for stellar brightness) had said the same in India.[1]

Two of the expeditions that went to look at the 1869 eclipse [5] focused on light from the solar corona (again then visible only during eclipses) and its spectrum. The new wavelength this time was 5303 Å, and the element supposed to produce it was, in due time dubbed coronium [6], by analogy with nebulium, helium, and various other elements, old and new. So, in chronological order and with some mild priority disputes and haggling over names, we acquired nebulium from Huggins [7], helium from Lockyer and Pogson [4], and coronium from Young and Harkness [5, 8], bitter enemies, who led two very separate expeditions.

The third favorable circumstance was that all these men were looking at spectra just before astronomers began switching from looking with their own eyes to recording images and spectra on photographic plates. Much of future astronomical spectroscopy would focus on the shorter wavelengths to which early silver halide emulsions were sensitive, while the discovery lines for all three astronomical elements lay in the yellow-green part of the spectrum, to which human vision is most sensitive.

15.3 Dethroning Nebulium and Coronium

When laboratory chemists began to fill in the gaps in Mendeleev's table with gallium (1875), scandium (1879), and germanium (1886), and Henry Moseley (and others) clarified a few ambiguities with X-ray emission spectroscopy [9–12], just before and after the First World War, it became clear that there was no space for either nebulium or coronium. And even helium required an eighth column to be added to the right side at the end of the nineteenth century, after William Ramsay found it in gas seeping from pitchblende, a uranium ore. That column now also accommodates Ne, Ar, Kr, Xe, Rn, and oganesson. A twenty-first-century eye (well, mine) looking at Hinrichs's periodic spiral [1, 13][2] feels an irresistible urge to add another spoke to the spiral between the one hosting F, Cl, Br, and I and the one populated by Li, Na, K, Rb, and Cs (Figs. 15.1 and 15.2). But I am not aware of any claim that he predicted the noble gases.

While nebulium and coronium are not real elements, their emission lines are, of course real, and remain strong in modern spectrograms. Eddington had guessed in around 1926 that nebulium might be some common element, highly ionized [14].[3]

[1]Although the helium line was observed in 1868, the proposal that it was due to a new element came later. In 1870, Lockyer began to consider whether the line could be the signature of a new element. The name first appears, however, in the inaugural address given by William Thomson (later Lord Kelvin) upon assuming the presidency of the British Association for the Advancement of Science on 2 August 1871.

[2]For more on Hinrichs, see Gregory Girolami's Chapter 6 in this volume.

[3]In Eddington's own words [14]: "The experimental physicists are at work trying more and more powerful means of battering atoms, so that one day a terrestrial atom will be stimulated to give nebulium light."

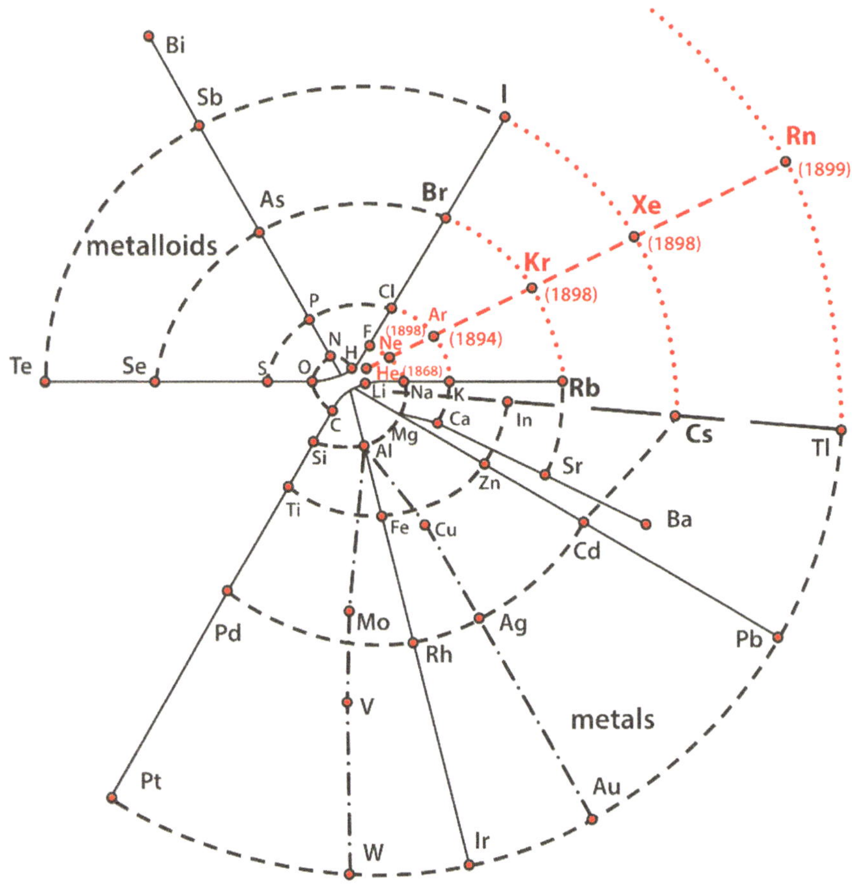

Fig. 15.1 Hinrichs's Chart of Elements from *Atomechanik oder die Chemie eine Mechanik der Panatome* (1867) [13], modified with addition of the inert gases by V. Trimble (from a version originally by JE Ybarra [17], based on the version in van Spronsen [18]), permission granted by Elsevier

Fig. 15.2 The inner core of Fig. 15.1, showing that both Ne and Ar fall on the extrapolated spirals

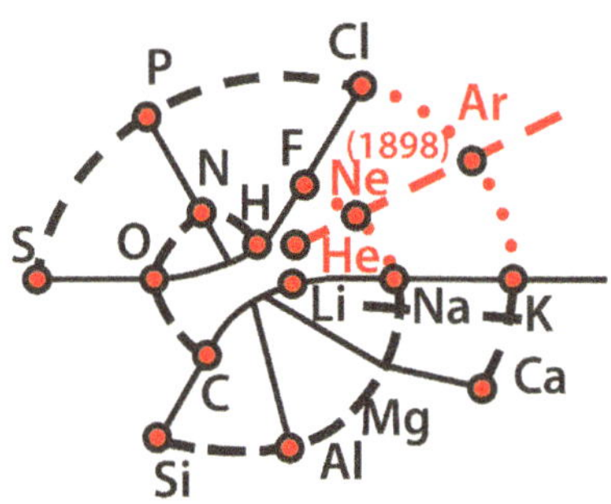

He was half right—common, but not highly ionized—and it was Ira Sprague Bowen the next year [15, 16] who showed that the strongest were twice-ionized oxygen, radiatively de-exciting from levels the atoms could reach only by collisional excitation. At sensible, terrestrial densities, collisional de-excitation would follow immediately. In the tenuous gasses of the Cat's Eye, NGC 7027, and all the rest, there was time for radiative de-excitation by photons carrying a unit of orbital as well as one unit of spin angular momentum. The transitions are called "forbidden," but obviously only relatively so, being less probable than "permitted" transitions by one or more powers of the fine structure constant,

$$\alpha = \frac{e^2}{\hbar c} or\ 1/137.$$

When I was a student back in the Mesozoic, the standard explanation was that given by Condon & Shortly [19].[4]

The decoding of coronium came during the Second World War, from Bengt Edlén [20][5] and Walter Grotrian [21], who considered the systematics of elements from Al up to Fe as they lost ever more electrons: neutral Al, Si missing one electron, P missing two electrons, S missing 3, Cl missing 4, Ar missing 5, up to Fe missing 13 (called Fe XIV, but that is just to make it more difficult) will each have the same electron configurations and so similar spectra, but shifted to higher and higher energies and shorter and shorter wavelengths, because the remaining electrons see ever-stronger positive charges from their nuclei.

The news was carried across WWII battle lines by Belgian Pol Swings [22]. Edlén's article was the written version of his 1945 George Darwin lecture and is a wonderful read though it must have made rather a dense talk. He clarified the ways in which the coronal lines are and are not like the nebular ones. Both are forbidden transitions, but the nebular are magnetic dipole transitions, with radiative lifetimes of hundreds to thousands or more seconds, while the coronal lines are electric quadrupole transitions with lifetimes of a fraction of a second (but strong electric dipole transitions have lifetimes less than microseconds). In addition, because sunlight is much brighter in the corona than is starlight in typical extended nebulae, some of the coronal excitation is probably radiative.

Lest I forget to mention it, a key step in solving the nebulium and coronium riddle was the recognition of many more lines of each type (that is ones with no laboratory counterparts) and improved wavelength measurements, so that the theorists had more numbers and more precise numbers to match. Data and ideas either march step by step together or circle repeatedly around some central truth, depending on your mental model of how science works.

---

[4]The nebulium lines have non-zero intensities mostly due to magnetic dipole contributions, which arise because the electrons are in motion.

[5]This is not the "discovery" paper, but it is exceedingly clear and well-written.

In this regard, the success of Bernard Lyot in constructing a coronagraph in 1930, with later improvement [23] so that he didn't have to wait for eclipses, was of immense importance as was the trick used by Lockyer and Janssen to catch chromospheric spectra any old day.

In due course, the chemists had to invite the astronomers back in when they wanted to know how much of each element was to be found in the cosmos and where it all came from. These investigations are still work in progress. There is, however, a fine old custom in astronomy that, whenever someone thinks of a new way of doing something or making something (cosmic rays; X-rays from binary stars; and also, for instance, gold, silver, and rare earths made by neutron star mergers) that someone is inclined to claim that the new mechanism is the only important one.[6] And if each mechanism were as powerful as the inventors claimed, we would be blinded by the X-rays and drowning in gadolinium.

15.4 How Much Is There of What?

Since our earth is the most readily available sample of cosmic stuff, I had supposed that geologists would have been the first to address and track through the years the relative abundances of the elements. Eight introductory geology and geophysics texts lie on my office book shelf, left over from before we had an Earth Sciences department here at UCI and I taught "Geophysics, the Making of the Earth" as a physics course. None has a word to say about the issue (though much about granite vs basalt, perovskites, and so forth). Nevertheless, I. E. Kleiber (1885), responsible for the earliest attempt at cosmic abundances that I've found, was probably a geoscientist. He considered the earth and meteorites. He is credited as first in a fine survey of the history of solar system composition that makes up the opening pages of the Lodders and Fegley text, *Chemistry of the Solar System* [24]. I have also pontificated on the subject in somewhat less detail [25].

Early efforts to determine the mix of elements in the atmospheres of the sun and stars were bedeviled by many now known to be common (CNO, Ne, etc.) not having any strong lines, sometimes not any lines at all, from gas at 5500 K in the visible part of the spectrum. But the most divisive, disputed, misunderstood issue for several decades following 1925 was the ratio of hydrogen to the sum of everything else from carbon to lead.

When it became clear that contraction from a gas cloud to the present density of the sun, about 1.4 g cm^{-3}, could not supply enough energy to power the sun for the then-known minimum life of the earth (1–2 Gyr), Perrin, Eddington [26, 27] and perhaps others suggested that fusion of hydrogen to helium might suffice, given that one helium atom is about 0.8% less massive than four hydrogen atoms, and $E = mc^2$. Necessarily, there had to be some hydrogen to start and should be some

[6]The situation in geophysics when plate tectonics was first accepted by the community and then blamed for everything on earth was perhaps similar. Advocates of wind, solar, geothermal, nuclear, etc., power perhaps also overestimate the contribution their preferred source can make.

helium around now. But Eddington, who was then a person of very great influence in the astronomical communities of the UK and US and, to a lesser extent in continental Europe, was firmly of the opinion that he could work out a relationship among the masses, luminosities, and radii of the stars which required radiation pressure to be comparable to gas pressure and could come out right only if hydrogen at most was a rather minor constituent, not more than about 7% [28]. This was a blatant contradiction to the analysis of the spectra of a number of stars, mostly both brighter and cooler than the sun, whose line strengths strongly suggested that most of the mass and, especially, most of the atoms were H and He [29]. It did not help to make the contest more equal that the author of the latter paper was a brand new PhD and a woman, Cecilia Payne, who had worked a bit with Eddington in Cambridge UK, though her thesis came from Cambridge MA. Indeed you have to read rather deeply into her book to realize what she had really learned, because both her advisor, Harlow Shapley, and the dean of American astronomers, Henry Norris Russell, insisted that she ascribe the strength of the hydrogen lines, even at rather low temperatures, to "anomalous excitation" (meaning more atoms in the $n = 2$ Bohr orbit than you would expect, where they could absorb the Balmer lines seen in the stellar spectra). And "of course" she was looking only at the stellar atmosphere: hydrogen is light; and it would surely float. Yes, yes, and not really, because stars with surface temperatures at and below that of the sun have convective (well-mixed) atmospheres.

Thus the standard text from which nearly all American astronomers got their early training for the next 20 years, said that "Miss Payne" had established that nearly all stars had very nearly the same atmospheric composition [30][7] but about absolute abundances of hydrogen and helium they express mostly puzzlement. Payne's thesis necessarily had to meet with the approval of Shapley, since he was her official advisor, and without his imprimatur, she could not receive her degree or publish. Assent from Henry Russell of Princeton was also essential, both because of his prominent place in American astronomy and because he was the official external examiner for her Harvard thesis. The result is that you have to dig fairly deep into Payne [29], in fact to Table XVIII to find the numbers she calculated for relative abundances of elements, including both hydrogen and helium, the latter coming from the few hot stars in her sample of mostly K (cool) giants. And helium led all the rest, though it was the dominance of hydrogen over metals that provoked the strongest disbelief from all around, including Eddington, who had the opportunity to tell her to her face "on the stars; not in the stars" when she returned to Cambridge after receiving her PhD. D. H. DeVorkin's [31] investigation of the interaction considers Russell's advice to other young astronomers as well, concluding that it was her youth, not her gender, that persuaded him she should not try to establish a result too far from the contemporary norm.

[7]This is in a chapter called "The Constitution of the Stars," which, however, deals primarily with T and P in atmospheres and interiors, not with chemical composition. They recognize spectral types K, M vs R, N, S as differences in the carbon to oxygen ratio, but are totally puzzled by strengths of H and He lines, even allowing for floating.

Time marched on, and Eddington reconsidered the best mix of H and heavier elements to make his relationships of stellar mass, luminosity, and temperature come out to match data. His best values for X (the initial hydrogen abundance by mass) drifted up to about one-third in 1932, with concurrence from Bengt Strömgren the same year [32, 33]. Chandrasekhar [34] therefore went to press just before WWII broke out with $X = 0.36$ for the sun and numbers ranging from 0.10 to 0.48 for other stars.

Let's back off for just a second to see what might go wrong with treatments either of spectra or of properties of the star as a whole. Gas pressure depends on temperature and the number of particles. For fully ionized gases (not a bad approximation inside stars) each atom of hydrogen has a mass of 1/2 nucleon mass per particle, and one nucleon mass per electron. A typical mix of heavy elements ($Z = 6$–83) has a bit less than two nucleon masses per particle and a bit more than two per electron. If there is too much hydrogen, pressure won't balance gravity and the star is in trouble. But the atmosphere is what lets radiation leak out, or not, depending on the opacity of the gas. Opacity comes from electron scattering, bound-bound, free-free, and bound-free transitions of electrons relative to their nuclei, and is obviously going to be an extraordinarily complex calculation for any honest mix of all the stable elements. There are, of course, approximation methods, some better than others, but the critical issue is the ratio of all the heavies ("metals" we astronomers call them) to hydrogen, as Aller pointed out in 1963 [35] and in the earlier 1953 edition of the book [36].

We now hop back into the "how big is X?" fray enlightened by Rupert Wildt [37], who recognized in 1939 that a proton can actually bind two electrons (though the second by only about 0.75 eV). It gets called H-minus or a negative hydrogen ion, and provides lots of opacity for red and near infrared light if there are extra electrons around to be embraced. Those have to come from easily ionized elements like carbon. Thus these contribute twice to opacities, through their own electronic transitions and through providing H^- ions.

Once again, many people, many papers, but when the sun has properly risen again after WWII, Martin Schwarzschild's 1958 *Structure and Evolution of the Stars* [38] (another of those volumes at the educational core of the astronomical curriculum for about 20 years) has a sun that is, by numbers of atoms, overwhelmingly hydrogen, and even by mass about 74% H, 24% He, and 2% everything else, with somewhat more of the volatiles, CNO and Ne, than the old Russell mixture [39].

Eddington truly never said the sun and stars were made mostly of iron (though one early attempt at structure calculation took the mean molecular weight to be 56), but in any case, H atoms in the sun now outnumber the Fe ones by a factor of something like 40,000 and even outweigh them by 70-something. See Fig. 15.3, which comes from a review based on a two-week, NATO-sponsored summer school [40].

The basic picture has changed rather little from the 1956 compilation (significance next section) by Suess and Urey [41] to current numbers for the solar system [42]. It was, however, already recognized by Russell, Dugan, and Stewart [30] that there were some real star-to-star variations; that these were correlated with other properties of the stars (NG Roman, W Iwanowska, and M and B Schwarzschild

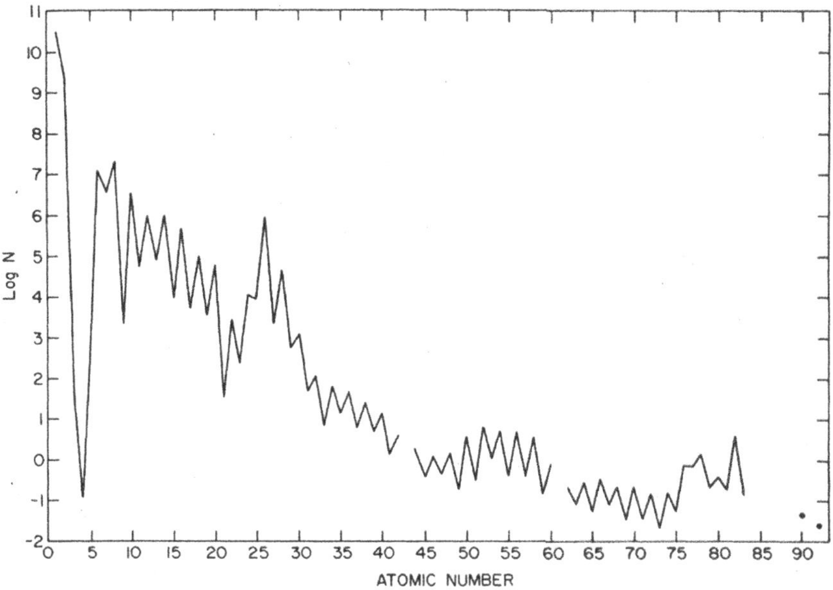

Fig. 15.3 The abundances of the chemical elements in Solar System material [40]. Used with permission of V Trimble

[43–47]), and that there have been, not surprisingly, systematic changes over the history of the universe, because, as Carl Sagan said, "We are made of star stuff [48]." Or, as we will see shortly, the early universe made all the elements up to helium, and Burbidge, Burbidge, Fowler, and Hoyle (B^2FH) [49] made all the rest.

15.5 Why?—Nuclear Physics

The chemical behavior of the elements is more or less described ("explained" is a strong word!) by the distribution of the orbital electrons into shells and subshells, with "magic numbers" at the closed shells $N = 2, 10, 18, 36, 54$, and 86. Now, what about the nuclei? Let us take as given the 1932 recognition of the existence of things to be called neutrons—a bit more massive than a proton plus an electron, and so in isolation prone to decay into the pair (plus a neutrino) in a bit more than 10 min. Thus we are spared the problem faced by Eddington in 1926 [28] of having to cram four protons and two electrons together to make a helium nucleus.[8] Neutrons, being

[8]This was part of his reason for some reluctance in putting forward 4H $\rightarrow$ He as the main solar and stellar energy source. Use of existing heavier atoms as catalysts came first from Atkinson and Houtermans [50]. That a helium nucleus was to be thought of as "made up of four positive electrons united with two negative electrons" came from Ernest Rutherford writing in *Popular Science Monthly* [51]. Positive electrons were what we now call protons or hydrogen nuclei.

charge = 0, spin = ½ fermions, made explaining the behavior of nuclei (such as their spins) seem a good deal simpler after 1932.

The nuclear shell model [52, 53] for which Maria Goeppert-Mayer and Hans Jensen shared the 1963 Nobel Prize with Eugene Wigner[9] (recognized for other contributions to nuclear physics); and of course, later input from many folks who did not receive Nobel Prizes, is the idea that you can assign each proton and neutron (separately) in a nucleus a principal quantum number $n = 1, 2, 3$, etc., an angular momentum number, l, from zero up to $n-1$, an angular momentum projection number m_l, from $m_l = -l$, by integers, up to $+l$, spin ½, and spin projection number $m_s = +½$ or $-½$. Right there we can see an obvious aspect of the abundance diagram mirrored—the odd-even effect, such that elements with Z = an even number (an outer pair of protons aligned spin up/spin down) are more abundant than the Z = odd numbers on either side. Take a look into isotope abundances in Fig. 15.4 to persuade yourself that an even number of neutrons is also favored.

Next, and this was the key idea, the spin-orbit coupling is much stronger for nucleons than for electrons, crudely because the orbits are much smaller and so closer to the spins. The resulting magic (strongly bound) numbers are 2, 8, 20 (28 somewhat), 50, 82, and 126. Pronounced abundance peaks are therefore found at ^{16}O ($N = Z = 8$); ^{40}Ca ($N = Z = 20$); ^{118}Sn ($Z = 50$, indeed tin has the largest number of stable isotopes of any element); ^{88}Sr, ^{89}Y, ^{90}Zr ($N = 50$); ^{138}Ba, ^{139}La, ^{140}Ce ($N = 82$); and at ^{208}Pb (doubly magic at $Z = 82$, $N = 126$).

A couple of other features are closely related, but involve the processes by which the nuclides were formed in stars and explosions. The peak around ^{56}Fe was ^{56}Ni ($Z = N = 28$) when it formed, which then beta decays back to Fe^{56}. And the double peaks around barium and lead reflect neutron captures that built up $N = 82$ and 126, but then also partly beta decayed back a few elements. Various nuclear processes also reflect excited states of a neutron or proton, and shell effects appear, as they do for atomic transitions, in reaction rates or cross sections. An enormous literature of successes and failures of the shell model, its competition with a liquid drop model (good for fission) and so forth, exists. Prejudiced, I would like to cite only Goeppert-Mayer and Jensen [54], though by then Aage Bohr, Ben Mottelson, and James Rainwater (Physics Nobel 1975) were also in the game, with papers starting in 1950. Some of us ignorantly reacted to this saying "Bohr, Mottelson, and Who??" doing an injustice to the first (and I suspect only) Nobel Prize winner with Native American ancestry. Wigner's work had started before WWII and continued after it.

On beyond the iron peak, there is another way to describe the systematics—the abundance of a nuclide is inversely proportional to its cross section for capturing one additional neutron, and it was this aspect that led to an early view of how astrophysics could account for the pattern, via synthesis of all the elements in a Big Bang universe. And, if you have been keeping count, you will have noticed that, as we go to heavier nuclides it takes more and more neutrons, exerting attractive

[9]The author is probably the only living Fellow of the American Physical Society to have danced a Viennese waltz with Eugene Wigner.

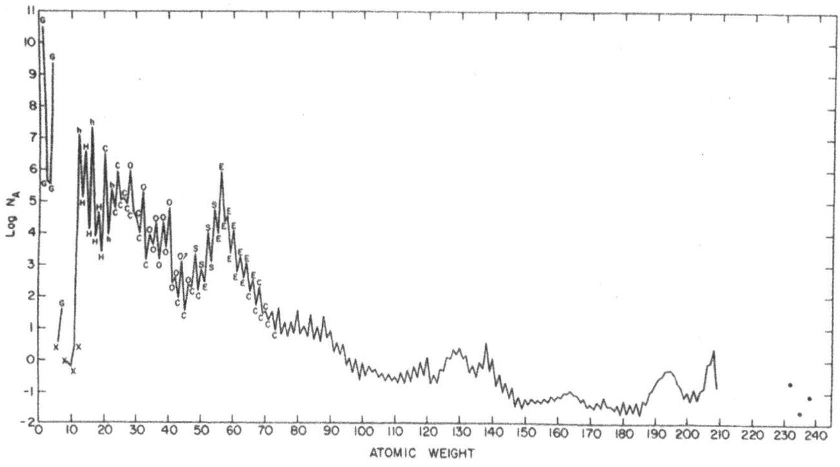

Fig. 15.4 The abundances by number of the nuclides in the Solar System material [40]. Lots of nuclear physics is reflected in this plot. For instance, around A = 70–90, higher and lower peak in the zig-zag alternate, showing that the most tightly bound, and therefore most abundant, nuclides are the ones with even numbers of both protons and neutrons. The small symbols attached to nuclides with atomic weights 1 to 70 indicate the dominant processes by which each was formed. Used with permission of V Trimble

nuclear forces, to overcome the repulsive Coulomb forces exerted by the protons. Beyond Pb and Bi, Coulomb always wins, and no element has a stable isotope, though some are quite long lived.

15.6 Why?—Astrophysics (Noise)

Astronomy entered the politically turbulent 1930s with enhanced agreement by assorted high-profile pundits that there was a lot of hydrogen in the sun and stars [39, 55–57] and at least one sort of mechanism [50, 58] by which hydrogen atoms might be built into heavier elements, releasing the 0.8% of their rest mass available according to measurements by Francis W. Aston [59, 60]. The mechanism was successive captures of protons and electrons by a few previously-existing atoms of lithium and such. The two main refinements (Bethe [61, 62], von Weizsäcker [63, 64]) were the replacement of lithium as the initial catalyst by carbon, with nitrogen and oxygen also in the cycle, and inverse beta decays of protons to neutrons inside the nuclei so that no electron captures were needed.

And then there was a war (WWII). The human costs were tragic beyond words. But astronomy benefitted. The development of radar led to radio astronomy. Captured V2 rockets and rapid improvements on them led to X-ray, gamma-ray, and ultraviolet astronomy. Many of the relocated scientists probably contributed more in their lands of refuge than they would have in their native countries. (This is impossible to prove, though it at least put them in countries less devastated by the war, on average, and therefore with increased access to equipment and funding.)

In contrast to the arguably positive astronomical effects of WWII, I think the Steady State universe, homogeneous in time as well as in space, was one of the less happy ones, although no one else seems to have described it that way. It was, however, wartime internment and then government service collaboration that brought Hermann Bondi (born in Austria), Thomas Gold (ditto), and Fred Hoyle (born in Yorkshire) together and gave them the opportunity to start thinking what the universe might be like on the largest scales. Steady State is the idea that the universe should be homogenous in time as well as in space, and it eventually failed a countably infinite number of observational tests starting in the 1950s [65] and continuing on to the present, with observations of distant galaxies showing that they are different from, and much younger than nearby galaxies today. A steady state universe requires constant input of new material to keep its density constant while it expands. The proponents chose to have material appear as hydrogen, with everything else produced in stars. Turning one-quarter into helium makes galaxies 10 times too bright, another of the failures. Thus we return in the next section to evolutionary universe models, with ages, densities, contents, and all in very close agreement with all sorts of observations [66, 67].

15.7 Why?—Astrophysics (Signal)

Now that those hills are out of the way, let's tackle the mountains of how, when, and where the distribution of elements we see came about. The starting point is a hot, dense early universe about 13.8 Gyr ago [67] There is just now (February 2020) some disagreement about exact numbers for the cosmic expansion rate—Hubble parameter—but the Planck satellite numbers for that, the density of matter, and so forth are more than good enough for our purposes here. I have pontificated on these issues before [25, 68–70], and it is perhaps a hopeful sign that the most recent of these is the shortest.

Early on, there were two schools of thought: Everything was made in the hot dense early universe or big bang [71–74] or everything was made in stars [49, 75, 76]. The best bet we have gradually concluded is that some of each has occurred. Nearly all the elements beyond helium (but only 1–2% of the mass of stars and all) come from nuclear reactions in stars, while very few elements (H, He, and a scrap of Li), and 98% of the baryonic mass, are left from Big Bang Nucleosynthesis.

George Gamow had actually started thinking about nucleosynthesis around 1935, but the key papers date from 1949–1950 [71, 72]. Discussion of them tends to get tangled up with the extent to which they did or did not predict relict radiation of the sort we now call the cosmic microwave background. Their ylem or primordial substance was pure neutrons. Those would begin to decay, releasing the difference in mass between a neutron and a proton plus electron (about

0.783 MeV), which, shared among the proton, electron, and a neutrino would take the gas up to close to 10^9 K, more than enough for fusion to turn some of the particles into helium atoms, ready to capture more neutrons and build up to the heaviest elements.

Actually no—the fatal flaw is the complete absence of any stable nuclides containing either 5 or 8 particles, though on several occasions both experimenters and theorists thought they had found one or more. Thus, neither two helium nuclei (alpha particles) nor one proton and one alpha can form anything. The scrap of Li-7 comes from He-4 plus either H-3 or He-3. He-3 is stable but quickly burns through to He-4, and H-3 lives about 12 years if left alone, but also burns through quickly. The process hangs up at helium. That the proper initial conditions would be a thermal equilibrium distribution of protons, neutrons, electrons (but not their antiparticles, or you get nothing but gamma rays) was first published by Chushiro Hayashi [77]. He is better known among astronomers for the first correct treatment of the phase in star formation when a cloud of gas is contracting under the force of gravity and energy is carried out by convection, called Hayashi tracks.

Hayashi's initial conditions led to better estimates of the production of the four stable H and He isotopes [78, 79] with temperature and density conditions chosen to evolve to the present numbers over the age of the universe. No $A = 5$ or 8, however, remained a fatal objection to producing everything in the early universe.

But where, you are asking, are dark matter, dark energy, and general relativity? Well, dark energy, otherwise known as the cosmological constant is so nearly constant that its energy density long ago was tiny compared to that of radiation and matter. Yes quite a lot of the matter was (and is) dark, but it simply didn't interact with the baryons, except gravitationally (and perhaps weakly) while they were busy making helium and such. The rate of expansion early on, and therefore the time available for Big Bang Nucleosynthesis did depend on the total density, according to well-known equations of general relativity. And the "consensus" or "standard" model of cosmology has the initial amounts of dark matter, baryonic matter, and radiation just right to end up now with exactly what we see when the temperature has dropped to 2.7 K.

The story of the alternative, everything being made in stars, customarily begins with Fred Hoyle in 1946 [80], though a handful or more of other physicists and astronomers had contemplated parts of the idea earlier [25]. Stars at their centers are obviously also hot and dense, and because they are not continuously expanding, can overcome the problem at $A = 5$ and 8 with three-body interactions. Bringing three helium nuclei together is called the triple-alpha process. The official "prediscovery" came from Ernst Öpik [81], and credit goes to Edwin Salpeter [82] for persuading the community that it would be important. Further contraction of the cores of massive stars lifts the temperature and density so that further, exoergic fusions continue on up to nickel-56, which beta decays to iron-56. The details of both the stellar evolution and the nuclear physics become exceedingly complex, and references 49 and 75 did not get everything quite right, but the principle holds, and computer simulations routinely replicate the observed evolution of stars and their elemental abundances [83].

It is, however, fair to say that the real attention-grabber in B^2FH was how they dealt with the abundance patterns on beyond the iron peak. ^{56}Fe is the most tightly bound of all the nuclides. Thus, if you want to go heavier, you must have spare energy to add. And the Coulomb barrier is so high that, if you merely smash one heavy nuclide into another, they mostly fragment rather than fusing. The best bet is to add neutrons. The Cameron and B^2FH insight was to interpret known abundances as showing this must have happened in two different ways and so probably in two different sorts of sites. The signature is the double humps in Figs. 15.3 and 15.4 (by atomic number and atomic weight, respectively) at around $Z = 50–55$ and $Z = 75–85$, more clearly at $A = 130–140$ and $A = 190–210$.

The "left hump" includes nuclides formed by addition of neutrons slowly, one by one (s-process) with time for unstable nuclides to decay before another neutron is added. This takes one up "the valley of beta stability" to magic numbers 50 and 82. The "right hand hump" comes from neutrons being piled on rapidly (r-process) until the next one isn't even bound. The products then decay back to nuclides with more protons and fewer neutrons than when they formed. Figure 15.5 shows a bit of the two processes from Yb to Os. Notice that there are a few "orphans," such as W-180 and Os-184, not reached by either s or r. B^2FH thought these might be due to proton captures—hence p-process. Removal of neutrons by energetic photons now seems more likely. All p-process nuclides are very rare; no element is dominated by them and we have no data on their abundances outside the solar system.

We still have Li, Be, and B to take care of, but let's sort out sites for s and r first. Both iron seeds and stray neutrons are available when stars like the sun and a bit more massive are fusing H to He and He to C and O in two thin shells around an inert C-O core. The iron had to be there from when the star formed. Thus the s-process nuclides can be made only in second generation or later stars. Assorted sources for the neutrons were suggested by Jesse L. Greenstein [84] and Alastair G. W. Cameron [85]. These involve alpha particles hitting things like C-13 and N-14, hence the s-process must happen in stars where both the CNO cycle hydrogen burning and helium fusion are occurring. The products get out with winds, formation of planetary nebulae, and perhaps some nova explosions of parent binary stars.

The r-process is a bit trickier. Yes, the collapse of an iron core of a massive star to make a type II supernova and a neutron star will have lots of both Fe and neutrons around, but it is not so easy for the products to get out without being photodissociated back to heaven knows what. My own favorite r-process site is David N. Schramm's "tube of toothpaste" effect, when an orbiting pair consisting of a neutron star plus a black hole spiral together and the black hole tries to swallow the neutron star [86, 87].[10] The name comes from what is likely to happen if you drop an open, nearly full tube of toothpaste on the bathroom floor and then step on it (well it was a difficult night-before). No certain examples of either the progenitor systems or the events have been seen.

[10]Wheeler credits Scientist X (Richard Feynman) for the idea and perhaps the name.

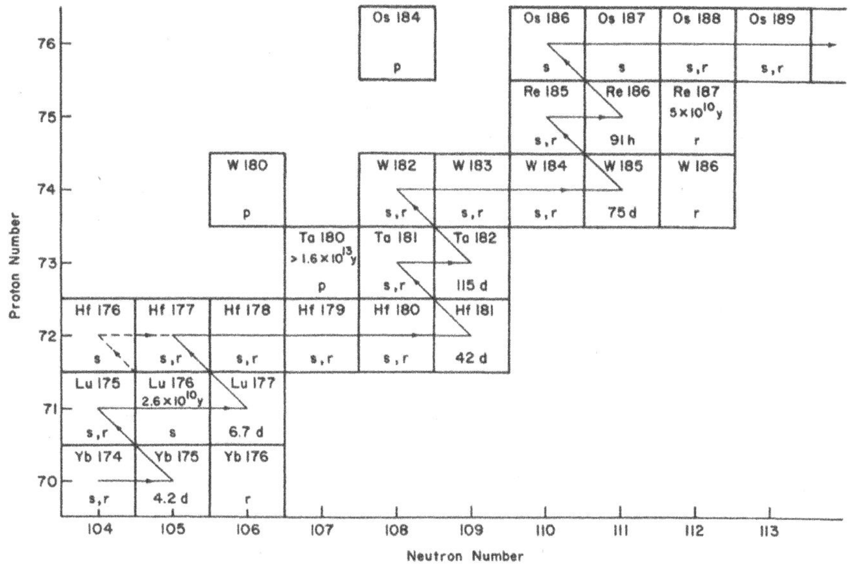

Fig. 15.5 The path of the s-process (capture of neutrons on a slow time scale) through the region A = 174–189 [40]. The r's indicate nuclides made by rapid neutron capture; some are made by both, and some by neither, the "orphan" nuclides marked p, for proton capture (or neutron removal). Used with permission of V Trimble

What has been seen was one LIGO event, a burst of gravitational waves from a merging pair of neutron stars, with an optical counterpart whose spectrum has been interpreted [88][11] as showing evidence for blended features due to a whole bunch of r-process products. Since then, a much smaller group [89] has done model-fitting to the near IR spectra and reported large quantities of a specific r-product (strontium). On the push-back side, another small group [90] has pointed out that some of the previously suggested processes must also occur and make an r-contribution.

Were you wondering where the iron came from? Yes, some from massive stars that give rise to supernovae and pulsars, but most of it from explosive burning of carbon and oxygen in what are called Type Ia supernovae (the ones used to calibrate cosmological distance scales) when a white dwarf made of carbon and oxygen is driven above the maximum possible stable (Chandrasekhar) mass either by accretion from a companion or by merger with another similar white dwarf. Hang on to that thought; you will need it in a few paragraphs.

Now about lithium, beryllium, and boron. These are remarkably sparse in the universe (Figs. 15.3 and 15.4) because they are very fragile and normally burn through to other stuff in stars. B^2FH said "x-process." The same can be said for deuterium, bound by only 2.2 MeV, but we now know that there is some/enough

[11]This is the LIGO visual follow-up.

left from the early universe. Remember B^2FH were "steady-state people" for whom there had never been a hot, dense, big bang. They suspected something involving cosmic rays might be relevant, and this is now known to be the rest of the answer. Very high energy protons whizzing through the gas of the interstellar medium from time to time hit CNO nuclei and break them up, with Li, Be, and B among the products.

Other rare, odd *A* or *Z* nuclides will also be made this way from more abundant ^{28}Si, ^{56}Fe, and all. For this we have fairly direct evidence: from the point of view of a cosmic ray oxygen nucleus, for instance, an interstellar proton looks like an incredibly fast particle aimed right at its heart, ready to break it up. And indeed LiBeB and some of the other odd nuclides are commoner in the cosmic rays than in the sun and such.

15.8 Why?—Astrophysics: A Bit of the Evidence and Missing Pieces

In the years after 1925, when Cecilia H. Payne had shown [29], and Russell, Dugan, and Stewart had accepted [30] that (nearly) all the stars had the same composition, it would not have been reasonable to look for evidence of on-going processes making more of the heavy elements, so that young stars would be more metal-rich than old ones. There was a period from around 1930 until after the Second World War when most astronomers thought all stars were the same age and had to be persuaded that star formation was also part of an on-going process. That is part of another story [91], but it didn't help in the present context. In any case, when the evidence first surfaced that Population II stars were metal poor [43, 47] it was thought either not important or not true or both. The version of that story I heard came from Lawrence H. Aller, who said what a "deuce of a time" (his words) he and Joe Chamberlain [92] had in getting their paper accepted for publication and that they had been required to scale back their factor of 100 deficiency of heavy elements to a factor of 10, just a bit more than 25 years after Cecilia had been required to scale back her "overabundance" of hydrogen and helium in stars!

It is now firmly established that old stars (as well as ones in other galaxies where there has been less star formation, evolution, and death) are generally deficient relative to our sun in all heavy elements, more so in nitrogen and s-process (secondary) nuclides and in iron, most of which comes from type Ia supernovae which start blowing up later in stellar evolution than core collapse (Type II etc.) supernovae [42]. Thus, we have turned once more to astronomical data to conclude that the synthesis of the heavy elements occurs in stars, continuously over the history of our galaxy and all the others.

15.9 Conclusions

Arthur Stanley Eddington, whom you met in Sect. 15.4 because he established much of early twentieth-century understanding of stellar structure without accepting that stars are made mostly of hydrogen and helium, had only one female co-author [93], Allie Vibert Douglas, who wrote his biography at the request of his surviving sister [94]. Safely back in Canada, where she was born (1894), died (1988), and spent most of her career apart from two years in Cambridge (1921–1923) she wrote, "You cannot solve the riddles of the stars without invoking aid of the atom, nor can you fully comprehend the atom without the aid of the stars" [95]. If you will let the stars stand for all of astronomy and astrophysics, and the atom stand for the periodic table and some additional bits of chemistry and physics, then her statement is essentially the conclusion here.

There was a good deal of confusion from the 1860s until well into the twentieth century over which spectral features corresponded to previously unknown chemical elements and which merely represented atoms of known elements under unusual conditions, as suggested by Eddington and by Russell et al. [30] (Sects. 15.2 and 15.3). Then came the era of sorting out the relative abundances of all the stable elements in the sun, stars, and nebulae, with a good deal of help from meteorites for isotopic ratios (Sect. 15.4), followed by the path-breaking work by Cameron (1957) [75, 76] and by Burbidge, Burbidge, Fowler, and Hoyle (1957) [49] (Sect. 15.5). They assigned virtually every stable isotope of every element (and even the unstable technetium, found in a few evolved stars) to one or more of about eight nuclear processes and sites (Sect. 15.7). Most of these abundances and sites have been fairly stable in our understanding down to the present time. Later, and still very much work in progress, is tracing the abundances and their variations through time in various populations of stars as a function of stellar age and formation environment. The classic beginning came from Olin C. Eggen, Donald Lynden-Bell, and Allan R. Sandage in 1962 [96]. They imagined a single giant cloud of hydrogen and helium gradually collapsing (conserving angular momentum), forming stars as it went from a sphere down to a halo and disk. This is not the current picture of galaxy formation, which involves many small entities merging to make large galaxies like the Milky Way, but the effect on stellar populations and the correlations of star composition, age, location, and kinematics is much the same (Sect. 15.8).

Is this the whole story? Of course not! There are long review articles, often whole conference proceedings, addressing practically every, I hope, factoid mentioned here. Is it all history? That probably depends on your age. History ends and current events begin when you start reading the literature for yourself. That was 1965 for me (summer after the first year of graduate school, and practically the first thing to be learned was that the universe was at a temperature of about 3 K). Somewhere in the world there are probably folks for whom it is even earlier (who don't come to as many conferences as they used to); and there are many for whom the line has to be drawn much later. For you, I have one thought: pay attention to what is going on around you, because some day you will be the only one who remembers.

Acknowledgements I am deeply grateful to the history division of the American Chemical Society for accepting an aging astronomer as almost one of their own! For what it is worth, my father was a chemist, inducted into Alpha Chi Sigma in the 1930s. Immeasurable thanks go to Vera Mainz for turning typewritten chaos into an elegant electronic format suitable for publication. A number of corrections and additions provided by Gregory S. Girolami and Carmen Giunta are also very much appreciated. Reviewer David H. DeVorkin of the Smithsonian Institution expertly added information and subtracted mis-information, for which I am much obliged!

References

1. Scerri ER (2007) The periodic table. Oxford University Press, Oxford, UK
2. Becker BJ (2011) Unravelling starlight: William and Margaret Huggins and the rise of the new astronomy. Cambridge University Press, Cambridge, UK
3. Nicholson JW (1911) The spectrum of nebulium. Mon Not R Astron Soc 72:49–64
4. Nath BB (2013) The story of helium and the birth of astrophysics. Springer, New York
5. Cottam S, Orchiston W (2015) Eclipses, transits, and comets of the nineteenth century: how America's perception of the skies changed. Astrophysics and Space Science Library, Volume 406. Springer, Cham
6. Gruenwald A (1887) On the remarkable relations between the spectrum of watery vapour and the line spectra of hydrogen and oxygen. Chem News 56:232
7. Huggins ML (1898) … Teach me how to name the … light. Astrophys J 8:54
8. Hufbauer K (1991) Exploring the sun. The Johns Hopkins Press, Baltimore, p 112
9. Heilbron JL (1974) H. G. J. Moseley: the life and letters of an English physicist, 1887–1915. University of California Press, Berkeley and Los Angeles
10. Macleod RM, Egdell RG, Bruton E (eds) (2018) For science, king, and country: the life and legacy of Henry Moseley. Uniform Press, London
11. Frederick-Frost KM (2017) For the love of a mother—Henry Moseley's rare earth research. Historical Stud Nat Sci 47:529–567
12. Trimble V, Mainz VV (2017) Who got Moseley's prize? In: Strom ET, Mainz VV (eds) The posthumous Nobel prize in chemistry, vol 1. Correcting the errors and oversights of the Nobel prize committee, American Chemical Society, Washington DC
13. Hinrichs G (1867) Programme der Atomechanik oder die Chemie eine Mechanik der Panatome. Privately printed, Iowa City, IA
14. Eddington AS (1927) Stars and atoms. Yale University Press, New Haven, CT
15. Bowen IS (1927) The origin of the nebulium spectrum. Nature 120:473
16. Bowen IS (1928) The origin of the nebular lines and the structure of the planetary nebulae. Astrophys J 67:1–15
17. Trimble V (2019, March) Astronomy Meets the Periodic Table. In: This Month in Astronomical History. American Astronomical Society Historical Astronomy Division. https://aas.org/posts/news/2019/02/month-astronomical-history-march Accessed 18 May 2020
18. van Spronsen JW (1969) The periodic system of the chemical elements, the first one hundred years. Elsevier, Amsterdam, p 121
19. Condon EU, Shortley GH (1964) The theory of atomic spectra. Cambridge University Press, Cambridge, p 282
20. Edlén B (1945) The identification of the coronal lines. Mon Not R Astron Soc 105:323–333
21. Grotian W (1939) Zur Frage der Deutung der Linien im Spektrum der Sonnenkorona. Naturwissenschaften 27:214
22. Swings P (1943) Edlén's identification of the coronal lines with forbidden lines of Fe X, XI, XIII, XIV, XV; Ni XII, XIII, XV, XVI; Ca XII, XIII, XV; AX, XIV. Astrophys J 98:116–128
23. Lyot B (1939) The study of the solar corona and prominences without eclipses. Mon Not R Astron Soc 99:580–594

24. Lodders K, Fegley B Jr (2011) Chemistry of the solar system. Royal Society of Chemistry Publishing, Cambridge, p 6
25. Trimble V (2010) The origins and abundances of the chemical elements before 1957: from Prout's law to Pasadena. Eur Phys Journal H 35:89–109
26. Perrin JB (1920) Atomes et lumière. Revue du Mois 21:113–166
27. Eddington AS (1920) The internal constitution of the stars. British Association for the Advancement of Science report of the eighty-eighth meeting 34–49 at 45
28. Eddington AS (1926) The internal constitution of the stars. Cambridge University Press, Cambridge, UK, pp 296, 314
29. Payne CH (1925) Stellar atmospheres: a contribution to the observational study of high temperature in the reversing layers of stars. PhD Dissertation, Harvard Monograph No. 1, Cambridge, MA
30. Russell HN, Dugan RS, Stewart JQ (1927) Astronomy: a revision of Young's manual of astronomy. Ginn and Company, Boston
31. DeVorkin DH (2010) Extraordinary claims require extraordinary evidence: CH Payne, HN Russell and standards of evidence in early quantitative spectroscopy. J Astronomical History Heritage 13:139–144
32. Eddington AS (1932) The hydrogen content of the stars. Mon Not R Astron Soc 92:471–481
33. Strömgren B (1932) The opacity of stellar material and the hydrogen content of the stars. Z Astrophysik 4:118–152
34. Chandrasekhar S (1939) Stellar structure. University of Chicago Press, Chicago, p 489
35. Aller LH (1963) The atmospheres of the sun and stars, 2nd edn. Ronald Press, New York, pp 54–55
36. Aller LH (1953) Astrophysics. Ronald Press, New York
37. Wildt R (1938) Negative ions of hydrogen and the opacity of stellar atmospheres. Astrophys J 90:611–619
38. Schwarzschild M (1958) Structure and evolution of the stars. Princeton University Press, Princeton, NJ
39. Russell HN (1929) On the compositions of the sun's atmosphere. Astrophys J 70:11–82
40. Trimble V (1975) The origin and abundances of the chemical elements. Rev Mod Phys 47:877–976
41. Suess HC, Urey HC (1956) Abundances of the elements. Rev Mod Phys 28:53–74
42. Lodders K (2019) Some elemental abundances. arXiv 1912.00844 https://arxiv.org/abs/1912.00844. Accessed 17 Feb 2020
43. Roman NG (1955) A catalogue of high-velocity stars. Astrophys J Suppl 2:195–223
44. Roman NG (1952) The spectra of the bright stars of types F5-K5. Astrophys J 116:122–143
45. Iwanowska W (1950) The spectrophotometric study of some high velocity stars. Bull Astron Obs U Toruń 9:25–38
46. Iwanowska W (1953) A spectrophotometric study of RR Lyrae type variables. Bull Astron Obs U Toruń 11:1–18
47. Schwarzschild M, Schwarzschild B (1950) A spectroscopic comparison between high-velocity and low-velocity stars. Astrophys J 112:248–265
48. Sagan C (1973) The cosmic connection: an extraterrestrial perspective. Anchor Press/Doubleday, Garden City, NY, pp 189–190
49. Burbidge EM, Burbidge GR, Fowler WA, Hoyle F (1957) The synthesis of the elements in stars. Rev Mod Phys 29:547–650
50. Atkinson RdE, Houtermans FG (1929) Zur Frage der Aufbaumöglichkeit der Elemente in Sternen. Z Physik 54:656–665
51. Rutherford E (1915) The constitution of matter and the evolution of the elements. Popular Science Monthly 87:105–141
52. Goeppert-Mayer M (1964) The shell model. Science 145:999–1006. This is a reprint of her Nobel Prize address
53. Goeppert-Mayer M (1948) On closed shells in nuclei. Phys Rev 74:235–239

54. Goeppert-Mayer M, Jensen JHD (1955) Elementary theory of nuclear shell structure. Wiley, New York
55. Unsöld A (1938) Über die Struktur der Fraunhoferschen Linien und die quantitative Spektralanalyse der Sonnenatmosphäre. Z Phys 46:765–781
56. Unsöld A (1938) Physik der Sternatmosphären: mit besonderer Berücksichtigung der Sonne. Verlag von Julius Springer, Berlin
57. McCrea WH (1931) Model stellar atmospheres. Mon Not R Astron Soc 91:836–857
58. Atkinson Rd'E (1931) Atomic synthesis and stellar energy II. Astrophys J 73:308–347
59. Aston FW (1920) LIX. The mass-spectra of the chemical elements. Philos Mag 30:611–625
60. Aston FW (1927) Bakerian lecture—a new mass spectrograph and the whole number rule. Proc R Soc Lond A 115:487–514
61. Bethe HA (1939) Energy production in stars. Phys Rev 55:434–456
62. Bethe HA, Critchfield CL (1938) The formation of deuterons by proton combination. Phys Rev 54:248–254
63. von Weizsäcker CF (1937) Über Elementumwandlungen im Innern der Sterne I. Phys Z 38:176–191
64. von Weizsäcker CF (1939) Über Elementumwandlungen im Innern der Sterne II. Phys Z 39:633–646
65. Kragh H (1996) Cosmology and controversy: the historical development of two theories of the universe. Princeton University Press, Princeton, NJ
66. Planck Collaboration (2018) Planck 2018 results VI. cosmological parameters arXiv 1807.06209 https://arxiv.org/abs/1807.06209. Accessed 17 Feb 2020
67. Planck Collaboration (2018) Planck 2018 results I. overview and the legacy of Planck arXiv 1807.06205 https://arxiv.org/abs/1807.06205. Accessed 23 Mar 2020
68. Trimble V (1991) Origin and abundances of the chemical elements revisited. Astron Astrophys Rev 3:1–46
69. Trimble V (1997) Origin of the biologically important elements. Orig Life Evol Biosph 27:3–21
70. Trimble V (1996) Cosmic abundances: past, present, and future. In: Holt SS, Sonneborn G (eds) Cosmic abundances ASP Conf Ser, vol 99. Astronomical Society of the Pacific, San Francisco
71. Gamow G (1946) Expanding universe and the origin of elements. Phys Rev 70:572–573
72. Alpher RA, Bethe HA, Gamow G (1948) The origin of chemical elements. Phys Rev 73:803–804
73. Alpher RA, Herman RC (1950) Theory of the origin and relative abundance distribution of the elements. Rev Mod Phys 22:153–212
74. Alpher RA, Herman RC (2001) Genesis of the big bang. Oxford University Press, New York
75. Cameron AGW (1957) Chalk River report CRL-41. Atomic Energy of Canada Ltd., Chalk River, Ontario
76. Cameron AGW (1957) Nuclear reactions in stars and nucleogenesis. Publ Astron Soc Pac 69:201–222
77. Hayashi C (1950) Proton-neutron concentration ratio in the expanding universe at the stages preceding the formation of the elements. Prog Theor Phys 5:224–235
78. Wagoner RV, Fowler WA, Hoyle F (1967) On the synthesis of elements at very high temperatures. Astrophys J 148:3–49
79. Cyburt RH et al (2016) Big bang nucleosynthesis: present status. Rev Mod Phys 88:015004-1–015004-22
80. Hoyle F (1946) The synthesis of the elements from hydrogen. Mon Not R Astron Soc 106:343–383
81. Opik E (1951) Stellar models with variable composition. II. sequences of models with energy generation proportional to the fifteenth power of temperature. Proc R Irish Acad A 54:49–77
82. Saltpeter EE (1952) Nuclear reactions in stars without hydrogen. Astrophys J 115:326–328

83. Maeder A, Meynet G (2012) Rotating massive stars: from first stars to gamma ray bursts. Rev Mod Phys 84:25–63
84. Greenstein JL (1954) Modern physics for the engineer. Ridenour LN (ed) McGraw-Hill, New York, p 235
85. Cameron AGW (1960) New neutron sources of possible astrophysical importance. Astrophys J 65:485
86. Lattimer JM, Schramm DN (1974) Black-hole—neutron-star collisions. Astrophys J Lett 192: L145–L147
87. Wheeler JA (1971) Mechanisms for jets. Pont Acad Scientiarum Scripta Varia 35:538–567 at 544–545. See section “Squeezed Tube of Toothpaste”
88. Abbott BP et al (2017) Multi-messenger observations of a binary star merger. Astrophys J Lett 848:L12 (59 pages)
89. Watson D et al (2019) Identifying strontium in the merger of two neutron stars. Nature 574:497–500
90. Siegel DM, Barnes J, Mezger BD (2019) Collapsars as a major source of r-process elements. Nature 569:241–244
91. Trimble V (2015) The fourth day of creation: the proto-history of young stars, star streams and exoplanets. In: Kasner JH et al (eds) Proc IAUS, vol 314. Cambridge University Press, pp 1–7 and references therein
92. Chamberlain JW, Aller LH (1951) The atmospheres of A-type subdwarfs and 95 Leonis. Astrophys J 114:52–72
93. Eddington AS, Douglas AV (1923) The progression of stellar velocity with absolute magnitude. Mon Not R Astron Soc 83:112–118
94. Douglas AV (1956) The life of Arthur Stanley Eddington. Thomas Nelson & Sons Ltd., London
95. Douglas AV (1930) From atoms to stars and the mystery of motion. McGill University Publications, Montreal, p 165
96. Eggen OC, Lynden-Bell D, Sandage AR (1962) Evidence from motions of old stars that the galaxy collapsed. Astrophys J 136:748–766

The Impact of Twentieth-Century Physics on the Periodic Table and Some Remaining Questions in the Twenty-First Century

16

Eric R. Scerri

Abstract

The historical development of the theoretical account of the periodic table provided by theoretical physics is reviewed, beginning with discoveries made at the start of the twentieth century. The article highlights the attempts to theoretically explain several features of the periodic table including the well-known period doubling or Madelung rule of orbital occupation. The account includes more recent group theoretical approaches which go beyond quantum mechanics and seek an explanation based in the underlying symmetry of the periodic table and how this symmetry is broken to produce the diversity of atoms that we are familiar with. The approach taken is one of seeking a global solution to such questions rather than merely solving the equations of quantum mechanics for each individual case.

16.1 Early Twentieth Century Developments

The twentieth century began with several influential discoveries that would have a large impact on the development of the periodic table and the study of the elements and that were made over a period of three consecutive years immediately preceding the turn of the twentieth century.[1]

[1]Although I will use the term periodic table in many cases, the more abstract concept of the periodic system may be more appropriate in some instances. Only in cases having to do with just the representation of the periodic system, is the term "table" entirely appropriate.

E. R. Scerri (✉)
Department of Chemistry& Biochemistry, University of California Los Angeles, Los Angeles, CA 90095, USA
e-mail: scerri@chem.ucla.edu

C. J. Giunta et al. (eds.), *150 Years of the Periodic Table*, Perspectives on the History of Chemistry, https://doi.org/10.1007/978-3-030-67910-1_16

First, Röntgen discovered X-rays in 1895 [1]. In addition to the well-known medical applications of this newly discovered form of electromagnetic radiation, they were soon used to study the structure of crystals and of matter in general. Between 1913 and 1914, Moseley discovered that the frequencies of certain X-ray lines showed a simple relationship with atoms of any particular element [2]. Moseley also succeeded in placing the elements into a more correct sequence that had been available to the discoverers of the periodic system, who had utilized the sequence of increasing atomic weights.

In 1896, just one year after Röntgen discovered X-rays, Becquerel discovered the phenomenon of radioactivity [3, 4] whereby certain unstable atoms decayed, while emitting particles of radiation including α, β and γ particles as they were subsequently identified by Rutherford. The very next year, Thomson discovered that the atom was not the featureless sphere that Dalton and others had imagined, because it seemed to contain some smaller particles [5] that were dubbed "electrons" by Stoney. While Thomson believed that these electrons were somehow embedded into the positive charge of the atom, his onetime student Rutherford led a team that fired beams of α particles at a thin gold foil and found a scattering pattern that could only be explained by assuming the presence of a dense small and positive nucleus at the center of any atom [6].

But even before the birth of Rutherford's nuclear atom, Thomson made one of the first attempts to explain the periodic table on the basis of his own model of the atom [7]. Most textbooks typically focus on the fact that Thomson regarded electrons as being embedded in the atom. These accounts typically fail to mention that Thomson also suggested that electrons were arranged as a series of concentric rings and that analogous arrangements of such rings could explain the similar properties among elements residing in the same groups of the periodic table. The essential physical explanation for the existence of chemical periodicity has therefore been in existence for well over 100 years, even if Thomson was mistaken as to how many electrons the atoms of each element possessed and the manner in which they moved. Among other notable contributors to early atomic theory one must mention the Curies who took up the exploration of Becquerel's radioactivity and succeeded in discovering the two elements of polonium and radium.

Meanwhile, starting in 1900 Planck unwittingly initiated the quantum revolution while explaining the radiation emanating from incandescent bodies such as light bulb filaments [8]. What emerged from his study was the counter-intuitive notion that black-body radiation was emitted only in discrete packets. The first significant application of this concept was made by Einstein, while he was in the process of explaining the photo-electric effect [9]. When light strikes a metal, with a sufficient energy, that depends of the metal in question, electrons are released and gain kinetic energy. This kinetic energy does not depend on the intensity of the light, as one might expect on the understanding that light is a wave phenomenon. Instead, the kinetic energy depends on the frequency of the light, but only if a certain threshold frequency is exceeded. Einstein resolved these apparent anomalies by suggesting that Planck's energy quantization applied to light in general rather than just black-body radiation. Einstein further postulated the existence of quanta, or particles, of light energy so that one quantum is required to dislodge each single electron before imparting kinetic energy to it.

In 1913 Bohr made another significant application of Planck's old quantum theory when he introduced it to the structure of the hydrogen atom [10]. By means of an ad hoc argument, Bohr asserted that electrons were confined to specific quantum orbits and that they could only undergo transitions to other fixed orbits. In the process of such transitions, as Bohr maintained, electrons could only absorb or emit specific quanta of energy which were reflected in the discrete spectra that had been obtained for any particular atom in the chemist's periodic table.[2]

Moreover, Bohr was able to provide more accurate electronic arrangements (now termed configurations), than Thomson had been able to, since the correct number of electrons in the atoms of all the elements had been deduced by then. As is well known, Bohr's model was extended by Sommerfeld in 1916 by appealing to the special theory of relativity and by assuming that Bohr's orbits were elliptical rather than only circular, thus effectively introducing a second degree of freedom to each electron [11]. This in turn resulted in the need to quote two quantum numbers in order to identify any particular electron, one more than in Bohr's original model with its one quantum number that corresponds to the main shell number.

A completely unknown Cambridge graduate student, Stoner [12], then introduced the use of a third quantum number, quickly followed by Pauli who, in 1925, added yet a fourth degree of freedom and accompanying fourth quantum number for each electron. Pauli announced his Exclusion Principle which would shortly be interpreted to require that the wavefunction of an atom is anti-symmetrical on the interchange of any two electrons [13].

Meanwhile, from 1923, a separate development was taking place in the context of wave-particle duality that Einstein had initiated. De Broglie proposed that wave-particle duality might work in both directions as it were [14]. Just as light waves had been shown to behave as particles, De Broglie proposed that particles, like electrons, might possess an intrinsic wave nature. The experimental confirmation of this proposal was achieved soon afterwards by Davisson and Germer who obtained an interference pattern when they fired a beam of electrons at a small crystal [15].

Schrödinger, working purely theoretically in 1926, proposed a wave mechanical equation to describe the motion of electrons in an atom [16]. Unlike Bohr's ad hoc quantization of angular momentum of electrons, Schrödinger succeeded in deriving quantization and in obtaining a more detailed description of the allowed motion of

[2]Although Bohr's theory was only quantitatively successful for one electron or hydrogenic systems, he applied it in a qualitative fashion in order to understand the periodic system and with a considerable degree of success.

electrons. Such motions were subsequently called orbitals, by contrast with Bohr's deterministic orbits. Solving Schrödinger's equation, by applying suitable boundary conditions, results in an infinite number of solutions which are characterized by three quantum numbers. These numbers corresponded to the quantum numbers mentioned above that had been gradually arrived at by Bohr, Sommerfeld and Stoner in more semi-empirical ways. However, Schrödinger's original treatment did not invoke Pauli's fourth quantum number, which to this day is "tagged on" because it is required by the spectral evidence.[3]

One of the most remarkable aspects of Schrödinger's model, when it is augmented thus with a fourth quantum number, is that it gives an almost complete explanation of the periodic table/system in a way that had eluded Thomson and even Bohr. If one combines together the allowed values of the four quantum numbers, in what I will call the Schrödinger-Pauli model, one can rigorously deduce that subsequent electron shells should contain 2, 8, 18 or 32 electrons in perfect agreement with the various possible period lengths that are found in the modern periodic table (Fig. 16.1).[4]

What makes this result even more significant is that it applies to the humble hydrogen atom with its one electron. And yet this approach provides the outline of an explanation of the capacity for each electron shell and for the possible period lengths for all the current 118 elements. I will be returning to this point in due course because it also contains a hint concerning more recent work on the explanation of the periodic table by appeal to the hydrogen atom.

Textbooks frequently rest contented with the Schrödinger-Pauli explanation for the periodic table, although it is clear that it can only be regarded as a stepping stone toward a fuller explanation which is not yet available, even after more than 150 years after the discovery of periodicity. What is still lacking is an equally rigorous, or global, explanation of why the sequence of period lengths in the periodic system is 2, 8, 8, 18, 18, 32, 32 (period doubling), instead of 2, 8, 18 and 32 as might be expected from the Schrödinger-Pauli approach.[5]

The order of filling of orbitals does not involve a strictly sequential filling of each shell but more of a filling in a "diagonal manner." This order is often displayed in a mnemonic shown in Fig. 16.2, and called the Madelung or $n + \ell$ rule [17].[6] On

[3]Dirac's more general theory does predict electron spin and hence a fourth quantum number even if Schrödinger's earlier non-relativistic theory does not.

[4]The form of the periodic table best suited to making this point is the 32-column or long form table.

[5]To be clear, quantum mechanics can exhaustively calculate the experimentally observed configuration of each atom provided that sufficient flexibility is built into the wavefunction. Consequently, it can reproduce the order of shell filling that agrees with the period doubling and the Madelung rule. What is intended by a "rigorous derivation" here, is one that would represent a global solution for the entire periodic table at once, as it were, without having to carry out calculations for every single atom.

[6]The rule appears to have been rediscovered a number of times. It is also found associated with the names of Karapetoff, Janet, Bose, Goudsmit, Klechkowski and Keller, somewhat depending on the national origin of the textbook that one examines. Nevertheless, the Francophone world seems to assign the rule to the Russian physicist Klechkowski.

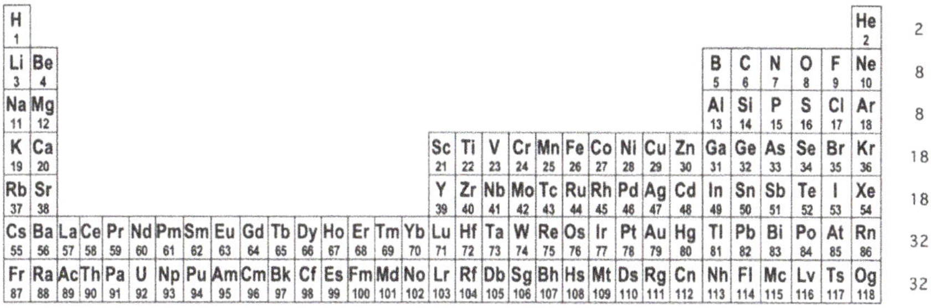

Fig. 16.1 32-column periodic table which most clearly displays the variation in period lengths, given by the formula $2n^2$

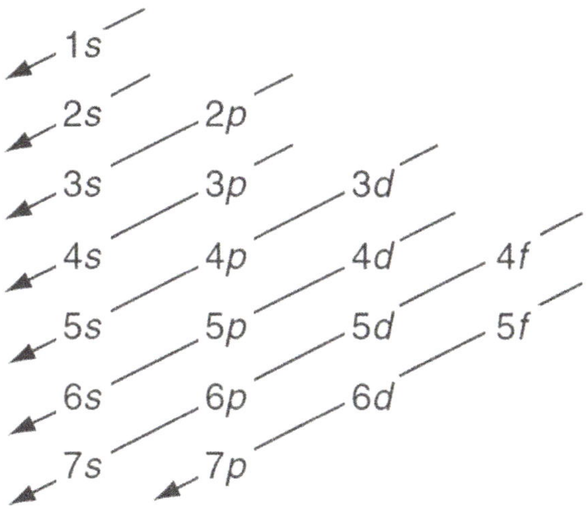

Fig. 16.2 Madelung or $n + \ell$ rule purporting to show order of orbital occupation

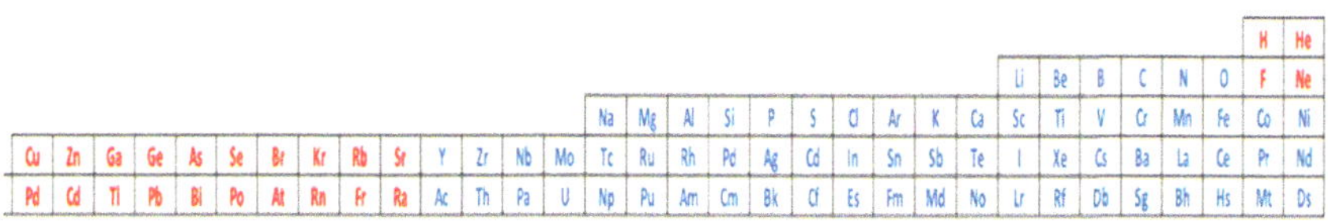

Fig. 16.3 This is how the periodic table would look if shells were to be filled in strict sequence, of only starting a new shell after the previous one is complete, beginning with the innermost shell. The familiar grouping of elements is lost, apart from the elements shown in red, a feature that appears to be coincidental. The fifth period of 50 elements has been truncated for the sake of convenience

the other hand, if the shells were to fill in a strictly sequential manner, meaning that each shell would fill completely before moving onto the next one, the result would be a rather unusual table as shown in Fig. 16.3 which would not reflect the chemical resemblances within groups in the conventional periodic table.[7]

16.2 Developments in the Second Half of the Twentieth Century

The unfortunate fact, for those who claim that the periodic system has been fully reduced, or explained by quantum mechanics, is that the Madelung rule itself has not yet been derived from first principles. This situation was highlighted by the theorist Löwdin while speaking at a conference to mark the 100th anniversary of the periodic table [18]:

> The energy rule for the neutral atoms was obviously in contradiction to Bohr's calculation on the hydrogen atom, which indicated that the energies should be increasing with increasing n. It is typical of the nature of "frontier-research" that Bohr abandoned this rule for the higher atoms, since it led to the wrong structure of the periodic system, and the modified rule $[(n + \ell, n)]$ seems to have been obtained in a more intuitive way. Bohr himself was never too explicit about his "Aufbau" principle, and [the rule] is sometimes referred to as the Goudsmit-rule or the Bose-rule. It is perhaps remarkable that, in axiomatic quantum theory, the simple energy rule has not yet been derived from first principles. [p 332]

Although many attempts have been made to explain what is sometimes referred to as the doubling of period lengths (except for the first period), none of them have been successful [19–23].

Moreover, some authors have sought to eliminate the Madelung rule because it fails to provide the precise order of orbital occupation for any particular atom starting with that of scandium [24–27].[8]

> Indeed, the $n + \ell$ rule has little meaning in chemistry. However, since the rule occurs in all textbooks and is absorbed by all students and teachers, it will die out only very slowly. Only a few chemists need correct details about the electronic structure of the chemical transition elements; their re-education in graduate courses is not too difficult. Most other undergraduate students will not need the $n + \ell$ blunder in their future career. [25]

These claims have been countered by various authors who point out that the Madelung rule remains valid when considering the nature of the differentiating electron, meaning the electron that makes a difference between any particular atom and the subsequent one [28–30].

---

[7]The order of occupation of orbitals in the ions of each atom is a different matter and is not given by the Madelung Rule.

[8]The configuration predicted for scandium by the Madelung rule is [Ar] $4s^2$ $3d^1$. However spectral data shows it to be [Ar] $3d^1$ $4s^2$. Similar apparent violations of the rule occur for all transition metals and those from the f-block of the periodic table.

Schwarz is correct in saying that the Madelung rule is violated when it comes to the progressive occupation of orbitals in any particular atom. But it is still true that the electron that differentiates an element from the previous one in the table follows Madelung's rule. In the case of potassium and calcium, the "new electron" relative to the previous atom is a 4s electron. But in scandium, the electron that differentiates it from calcium is a 3d one, even though it is not the final electron to enter the atom as it builds up. In other words, the simple approach to using the aufbau principle and the Madelung rule remains valid for the periodic table viewed as a whole. It only breaks down when considering one specific atom and its occupation of orbitals and ionization energies. The challenge of trying to derive the Madelung rule therefore remains, pace Schwarz. [28]

I would now like to turn to one promising line of research which has been in the making for more than 50 years and which brings us back to the hydrogen atom and its potential to generate the entire periodic table.[9] I am referring to the group theoretical approach to explaining the precise lengths of all the periods in the periodic table on the basis of the special symmetry of the hydrogen atom.

16.3 The Special Dynamical Symmetry of the Hydrogen Atom

The energy levels in the hydrogen atom that share a common n quantum number value all have the same energy.[10] This property, that is given the name of degeneracy, is rather difficult to explain fully using quantum mechanics. The first person to do so was Pauli who drew on the dynamical symmetry of the hydrogen atom, as will be explained. It is important to appreciate that this symmetry goes beyond the spherical shape of the electron distribution around the nucleus of the hydrogen atom.

The hydrogen atom features just one electron of course, that experiences a Coulombic force of attraction. This highly symmetrical scenario is responsible for the degeneracy among energy levels that share the same n and ℓ quantum numbers. For example, the three 2p orbitals in the hydrogen, and indeed any atom, are known to possess the same energy in the absence of perturbations. However, the hydrogen

[9]Cosmologists also consider hydrogen to have been the progenitor of all other elements in the sense that the first element to form after the Big Bang was indeed hydrogen. This can be considered as a modern-day version of Prout's hypothesis whereby all elements were regarded as composites of hydrogen, which of course they are if one focuses on the number of protons in the nucleus of the atom of any particular element.

[10]This is no longer the case if one considers the fine structure among spectroscopic levels or the even smaller splitting due to the Lamb shift. The approximate degeneracy being referred to is a feature of the non-relativistic Schrödinger treatment of the hydrogen atom. More strictly speaking, there is already nonrelativistic lifting of degeneracy at the Schrödinger level due to the finite size of nuclei which are not point charges without structure, but spatially extended objects. Further contributions to fine structure also arise from electron spin and the mass of the electron. I am grateful to a reviewer for suggesting this clarification.

atom alone possesses another more remarkable degeneracy, namely the fact that all the orbitals that share the same *n* quantum number have the same energy such as in the case of the 3s, 3p and 3d orbitals in the 3rd main shell. This degeneracy has long been known as an "accidental degeneracy" because its origin was unknown. Nowadays it should no longer be referred to as such since its cause has been fully explained. A full understanding of this issue involves some rather complicated theoretical physics and mathematics and will not be attempted in this article. (Further information on this question can be found in Blinder [31] and Thyssen and Ceulemans [32].) It is important to gather a broad understanding of the question however, because of its importance to recent attempts to obtain a global understanding of the periodic table that involves group theory and that go beyond the use of quantum mechanics in some respects.

As already briefly mentioned, the hydrogen atom possesses additional degeneracy than that associated with its rotational symmetry. This additional symmetry, sometimes described as being a hidden symmetry, is termed "dynamical symmetry" in order to distinguish it from the better-known geometrical symmetry of the spherical potential experienced by the electron.

16.4 The Laplace-Runge-Lenz or LRL Vector

The mathematical key to understanding the degeneracy of the hydrogen atom lies in drawing an analogy to a classical mechanical problem that was analyzed long ago by some well-known mathematicians including, Hermann, Bernoulli, Laplace and Hamilton, each of whom re-discovered a certain vector that has been given various names. Mechanical systems that belong to a class having central forces show a conservation of angular momentum and in addition the conservation of a vector that is known as the Laplace-Runge-Lenz vector since Wolfgang Pauli made use of it in quantum mechanics. The LRL vector is constant in both magnitude and direction at any point during the course of a planet's elliptical motion around the sun, or in the case of the atom, an electron's motion around the nucleus. More specifically, there are three components of the LRL vector which all represent constants of the motion for the planetary system or the classical understanding of the hydrogen atom. The LRL is therefore the key to discovering an extra constant of motion that is associated with the additional dynamical symmetry that exists in the hydrogen atom.

In 1926 Pauli obtained the quantum mechanical version of the LRL vector and was able to explain the formerly known accidental degeneracy of the hydrogen atom. In the same article Pauli used the LRL vector, within Heisenberg's matrix mechanical theory, to obtain the first solution of the energy levels of the hydrogen atom, ahead of Schrödinger's treatment which soon followed [33].[11]

[11]The reason why the vector bears the names of Laplace, Runge and Lenz is due to Pauli who learned of the existence of this vector while he was an assistant to Lenz, who in turn referred to the work of Runge and Laplace on the same vector.

Nevertheless, it appears that Pauli lacked the necessary knowledge of group theory that would have enabled him to grasp the full implications of the LRL vector for the question of the symmetry of the hydrogen atom. The latter feat was achieved by the Soviet physicist Vladimir Fock, who is perhaps better known for having modified Hartree's method of approximating the orbitals for many-electron atoms in order to comply with the requirement of the anti-symmetry of the wavefunction in the Hartree-Fock approach to computational chemistry and physics [34].

Returning to the hydrogen atom, in 1935 Fock discovered that the additional degeneracy could be rationalized by appeal to a form of supersymmetry that is hidden within a fourth spatial dimension [35]. Fock's hydrogenic wavefunctions were projections from the familiar three-dimensional space onto the surface of a four-dimensional hypersphere that is technically denoted as a 3-sphere. While the familiar sphere in three dimensions represents the symbol of perfect symmetry, physical phenomena such as the hydrogen atom have led to the realization that there exist symmetries even more perfect in higher dimensional spaces.

Here is how the Mexican group theorist Octavio Novaro explained the situation in a volume consisting of articles presented during the second international conference on the periodic table held in 2003 [36–40].[12]

> Let us reflect upon the depth of Fock's achievement: he identified two apparently unrelated systems, the three-dimensional Coulomb potential of the nucleus acting on the electron, and a forceless punctual mass constrained to move on the surface of a hypersphere in four dimensions. He therefore obtained a full group-theoretical explanation of the "accidental degeneracy" of the hydrogen atom. In fact, Fock also provided the closed-shell occupation numbers for model systems consisting of many non-interacting particles captured in an attractive Coulomb potential. These are the so-called "magic numbers" (2, 8, 18, 32, 50 etc.) which in Fock's approach are not or mysterious at all, as they correspond to the irreducible representations of the group O(4)...Elaborating on these results, Bargman demonstrated that the O(4) symmetry of the hydrogen atom stems from the conservation of two constants of motion: the angular momentum in three dimensions and the Runge-Lenz vector and that these are precisely the generators of this group.

This work provided the first significant connection between symmetry principles and attempts to fully explain the periodic table.[13] However, it did not give an explanation of period doubling in the periodic table or the Madelung rule that governs the occupation of atomic orbitals.

[12]There have been four international conferences on the periodic table up to the present time. The first was held in the Vatican City to commemorate the centenary of Mendeleev's 1869 article, in which he announced his periodic table, and included presentations from physicists John Wheeler and Emilio Segrè. The second conference took place in Banff, Canada in 2003. The city of Cusco in Peru was the location of the third international conference while the fourth was held in St. Petersburg to commemorate the 150th anniversary of Mendeleev's paper of 1869. Proceedings for all of these meetings have been published [37–40].

[13]Novaro's claim that degeneracy was rendered non-accidental as a result of Fock's work is historically inaccurate. As was mentioned above, this development was due to Pauli.

16.5 An Alternative Philosophical Approach of Moving Beyond Particles and Individual Elements: Heisenberg and Isospin[14]

In order to understand the subsequent developments in the still unfolding story, we must go backwards historically and consider the work of another luminary figure among the founders of quantum mechanics, namely Werner Heisenberg. This author was deeply influenced by ancient Greek philosophy and more specifically the writings of Plato. For example, in a book written in 1971 Heisenberg states [41],

> The elementary particles in Plato's Timaeus are finally not substance but mathematical forms. [p 8]

and

> So far we had always believed in the doctrine of Democritus, which can be summarised by: "In the beginning was the particle." We had assumed that visible matter was composed of smaller units, and that, if only we divided these long enough, we should arrive at the smallest units, which Democritus had called "atoms" and which modern physicists called "elementary particles." But perhaps this entire approach has been mistaken. Perhaps there was no such thing as an indivisible particle. In the beginning was symmetry! [p 133]

Working in the 1930s Heisenberg set out to understand the relationship between the proton and the neutron, two fundamental particles with almost identical masses. These near identical masses implied a degeneracy, which in turn suggested that there might be a form of symmetry that allows for interconversion among these particles. Such transformations are known to occur physically, such as when a neutron is transformed into a proton plus a β particle along with a neutrino [42].

By analogy with electrons which possess two spin states with very similar energies, Heisenberg postulated a property that he termed isospin which characterized the proton and neutron as showing alternative isospin states of the same fundamental particle. While the mathematical symmetry group that allows the spin states of electrons to transform into each other is SU(2), Heisenberg found that the same symmetry group transforms a proton into a neutron and vice versa.

16.6 Gell-Man and the Eight-Fold Way

Heisenberg's approach to the transformation of the proton into the neutron lay dormant for a period of about 30 years before it was revived by the physicist Murray Gell-Mann [43]. By this time the number of elementary particles identified by physicists had greatly increased, so much so that the situation became described as the "particle zoo." This state of affairs was seen to be analogous to the state of chemistry with its sixty or so elements that were recognized before the advent of the periodic table in the 1860s.

[14] the remaining parts of the present article draw heavily from the work of Thyssen and Ceulemans [32].

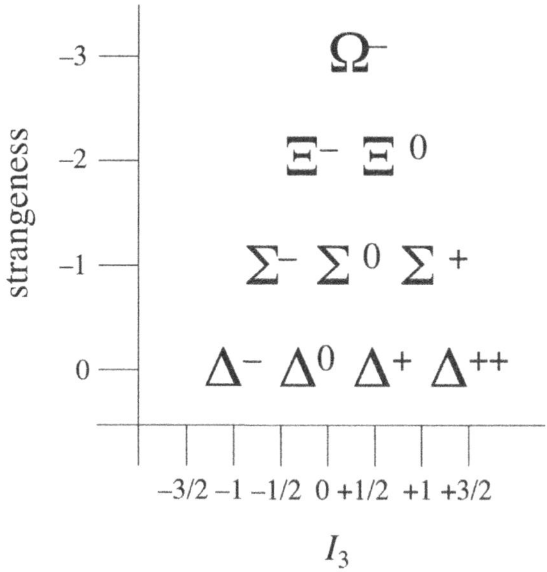

Fig. 16.4 Gell-Mann's eight-fold way allowed him to predict the existence of the Ω^- particle that was experimentally confirmed in 1964

Particle physics began seeking the underlying symmetry which could provide the connection, and indeed the interconversion, of elementary particles within a number of families of particles such as the leptons and hadrons. It was at this point in the development of the field that Gell-Mann found that the SU(3) symmetry group was what was required to bring about transformations among the particles in the hadron family [44].

In addition to providing a fundamental connection between these seemingly distinct particles Gell-Mann's scheme famously allowed him to make a prediction of the existence of a particle that represented a missing gap in his diagram, in much the same way that Mendeleev had left empty spaces and had successfully predicted several new elements when he had tamed the element zoo one hundred years previously in the 1860s. Gell-Mann not only imposed order on the particle zoo, he also successfully predicted the existence of the Ω^- that was experimentally confirmed in 1964 [44] (Fig. 16.4).

The success of this approach was rewarded by the Nobel Prize to Gell-Mann in 1969, precisely 100 years after Mendeleev's discovery of the periodic table. This work served to strengthen the philosophical Plato's view that form is more fundamental than substance, and in the terms of Heisenberg, that symmetry is more fundamental than particles. In the latter case, all the members of a family of fundamental particles could be regarded as manifestations of a super-particle whose properties are governed by the underlying symmetry group.[15]

[15]The even more recent work on the postulation of the Higgs particle and its experimental verification in the twenty-first century lends further support for the view that symmetry is more fundamental than particles and that the existence of individual particles results from the breaking of symmetry.

Author	*Year*	*Chain of subgroups*
Barut	1972	$SO(4,2) \supset SO(3,2) \supset SO(3) \otimes O(2)$
Odabaşi	1973, 2010	$SO(4,2) \supset SO(3,2) \supset SO(3) \otimes O(2)$
Ostrovsky	1980, 1981, 2006	$O(4,2) \otimes SU(2)_S \otimes SU(2)_T \supset O(4,2) \otimes SU(2)_S \supset O(4) \otimes SU(2)_S \supset O(3)$
Novaro and Berrondo	1972	$SU(2) \otimes SU(2) \otimes SU(2) \supset O(4) \supset SO(3)$
Berrondo and Novaro	1973, 1989, 2006, 2010	$E(4) \supset SU(2) \otimes SU(2) \otimes SU(2) \supset O(4) \supset SO(3)$
Rumer and Fet	1971	$Spin(4) \supset SU(2)_M, SU(2)_C$
Konopel'chenko	1972	$SO(2,4) + R \supset SO(2,4) \supset SO(4) \supset SO(3) \supset SO(2)$
Fet	1974, 1975	$SO(4,2) \otimes SU(2) \supset Spin(4) \otimes SU(2) \supset SU(2)_C \otimes SU(2)$
Byakov et al.	1976, 1977	$SO(4,2) \otimes SU(2) \supset Spin(4) \otimes SU(2) \supset SU(2)_C \otimes SU(2)$
Fet	1979, 1980, 1981	$O(4,2) \otimes SU(2) \otimes Z_2$
Fet	1989, 1992, 1996, 2010	$O(4,2) \otimes SU(2) \otimes SU(2)$
Kibler and Negadi	1989, 2004, 2006, 2007	$SO(4,2) \otimes SU(2) \supset SO(4) \otimes SU(2) \supset SO(3) \otimes SU(2) \supset SU(2)$
Thyssen and Ceulemans	2013	$SO(4,2) \otimes SU(2) \supset SO(3,2) \otimes SU(2) \supset SO'(4) \otimes SU(2)$

[a] P. Thyssen, "Symmetry and Symmetry Breaking in the Periodic Table: Towards a Group-Theoretical Classification of the Chemical Elements." PhD diss., Katholieke Universiteit Leuven, 2013.

Fig. 16.5 A table reproduced from Thyssen and Ceulemans [32] showing various ways in which group theorists have examined the breaking of SO(4,2) symmetry in an attempt to explain the period doubling as well as the Madelung rule associated with the periodic table. Reproduced with permission [32]

16.7 Fet, Barut and Others on Super-Elements

In the 1970s, several physicists, working independently, adopted a group theoretical approach to trying to explain the periodic system of the elements in a more fundamental, or global manner, than quantum mechanics had succeeded in doing [45, 46].[16] Authors including Fet in the then Soviet Union and Barut in the US shared the view of Plato, Heisenberg and Gell-Mann that symmetry operates at a more fundamental level than matter or particles [47, 48]. They applied this way of thinking to the periodic table of the elements and postulated the existence of a "super-element" which gives rise to all the individually known elements when symmetry is broken. Just as in the work of Heisenberg and Gell-Mann, it became a question of identifying what symmetry was responsible for the interconversion of any element into any other particular element in this case. Stated otherwise, they sought the particular form of symmetry which when broken could give rise to all the known individual elements.

The required symmetry was identified by various physicists as being SO(4,2). However, there is much disagreement as to how this underlying symmetry should be broken via a series of reductions, to a chain of sub-groups, in order to recover the characteristic period doubling of the periodic table and its associated Madelung rule. As shown in Fig. 16.5, many competing schemes have been proposed. Indeed, there is even disagreement as to whether the Madelung rule requires any explanation at all. Many of these group theoretical approaches are concerned with the super-element as discussed above. Consequently, some of these authors no longer

[16] An alternative atomic physics approach was simultaneously explored by other authors including Demkov and Ostrovsky [21], Ostrovsky [45], Novaro [36] and Kibler [46].

feel restricted by having to recover the Madelung rule in particular, which they believe is only relevant when one considers the elements separately as opposed to collectively.

16.8 Conclusion

Attempts to explain the periodic table have been a great driving force for physicists throughout the twentieth century and also into the twenty-first century. Although quantum mechanics provides an ab initio explanation for the lengths of periods, it has not yet explained the phenomenon of period doubling or the Madelung rule which governs the manner in which atoms are built-up as one traverses the periodic table. Beginning in the 1970, but drawing of earlier work stretching back to classical mechanics of a two-body system, group theorists have moved beyond quantum mechanics in order to seek the symmetry that underlies the periodic system. Although considerable progress has been achieved, such as the recognition of the symmetry group that underlies the periodic table, this project has not yet been entirely successful. What remains to be carried out is to discover precisely how the underlying SO(4,2) symmetry is broken to produce the well-known aspects of the table whose earliest version was published by Mendeleev just over 150 years ago.

References

1. Röntgen WC (1895) Ueber eine neue Art von Strahlen. Sitzungs-Berichte Physik-medic Gesellschaft Würzburg, 132–141
2. Moseley HGJ (1913) The high-frequency spectra of the elements. Philos Mag 26:1024–1034
3. Becquerel AH (1896) Sur les radiations émises par phosphorescence. C R Séances Acad Sci 122:420–421
4. Becquerel AH (1896) Sur les radiations invisibles émises par les corps phosphorescents. C R Séances Acad Sci 122:501–503
5. Thomson JJ (1897) Cathode rays. The Electrician 39:104–109
6. Rutherford E (1911) The scattering of α and β particles by matter and the structure of the atom. Philos Mag 21:669–688
7. Thomson JJ (1904) On the structure of the atom: an investigation of the stability and periods of oscillation of a number of corpuscles arranged at equal intervals around the circumference of a circle; with application of the results to the theory of atomic structure. Philos Mag 7:237–265
8. Planck M (1900) Zur Theorie des Gesetzes der Energieverteilung im Normalspectrum. Verhandl Dtsch Phys Ges 2:237–245
9. Einstein A (1905) Über einen Erzeugung und Verwandlung des Lichtes betreffenden heuristischen Gesichtspunkt. Ann Phys 17:132–148
10. Bohr N (1913) On the constitution of atoms and molecules, part III: Systems containing several nuclei. Philos Mag 26:857–875
11. Sommerfeld A (1930) Wave mechanics. EP Dutton & Co, New York
12. Stoner E (1924) The distribution of electrons among atomic levels. Philos Mag 48:719–736
13. Pauli W (1925) Über den Zusammenhang des Abschlusses der Elektronengruppen im Atom mit Komplexstruktur der Spektren. Z Phys 31:765–783

14. De Broglie L (1925) Recherches sur la théorie des quanta. Ann Phys 10(3):22–128
15. Davisson C, Germer LH (1927) Diffraction of electrons by a crystal of nickel. Phys Rev 30:705–741
16. Schrödinger E (1926) Quantisierung als Eigenwertproblem. Ann Phys 384(4):273–376
17. Madelung E (1936) Die mathematischen Hilfsmittel des Physikers, 3rd edn. Springer, Berlin, p 359
18. Löwdin PO (1969) Some comments on the periodic system of the elements. Int J Quantum Chem 3S:331–334
19. Allen LC, Knight ET (2002) The Löwdin challenge: origin of the (Madelung) rule for filling the orbital configurations of the periodic table. Int J Quantum Chem 90:80–82
20. Bent HA, Weinhold F (2007) News from the periodic table: an introduction to periodicity symbols, tables and models for higher order valency and donor-acceptor kinships. J Chem Educ 84:1145–1146
21. Demkov YN, Ostrovsky V (1972) $n + \ell$ filling rule in the periodic system and focusing potentials. Soviet Physics JETP 35:66–69
22. Pan Wong D (1979) Theoretical justification of Madelung's rule. J Chem Educ 56:714–717
23. Scerri ER (2009) The dual sense of the term "element", attempts to derive the Madelung rule and the optimal form of the periodic table, if any. Int J Quantum Chem 109:959–971
24. Schwarz WHE, Rich RL (2010) Theoretical basis and correct explanation of the periodic system: review and update. J Chem Educ 87:435–443
25. Wang SG, Schwarz WHE (2009) Icon of chemistry: the periodic system of chemical elements in the new century. Angew Chem Int Ed 48:3404–3415
26. Schwarz WHE (2010) The full story of the electron configurations of the transition elements. J Chem Educ 87:444–448
27. Schwarz WHE, Wang SG (2010) Some solved problems of the periodic system of chemical elements. Int J Quantum Chem 110:1455–1465
28. Scerri ER (2019) Can quantum ideas explain chemistry's greatest icon? Nature 565:557–559
29. Scerri ER (2019) Happy 150th birthday to the periodic table. Chem A Eur J 25:7410–7415
30. Thyssen P, Ceulemans A (2020) Particular symmetries: group theory of the periodic system. Substantia 4(1):7–22
31. Blinder SM (2001) Quantum alchemy: transmutation of atomic orbitals. J Chem Educ 78:391–394
32. Thyssen P, Ceulemans A (2017) Shattered symmetry. Oxford University Press, New York
33. Pauli W (1926) On the hydrogen spectrum from the standpoint of the new quantum mechanics. Z Phys 36:336–363. English translation in van der Waerden, B. L. (1968) Sources of quantum mechanics. Dover, New York, p 387ff
34. Fock V (1930) Näherungsmethode zur Lösung des quantenmechanischen Mehrkörperproblems. Z Phys 61:126–148
35. Fock V (1935) Zur Theorie des Wasserstoffatoms. Z Phys 98:145–154
36. Novaro O (2006) Group theory and the periodic table. In: Rouvray DH, King RB (eds) The mathematics of the periodic table. Nova Science Publishers, New York, pp 217–234
37. Verde M (ed) (1971) Atti del convegno Mendeleeviano: periodicità. e simmetrie nella struttura elementare della materia. Accademia delle Scienze di Torino, Turin and Rome
38. Rouvray DH, King RB (eds) (2004) The periodic table: into the 21st century. Research Studies Press Ltd., Baldock, Hertfordshire, UK
39. Scerri ER, Restrepo G (eds) (2018) Mendeleev to Oganesson. Oxford University Press, New York
40. Burrows HD, Stohner J (2019) Road to Mendeleev 150: 4th International Conference on the Periodic Table. Pure Appl Chem 91. https://doi.org/10.1515/pac-2019-1008
41. Heisenberg W (1971) Physics and beyond: Encounters and conversations (Vol 42). Harper and Row, New York
42. Heisenberg W (1932) Über den Bau der Atomkerne. Z Phys 77:1–11

43. Gell-Mann M (1961) The eightfold way: a theory of strong interaction symmetry. California Inst of Tech Synchrotron Laboratory report TID-12608, Pasadena, CA, March 15, 1961
44. Barnes VE et al (1964) Observation of a hyperon with strangeness minus three. Phys Rev Lett 12(8):204–206
45. Ostrovsky VN (2006) Group theory applied to the periodic table of the elements. In: Rouvray DH, King RB (eds) The mathematics of the periodic table. Nova Science Publishers, New York, pp 265–312
46. Kibler MA (2006) Group-theoretical approach to the periodic table: old and new developments. In: Rouvray DH, King RB (eds) The mathematics of the periodic table. Nova Science Publishers, New York, pp 237–263
47. Fet AI (2016) Group theory of chemical elements. De Gruyter, Berlin
48. Barut AO (1972) Group structure of the periodic system. In: Wybourne BG (ed) The structure of matter: Rutherford Centennial Symposium. University of Canterbury Press, Christchurch, New Zealand

An Essay on Periodic Tables

17

Pekka Pyykkö

Abstract

After a compact history of the PT, from Döbereiner's triads to the theoretical predictions up to element 172, a number of particular issues is discussed: Why may $Z = 172$ be a limit for stable electron shells? What are the expected stability limits of the nuclear isotopes? When are formally empty atomic orbitals used in molecular electronic structures? What is 'Secondary Periodicity'? When do the elements (Ir, Pt, Au), at the end of a bond, simulate (N, O, I), respectively? Some new suggestions for alternative PTs are commented upon. As a local connection, Johan Gadolin's 1794 analysis of the Ytterby mineral is mentioned.

Keywords

Periodic tables · Secondary periodicity · Superheavy elements · Chemical bonding · Analogies between elements · Lanthanides

17.1 Historical Introduction

A Periodic Table of Elements (PT) arranges chemical elements as a function of their properties—how so? Any student might answer: by their nuclear charge, Z. Currently the elements with $Z = 1$–118 have been found in nature or artificially pro-

---

This chapter has been reproduced with permission from Ref. [1]. https://doi.org/10.1515/pac-2019-0801. Copyright 2019 IUPAC and De Gruyter.

A collection of invited papers based on presentations at "Mendeleev 150": 4th International Conference on the Periodic Table endorsed by IUPAC, Saint Petersburg (Russia), 26–28 July 2019.

P. Pyykkö (✉)
Department of Chemistry, University of Helsinki, POB 55 (A. I. Virtasen Aukio 1), 00014 Helsinki, Finland
e-mail: Pekka.Pyykko@helsinki.fi

C. J. Giunta et al. (eds.), *150 Years of the Periodic Table*, Perspectives on the History of Chemistry, https://doi.org/10.1007/978-3-030-67910-1_17

duced. What did people use in the nineteenth century, before they knew about nuclei or nuclear charges? Atomic weights, *m*. Except for a few anomalies, arrangement by *m* gave the same running order toward heavier elements, as by *Z*.

History in a nutshell. The periodic behavior of chemical elements slowly became apparent in fragments, perhaps beginning with Döbereiner's *triads*, such as (Ca, Sr, Ba) in 1817 or (Li, Na, K), (S, Se, Te) and (Cl, Br. I) in 1829.[1] In each of these triads, the *m* of the second element is approximately equal to the average of the first and third.

In 1843, Gmelin had a table of 55 elements, with oxygen in the correct place. Following the spiralized "telluric screw" (de Chancourtois 1862) and the Law of "octaves" (Newlands 1863, 1865), Meyer constructed a square table of 28 elements (with gaps) in 1864. In 1869, Mendeleev wrote two articles (one in Russian and the other—a short summary—in German) explicitly predicting the existence of three missing elements with the atomic weights 45, 68, and 70. These were discovered in 1879, 1875, and 1886, and are now known as scandium (Sc; atomic weight 44.956), gallium (Ga; atomic weight 69.723), and germanium (Ge; atomic weight 72.640), respectively.[2]

Mendeleev's articles (1869) also overlapped with the writing of his textbook *Fundamentals of Chemistry* (see Kaji [2]).

In 1900 Ramsay suggested that the new (nearly) noble gases should form a separate group, which is currently referred to as Group 18.[3] Similarly, in 1945, Seaborg proposed that the newly discovered actinides should form their own row below the lanthanides. He purposely chose the elemental names europium, americium (Eu, Am), gadolinium, curium (Gd, Cm), and terbium, berkelium (Tb, Bk) to emphasize the (4f, 5f) analogy by selecting for these actinides names of a continent, a celebrated scientist, and a town.[4]

Accelerator experiments have now completed the 6d series Rf-Cn and the 7p series Nh-Og, resulting in the currently accepted PT (highlighted in yellow in Fig. 17.1).

Note that supercritical collision systems can also be created [6]. The theoretical predictions (shown in white in Fig. 17.1) by the present author [4, 5, 7] support the idea of two 8s elements (E119 and 120), an overlapping "grey" zone of additional shells (8p, 7d, and 6f) at E121–124, and then a systematic sequence of increasing 5g occupation numbers for the elements E125 onwards. Nominally, all of the elements E121–138 are assigned to a "5g" series. As a parallel case, recall that we regard Th as an actinide, although the free thorium atom has no 5f orbital occupation. Accordingly, a name *pre-f* was recently introduced [7].

[1]For more information on triads, see Chapter 3 of this volume—Ed.

[2]Other chapters in this volume treat in greater detail Beguyer de Chancourtois (5), Newlands (6), Meyer (8 and 9), Mendeleev (2), and the discovery of predicted elements (10)—Ed.

[3]Chapter 12 of this volume recounts the discovery of the noble gas elements—Ed.

[4]That is, Seaborg chose names for the actinides by analogy to the lanthanides' names—Ed.

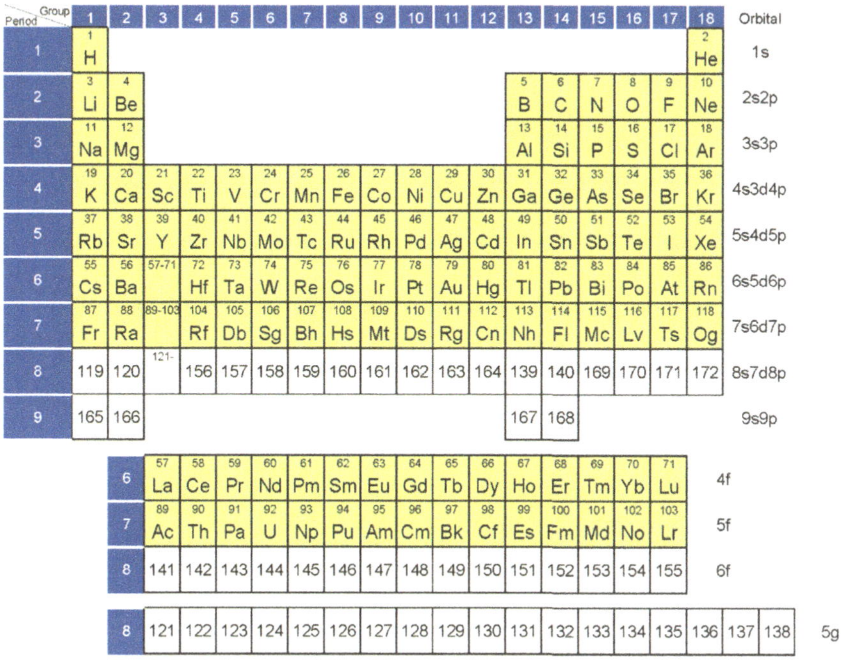

Fig. 17.1 The present Periodic Table (yellow) and possible assignments of the future elements E119–E172 (white). Picture reproduced from Haba [3]. Table reproduced from Pyykkö [4, 5]. Note the p-orbital spin-orbit-induced anomalies at E139–140 and E167–168, and the 9s-orbital-induced location of E165–166

As a further data point supporting Fig. 17.1, Indelicato et al. [8] find an $8s^2 8p^2 5g^{18}$ configuration for the cations $E143^{3+}$ to $E148^{8+}$. The atomic calculations on the superheavy elements (SHE) range from the Dirac-Slater (DS) ones by Fricke et al. [9], and the Dirac-Fock (DF) ones by Desclaux [10] to massive multiconfiguration DF (MCDF) approaches, such as [9], and sophisticated many-body approaches [11, 12]. As a simple example, see Fermi [13]. As to molecules, theoretical calculations on hypothetical octahedral hexafluorides, MF_6, support the onset of the expected 5g occupation from E125 to at least E129. Note that the metal atom, M, delivers six electrons to the six fluorides. The other valence electrons go to the 5g shell. These molecular relativistic density-functional calculations were reported by Dognon and Pyykkö [14].[5]

For more comprehensive treatises on the history of the PT, see Gordin [15], Kaji [2] or Scerri [16].

[5] The very compact size of the 5g shell would make these elements 'superlanthanides'.

17.2 Technical Details

17.2.1 Why Must Z Be ≤ 172?

The current border of the PT at $Z = 118$ is set by the nuclear instability of the existing isotopes and by their small nucleosynthetic cross sections. For discussing chemical properties, even if the nuclei existed, there may also be limits, arising from the chemical reactivity of the vacuum in strong Coulomb fields, due to quantum electrodynamics, QED.

For any elements beyond E172, or so, the lowest or 1s shell would dive to the lower, positron-like continuum of the Dirac equation. It is not yet fully understood what would physically happen. Another way to study the question is to consider heavy-atom collisions [6]. For one, point-like nucleus this diving would already take place at $Z = 137$. For the earlier literature on this question, see [5, p. 162].

17.2.2 Relativity Versus QED

It is well-known that the (Dirac) relativistic effects contract and stabilize the ns and np* (= $np_{1/2}$) shells while the ensuing indirect relativistic effects expand and destabilize the d and f shells. As previously discussed [7, 8], the QED effects, dominated by the vacuum fluctuations (the zero-point oscillations of the electromagnetic field), cancel about –1% of the previous effect, for the heavier elements. The other lowest-order contribution, of opposite sign, is vacuum polarization. One could say that the Dirac-Fock-Breit Hamiltonian is "101 percent correct" (cp. [8]). The QED effects can be seen in accurate quantitative comparisons but have so far not led to qualitative chemical changes.

17.2.3 Which Orbitals to Use in Chemistry?

The chemical behavior of the elements in Fig. 17.1 is mostly driven by the orbitals, occupied in the atomic ground state, and given in the right-hand marginal. Sometimes also other orbitals, which are unoccupied in the atomic ground state but energetically accessible for bond formation, can participate. Thus we can have the predicted [17] *pre-s* Og^- anion, the *pre-p* Be, Mg; Zn, making bonds with their ns + np orbitals, the *pre-d* Ca, Sr, Ba; Cs, and the *pre-f* Th [7].

A recent example was the synthesis of $[Ba(CO)_8]$, which fulfills an 18-electron rule by using the originally empty 5d shell of the central barium atom [18]. The atomic ground state does not always explain the molecular outcome. Moreover, remember that electron correlation can make the concept of electron configurations diffuse.

17.2.4 "Secondary Periodicity"

Biron [19] pointed out in 1915 that every second period has specific properties. Taking Group 15 as an example, along the series (N, P, As, Sb, Bi) (Biron, p. 971), the dominant oxidation states are III, V, III, V, III, respectively. The anomaly at As can be attributed to partial screening by the filled 3d shell. The anomaly at Bi is due to both an analogous partial screening by the 4f shell and to relativistic effects [20].

Another vertical anomaly is the small radius of every atomic shell (1s, 2p, 3d, 4f, 5g) with a new orbital angular momentum quantum number, *l*. The author [21] used the name "primogenic repulsion" for its effect on the higher shells. In Russian literature the term "kainosymmetric" is often used for these "first-born" shells [22, 23].

17.2.5 The Inert-Pair Effect

Sidgwick [24] called attention in 1933 to a decrease of the main oxidation state by two units for 6th-Period elements, take Pb(II) as an example. A first explanation would be the relativistic 6s stabilization. Closer studies involve the hybridization of the metal (6s, 6p) orbitals with the ligand np orbitals [25].

17.2.6 "False Friends"

The gold atom is almost as electronegative as iodine; we can see its outermost shell as either a $6s^1$ electron or a $6s^{-1}$ hole. A wide chemistry of the auride ion, Au^- is known [26]. For a comparison of aurides with other "halides," see also [27].

Going one step left from gold, solid Cs_2Pt and other $Pt^{(-II)}$ compounds were studied in the group of Jansen [28]. In molecules, in addition to σ bonding, also analogous $2p\pi$ and $5d\pi$ bonding was identified between OCO and PtCO, respectively, leading to multiple bonding [29].

In the uranyl-like isoelectronic series, OUN^+ and $OUIr^+$ were found to have similar triple bonds [30] and the latter species was later produced in mass-spectroscopy [31]. These later chemical analogies were initially unexpected.

17.2.7 Nuclear Stability

The chemical predictions quoted here are based on theoretical, relativistic quantum chemical calculations using established electronic Hamiltonians. The nuclei are simply assumed to exist, with a realistic, finite nuclear size. The synthesis of heavier nuclei, up to E118 (Oganesson), is demanding. The most recent nuclear syntheses were completed in a friendly collaboration between laboratories in Oak Ridge and Dubna. The lifetimes of these nuclides are short; for example, the present Og isotopes have lifetimes below a millisecond. The most challenging production bottleneck is, however, not the short lifetime but the small nucleosynthetic cross-section for these elements. If the experiment runs for a year and yields less than a handful of desired product nuclei, this creates an obvious problem, even with nearly 4π detection (all scattering directions seen) and an almost noiseless apparatus. The current situation on superheavy elements is discussed by Giuliani et al. [32].

A quantum chemist can always assume a finite nucleus of realistic size for any $Z < 173$ and do ab initio calculations for that theoretical model, whether or not such nuclei or their compounds are ever made. One even could claim that some general conclusions could possibly be drawn, such as the possible existence of a 5g series or the vast spin-orbit effects.

17.3 In Defence of the Current PT: What Are the Choices?

Figure 17.1 illustrates a recent choice of IUPAC PT layout (highlighted in yellow). Note the placement of all lanthanides and actinides in Group 3. One could, however argue certain points:

(a) H is now in Group 1, because it is often manifested as H^{+1}, or neutral. If an emphasis was placed on hydrides (H^{-1}), one could argue for having hydrogen also in Group 17.

(b) The noble gas He is now in Group 18 with the other noble gases. Apart from spectroscopic species, like the astrophysical, diatomic HeH^+, or high-pressure compounds, like Na_2He [33], helium is still a very noble gas. As stressed by the authors, this compound is a sodium electride containing He(0). If one would like to emphasize helium's $1s^2$ electronic structure, one could also have it in Group 2. Its chemical behavior is, however, not that of an alkaline earth. Note the easy ns-np-(n–1)d hybridization among heavier Group-2 elements, lacking for helium.

(c) How long f-element rows? One now has a 15-element lanthanide (Ln) row from La to Lu. All of these elements are (mostly) trivalent. Their ionic radii or ionization potentials exhibit a systematic relationship along the series. It is entirely plausible to count from $4f^0$ to $4f^{14}$, and to leave a hole in Group 3 of Period 6. Moreover this completely avoids the heated argument on which end should one cut off—La or Lu. A clear advantage is then having all these, mostly trivalent, rare earths in the single Group 3, corresponding to three valence electrons.

(d) Impressive experiments and computations [34] have recently verified that a free Lr atom has a $7s^27p^1$ configuration, different from the $6s^25d^1$ for a Lu atom. Computations for a handful of *molecules*, however, find a complete analogy between Lu and Lr [35], see Fig. 17.2. That said, if our PT is to be driven by chemistry, there are no reasons to change Fig. 17.1.

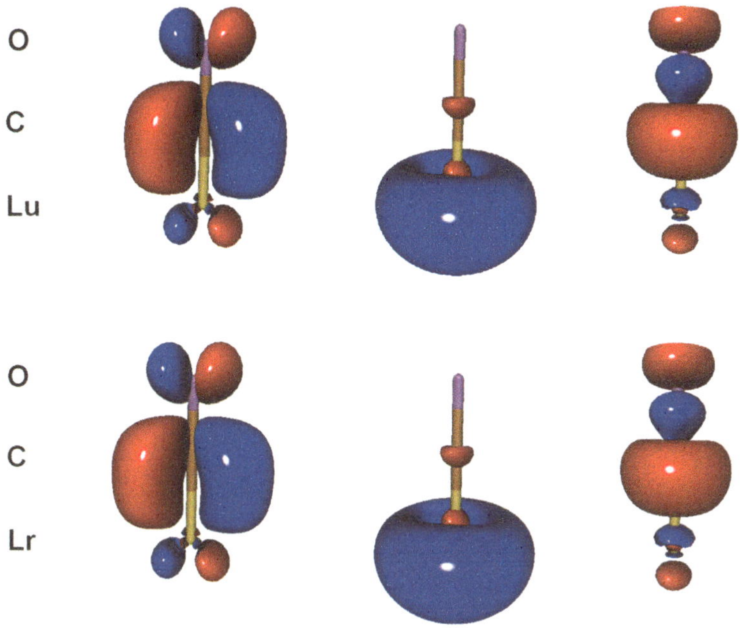

Fig. 17.2 The valence molecular orbitals (MO) of theoretical LuCO and LrCO molecules [36]. Note the similarity. The LuCO is experimentally known, see [37]. As discussed there, the main Ln bonding orbitals are the 6s and 5d. The MO:s from left to right are the M-C pi bond (a donation from M to the CO pi* MO), a sigma lone pair, and the M-C sigma bond. The similarity between Lu and Lr is not a mistake, but the message

(e) The 32-column option? Some desirable properties of a PT could be:

1. The column (or "Group") number *g* (or *g-10*) is the maximum number of valence electrons. The oxidation state is then the same number or, counting holes in the spirit of Abegg's contravalence [36], the negative number, *g-18*.
2. The valence atomic orbitals along a row are constant in a block and equal to those in the right-hand margin.
3. The nuclear charge, *Z*, increases systematically toward the right.
4. The shape should be typographically convenient.

Of these properties, Fig. 17.1 fulfills 1, 2, and 4 but violates 3 for certain superheavy elements when relativistic effects so require. Conversely, the 32-column "long-form" PT favored by Scerri [38] (see his Fig. 17.1) violates (1) by having very many potentially Group-3 columns and may also violate (4). It does satisfy (3).

17.3.1 The "Madelung Rule"

The order of filling shells in a neutral atom is approximately the one given in Fig. 17.3. It was extended up to $Z = 172$ in [6]. This mnemonic device is usually called the Madelung or *(n + l, n)* rule [39], although its shape appears to be first presented by Janet [40, 41] or actually Sommerfeld [42, 43].[6]

17.3.2 Models for Reproducing the PT

How to explain this approximate order of level filling in neutral, or nearly neutral atoms? For an electron in a Coulomb field, each new n introduces a new l_{max} and this degeneracy of all levels with the same n was discussed by Fock [44] using momentum-space wave functions in a four-dimensional space. Ostrovsky [45] used coordinate space.

Concerning the physics of many-electron atoms, the Dirac-Fock-Breit (DFB) Hamiltonian, supplemented with some estimate of leading quantum electrodynamic (QED) effects, gives an excellent description. As an example, Pašteka et al. [46] calculated the ionization potential and electron affinity of a gold atom with milli-electronvolt (meV) accuracy. The bottleneck rather was in the handling of electron correlation. Coupled-cluster methods with up to pentuple excitations were used. Thus no surprises are expected here. Atoms follow Physics.

A much simpler task is the question of the filling order of one-electron levels in some effective-potential model for a many-electron atom, such as a Thomas-Fermi one. This has been tried since Fermi [13] or Goeppert Mayer [47], and various versions have been included in textbooks, such as Sommerfeld [48], Gombás [49], or Landau and Lifshitz [50]. Some examples of these studies are [51–53]. The predictions for filling new shells are quite similar, see Table 17.1. In a screened-Coulomb potential, the attraction must be sufficient to balance the centrifugal potential for l. This type of reasoning was used by Goeppert Mayer [48] to discuss the Z where 4f and 5f states are first occupied. For reviews see Ostrovsky below. The T-F treatment yields the Z values for a new l at

$$Z = 0.17(2l + 1)^3 \tag{17.1}$$

with the pre-coefficient chosen by Landau and Lifshitz [50]. The results in Table 17.1 are in a surprisingly close agreement with more exact results. Note that the T-F potential is just one of the various screened-Coulomb potentials, which do the job. The literature on which T-F potential at which Z starts a given l is broad, see Table 17.1. Essén [54] gives for filling the first state of l the nuclear charge

[6]See Chapter 16 of the present volume for further information on the Madelung rule and related topics—Ed.

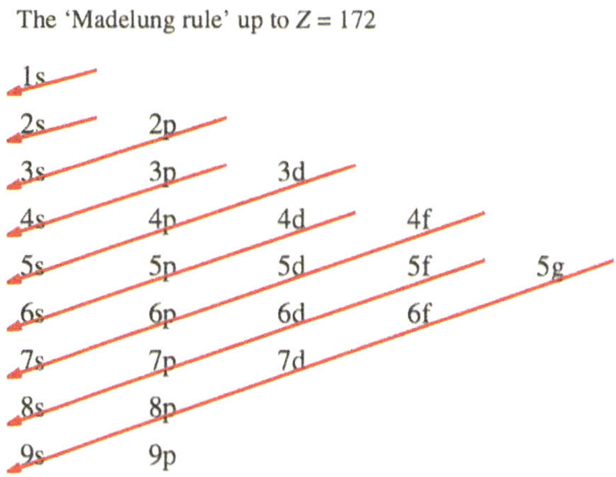

Fig. 17.3 The "Madelung rule" for filling atomic orbitals up to Z = 172, corresponding to Fig. 17.1. Reproduced from Pyykkö [4]

Table 17.1 The first nuclear charges, *Z*, where the atomic orbitals *l* are occupied for a Schrödinger equation in a Thomas-Fermi potential (from Fermi [13], Goeppert Mayer [47], Iwanenko and Larin [51], Landau and Lifshitz [50] (present Eq. 17.1), Essén [54]) and from Dirac-Fock-level relativistic calculations [8]

l	Fermi [13]	Goeppert [47]	Iwanenko [51]	Landau [50]	Essén [54]	DF [8]
1	5		5	5	5	5
2	21		22	21	21	21
3	≥ 55, 58	≥ 57; ≥ 91	58	58	57	58
4				124	121	~ 125

$$Z(l) = 2\left(\sum_{n=0}^{l} 2n^2\right) + 1 \tag{17.2}$$

The doubling of periods with two periods of lengths 8, 18, and 32 each has been noticed by many authors, Löwdin [55], Demkov and Ostrovsky [56, 57], Odabasi [58], Ostrovsky, Essén [54], Katriel [59], Kitagawara and Barut [60], Novaro [61] and Kibler [62] have discussed possible underlying dynamical symmetries. An alternative is that there is no deep symmetry reason and that one only needs the quoted "nuts and bolts" of the DFB Hamiltonian [46] +QED.

For later reviews on the doubling question, see Ostrovsky [63, 64].

Concluding, of the existing literature on the PT, we would still like to remind the reader of the articles by Schwarz [65–67] and of a factor-analytical search for chemical similarities by Leal and Restrepo [68].

17.4 Did Our Department Contribute to the Story?

The new Chemical Laboratory building of the old Kungliga Åbo Akademi (our direct legal predecessor before the removal from Turku to Helsinki in 1828) was inaugurated on April 13, 1764 under Professor Pehr Gadd. The 250th anniversary was celebrated in 2014 by having business as usual.

Gadd's successor, Johan Gadolin (1760–1852) [69] published the first chemical analysis of a black mineral from Ytterby (near Stockholm) in 1794 (German version published in 1796). The paper was translated to modern chemical terminology by Pyykkö and Orama [70].

Starting from this mineral, which was essentially $FeBe_2Y_2Si_2O_{10}$, he obtained a new "earth," *i.e.* oxide, essentially Y_2O_3, possibly with traces of other rare earths. The mineral was later named *gadolinite*. Although one cannot necessarily claim that Gadolin found a specific element, he did contribute to finding an entire family of 16 elements. Organizing these kept chemists busy for another century. In 1886 one of them was given the name *gadolinium (Gd)*.

Circumstantial evidence from the parallel case of samarium (Sm) by the same authors suggests that Lecoq de Boisbaudran and Marignac considered both the mineral and the eponym, or the person behind its name. For more details on Gd, see [71] and supplements to it. In 1994 a bicentennial conference, 2-ICFE, was held in Helsinki, with a conference excursion to Ytterby, see Fig. 17.4.

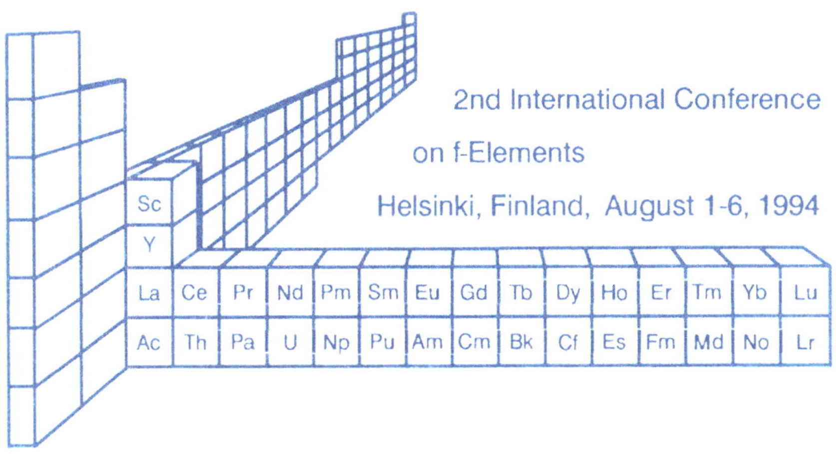

Fig. 17.4 In 1994 an International Conference on f-Elements was held in Helsinki to celebrate the bicentennial of Johan Gadolin's 1794 analysis [70] (An earlier version of the present article was published also in the 2019 brochure ChemistryNews of the author's Department)

References

1. Pyykkö P (2019) An essay on periodic tables. Pure Appl Chem 91:1959–1967
2. Kaji M (2002) D I Mendeleev's concept of chemical elements and the principles of chemistry. Bull Hist Chem 27:4–16
3. Haba H (2019) A new period in superheavy-element hunting. Nature Chem 11:10–13
4. Pyykkö P (2016) Is the periodic table all right ("PT OK")? Eur Phys J, Web of Conf 131:01001-1–01001-6
5. Pyykkö P (2011) A suggested periodic table up to $Z \leq 172$, based on Dirac-Fock calculations on atoms and ions. Phys Chem Chem Phys 13:161–168
6. Aleksandrov IA, Plunien G, Shabaev VM (2019) Locally-constant field approximation in studies of electron-positron pair production in strong external fields. Phys Rev D 99:016020-1–016020-12
7. Pyykkö P (2012) The physics behind chemistry and the periodic table. Chem Rev 112:371–384
8. Indelicato P, Bieroń J, Jönsson P (2011) Are MCDF calculations 101% correct in the super-heavy elements range? Theor Chem Acc 129:495–505
9. Fricke B, Greiner W, Waber JT (1971) The continuation of the periodic table up to Z = 172. The chemistry of superheavy elements. Theor Chim Acta 21:235–260
10. Desclaux JP (1973) Relativistic Dirac-Fock expectation values for atoms with Z = 1 to Z = 120. At Data Nucl Data Tables 12:311–406
11. Eliav E, Fritzsche S, Kaldor U (2015) Electronic structure theory of the superheavy elements. Nucl Phys A 944:518–550
12. Porsev SG, Safronova MS, Safronova UI, Dzuba VA, Flambaum VV (2018) Nobelium energy levels and hyperfine-structure constants. Phys Rev A 98:052512-1–052512-6
13. Fermi E (1928) Eine statistische Methode zur Bestimmung einiger Eigenschaften des Atoms und ihre Anwendung auf die Theorie des periodischen Systems der Elemente. Z Phys 48:73–79 (an English abstract is available in (1928) Nature 121:502)
14. Dognon JP, Pyykkö P (2017) Chemistry of the 5g elements: relativistic calculations on hexafluorides. Angew Chem Int Ed 56:10132–10134
15. Gordin MD (2018) A Well-ordered thing: Dmitrii Mendeleev and the shadow of the periodic table, rev edn. Princeton University Press, Princeton, NJ
16. Scerri ER (2007) The periodic table: its story and its significance. Oxford University Press, Oxford
17. Eliav E, Kaldor U, Ishikawa Y, Pyykkö P (1996) Element 118: the first rare gas with an electron affinity. Phys Rev Lett 77:5350–5352
18. Wu X, Zhao L, Jin J, Pan S, Li W, Jin X, Wang G, Zhou M, Frenking G (2018) Observation of alkaline earth complexes $M(CO)_8$ (M = Ca, Sr, or Ba) that mimic transition metals. Science 361:912–916
19. Biron EV (1915) Secondary periodicity. Zh Russk Fiz-Khim Obshch 47:964–988. See p 975
20. Pyykkö P (1979) Interpretation of secondary periodicity in the periodic table. J Chem Res Synop 380–381
21. Pyykkö P (1979) Dirac-Fock one-centre calculations part 8. The 1Σ states of ScH, YH, LaH, AcH, TmH, LuH and LrH. Phys Scr (Sweden) 20:647
22. Shchukarev SA (1977) New Understanding of DI Mendeleev's system. 1. Periodicity in stratigraphy of atomic electron shells in the system and concept of kainosymmetry. (1977) J Gen Chem (USSR) 47:227–238 (Russian language version available at (1977) Zh Obshch Khim 47:246–259)
23. Imyanitov NS (2011) The periodic law. Formulations, equations, graphic representations. Russian J Inorg Chem 56:2183–2200
24. Sidgwick NV (1933) The covalent link in chemistry. Cornell University Press, Ithaca, NY, p 189

25. Pyykkö P (1988) Relativistic effects in structural chemistry. Chem Rev 88:563–594. See pp 578–580, 585. The later literature would require its own review
26. Jansen M (2008) The chemistry of gold as an anion. Chem Soc Rev 37:1824–1835
27. Pyykkö P (2002) Relativity, gold, closed-shell interactions, and CsAu•NH_3. Angew Chem Int Ed 41:3573–3578 (German language version available at (2002) Ang Chem 114:3723–3728)
28. Karpov A, Nuss J, Wedig U, Jansen M (2003) Cs_2Pt: A platinide (–II) exhibiting complete charge separation. Angew Chem Int Ed 42:4818–4821 (German language version available at (2003) Angew Chem 115:4966–4969)
29. Patzschke M, Pyykkö P (2004) Darmstadtium carbonyl and carbide resemble platinum carbonyl and carbide. Chem Comm 1982–1983
30. Gagliardi L, Pyykkö P (2004) Theoretical search for very short metal-actinide bonds: NUIr and isoelectronic systems. Angew Chem Int Ed 43:1573–1576 (also available at (2004) Angew Chem 116:1599–1602)
31. Santos M, Marçalo J, Pires de Matos A, Gibson JK, Haire RG (2006) Actinide-transition metal heteronuclear ions and their oxides: $\{IrUO\}^+$ as an analogue to uranyl. Eur J Inorg Chem 3346–3349
32. Giuliani SA, Matheson Z, Nazarewicz W, Olsen E, Reinhard PG, Sadhukhan J, Schuetrumpf B, Schunk N, Schwerdtfeger P (2019) Colloquium: Superheavy elements: Oganesson and beyond. Rev Mod Phys 91:011001-1–011001-12
33. Dong X, Oganov AR, Goncharov AF, Stavrou E, Lobanov S, Saleh G, Qian GR, Zhu Q, Gatti C, Deringer VL, Dronskowski RA, Zhou XF, Prakapenka VB, Konopkova Z, Popov IA, Boldyrev AI, Wang HT (2017) A stable compound of helium and sodium at high pressue. Nature Chem 9:440–445
34. Sato TK, Asai M, Borschevsky A, Stora T, Sato N, Kameya Y, Tsukada K, Düllmann ChE, Eberhardt K, Eliav E, Ichikawa S, Kaldor U, Kratz JV, Miyashita S, Nagame Y, Ooe K, Osa A, Renisch D, Runke J, Schädel M, Thörle-Pospiech P, Toyoshima A, Trautmann N (2015) Measurement of the first ionization potential of lawrencium, element 103. Nature 520:209–211
35. Xu WH, Pyykkö P (2016) Is the chemistry of lawrencium peculiar? Phys Chem Chem Phys 18:17351–17355
36. Abegg R (1904) Die Valenz und das periodische System. Versuch einer Theorie der Molekularverbindungen. Z Anorg Chem 39:330–380. See p 344
37. Xu WH, Jin X, Chen MH, Pyykkö P, Zhou MF, Li J (2012) Rare-earth monocarbonyls MCO: comprehensive infrared observations and a transparent theoretical interpretation for M = Sc; Y; La-Lu. Chem Sci 3:1548–1554
38. Scerri ER (2019) Five ideas in chemical education that must die. Found Chem 21:61–69
39. Madelung E (1936) Die mathematischen Hilfsmittel des Physikers, 3rd edn. Springer, Berlin, p 359
40. Janet C (2010) La classification hélicoidale des éléments chimiques. Impr. Dép. de l'Oise, Beauvais
41. Stewart PJ (2010) Charles Janet: unrecognized genius of the periodic system. Found Chem 12:5–15
42. Allen LC, Knight ET (2002) The Löwdin challenge: Origin of the $n + \ell$, n (Madelung) rule for filling the orbital configurations of the periodic table. Int J Quantum Chem 90:80–88 (quote for the first version of the Madelung rule)
43. Sommerfeld A (1926) Electronic structure of the atom and the quantum theory. Mem Proc Manchester Lit Phil Soc 70:141–151
44. Fock VA (1936) Zur Theorie des Wasserstoffatoms. Z Phys 98:145–154
45. Ostrovsky VN (1981) Dynamic symmetry of atomic potential. J Phys B 14:4425–4439
46. Pašteka LF, Eliav E, Borschevsky A, Kaldor U, Schwerdtfeger P (2017) Relativistic coupled cluster calculations with variational quantum electrodynamics resolve the discrepancy between experiment and theory concerning the electron affinity and ionization potential of gold. Phys Rev Lett 118:023002-1–023002-5

47. Goeppert Mayer M (1941) Rare-earth and transuranic elements. Phys Rev 60:184–187
48. Sommerfeld A (1951) Atombau und Spektrallinien, vol 2: Wellenmechanischer Ergänzungsband (4th edn). F Vieweg & Sohn, Braunschweig, p 698. See Chapter 10, Section 4, equation 6
49. Gombás P (1956) Statistische Behandlung des Atoms. Handbuch der Physik 36:109–231. See p 181
50. Landau LD, Lifshitz EM (1956) Quantum mechanics. Non-relativistic theory (2nd edn). Pergamon Press, Oxford. See §70, 73
51. Iwanenko D, Larin S (1953) Theory of the periodic system of the elements. US Atomic Energy Commission, Technical Information Service, Washington DC (original article in Russian at (1953) Dokl Akad Nauk SSSR 88:45)
52. Klechkovskii VM (1951) Gruppy v posledovatel'nom zapolnenii elektronnykh konfiguratzii atomov [(n + ℓ) groups in successive filling of atomic electron configurations]. Dokl Akad Nauk SSSR 80:603–606
53. Klechkovskii VM (1962) Justification of the rule for successive filling on (n + ℓ) groups. Soviet Phys J Exp Theor Phys 14:334–335 (for the original Russian article, see (1962) Zh Eksp Teor Fiz 41:465–466)
54. Essén H (1982) Periodic table of the elements and the Thomas-Fermi atom. Int J Quantum Chem 21:717–726
55. Löwdin PO (1969) Some comments on the periodic system of the elements. Int J Quantum Chem Symp 3:331–334
56. Demkov YN, Ostrovsky VN (1971) Intrinsic symmetry of the Maxwell 'fish-eye' problem and the Fock group for the hydrogen atom. Soviet Phys J Exp Theor Phys 13:1083–1087 (for the original Russian article, see (1971) Zh Eksp Teor Fiz 60:2011–2018)
57. Demkov YN, Ostrovsky VN (1972) n + ℓ Filling rule in the periodic system and focusing potentials. Soviet Phys J Exp Theor Phys 35:66–69 (for the original Russian article, see (1972) Zh Eksp Teor Fiz 62:125–132)
58. Odabasi H (1973) Some evidence about the dynamical group *SO* (4, 2) symmetries of the periodic table of elements. Int J Quantum Chem Symp 7:23–33
59. Katriel J, Jørgensen CK (1982) Possible broken supersymmetry behind the periodic table. Chem Phys Lett 87:315–319
60. Kitagawara Y, Barut AO (1983) Period doubling in the n + l filling rule and dynamical symmetry of the Demkov-Ostrovsky atomic model. J Phys B 16:3305–3327
61. Novaro O (1989) Group theoretical aspects of the periodic table of the elements. J Mol Struct: THEOCHEM 199:103–118
62. Kibler M (1989) The periodic system of chemical elements: old and new developments. J Mol Struct: THEOCHEM 187:83–93
63. Ostrovsky VN (2001) What and how physics contributes to understanding the periodic law. Found Chem 3:145–181
64. Ostrovsky VN (2003) Physical explanation of the periodic table. Ann N Y Acad Sci 988:182–192
65. Wang SG, Schwarz WHE (2009) Icon of chemistry: the periodic system of chemical elements in the new century. Angew Chem Int Ed 48:3404–3415 (the original German article is available at (2009) Angew Chem 121:3456–3467)
66. Schwarz WHE, Rich RL (2010) Theoretical basis and correct explanation of the periodic system: review and update. J Chem Educ 87:435–443
67. Schwarz WHE (2010) The full story of the electron configurations of the transition elements. J Chem Educ 87:444–448
68. Leal W, Restrepo G (2019) Formal structure of periodic system of elements. Proc Roy Soc A 475:20180581-1–20180581-20

69. Gadolin J (1794) Undersökning av en svart tung stenart ifrån Ytterby stenbrott i Roslagen. Kungl Svenska Vetenskapsak Handl 15:137–155 (German version available at (1796) Crell's Ann 313–329)
70. Pyykkö P, Orama O (1996) What did Johan Gadolin actually do? In: Evans CH (ed) Episodes from the history of the rare earth elements. Kluwer, Dordrecht, pp 1–12
71. Pyykkö P (2015) Magically magnetic gadolinium. Nature Chem 7:680–680

The Periodic Table at 150: A Philatelic Celebration

18

Daniel Rabinovich

Abstract

The International Year of the Periodic Table of the Chemical Elements (IYPT), which marks the sesquicentennial of the iconic chart's introduction by Dmitri Mendeleev in 1869, provides a unique opportunity to bring its colorful history to life. Postage stamps, released by many countries to recognize scientific achievements or to honor well-known scientists, are used in this article to celebrate the IYPT. In particular, a detailed description of all the relevant stamps issued during 2019, some of which highlight important milestones in the development of the modern periodic table and the discovery of certain chemical elements, is presented.

18.1 Introduction

The United Nations (UN) and its Educational, Scientific and Cultural Organization (UNESCO) have periodically sponsored the observation of yearlong themes to promote, through awareness and action, their stated goals to increase universal respect for justice, the rule of law, human rights, and international cooperation. For instance, the International Years of Physics (2005), Astronomy (2009), Chemistry (2011), Crystallography (2014), and Light and Light-based Technologies (2015), underscored the importance and contributions of these disciplines to society.

D. Rabinovich (✉)
Department of Chemistry, The University of North Carolina at Charlotte,
9201 University City Boulevard, Charlotte, NC 28223, USA
e-mail: drabinov@uncc.edu

C. J. Giunta et al. (eds.), *150 Years of the Periodic Table*, Perspectives on the History of Chemistry, https://doi.org/10.1007/978-3-030-67910-1_18

A resolution adopted by the UN General Assembly on 20 December 2017 proclaimed 2019 as the International Year of the Periodic Table of the Chemical Elements (IYPT), which commemorates the 150th anniversary of the publication of the original elemental chart by the renowned Russian chemist Dmitri Mendeleev (1834–1907). The resolution was endorsed by the International Union of Pure and Applied Chemistry (IUPAC), which was celebrating its centennial in 2019, and several related organizations, including the International Union of Pure and Applied Physics (IUPAP) and the International Union of History and Philosophy of Science and Technology (IUHPST). Not surprisingly, the IYPT was recognized throughout the world in multiple ways, including the organization of thematic conferences and symposia, special activities for children and online games (e.g., IUPAC's Periodic Table Challenge), the publication of a myriad of articles in magazines and technical journals, and the release of postage stamps by several countries.

The use of postage stamps as didactic tools in science communication and teaching is well documented [1–12], and selected publications have explicitly focused on the periodic table and the discovery of chemical elements [13–15]. In honor of the IYPT, this article presents in chronological order all the relevant stamps released during 2019, with brief descriptions that highlight the creativity or idiosyncrasies of the individuals that designed them or the postal authorities that issued them [16–18].

18.2 IYPT Stamps

18.2.1 Algeria and the IYPT Logo

The first postage stamp honoring the IYPT was issued right after New Year's Day (Fig. 18.1). On 2 January 2019, *Algérie Poste* placed into circulation a rather simple stamp featuring the IYPT logo, which includes a depiction of Mendeleev and a globe incorporating the chemical symbols of mendelevium (Md) and the four most common elements found in living organisms (hydrogen, carbon, nitrogen, and oxygen). The stamp, with a face value of 25 Algerian dinars, includes the translation of "International Year of the Periodic Table of the Chemical Elements" into Algeria's two official languages, namely Arabic and Kabyle, the latter being one of the Berber languages spoken mainly by people in the north and northeast parts of the country.

18.2.2 Spanish Chemical Pride

On 9 January 2019, the Spanish postal service (*Correos*) issued a meaningful stamp that prominently displays the chemical symbols of vanadium, tungsten, and platinum (Fig. 18.2). Even though it may not be evident from the stamp's design why these particular elements were selected to represent the periodic table (the Spanish

Fig. 18.1 IYPT stamp from Algeria issued on 2 January 2019

flag in the lower right corner of each periodic table "tile" may provide a clue), it turns out that they are the three chemical elements discovered by Spaniards [19].

The discovery of vanadium is usually ascribed to the Spanish mineralogist Andrés Manuel del Río (1764–1849), who was born in Madrid but spent most of his professional life (and died) in Mexico [20, 21]. In 1801 he analyzed some lead-containing minerals (one of which is now known as vanadinite) sent to him from the Purísima del Cardenal mine in the State of Hidalgo and established the presence of a new metallic element, which he initially named panchromium (based on the vivid colors of the compounds he prepared with it) and later erythronium (since the color of the compounds changed to red upon heating). The identity of the new element was only confirmed in 1830 by the Swedish chemist Nils Gabriel Sefström (1787–1845), who changed its name yet again to vanadium in honor of Vanadis, the Scandinavian goddess of love and beauty. Incidentally, vanadium is the only chemical element to have been discovered in Mexico.

Tungsten, the only element discovered in the Iberian Peninsula, was isolated in 1783 by the Spanish chemist Juan José Delhuyar[1] (1754–1796) and his younger brother Fausto (1755–1833) while working at the Royal Seminary of Vergara in the province of Gipuzkoa (Basque Country) in the north of Spain [22]. The current symbol of the element (W) is derived from wolfram, the original name proposed by the Delhuyar brothers, who isolated it from a sample of the mineral wolframite, a mixed tungstate (or wolframate?) of iron and manganese, with chemical formula $(Fe,Mn)WO_4$. Interestingly, wolfram is still today the preferred name for element 74 in Spain, Scandinavia, Russia, Turkey, Germany, Greece, and other countries, even though tungsten is recommended by IUPAC and widely used in Latin America and most English- and French-speaking countries [23, 24].

Platinum, a fairly rare and expensive element usually found associated with nickel and copper ores, was already known to pre-Columbian natives in the region of present-day Colombia and Ecuador. However, the "European" or "modern" discovery of platinum is usually attributed to the Spanish scientist and explorer

[1]Other known spellings of the last name include Elhuyar, D'Elhuyar, and de Luyart (in the original publication from 1783), but a majority of modern sources seem to prefer Delhuyar. See Caswell [22].

Fig. 18.2 IYPT stamp from Spain issued on 9 January 2019

Antonio de Ulloa (1716–1795) [25]. He was a member of the French Geodesic Mission to South America (1735–1744) organized by the French Academy of Sciences and led by the renowned explorer and geographer Charles Marie de La Condamine (1701–1774). Upon his return to Spain, de Ulloa wrote extensively about the expedition and described for the first time in 1748 some of the physical and chemical properties of the noble metal.

18.2.3 Kyrgyzstan and Mendeleev

On 12 April 2019, a stamp commemorating the IYPT was issued in the Kyrgyz Republic, a landlocked country in Central Asia, bordering Kazakhstan, Tajikistan, Uzbekistan, and China (Fig. 18.3).[2] In addition to the UNESCO and IYPT logos in the upper right corner, the colorful stamp, with a face value of 100 Kyrgyzstani som, shows a portrait of Mendeleev and the chemical symbols of mendelevium, zinc, copper, helium, argon, gold, cadmium, and xenon, among other elements.

18.2.4 Moldova and Rubik's Cube

The Republic of Moldova, a country in Eastern Europe bordered by Romania and Ukraine, recognized the IYPT on 23 May 2019 (Fig. 18.4). The attractive 15.50-Moldovan leu stamp includes the IYPT and UNESCO logos and a picture of a Rubik's Cube with different chemical symbols in each of its six faces. I have not been able to ascertain the connection between the periodic table and the famous puzzle invented in 1974 by the Hungarian architect Ernö Rubik (b. 1944). I can only speculate that the idea behind the design is a subtle tribute to the Russian chemist who solved the puzzle of sorting the chemical elements based on their recurring properties (and, importantly, predicted the existence of unknown elements) 150 years ago.

[2]The suffix -stan means "place of" or "country of" in ancient Persian, hence the country names refer to the land of the Kyrgyz, Kazakhs, Tajiks, Uzbeks, and so forth.

Fig. 18.3 IYPT stamp from Kyrgyzstan issued on 12 April 2019

Fig. 18.4 IYPT stamp from Moldova issued on 23 May 2019

18.2.5 Mendeleev's Final Draft of the PT

Magyar Posta, the Hungarian postal service, issued on 3 June 2019 an IYPT stamp that includes a portrait of Mendeleev, a small reproduction of the final draft of his manuscript of the periodic table (i.e., the one dated 17 February 1869), and, to the left, a portion of a modern periodic table (containing the elements of groups 1–7 and the first half of the f-block) (Fig. 18.5). The inscription in Hungarian ("a periódusos rendszer megalkotója") near the upper left corner, below Mendeleev's full name and the dates of his birth and death (1834–1907), translates literally to "the creator of the periodic table."

18.2.6 The 7th Period Is Complete

On 24 June 2019, Bulgaria issued a postage stamp that includes the symbols of all 118 chemical elements in the periodic table and a diffuse portrait of Mendeleev in the background (Fig. 18.6). Although there are many stamps that show chemical symbols or portions of the periodic table, this may well be the first stamp to display a complete table (vide infra). As such, it includes the symbols of the four most

Fig. 18.5 IYPT stamp from Hungary issued on 3 June 2019

recent elemental additions to the beloved chart, whose names were approved by IUPAC in November 2016, namely nihonium (Nh), moscovium (Mc), tennessine (Ts), and oganesson (Og).

18.2.7 Portugal: Four to 118 in 2,500 Years

Portugal went all out to honor the IYPT and, on 24 July 2019, released not one but two stamps plus a so-called souvenir sheet (Figs. 18.7 and 18.8). One of the stamps, with a face value to cover the cost of domestic postage for letters weighing up to 20 g ("N" is for "national") displays the symbol of hydrogen, the most common element in the Universe, and graphical representations of the four classical elements from Ancient Greece (earth, water, air, and fire) originally envisioned by the philosopher Empedocles (ca. 490–430 BCE). The other stamp, to cover the cost of international ("I") postage for letters weighing up to 20 g, displays an elegant black-and-white portrait of Mendeleev and the chemical symbol, name, atomic number ($Z = 101$), and atomic mass ($A = 258$) of the most stable isotope ($t_{1/2} = 51.5$ d) of the element mendelevium.

The souvenir sheet, with a face value of €2.00, features a complete periodic table and uses carbon as an example to explain the information included for each element. In contrast to the stamp from Bulgaria shown above, the fairly large size of the souvenir sheet (13.5 cm × 12.5 cm, with a total area nearly four times that of a typical credit card) allows for the inclusion of the name, symbol, atomic number, and atomic mass for each and every element. Postally used examples of this sheet must be pretty uncommon, but wouldn't we be thrilled to get a large envelope by mail with a periodic table used for postage?

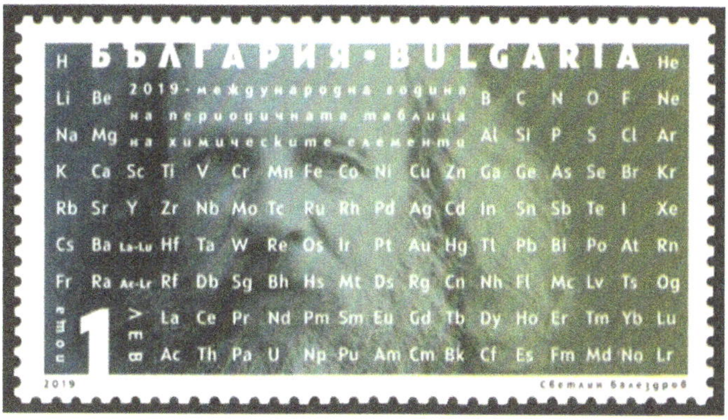

Fig. 18.6 IYPT stamp from Bulgaria issued on 24 June 2019

Fig. 18.7 IYPT stamps from Portugal issued on 24 July 2019

18.2.8 Sri Lanka's Colorful Table

The Philatelic Bureau of Sri Lanka Post released on 6 October 2019 a 45-rupee stamp that includes a full periodic table but unfortunately only the chemical symbol of each element is legible (Fig. 18.9). In addition, the stamp's design incorporates the IYPT logo and a portrait of Mendeleev that looks remarkably similar (but not identical) to the one used in the stamp from Kyrgyzstan (Fig. 18.3). The choice of colors in the periodic table is also somewhat unusual: while distinct hues are used for the alkali metals, the alkaline earths, the halogens, and the noble gases, the metalloids (B, Si, Ge, As, Sb, Te) get their own dark green color and the remaining elements of groups 13–16 are split into metals (lavender) and nonmetals (pale green).

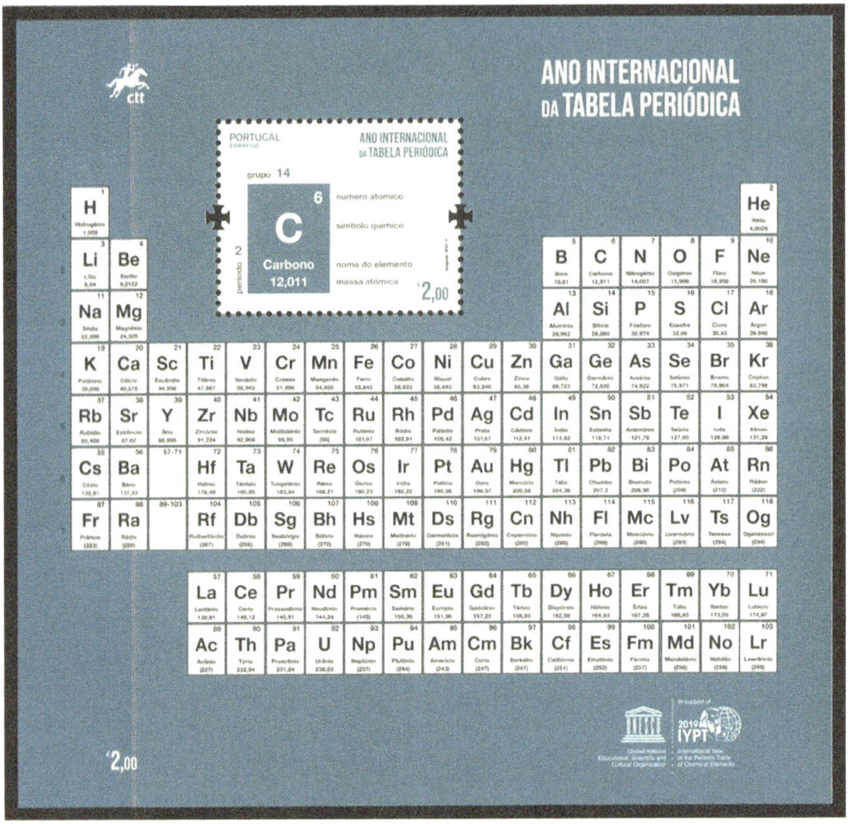

Fig. 18.8 IYPT souvenir sheet from Portugal issued on 24 July 2019

Fig. 18.9 IYPT stamp from Sri Lanka issued on 6 October 2019

Fig. 18.10 IYPT stamp from North Macedonia issued on 9 October 2019

Fig. 18.11 Personalized stamps prepared for hypothetical superheavy elements

18.2.9 North Macedonia's Groovy PT

Last but not least, the Republic of North Macedonia[3] in the Balkan Peninsula issued on 9 October 2019 an eye-catching 50-denar stamp that highlights the relative abundance in the Earth's crust of the 90 naturally occurring chemical elements (Fig. 18.10). Thus, the area occupied by each element is proportional to its abundance using a logarithmic scale since the rarer elements would otherwise be almost invisible. The image is based on a periodic table developed by the European Chemical Society (EuChemS) and is readily available online [26].

Significantly, the captions on the left, in Macedonian and Albanian, can be translated to "150 years – periodic *system* of the elements," which is the original term used by Mendeleev, not "periodic *table*." The stamp also includes in the upper left corner very small logos for both UNESCO and the IYPT.

[3]The country formerly known as the Republic of Macedonia officially (albeit reluctantly) changed its name to the Republic of North Macedonia in February 2019 after a protracted and bitter dispute with Greece, which also has a region named Macedonia.

18.3 Concluding Remarks

What's in the future of the periodic table? There are already concerted efforts by nuclear physicists to generate one or more atoms of elements 119 or 120, mainly at the RIKEN's Nishina Center for Accelerator-Based Science in Japan and the brand-new Superheavy Element Factory at the Flerov Laboratory of Nuclear Reactions in Dubna, Russia [27–29]. Are we ever going to get to the legendary Island of Stability and find superheavy isotopes with more manageable half-lives [30, 31]? What is the limit of the periodic table? Pekka Pyykkö, a computational chemistry professor at the University of Helsinki, has done sophisticated calculations to predict the electron configurations of elements with atomic numbers up to 172 and their positions on a future expanded periodic table [32, 33].[4]

One thing is certain: the synthesis and investigation of new chemical elements is becoming increasingly difficult and requires a massive investment of resources. In this regard, Peter Elias, an engineer and philatelist with a keen sense of humor, has recently prepared a pair of "personalized" stamps[5] that underscore that such elements may just be too expensive to synthesize or simply won't exist at all (Fig. 18.11)!

Acknowledgements I sincerely thank Prof. Javier García-Martínez (University of Alicante, Spain), Dr. Christo Balarew (Bulgarian Academy of Sciences, Sofia), Dr. Fabienne Meyers (IUPAC), Prof. Zoran Zdravkovski (Ss. Cyril and Methodius University, Skopje, North Macedonia), and Mr. Peter C. Elias (PC and Associates, Plano, Texas) for helpful discussions and for providing copies of some of the stamps illustrated in this article.

References

1. Heilbronner E, Miller FA (1998) A philatelic ramble through chemistry. Wiley-VCH, Weinheim
2. Yardley CB (2015) The representation of science and scientists on postage stamps. ANU Press, Canberra, ACT, Australia
3. Habashi F, Hendricker D, Gignac C (1999) Mining and metallurgy on postage stamps. Métallurgie Extractive Québec, Saint Foy, QC, Canada
4. Habashi F (2019) Postage stamps: metallurgy, art, history part 2. Métallurgie Extractive Québec, Saint Foy, QC, Canada
5. Schaeffer HF (1934) Philately serves chemistry. J Chem Educ 11:259–266
6. Miller FA (1983) The history of spectroscopy as illustrated on stamps. Appl Spectrosc 37:219–225
7. Miller FA (1986) A postage stamp history of chemistry. Appl Spectrosc 40:911–924
8. Schreck JO (1986) Postage stamps as a teaching tool in chemistry. J Chem Educ 63:283–287
9. Rappoport Z (1992) Chemistry on stamps (Chemophilately). Acc Chem Res 25:24–31
10. Norkus P, Norkus E, Vaitaitis AP (2007) Chemistry in philately 1. Symbols of chemical elements. Chemija 18(4):21–33

[4]See also Pekka Pyykkö's Chapter 17 in this volume, reprinted from *Pure and Applied Chemistry*.
[5]Personalized or customized stamps, often used in birth announcements or wedding invitations, are privately designed but printed by companies authorized by postal authorities (such as the U.S. Postal Service) and are valid for postage.

11. Nawlakhe UA, Nawlakhe AM (2011) IYC stamps as science communicators. Philat Chim Phys 33:122–129
12. Rabinovich D (2011) 2011: a stamp odyssey. Chem Int 34(3):36–37
13. Garrigós L, Ferrando F, Miralles R (1987) A simple postage stamp periodic table. J Chem Educ 64:682–685
14. García Martínez J, Salas Peregrín JM (2007) La química a través de sus sellos: Una revisión comparativa de la filatelia dedicada a Mendeléiev. An Quím 103(1):50–57
15. García Martínez J, Román Polo P (2007) 2007, el año de Mendeléiev: Una iniciativa para la divulgación de la química. An Quím 104(1):50–55
16. Rabinovich D (2019) International year of the periodic table (IYPT): A midyear philatelic report. Philat Chim Phys 40(2):56–65
17. Rabinovich D (2019) IYPT and the mother of all tables. Chem Int 41(4):60–61
18. Pinto Cañón G, Martín Sánchez M, Prolongo Sarria M (2020) El año internacional de la tabla periódica desde la filatelia. An Quím 116(3):164–172
19. Elguero Bertolini J (2007) España y los elementos de la tabla periódica. An Quím 103(1): 70–76
20. Weeks ME (1935) The scientific contributions of Don Andrés Manuel del Río. J Chem Educ 12(4):161–166
21. Caswell LR (2003) Andrés del Río, Alexander von Humboldt, and the twice-discovered element. Bull Hist Chem 28(1):35–41
22. Caswell LR (1999) The Delhuyar brothers, tungsten, and Spanish silver. Bull Hist Chem 23:11–19
23. Román Polo P (2005) Wolframio, sí; tungsteno, no. An Quím 101:42–48
24. Jensen WB (2008) Why tungsten instead of wolfram? J Chem Educ 85(4):488–489
25. Pinto G (2017) Antonio de Ulloa and the discovery of platinum: an opportunity to connect science and history through a postage stamp. J Chem Educ 94:970–975
26. Cole-Hamilton DJ (2019) Elements of scarcity. Chem Int 41(4):23–28
27. Haba H (2019) A new period in superheavy-element hunting. Nat Chem 11:10–13
28. Ball P (2019) On the edge of the periodic table. Nature 565:552–555
29. Kean S (2019) The quest for superheavies. Science 363:466–470
30. Oganessian Y (2012) Nuclei in the "island of stability" of superheavy elements. J Phys Conf Ser 337:012005
31. Chapman K (2020) The transuranic elements and the island of stability. Phil Trans R Soc A 378:20190535
32. Pyykkö P (2011) A suggested periodic table up to $Z \leq 172$, based on Dirac-Fock calculations on atoms and ions. Phys Chem Chem Phys 13:161–168
33. Pyykkö P (2016) Is the periodic table all right ("PT OK")? EPJ Web Conf 131:01001

Index

A

Abelson, Phillip, 332
abundance (elemental), 8, 392, 394–397, 399, 400, 403
Académie des Sciences, 17, 61, 65, 67, 69, 71, 72, 75, 76, 89
actinides, 159, 252, 267, 270, 336, 337, 341, 426, 430
Algeria, 8, 440, 441
alkali metals, 54, 56, 100, 137, 146, 147, 233, 250, 293, 365, 445
alkaline earths, 45, 53, 57, 100, 103, 147, 233, 265, 292, 293, 365, 445
Aller, Lawrence H., 394, 402
alloy, 72, 237, 241, 242, 277
aluminum, 9, 45, 53, 56, 58, 67, 71, 95, 103, 159, 161, 206, 219, 233, 235, 236, 242, 260, 278
Alvarez, Luis, 331
American Chemical Society (ACS), 3, 4, 334, 336, 360, 362, 363
americium, 9, 336, 337, 426
ammonium, 57, 65, 66, 71, 87, 163, 262, 272, 273, 278, 280
Ampère, André-Marie, 17, 72
analogies (between elements), 17, 21, 27, 99, 118, 246, 429
anode, 237
antimony, 9, 23, 50, 52, 57, 71, 72, 118, 146, 235, 245, 246
argon, 7, 9, 309, 312, 316, 318, 442
argyrodite, 244–247
Arnoult, Eugène d', 142
Arrhenius, Carl Axel, 260
arsenic, 9, 23, 50, 52, 57, 72, 73, 95, 118, 235, 238, 245, 247
artiad elements, 54, 56, 57
Aston, Francis W., 397
astronomy, 4, 5, 7, 8, 44, 387, 388, 392, 393, 397, 403, 439
atogram, 145, 148, 149, 151, 155, 157
atomare, 145, 155
Atomechanics, 139, 141, 150, 163, 164
atometer, 145, 146
atomic mass, 198, 237, 242, 248, 259, 262, 263, 265, 266, 444
atomic number, 6, 27, 38–40, 50, 62, 118, 153, 228, 229, 232, 233, 238, 244, 248, 251, 253, 268, 270, 272, 273, 289, 291, 316, 330, 336, 400, 444, 448
atomic volume, 6, 18, 26, 27, 134, 162, 209, 210, 228, 245
atomic weight, 4, 5, 7, 8, 15, 17–27, 29, 35, 36, 38–42, 47–58, 62, 63, 65, 69, 70, 72, 73, 76, 77, 79, 81, 83, 84, 86, 87, 89, 96, 97, 99–107, 110, 111, 113–122, 124, 126, 127, 134, 137, 138, 141–146, 148–151, 153, 154, 157, 159–164, 198, 201, 203–212, 216, 218, 219, 228, 230, 234, 239, 240, 246, 248, 249, 252, 265, 266, 303, 306, 311–316, 388, 397, 400, 410, 426
atomicity, 96, 106, 155, 312. *See also* quantivalence, valence
Auer von Welsbach, Carl, 249, 262, 263, 266
Avogadro's hypothesis, 17, 51, 72, 198, 220
axioms of chemistry, 49

B

Bad Dürkheim, 249
Bahr, Johan Fredrik, 217, 239
Bartlett, Neil, 318, 319, 324
Barut, Asim Orhan, 420, 433
base, 45, 49, 203, 235, 239
Bassett, Henry, 266
bastnasite, 278

C. J. Giunta et al. (eds.), *150 Years of the Periodic Table*, Perspectives on the History of Chemistry, https://doi.org/10.1007/978-3-030-67910-1

Baumhauer, Heinrich, 25
bauxite, 235
Becquerel, Henri, 410
Béguyer de Chancourtois, Alexandre-Émile, 4, 17, 61–63, 65, 67–73, 75–79, 81, 83–89, 126, 134, 138, 426
Bémont, Gustave, 250
Berg, Otto, 250
Bergakademie Freiberg, 242, 243, 247
Berkeley, University of California, 331–341
berkelium, 9, 337, 338, 426
Berlin, Nils Johan, 239
Berthelot, Marcellin, 77, 240, 316
Berthollet, Claude Louis, 198, 203, 220, 221, 223
beryllium, 9, 23, 24, 45, 117, 219, 239, 260, 330, 401
Berzelius, Jöns Jacob, 48, 49, 142, 148, 198, 203, 220, 238, 242, 260, 261, 265, 365
Bethe, Hans, 397
big bang, 396, 398, 399, 402, 415
bismuth, 9, 23, 50, 52, 57, 65, 235, 245, 246, 334
blende, 234
Boerhaave, 44
Bohr, Niels, 8, 411, 412, 414
Bondi, Hermann, 398
bonding, 155, 223, 288, 429, 431
borax, 56, 245
boron, 9, 45, 53, 56, 66, 70, 72, 84, 87, 103, 117, 121, 146, 161, 235, 401
Boscovich, Roger, 44
Bowen, Ira Sprague, 391
Brauner, Bohuslav, 248, 249, 266, 267
Breithaupt, August, 243, 244
Breslau, University of, 198, 201, 216
Brezina, Aristides, 140
British Association for the Advancement of Science, 5, 50, 51, 125, 126, 305, 309, 313, 389
Brock, William Hodson, 44, 373, 374
Brodie, Benjamin, 95
Broglie, Louis de, 411
Browne, Charles Albert, 360
Brunck, Otto, 244
Bulgaria, 443–445
Bunsen burner, 233, 234, 239, 249, 250, 252
Bunsen, Robert, 197, 215, 233, 239, 249, 250, 252, 261, 262, 388
Burbidge, Burbidge, Fowler and Hoyle (B^2FH), 395, 400, 401, 403
Bürger, Karl, 242, 251
Butlerov, Alexander, 99

C

Cady, Hamilton Perkins, 350, 351, 359
calcoids, 147
californium, 9, 338, 340
Cameron, Alastair, 400, 403
Cannizzaro, Stanislao, 17, 18, 20, 45, 51, 69, 70, 74, 76, 81, 97, 100, 101, 103, 111, 113, 115, 124, 137, 146, 198, 216
carbon, 9, 20, 45, 72, 96, 100, 106, 121, 146, 156, 323, 368, 369, 392–394, 397, 401, 440, 444
Cat's Eye nebula, 388
Cavendish, Henry, 304, 305, 307, 309
cerium, 9, 110, 260, 261, 265–267, 272, 273, 277, 281
cesium, 9, 219, 249, 250, 388
Ceulemans, Arnout, 416, 418, 420
Chadwick, James, 330
chalcogen, 147, 151–155, 159, 237
Chamberlain, Joseph, 402
Chandrasekhar, Subrahmanyan, 394, 401
Charles James, 263
chart (of the elements), 4, 5, 26, 62–65, 67–76, 78, 79, 84, 85, 87–89, 114, 133, 134, 149–151, 153, 154, 157, 158, 160, 162, 163, 212, 390, 440, 444
chemical analysis, 50, 69, 113, 234, 244, 249, 333, 346, 434
chemical atoms, 45, 46, 50, 204, 216
Chemical News, 18, 19, 79, 93, 97, 113, 114, 117–119, 121, 122, 125, 126
chemical properties, 316
Chemical Society (London), 19, 95, 108, 111, 113, 121–124, 126, 127, 201, 232
chemistry, agricultural, 241
Chemnitz, 208, 243
chloride, 56, 57, 156, 163, 234, 237, 242, 245, 261, 272, 275, 278, 280, 290, 292, 293
chloroids, 147, 148, 153
classification, classify (the elements), 3, 5, 6, 14, 15, 17–20, 29, 36, 57, 61, 67–69, 72, 76, 77, 81, 85, 86, 93, 95, 96, 98, 99, 106, 111, 113, 121, 122, 124–127, 134, 138, 140, 145, 146, 148–150, 154, 159, 162, 201, 212, 216, 222, 233, 236, 252, 312
Claus, Carl Ernst, 148
Cleve, Per Theodor, 22, 240, 242, 252, 262, 266
cleveite, 313
CNO cycle, 400
Cognac (France), 232

collaboration, collaborator, 113, 238, 250, 305, 306, 332, 341, 348, 360, 371, 398, 429
color, 52, 196, 233, 239, 249, 261, 277, 304, 339, 388, 441, 445
complex, complexation, 270, 273–276, 281–288, 290, 293
composition, 47, 50, 52, 53, 67, 137, 218, 244, 260, 319, 393, 402, 403
Comptes Rendus, 17, 22, 62, 67, 70–73, 75, 79, 240
Compton, Arthur, 333
Conant, James, 333
Cooke, Josiah Parsons Jr., 5, 16, 17, 20, 43, 50–58, 77, 137, 138
coordination number, 7, 283, 285–288, 290, 293, 294
Copenhagen, 135, 140
co-precipitation, 235
coronagraph, 392
coronal lines, 391
coronium, 252, 389, 391
Couper, Archibald Scott, 99
Course of Practical Chemistry, A, 18, 96, 107, 108, 110
Cronstedt, Axel Fredrik, 260
Crookes, William, 79, 89, 309, 310, 313, 388
crystal forms, 47, 140, 157
crystal structure, 47, 241, 284, 286, 288, 292, 293
crystallization, fractional, 261–263, 271, 272
Cunningham, Burris, 337
cuproids, 147, 154
Curie, Marie, 253, 336, 410
Curie, Pierre, 336, 410
curium, 9, 336, 337, 426
cyanide, 156, 163

D
Dains, Franklin Burnett, 352–355, 357, 360, 365
Dalton, John, 14, 20, 44–47, 49, 58, 84, 95, 410
Dana, James Dwight, 144
Darwin, Charles, 139, 164, 346
Daubeny, Charles, 143
Davis, Tenney Lombard, 371–373
Davisson, Clinton, 411
Davy Medal, 6, 18, 124, 125, 201, 211, 228, 236
Davy, Humphry, 44, 46, 47, 49, 56, 57, 365
degeneracy, 415–418, 432
Delafontaine, Marc, 262
Delhuyar, Fausto, 441
Delhuyar, Juan José, 365, 441
Demarçay, Eugène-Anatole, 262
density, 24, 103, 157, 162, 209, 235, 236, 242, 250, 306–311, 313, 314, 392, 398, 399
detection, 232, 249–251, 253, 339, 430
DeVorkin, David H., 393, 403
Dewar, James, 309, 312
Dexter Award, 362, 363
didymium, 110, 243, 249, 262, 265, 266
Dirac equation, 428
discovery criteria, 242, 253
discovery, effective, 250
Discovery of the Elements, 4, 7, 232, 343, 354, 357, 363–374, 379
dispersion, chromatic, 247
distillation, 249
Döbereiner, Johann Wolfgang, 5, 15, 36–38, 41, 51, 99, 114, 150, 202, 203, 211, 212, 313, 426
Douglas, Allie Vibert, 403
Dresden, 243, 247
Dronsfield, Alan, 250
du Bois-Reymond, Emil, 142
Dubna (Russia), 316, 341, 429, 448
Dulong and Petit, law of, 70, 72, 103, 220, 223
Dumas, Jean Baptiste André, 51, 52, 75, 77, 78, 81, 103, 113, 114, 118, 119, 137, 138, 141, 163, 218, 221, 305
dvi-manganese, 252
dysprosium, 9, 234, 236, 262, 338

E
earth, ceric, 238
earth, yttric, 238
École des Mines, 62, 63, 67, 76, 77, 79
École Polytechnique, 62, 233
Eddington, Arthur Stanley, 389, 392–395, 403
Edlén, Bengt, 391
Eggen, Olin C., 403
Einstein, Albert, 357, 410, 411
einsteinium, 7, 9, 340, 341
eka-aluminum, 22, 41, 42, 76, 124, 232, 235, 237, 252
eka-antimony, 245
eka-barium, 252
eka-boron, 22, 41, 232, 237, 240, 242, 248, 252, 262
eka-cesium, 252
eka-iodine, 252
eka-lanthanum, 252
eka-manganese, 252
eka-silicon, 23, 41, 42, 232, 242, 245, 246, 248, 373
eka-tantalum, 252
electric discharge, 307

electrolysis, 223, 235, 241, 247, 340
Electromagnetische Telegraphie, Die, 135
electronegativity, 159, 292, 317, 372
element 43, 253
element 61, 253, 267–270, 273
element, highly ionized, 389
elements, classical, 444
elements, common, 235, 440, 444
elements, families or groups of, 44, 50–52, 56, 103, 105, 106, 116, 122, 147, 148, 152, 204, 218, 233
elements, missing, 228, 232, 234, 268, 426
elements, superheavy, 330, 338, 340, 341, 427, 430, 431, 447, 448
elements, synthetic, 316, 330, 340
elements, transuranium, 253, 316, 332, 333, 335, 337–339, 373. *See also* transuranides
Élie de Beaumont, Léonce, 62, 63, 67, 79, 88
Entwurf der Atomechanik, 141
equivalent, 16, 36, 45, 69, 75, 81, 95, 96, 107, 113–115, 117–120, 126, 143, 146, 198, 203, 236, 239
erbia, 239, 240, 261, 262
erbium, 9, 239, 240, 263, 265
erythronium, 441
Eschricht, Daniel Frederik, 135
EuChemS, 447
europium, 9, 261, 262, 267, 280, 282, 336, 426
euxenite, 238, 240, 262
extraction, solvent, 274, 278
eyselite, 247

F
Faraday, Michael, 47, 49, 94
Fermi, Enrico, 7, 330, 331, 334, 335, 338, 340, 427, 432, 433
fermium, 7, 9, 340, 341
ferroids, 147, 151, 154
Fet, Abram Ilyich, 420
Fischer, Werner, 242, 251
Flerov, Georgy, 336, 337, 341
Flügel, Felix, 139
fluorine, 9, 45, 48, 50, 56, 66, 73, 75, 100, 118, 141, 146, 148, 266, 324
Fock, Vladimir, 417, 427, 428, 432, 433
forbidden transitions, 388, 391
Forchhammer, Johan Georg, 135, 142
Foster, George Carey, 122, 127
fractionation, 239, 240, 314
Frankland, Edward, 94, 96, 305
Fraunhofer lines, 249, 388, 391
Freiberg, 241–244, 246
Fresenius, Carl Remigius, 139
Freund, Ida, 230
Frisch, Otto, 331
fusibility, 159

G
Gabelsberger, Franz Xaver, 141
Gadolin, Johan, 336, 434
gadolinite, 239, 240, 260, 262, 434
gadolinium, 9, 234, 236, 260–262, 277, 392, 426, 434
Gahn, Johan Gottlieb, 242
gallium, 4, 6, 9, 22, 23, 41, 56, 76, 79, 122, 124, 126, 161, 227, 231–237, 248, 251, 252, 389, 426
Gamow, George, 398
Gaudin, Marc Antoine, 84, 157
Gay-Lussac, Joseph-Louis, 49, 52
Geinitz, Hans Bruno, 139
Gell-Mann, Murray, 418–420
geometries (of molecules), 155, 286
Gerhardt, Charles, 20, 69, 96–98, 100, 221
germanium, 4, 6, 9, 23, 41, 117, 161, 227, 231, 232, 242, 244–248, 251, 373, 389, 426
Germer, Lester, 411
Ghiorso, Albert, 334, 335, 337, 339, 340
Gilbert, Davies, 46
Gilman, Albert Franklin, 346, 347
Gladstone, John Hall, 15, 17, 18, 77, 83, 99–101, 122, 137
Gmelin, Leopold, 15, 16, 159, 218, 426
Goeppert-Mayer, Maria, 338, 396, 432, 433
Goethe, Johann Wolfgang von, 36, 203
Gofman, John, 335
gold, 9, 52, 53, 69, 147, 235, 338, 364, 368, 392, 410, 429, 432, 442
Gold, Thomas, 398
Goldschmidt, Victor Moritz, 241, 289, 291
Gordin, Michael D., 3, 4, 10, 85, 125, 127, 212, 427
Gordon, Neil Elbridge, 352–354, 359, 361
Gothland Island, 237
Graham, Thomas, 48–50
graphic, 84
graphical representation (of the periodic law), 24, 25, 26, 29, 78, 81, 84, 85
Greenstein, Jesse, 400
Grieneisen, Hans, 242, 251
Grotrian, Walter, 391
group theory, 8, 416, 417
Guy's Hospital, 94, 95

H
Hageboeck, August, 140
Hahn, Otto, 331, 332

Haidinger, Wilhelm, 139, 142
halogens, 52, 56, 57, 134, 146–148, 151, 152, 155, 159, 204, 237, 275, 279, 280, 308, 314, 445
Hansen, Peter Andreas, 142
Hartog, Philip, 65, 78, 84, 87, 88
Hartree, Douglas, 417
Hayashi, Chushiro, 399
Hearn, Edward David, 139
Heidelberg, 197, 215, 217, 250
Heisenberg, Werner, 416, 418–420
helium, 4, 8, 9, 245, 251, 305, 313, 318, 336, 337, 350, 389, 392, 393, 395, 398–400, 402, 403, 430, 442
Helmholtz, Hermann, 81, 164
Henry, William, 44
Hermes Trismegistus, 366
Herschel, John F.W., 4, 5, 43, 48–51, 58
Hillebrand, William, 306, 313
Hinrichs, Gustavus Detlef, 4, 5, 17, 23, 133–164, 390
Hisinger, Wilhelm, 49, 261
history of chemistry, 3, 14, 16, 41, 46, 51, 67, 221, 344, 352–354, 357, 359, 361, 363, 372, 373, 379
Hofmann, August Wilhelm, 112, 139
holmium, 9, 240, 262, 290, 292
Hooker, Samuel, 359
Hoppe, Rudolf, 324, 319, 323–325
Hoyle, 398, 399
Huggins, William, 388, 389
Humboldt, Alexander von, 142, 164
Humpidge, Thomas Samuel, 24
Hungary, 8, 444
hydrogen, 7–9, 15, 25, 44, 46, 52–54, 56, 57, 65, 69, 96, 97, 117, 137, 141–144, 147, 152, 155, 160, 201, 206, 208, 209, 245, 305, 311, 339, 350, 368, 388, 392–395, 397, 398, 400, 402, 403, 411, 412, 414–417, 430, 440, 444
hydrogen atom, fraction of, 137, 143
hydrogen, ratio to other elements, 392
Hyman, Herbert, 319–323

I

Illinois, University of, 134, 138, 140, 141, 160, 161, 164, 268, 350
indium, 9, 107, 111, 118, 233, 235, 236, 243, 265
infrared, 247, 268, 394
International Year of the Periodic Table (IYPT), 1–4, 8, 44, 195, 196, 211, 440–447
iodine, 9, 15, 23, 45, 48, 50, 65, 72, 103, 119, 141, 148, 203, 205, 211, 429
ion-exchange, 270, 273, 274, 278
ionization potential, 280, 319, 430, 432
Iowa Weather and Crop Service, 135
Iowa, University of, 134–136
iridoids, 148
iron, 9, 53, 65, 69, 72, 99, 104, 147, 219, 260, 278, 313, 368, 388, 394, 396, 399–402, 441
island of stability, 316, 448
isolation (chemical), 232, 235, 242, 247, 251–253, 259, 261, 263, 266, 282, 309, 316, 319, 365
isomorphism, 69, 134, 155, 156, 163
isospin, 418
isotope, 57, 212, 237, 248, 253, 268, 316, 332, 339, 396, 399, 415, 428, 429, 448
isotope, stable, 242, 247, 268, 396, 397, 403, 444
IUPAC, 2, 8, 251, 316, 341, 425, 430, 440, 441, 444

J

James, Charles, 262
James, Ralph, 335
Janssen, Jules, 305, 388, 392
Jensen, Hans, 338, 396
Jensen, William B., 19, 35, 252
Joint Institute for Nuclear Research, 316, 341
Joliot-Curie, Frédéric, 330
Joliot-Curie, Irène, 330
Journal of Chemical Education, 7, 232, 352–354, 365, 366, 368, 373

K

kadmoids, 147, 151
kaloids, 147
Kansas, University of, 7, 344, 350–353, 355, 360, 371
Karlsruhe Congress, 16–18, 51, 81, 96, 112, 216
Karlsruhe Polytechnic, 200
Kaye, Albert, 318, 319
Kekulé, Friedrich August, 58, 99, 197, 198, 216, 217, 224
Kirchhoff, Gustav Robert, 216, 233, 249, 250, 252, 261, 388
Kirk, Paul, 335
Klaproth, Martin Heinrich, 261, 366
Kleiber, E.I., 392
Koenig, Frederick Otto, 372
Koksharov, Nikolai, 140

Kragh, Helge, 21, 305
Kremers, Peter, 77
Kresge, Sebastian Spering, 359
Kresge-Hooker Science Library, 359–361
Krieghbaum, Hillier, 372
Kristiania (Norway), 238
Krönig, August Karl, 142, 143
Krüss, Gerhard, 241
krypton, 9, 314, 316, 318
Kurrent script, 140
Kyrgyzstan, 442, 443, 445

L
Landolt, Hans, 197, 211, 216, 217
lanthanides, 3, 6, 7, 159, 239, 260–264, 266, 267, 270–279, 281–294, 315, 335, 336, 338, 426, 430
lanthanum, 9, 243, 260, 261, 265, 266, 272, 277, 280, 290, 291, 293, 337, 340
Laplace-Runge-Lenz (LRL) vector, 416, 417
Lapparent, Albert de, 79–81, 88, 89
Laszlo, Pierre, 317, 318
Laurent, Auguste, 69, 96, 97, 100, 221
Lavoisier, Antoine, 44, 220, 357
Lawrence, Ernest, 331–334, 337, 341
Lawrence Berkeley Laboratory, 316, 337
Lawrence Livermore Laboratory, 4, 316, 340
Lea, Mathew Carey, 137, 138
Lecoq de Boisbaudran, Paul-Émile, 22, 79–81, 88, 89, 232–236, 246, 248, 251, 262, 434
Leicester, Henry Marshall, 7, 362, 363, 365, 366, 368–370, 373
Lenssen, Ernst, 36, 38, 78, 102
Liebig, Justus von, 139, 228
line (spectral), 157, 234, 235, 240, 250, 305–307, 309, 314, 317, 388, 389, 391, 393, 410
Linnean nomenclature, 145
liquid range, 237
lithium, 9, 50, 54, 65, 73, 87, 117, 141, 203, 249, 365, 388, 397, 401
Lockyer, Norman, 305, 306, 313, 388, 389, 392
long form (periodic table), 26, 27, 159, 412, 431
Löwdin, Per-Olov, 414, 433
Ludwig, Carl, 196, 215
lutetium, 9, 262, 263, 266, 290, 292, 293
Lynden-Bell, Donald, 403
Lyot, Bernard, 392

M
Madelung rule, 9, 412, 414, 415, 417, 420, 421, 432, 433
magic numbers, 395, 400, 417
Manhattan project, 268, 273, 274, 334, 335, 340, 341
Marignac, Jean-Charles Galissard de, 143, 239, 262, 434
Mathews, Joseph Howard, 348, 349
Maumené, Edme-Jules, 143
Mayer, Julius Robert, 164
McCay, LeRoy Wiley, 243, 244
McMillan, Edwin, 332, 333, 338
Meitner, Lise, 3, 4, 331, 341
melting point, 157, 162, 241, 250, 292
Mendeleev, Dmitri Ivanovich, 2–8, 10, 13–15, 17–29, 35, 38, 41, 42, 58, 73, 76–79, 81, 85, 87, 93, 97, 99, 100, 105, 107, 110, 111, 122–127, 134, 158, 159, 162, 163, 198, 201, 208–212, 215, 227, 228, 230–232, 235–237, 240–242, 245–248, 251, 252, 259, 262, 264–266, 279, 303, 312–314, 343, 373, 387, 389, 417, 419, 421, 426, 440, 442–445, 447
mendelevium, 4, 9, 341, 362, 440, 442, 444
mercury vapor lamps, 242
Meroney, Virgil, 339
metals, 4, 15, 36, 45, 47, 53, 54, 56, 57, 99, 103–107, 134, 146, 150, 159, 160, 218, 219, 236, 237, 241, 247, 250, 266, 272, 273, 277, 278, 282–286, 294, 305, 308, 318, 393, 394, 402, 410, 442, 445
Mexico, 16, 441
Meyer, Julius Lothar, 3–6, 17–19, 22, 23, 25, 26, 73, 77, 79, 81, 85, 87, 94, 97, 100, 103, 110, 122–126, 134, 138, 162, 195–212, 215–217, 224, 228–230, 245, 248, 313, 314, 426
Miller, William Allen, 388
mineral, 22, 48, 50, 53, 56, 67, 69, 234, 238, 240, 243–245, 261, 278, 313, 441
mineralogy, 70, 72, 196, 198, 208, 243
mixture, 49, 53, 235, 239, 240, 242, 261, 262, 268, 270, 272, 273, 278, 304, 305, 309–313, 319
Modernen Theorien, 6, 18, 19, 126, 198, 203, 204, 206, 215, 216
Moissan, Henri, 56, 310, 316
Moldova, 443
molybdoids, 147, 154
monazite, 268, 278
Morgan, Leon, 335, 336
Mosander, Carl Gustav, 261
moscovium, 9, 444
Moseley, Henry Gwyn Jeffreys, 8, 50, 228, 248, 249, 253, 267, 269, 389, 410

N

name (of element), 69, 235, 236, 240, 245, 263, 270, 309, 313, 314, 316, 331, 333, 336–338, 340, 341, 389, 426, 434, 441, 444
Nantes, Edict of, 232
Naumann, Carl Friedrich, 140
nebula, 144, 388
nebular lines, 388, 391
nebulium, 388, 389, 391
neodymium, 248, 249, 261, 262, 268, 270, 272, 277, 285
neon, 9, 314, 318
neptunium, 4, 7, 9, 332, 333, 335, 338
Neubert, Eduard Wilhelm, 244
Neumann, Franz Ernst, 197, 208, 216
Neustadt-Eberswalde Forest Academy, 198
neutron, 4, 7, 268, 273, 330–332, 335, 338, 339, 392, 395–401, 418
neutron capture, 7, 247, 330, 331, 335, 338, 396, 401
Newlands, John A.R., 4, 5, 17–19, 21, 23, 73, 77–79, 85, 93, 103, 110–127, 134, 246, 251, 314, 426
newtonium, 252
night vision, 247
nihonium, 9, 444
Nilson, Lars Fredrik, 22, 24, 237–242, 248, 251, 252, 262
niton, 315
nitrogen, 9, 15, 44, 50, 52, 57, 72, 73, 100, 118, 134, 146, 147, 156, 159, 163, 197, 240, 294, 304, 306, 307, 309, 310, 368, 388, 397, 402, 440
Nobel Prize, 51, 331, 338, 372, 396, 419
nobelium, 9, 341
noble gases, 4, 7, 27, 38, 40, 122, 152, 303, 304, 314–317, 319, 323, 324, 389, 426, 430, 445
Noddack, Ida (Tacke), 3, 4, 268, 330
North Macedonia, 447, 448
Norway, 242
nuclear age, 329
nuclear weapons, 7, 333, 337
nucleosynthesis, 398, 399

O

Oak Ridge, 268, 335, 338, 429
octaves, law of, 18, 21, 77, 118, 119, 121–125, 127, 315, 426
Odling, William, 5, 17, 20, 22, 23, 73, 77, 93–108, 110, 111, 117, 121, 123–127, 134, 138
Oganessian, Yuri, 316, 341
oganesson, 4, 9, 316, 341, 389, 429, 444
Olszewski, Karol, 310
Öpik, Ernst, 399
ore, 233–235, 240, 243, 244, 268, 277, 278, 306, 389, 441
organizing principle, 53, 163, 228, 248
Oslo, 238
Osnovy Khimii, 109
Owens, Robert Bowie, 315
oxidation number or state, 7, 73, 237, 242, 247, 262, 279, 284, 291, 336, 429, 431
oxide, 6, 36, 45, 46, 52, 53, 54, 56, 57, 70, 72, 153, 203, 234, 237, 239, 240, 242, 245, 261, 262, 265, 266, 272, 274, 277–279, 281, 289, 290, 292, 304, 319, 434
oxygen, 9, 15, 45, 50, 52, 54, 56, 57, 63, 65, 69, 73, 96, 97, 100, 118, 134, 142, 146, 153, 155, 159, 163, 197, 198, 294, 306, 309, 391, 393, 397, 401, 402, 426, 440

P

panchromium, 441
Paneth, Fritz, 317
pantatom, 6, 144–148, 155, 156
pantogen, 140, 144, 149, 154, 160
pantoids, 147
Pauli, Wolfgang, 8, 411, 412, 416, 417
Pauling, Linus Carl, 317, 318, 334, 372
Payne, Cecilia H., 393, 402
periodic law, 14, 15, 17–19, 21–27, 29, 38, 41, 61, 62, 75–78, 81, 83–88, 94, 110–112, 122, 124, 125, 127, 216, 231, 232, 240, 343
periodic table, 2–3, 8, 10, 15, 20, 21, 24–26, 35, 36, 38–40, 44, 45, 56, 58, 67, 81, 85, 93, 106, 107, 110, 134, 150, 153, 159, 163, 195, 210, 215, 216, 230, 232, 239, 241, 252, 253, 259, 264–267, 291, 303, 305, 307, 308, 312–317, 331, 333, 335, 338, 340, 341, 344, 366, 403, 409-421, 425, 427, 430, 439, 440, 443–445, 447, 448
periodicity, 3–5, 13, 14, 22, 24–27, 29, 65, 75–77, 79, 81, 86–89, 93, 107, 111, 117–123, 127, 204, 209, 410, 412
perissad elements, 54, 56, 58
Perlman, Isadore, 334, 337, 340
Perrier, Carlo, 331, 340
Perrin, Jean, 392
Pettenkofer, Max von, 77, 78, 141, 218
Pettersson, Sven Otto, 24, 238, 239
philatelic, philately, 4, 7, 8, 439, 445
philosophy of science, 224
phlogiston, 220, 304, 305

phosphoids, 147
Pierrefitte mine, 233, 234
Planck, Max, 410, 411
platinum, 9, 53, 99, 104, 107, 268, 285, 291, 440, 441
plutonium, 7, 9, 333–335, 338, 339
pnictogens, 147, 151, 152, 155, 159
Poggendorff, Johann Christian, 139, 142
Pogson, Norman R., 389
polynomial formulas (for atomic weights), 137, 138
Portugal, 444–446
potassium, 7, 9, 45, 47, 48, 51, 54, 56, 65, 71, 73, 79, 87, 141, 147, 156, 163, 203, 249, 260, 261, 331, 365, 368, 415
Prandtl, Wilhelm, 268
praseodymium, 9, 249, 261, 262, 270, 272, 281
precipitation, 235, 261, 273, 350
precipitation, fractional, 245
prediction, 4–6, 14, 19, 22, 23, 25, 29, 37, 41, 79, 87, 102, 117, 161, 163, 211, 212, 230, 235, 236, 241, 249, 252, 303, 419, 426, 429, 432
Priestley, Joseph, 304
priority dispute, 79, 108, 164, 208, 210, 228, 247, 251, 253, 389
prism, 69, 137, 145, 146, 249
Programme der Atomechanik, 6, 139, 147, 149, 151, 152, 156
promethium, 6, 9, 263, 268, 271, 335, 340
properties of elements, 5, 86, 150, 314
properties, chemical, 15, 19, 23, 62, 68, 77, 138, 152, 162, 204, 240, 241, 264, 428, 442
properties, physical, 15, 22, 27, 62, 65, 107, 110, 230, 292, 315, 316
protactinium, 9, 251
Prout, William, 50, 141, 143, 201
Prout's hypothesis, 52, 65, 137, 143, 160, 202, 204, 212, 306, 415
purification, 240, 273

Q

quadrivalency, 246
quantivalence, 54, 56–58, 96, 123. *See also* atomicity, valence
quantum mechanics, 7, 414, 416, 418, 420, 421

R

radioactive, 56, 252, 263, 278, 315, 331, 334, 366
radioisotopes, 242, 247, 251, 332
radon, 9, 315, 318, 319, 330
Rammelsberg, Karl Friedrich August, 140
Ramsay, William, 7, 84, 306–310, 312–315, 317, 389, 426
Rancke-Madsen, Edmond, 250
rare earths, 3, 4, 6, 7, 99, 233, 239–241, 248–251, 253, 259–261, 264–266, 268, 270–274, 276–279, 285, 289, 294, 340, 392, 430, 434
Rayleigh, Lord (John William Strutt), 306–309, 312, 313, 315
Rayner-Canham, Geoffrey, 349
Rayner-Canham, Marelene, 349
Regnault, Victor, 51, 70
Reich, Ferdinand, 243
relativistic effects, 316, 428, 429, 431
Remelé, Adolf, 208
rhodoids, 148
Richards, Theodore, 51, 348
Richter, Benjamin, 45, 47, 49
Richter, Hieronymus Theodor, 243, 244
Río, Andrés Manuel del, 366, 441
Ripon College, 346–349
Robinson, Jimmy, 339
Rocke, Alan, 3, 4, 45, 96, 98, 99, 216, 224
Rolla, Luigi, 268
Röntgen, Wilhelm, 410
Royal Academy of Agriculture (Sweden), 241
Royal Institution of Great Britain, 46, 95, 98, 103, 108, 110, 312
Royal Society (London), 6, 18, 24, 46, 95, 110, 124, 201, 211, 228, 304, 309, 314
r-process, 400, 401
rubidium, 9, 73, 249, 250, 388
Rubik, Ernö, 443
Rudorf, George, 84
Russell, Henry Norris, 393, 394, 402, 403
Russian Chemical Society, 2, 110, 210, 232, 235
Rutherford, Daniel, 366
Rutherford, Ernest, 202, 315, 395, 410

S

Saint Bartholomew's Hospital, 94
Saint Louis University, 135
Sainte-Claire Deville, Charles Joseph, 67–69
Salpeter, Edwin, 399
samarium, 9, 234, 236, 248, 261, 262, 268, 280, 286, 290, 434
samarskite, 236, 262
Sandage, Allan R., 403
Saxony, 242
scandium, 4, 6, 9, 22, 24, 41, 161, 227, 232, 237–242, 248, 251, 252, 262, 266, 291–293, 389, 414, 415, 426

Scerri, Eric R., 4, 8, 10, 15, 112, 134, 213, 409, 427, 431
Scheerer, Theodor, 238
Schleswig-Holstein War, 135
Schottky diode, 247
Schrauf, Albrecht, 140
Schrobilgen, Gary, 317, 318
Schrödinger, Erwin, 8, 411, 412, 415, 416
Schrötter, Anton, 142
Schwarzschild, Barbara, 394
Schwarzschild, Martin, 394
Seaborg, Glenn, 252, 267, 332–341, 426
seaborgium, 9, 341
secondary periodicity, 429
Sefström, Nils Gabriel, 242, 441
Segrè, Emilio, 330–333, 340, 417
selenium, 9, 52, 57, 65, 118, 203, 230, 235, 238, 245
semiconductor, 237, 247
Setterberg, Carl, 250
Seubert, Karl, 81, 82, 196, 200, 201, 206-208
Shapley, Harlow, 393
short form (periodic table), 26–28, 109, 159
sideroids, 147
Sigma Xi, 355, 356
silicon, 9, 45, 53, 58, 70, 72, 87, 106, 117, 206, 245, 246
silver, 9, 48, 54, 56, 96, 100, 147, 235, 244, 317, 364, 368, 389, 392
Soddy, Frederick, 253, 315
sodium, 9, 47, 48, 50, 54, 56, 65, 73, 141, 203, 249, 305, 331, 348, 365, 368, 388, 389, 430
solar power, 247
solubility, 237, 239, 272
Sommerfeld, Arnold, 411, 412, 432
Spain, 441, 442
specific gravity, 162, 203, 237, 241, 242, 248
specific heat, 24, 69, 70, 157, 248, 266, 310, 312, 313
spectroscope, 249, 252, 261, 388
spectroscopy, 51, 234, 250, 251, 261, 306, 310, 313, 388, 389
spectrum, 234, 239, 242, 246, 249, 250, 262, 263, 268, 305, 313, 314, 388, 389, 392, 401
speculation, 65, 89, 204, 216, 222, 231, 241, 247–249, 251, 252
Spedding, Frank H., 273
Spence, Roderick, 339, 340
Speter, Max, 357, 359
sphalerite, 234, 235
spin-orbit coupling, 316, 396, 427, 430
spiral (periodic system), 5, 17, 25, 26, 29, 84, 134, 149, 150, 153, 155, 160, 163, 389, 390
Springer, Anna, 135
Springer, Auguste, 135
Sri Lanka, 8, 445, 446
stamp, 8, 440–447
Stas, Jean-Servais, 137, 143, 160
steady state universe, 398
stellar evolution, 399, 402
Stockholm, 237, 241, 340, 434
stoichiometry, 45, 47, 67, 70, 319
Stoner, Edmund, 411, 412
Stoney, George Johnstone, 410
Strecker, Adolph, 78, 107
Street, Kenneth, 340
Strömgren, Bengt, 394
structural theory, 81, 99
Studiosus, 114, 115, 117, 118
sulfide, 106, 234, 235, 238, 245, 261, 305
sulfur, 9, 45, 52, 57, 65, 69, 118, 203, 244, 245, 305, 368
sulphoids, 147
Svanberg, Lars Fredrik, 237
Sweden, 48, 237, 241, 242, 331
Swift, Doris Brewster, 350, 378
Swings, Pol, 391
symbol, 15, 96, 160, 233, 238, 244, 246, 333, 338, 440–445
symmetry, 8, 415–421, 433
synthetic elements, 7

T

tabular (forms), 17, 26, 27
technetium, 9, 331, 340, 403
telluric screw, 69, 426. *See also* vis tellurique
tellurium, 9, 23, 52, 57, 69, 103, 118, 119, 146, 203, 204, 211, 245
tennessine, 9, 338, 444
terbia, 261, 262
terbium, 9, 262, 281, 287, 426
tetragonoids, 144, 146, 147, 151, 152, 154
Thalén, Tobias Robert, 240, 242, 251
thallium, 9, 72–74, 107, 121, 204, 219, 235, 251, 388
theoretical chemistry, 198, 200, 224
thermometer, 237
Thompson, Stanley, 332, 334, 335, 337, 339, 340
Thomson, J.J., 8, 410–412
Thomson, Thomas, 45–48, 50
Thomson, William, 305, 389
thorium, 9, 239, 240, 265, 266, 278, 315, 426

thulium, 9, 240, 262, 263, 272
Thyssen, Pieter, 416, 418, 420
Tilden, William, 110
tin, 9, 53, 107, 116, 235, 240, 246, 368, 396
titanoids, 147, 151
transfermium, 341
transition elements, 35, 40, 53, 103, 104, 134, 150, 158–161, 163, 228, 241, 263, 288, 291–293, 323, 414
transuranides, 251. *See also* elements, transuranic
triads, 4, 5, 14–17, 35, 36, 38–42, 51, 52, 99, 100, 102, 103, 105, 107, 114, 116, 119, 202, 203, 212, 313, 426
trigonoids, 144, 146, 152, 154
tri-manganese, 252
Tübingen, University of, 200–202, 206
tungsten, 9, 52, 72, 335, 440, 441
Tyndall, John, 139, 142
type theory, 96–98

U
Ulloa, Antonio de, 442
UNESCO, 1, 2, 44, 439, 442, 443, 447
Uppsala University, 237
uraninite, 306, 313
uranium, 7, 9, 23, 53, 57, 65, 107, 110, 153, 267, 273, 306, 330–333, 335, 337–339, 389
Urbain, Georges, 262, 263, 285
urstoff, 137

V
valence, 23, 24, 27, 35, 69, 95–97, 133, 153, 154, 155, 162, 204, 205, 228, 239, 242, 248, 312, 372, 427, 430, 431. *See also* atomicity, quantivalence
van den Broek, Antonius, 228
van Spronsen, Johannes (Jan), 84–86, 110, 160, 213, 390
Vanadis, 441
vanadium, 9, 52, 72, 153, 155, 207, 242, 370, 440, 441
Vauquelin, Louis Nicholas, 50, 261
Venable, Francis, 15, 17, 81, 83
Virchow, Rudolf, 215
vis tellurique, 4, 5, 17, 61–89, 126. *See also* telluric screw
Visby, Sweden, 237
volatility, 159, 241, 245
von Richter, Victor, 23, 210, 245

W
Waage, Peter, 238
Walker, James, 160
Wappler, Rudolph Benno, 244
Watts's Dictionary, 100, 103–107, 110, 111, 121
Wayne State University, 354, 356, 359, 361, 363
Weeks, Mary Elvira, 7, 232, 343–357, 359–366, 368–374, 378, 379
Weinhold, Adolf Ferdinand, 208
Weisbach, Albin, 244
Weizsäcker, Carl Friedrich von, 397
Wells, H.G., 317, 330
Werner, Alfred, 266, 267, 284
Wigner, Eugene, 338, 396
Wildt, Rupert, 394
Will, Heinrich, 137, 139
Williamson, Alexander, 95, 115, 223
Winkler, Clemens Alexander, 23, 242–248, 251, 373
Winkler, Kurt Alexander, 242
Wisconsin, University of, 348, 349
wolfram, 441
wolframite, 441
Wollaston, William Hyde, 44, 46, 95, 96, 113
women in chemistry, 349, 371
Wurtz, Charles Adolphe, 69, 76, 77, 79, 210, 235, 236, 240
Würzburg, 196, 197, 215

X
xenon, 9, 315, 317–319, 323, 324, 442
x-ray diffraction, 285
x-ray spectra, 267
x-ray tubes, 237

Y
Yost, Donald, 317–319
ytterbia, 239, 240, 262
ytterbium, 9, 239, 240, 262, 263, 280
Ytterby (Sweden), 239, 260, 340, 434
yttrium, 6, 9, 45, 110, 159, 239, 260, 262, 263, 265, 283, 285, 291, 292

Z
zinc, 53, 57, 65, 72, 234, 235, 241, 273, 348, 442
Zschopenthal (Germany), 243
Zürich, University of, 196, 215

The manufacturer's authorised representative in the EU is Springer Nature Customer Service Centre GmbH, Europaplatz 3, 69115 Heidelberg, Germany. If you have any concerns regarding our products, please contact ProductSafety@springernature.com

Printed and bound by CPI Group (UK) Ltd, Croydon, CR0 4YY
30/09/2026
02225321-0020